Further Mechanics and Probability

L. Bostock, B.Sc.

formerly Senior Mathematics Lecturer, Southgate Technical College

S. Chandler, B.Sc.

formerly of the Godolphin and Latymer School

Stanley Thornes (Publishers) Ltd.

First published in 1985 by
Stanley Thornes (Publishers) Ltd
Old Station Drive
Leckhampton
CHELTENHAM GL53 0DN

British Library Cataloguing in Publication Data

Bostock, L.
Further mechanics and probability.
1. Mathematics – 1961 –
I. Title II. Chandler, S.
510 QA39.2

ISBN 0-85950-142-6

Typeset by Tech-Set, 15 Enterprise House, Team Valley, Gateshead, Tyne & Wear.
Printed and bound in Great Britain at The Bath Press, Avon.

PREFACE

Further Mechanics and Probability completes a set of four books designed to cover the work needed for Advanced level courses either in Mathematics and Further Mathematics or in Pure Mathematics and Applied Mathematics.

This volume is a continuation of the work covered in *Mechanics and Probability*. Rigid body dynamics is dealt with fully and certain topics from the first volume are developed further. The sections on Probability in the two volumes cover most of the work needed to satisfy those syllabuses where mechanics and probability are examined in one paper, e.g. the University of London Syllabus B in Further Mathematics.

Several topics in this book require a working knowledge of the solution of differential equations and a reasonable degree of competence in integration techniques. These sections contain much material which is suitable for mathematical models.

Many worked examples are incorporated in the text to illustrate each main development of a topic and a set of straightforward problems follows each section. A selection of more demanding questions is given in the miscellaneous exercise at the end of each chapter and multiple choice exercises are included when appropriate.

We wish to express our sincere thanks to those friends and colleagues whose suggestions, criticisms and calculations have been most helpful. We are grateful to the following Examination Boards for their permission to reproduce questions from past examination papers:

The Associated Examining Board
Joint Matriculation Board
University of London University Entrance and School Examination Council
University of Cambridge Local Examinations Syndicate
Oxford Delegacy of Local Examinations
Oxford and Cambridge School Examinations Board
Southern Universities' Joint Board
Welsh Joint Education Committee

Responsibility for the answers to these questions is the authors' alone.

L. Bostock
S. Chandler

CONTENTS

NOTES ON USE OF THE BOOK

1. Notation Used in Diagrams

Force
Velocity
Acceleration
Dimensions

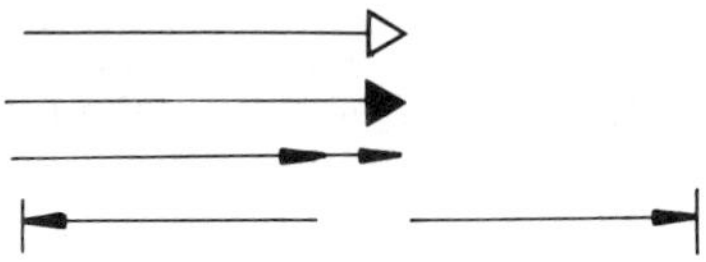

The Value of *g*

In this book, unless otherwise stated, the value of *g* is usually taken as $10\,\mathrm{m\,s^{-1}}$.

2. Instructions for Answering Multiple Choice Exercises

These exercises are at the end of some chapters. The questions are set in groups, each group representing one of the variations that may arise in examination papers. The answering techniques are different for each type of question and are classified as follows:

TYPE I

These questions consist of a problem followed by several alternative answers, only *one* of which is correct.
Write down the letter corresponding to the correct answer.

TYPE II

In this type of question some information is given and is followed by a number of possible responses. *One or more* of the suggested responses follow(s) directly from the information given.
Write down the letter(s) corresponding to the correct response(s).
e.g.: PQR is a triangle.
(a) $\widehat{P} + \widehat{Q} + \widehat{R} = 180°$.
(b) PQ + QR is less than PR.
(c) If $\widehat{P}$ is obtuse, $\widehat{Q}$ and $\widehat{R}$ must both be acute.
(d) $\widehat{P} = 90°$, $\widehat{Q} = 45°$, $\widehat{R} = 45°$.
The correct responses are (a) and (c).
(b) is definitely incorrect and (d) may or may not be true of the triangle PQR. There is not sufficient information given to allot a particular value to each angle. Responses of this kind, which require more information than is given, should not be regarded as correct.

TYPE III

Each problem contains two independent statements, (a) and (b).

1) If (a) implies (b) but (b) does not imply (a) write *A*.
2) If (b) implies (a) but (a) does not imply (b) write *B*.
3) If (a) implies (b) *and* (b) implies (a) write *C*.
4) If (a) denies (b) *and* (b) denies (a) write *D*.
5) If none of the first four relationships apply write *E*.

TYPE IV

A problem is introduced and followed by a number of pieces of information. You are not required to solve the problem but to decide whether:
1) the total amount of information given is insufficient to solve the problem. If so write *I*,
2) the given information is *all* needed to solve the problem. In this case write *A*,
3) the problem can be solved without using one or more of the given pieces of information. In this case write down the letter(s) corresponding to the item(s) not needed.

TYPE V

A single statement is made. Write *T* if the statement is true and *F* if the statement is false.

CHAPTER 1

VECTORS. SCALAR PRODUCT. APPLICATIONS

In our previous book, *Mechanics and Probability*, a variety of problems were solved by vector methods, particularly by using Cartesian components. Because much of that work was limited to two dimensions, this book begins with a reminder of these methods as they can be applied to three dimensions.

BASIC FACTS

1) Two vectors $\mathbf{a}$ and $\mathbf{b}$ are parallel if $\mathbf{a} = \lambda\mathbf{b}$.

2) A vector of magnitude V in a direction $\mathbf{d}$ is given by $V\hat{\mathbf{d}}$.

3) A line through a point with position vector $\mathbf{a}$, and in a direction $\mathbf{d}$ has a vector equation $\mathbf{r} = \mathbf{a} + \lambda\mathbf{d}$.

4) For a point P, which at time t has a position vector $\mathbf{r}$, the velocity $\mathbf{v}$ is $\dfrac{d\mathbf{r}}{dt}$ and the acceleration $\mathbf{a}$ is $\dfrac{d\mathbf{v}}{dt}$.

Conversely, $\mathbf{r} = \int \mathbf{v}\,dt$ and $\mathbf{v} = \int \mathbf{a}\,dt$.

5) The scalar product of two vectors $\mathbf{a}$ and $\mathbf{b}$ is $\mathbf{a}.\mathbf{b} = |\mathbf{a}||\mathbf{b}|\cos\theta$ where θ is the angle between the directions of $\mathbf{a}$ and $\mathbf{b}$. It follows that, if $\mathbf{a}$ and $\mathbf{b}$ are perpendicular then $\mathbf{a}.\mathbf{b} = 0$. In particular, if $\mathbf{a} = x_1\mathbf{i} + y_1\mathbf{j} + z_1\mathbf{k}$ and $\mathbf{b} = x_2\mathbf{i} + y_2\mathbf{j} + z_2\mathbf{k}$ then $\mathbf{a}.\mathbf{b} = x_1x_2 + y_1y_2 + z_1z_2$.

6) The work done by a constant force $\mathbf{F}$ in causing a displacement $\mathbf{r}$ is given by $\mathbf{F}.\mathbf{r}$.

7) The kinetic energy of a particle of mass m, moving with velocity $\mathbf{v}$ is given by $\frac{1}{2}m\mathbf{v}.\mathbf{v}$. (i.e. $\frac{1}{2}m\mathbf{v}^2$.)

EXAMPLES 1a

1) Two forces, $\mathbf{F}_1 = 3\mathbf{i}+4\mathbf{j}-2\mathbf{k}$ and $\mathbf{F}_2 = \mathbf{i}-5\mathbf{i}+3\mathbf{k}$, act through points with position vectors $-5\mathbf{i}+6\mathbf{j}-\mathbf{k}$ and $-\mathbf{i}+5\mathbf{j}$ respectively. Find the equations of their lines of action; show that they intersect and hence find the equation of the line of action of their resultant.

The direction vector of the line of action of $\mathbf{F}_1$ is $3\mathbf{i}+4\mathbf{j}-2\mathbf{k}$.
The equation of the line of action of $\mathbf{F}_1$ is therefore

$$\mathbf{r}_1 = -5\mathbf{i}+6\mathbf{j}-\mathbf{k}+\lambda(3\mathbf{i}+4\mathbf{j}-2\mathbf{k})$$

The direction vector of the line of action of $\mathbf{F}_2$ is $\mathbf{i}-5\mathbf{j}+3\mathbf{k}$.
The equation of the line of action of $\mathbf{F}_2$ is therefore

$$\mathbf{r}_2 = -\mathbf{i}+5\mathbf{j}+\mu(\mathbf{i}-5\mathbf{j}+3\mathbf{k})$$

If the forces are concurrent it will be at a point where $\mathbf{r}_1 = \mathbf{r}_2$
i.e. where

$$(3\lambda-5)\mathbf{i}+(4\lambda+6)\mathbf{j}+(-2\lambda-1)\mathbf{k} = (\mu-1)\mathbf{i}+(5-5\mu)\mathbf{j}+3\mu\mathbf{k}$$

Then $\quad 3\lambda-5 = \mu-1 \quad$ (coefficients of $\mathbf{i}$)

$\quad 4\lambda+6 = 5-5\mu \quad$ (coefficients of $\mathbf{j}$)

These give $\quad \lambda = 1$ and $\mu = -1$

With these values, the coefficients of $\mathbf{k}$ become:

$$\left.\begin{array}{ll}\text{from line of action of } \mathbf{F}_1 & -2\lambda-1 = -3 \\ \text{from line of action of } \mathbf{F}_2 & 3\mu = -3\end{array}\right\} \text{equal values}$$

So the forces are concurrent at a point P with position vector

$$\mathbf{r} = -5\mathbf{i}+6\mathbf{j}-\mathbf{k}+1(3\mathbf{i}+4\mathbf{j}-2\mathbf{k})$$

i.e. $\quad \mathbf{r} = -2\mathbf{i}+10\mathbf{j}-3\mathbf{k}$

Since $\mathbf{F}_1$ and $\mathbf{F}_2$ intersect, their resultant $\mathbf{F}$ will also pass through the point of intersection.

$$\begin{aligned}\mathbf{F} &= \mathbf{F}_1+\mathbf{F}_2 \\ &= 3\mathbf{i}+4\mathbf{j}-2\mathbf{k}+\mathbf{i}-5\mathbf{j}+3\mathbf{k} \\ &= 4\mathbf{i}-\mathbf{j}+\mathbf{k}\end{aligned}$$

The equation of the line of action of $\mathbf{F}$ is therefore

$$\mathbf{r} = -2\mathbf{i}+10\mathbf{j}-3\mathbf{k}+\lambda(4\mathbf{i}-\mathbf{j}+\mathbf{k})$$

2) Three forces $\mathbf{F}_1$, $\mathbf{F}_2$ and $\mathbf{F}_3$ act in a plane.

$\mathbf{F}_1 = \mathbf{i}+2\mathbf{j}+3\mathbf{k}$ and acts at the point with position vector $3\mathbf{i}+\mathbf{j}$,
$\mathbf{F}_2 = 4\mathbf{i}-\mathbf{j}+\mathbf{k}$ and acts at the point with position vector $13\mathbf{i}-6\mathbf{j}-3\mathbf{k}$,
$\mathbf{F}_3 = -5\mathbf{i}-\mathbf{j}-4\mathbf{k}$ and acts at the point with position vector $6\mathbf{i}-2\mathbf{j}-2\mathbf{k}$.

Show that the three forces are concurrent and deduce that they are in equilibrium.

The equation of the line of action of $\mathbf{F}_1$ is $\mathbf{r}_1 = 3\mathbf{i}+\mathbf{j}+\lambda(\mathbf{i}+2\mathbf{j}+3\mathbf{k})$.
The equation of the line of action of $\mathbf{F}_2$ is $\mathbf{r}_2 = 13\mathbf{i}-6\mathbf{j}-3\mathbf{k}+\mu(4\mathbf{i}-\mathbf{j}+\mathbf{k})$.
If these lines intersect it will be where

$$\left.\begin{aligned} 3+\lambda &= 13+4\mu \\ 1+2\lambda &= -6-\mu \end{aligned}\right\} \Rightarrow \quad \mu = -3, \quad \lambda = -2$$

These values of λ and μ give $\mathbf{r}_1 = \mathbf{i}-3\mathbf{j}-6\mathbf{k}$

$$\mathbf{r}_2 = \mathbf{i}-3\mathbf{j}-6\mathbf{k}$$

Hence the lines of action of $\mathbf{F}_1$ and $\mathbf{F}_2$ meet at the point which has position vector $\mathbf{i}-3\mathbf{j}-6\mathbf{k}$.
Now the equation of the line of action of $\mathbf{F}_3$ is

$$\mathbf{r}_3 = 6\mathbf{i}-2\mathbf{j}-2\mathbf{k}+\eta(-5\mathbf{i}-\mathbf{j}-4\mathbf{k})$$

This line also passes through $\mathbf{i}-3\mathbf{j}-6\mathbf{k}$ (using $\eta = 1$).

So all three lines of action are concurrent.

Also $\mathbf{F}_1+\mathbf{F}_2+\mathbf{F}_3 = (\mathbf{i}+2\mathbf{j}+3\mathbf{k})+(4\mathbf{i}-\mathbf{j}-\mathbf{k})+(-5\mathbf{i}-\mathbf{j}-4\mathbf{k}) = \mathbf{0}$.
Therefore the resultant of the three concurrent forces is zero and hence the forces are in equilibrium.

3) A particle P is moving with a constant speed of $6\,\mathrm{m\,s^{-1}}$ in a direction $2\mathbf{i}-\mathbf{j}-2\mathbf{k}$. When $t = 0$, P is at a point with position vector $3\mathbf{i}+4\mathbf{j}-7\mathbf{k}$. Find the position vector of P after (a) t seconds, (b) 4 seconds.

If $\mathbf{v}$ is the velocity vector of P then the direction of $\mathbf{v}$ is $2\mathbf{i}-\mathbf{j}-2\mathbf{k}$.

Hence $\hat{\mathbf{v}} = \frac{1}{3}(2\mathbf{i}-\mathbf{j}-2\mathbf{k})$

But $|\mathbf{v}| = 6$

Therefore $\mathbf{v} = |\mathbf{v}|\hat{\mathbf{v}} = 4\mathbf{i}-2\mathbf{j}-4\mathbf{k}$

If $\mathbf{d}$ is the displacement vector of P in time t,

$$\mathbf{d} = t(4\mathbf{i}-2\mathbf{j}-4\mathbf{k})$$

(a) P is initially at $3\mathbf{i}+4\mathbf{j}-7\mathbf{k}$, so if $\mathbf{r}$ is the position vector of P after time t then

$$\mathbf{r} = 3\mathbf{i}+4\mathbf{j}-7\mathbf{k}+t(4\mathbf{i}-2\mathbf{j}-4\mathbf{k})$$

(b) When $t = 4$

$$\mathbf{r} = 3\mathbf{i}+4\mathbf{j}-7\mathbf{k}+4(4\mathbf{i}-2\mathbf{j}-4\mathbf{k})$$

Hence $$\mathbf{r} = 19\mathbf{i}-4\mathbf{j}-23\mathbf{k}$$

4) Two particles A and B pass simultaneously through two points A_0 and B_0 with position vectors $\mathbf{a}_0$ and $\mathbf{b}_0$. A and B are moving with constant velocities $\mathbf{v}_A$ and $\mathbf{v}_B$.
If $\mathbf{a}_0 = \mathbf{i}+4\mathbf{j}-26\mathbf{k}$, $\mathbf{b}_0 = 16\mathbf{i}+\mathbf{j}-2\mathbf{k}$, $\mathbf{v}_A = 4\mathbf{i}+\mathbf{j}+5\mathbf{k}$ and $\mathbf{v}_B=-\mathbf{i}+2\mathbf{j}-3\mathbf{k}$, show that A and B will collide.
Find the time at which collision occurs and the position vector of the point of collision.

If at time t after passing through A_0 and B_0, the position vectors of A and B are $\mathbf{r}_A$ and $\mathbf{r}_B$

then
$$\begin{cases} \mathbf{r}_A = \mathbf{i}+4\mathbf{j}-26\mathbf{k}+t(4\mathbf{i}+\mathbf{j}+5\mathbf{k}) \\ \mathbf{r}_B = 16\mathbf{i}+\mathbf{j}-2\mathbf{k}+t(-\mathbf{i}+2\mathbf{j}-3\mathbf{k}) \end{cases}$$

or
$$\begin{cases} \mathbf{r}_A = (1+4t)\mathbf{i}+(4+t)\mathbf{j}+(-26+5t)\mathbf{k} \\ \mathbf{r}_B = (16-t)\mathbf{i}+(1+2t)\mathbf{j}+(-2-3t)\mathbf{k} \end{cases}$$

Now A and B will collide if there is a particular value of t for which $\mathbf{r}_A=\mathbf{r}_B$.

Equating coefficients of $\mathbf{i}$ gives $1+4t = 16-t$ so $t=3$
A collision will occur if, when $t=3$, the coefficients of $\mathbf{j}$ and $\mathbf{k}$ are also equal for A and B.

Using $t=3$, $$\left.\begin{aligned} \mathbf{r}_A &= 13\mathbf{i}+7\mathbf{j}-11\mathbf{k} \\ \mathbf{r}_B &= 13\mathbf{i}+7\mathbf{j}-11\mathbf{k} \end{aligned}\right\} \text{ equal coefficients of } \mathbf{i}, \mathbf{j} \text{ and } \mathbf{k}$$

Hence, when $t=3$, A and B collide at a point with position vector
$$13\mathbf{i}+7\mathbf{j}-11\mathbf{k}$$

5) Two particles are moving with constant velocities. At a particular instant the particle with velocity vector $3\mathbf{i}-4\mathbf{j}+7\mathbf{k}$ passes through a point with position vector $\mathbf{i}+2\mathbf{j}-3\mathbf{k}$ and the particle with velocity vector $2\mathbf{i}-6\mathbf{j}+5\mathbf{k}$ passes through the origin. Find the shortest distance between the particles in the ensuing motion and the time when they are closest together.

Let the two particles be A and B.

Then $$\mathbf{v}_A = 3\mathbf{i} - 4\mathbf{j} + 7\mathbf{k}$$

$$\mathbf{v}_B = 2\mathbf{i} - 6\mathbf{j} + 5\mathbf{k}$$

At time t $$\mathbf{r}_A = \mathbf{i} + 2\mathbf{j} - 3\mathbf{k} + t(3\mathbf{i} - 4\mathbf{j} + 7\mathbf{k})$$

$$\mathbf{r}_B = \mathbf{0} + t(2\mathbf{i} - 6\mathbf{j} + 5\mathbf{k})$$

So that $$\mathbf{r}_A - \mathbf{r}_B = \mathbf{i} + 2\mathbf{j} - 3\mathbf{k} + t(\mathbf{i} + 2\mathbf{j} + 2\mathbf{k})$$

The distance between A and B at time t is given by

$$l = |\mathbf{r}_A - \mathbf{r}_B|$$

$$l^2 = (t+1)^2 + (2t+2)^2 + (2t-3)^2$$

So that $$l^2 = 9t^2 - 2t + 14$$

The least value of l can now be found using either of the following methods.

(a) $$l^2 = 9t^2 - 2t + 14$$

Therefore $$\frac{l^2}{9} = t^2 - \frac{2}{9}t + \frac{14}{9}$$

$$= \left(t - \frac{1}{9}\right)^2 + \frac{14}{9} - \left(\frac{1}{9}\right)^2$$

The expression on the right is least when $\left(t - \frac{1}{9}\right)^2 = 0$

Hence l is least when $t = \frac{1}{9}$, and then

$$l^2 = 9\left(\frac{14}{9} - \frac{1}{81}\right) = \frac{125}{9}$$

So the shortest distance between A and B is $\frac{5\sqrt{5}}{3}$ units

and occurs when $t = \frac{1}{9}$.

(b) $$l^2 = 9t^2 - 2t + 14$$

$$\frac{d(l^2)}{dt} = 18t - 2$$

and $$\frac{d^2(l^2)}{dt^2} = 18$$

When l^2 is least, $\dfrac{d(l^2)}{dt} = 0$, therefore $t = \dfrac{1}{9}$

and since $\dfrac{d^2(l^2)}{dt^2}$ is positive, l^2 is minimum when $t = \dfrac{1}{9}$.

Then $$l^2 = 9\left(\frac{1}{9}\right)^2 - 2\left(\frac{1}{9}\right) + 14 = 14 - \frac{1}{9}$$

So the minimum value of l is $\dfrac{5\sqrt{5}}{3}$ units.

6) Two spheres A and B, with equal radii 3 units, are moving with constant velocity vectors $\mathbf{i}+5\mathbf{j}-3\mathbf{k}$ and $2\mathbf{i}+9\mathbf{j}-11\mathbf{k}$ respectively. Their centres pass simultaneously through two points with position vectors $6\mathbf{i}+3\mathbf{j}+\mathbf{k}$ and $\mathbf{i}+3\mathbf{j}+11\mathbf{k}$ respectively. Show that they collide and find the time which elapses before impact.

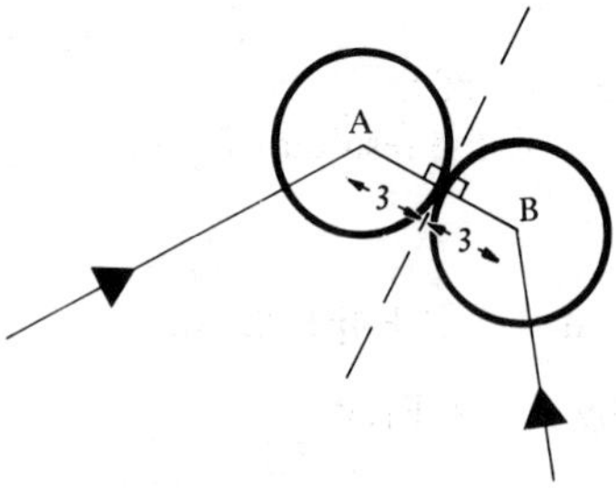

If the spheres collide, their centres will be exactly 6 units apart at the instant of impact.
Considering the motion of the centre of each sphere, their position vectors at time t are respectively

$$\begin{aligned}\mathbf{r}_A &= 6\mathbf{i}+3\mathbf{j}+\mathbf{k}+t(\mathbf{i}+5\mathbf{j}-3\mathbf{k})\\ &= (6+t)\mathbf{i}+(3+5t)\mathbf{j}+(1-3t)\mathbf{k}\\ \mathbf{r}_B &= \mathbf{i}+3\mathbf{j}+11\mathbf{k}+t(2\mathbf{i}+9\mathbf{j}-11\mathbf{k})\\ &= (1+2t)\mathbf{i}+(3+9t)\mathbf{j}+(11-11t)\mathbf{k}\end{aligned}$$

The distance l between the centres at any time t is given by

$$l = |\mathbf{r}_A - \mathbf{r}_B|$$

But $$\mathbf{r}_A - \mathbf{r}_B = (5-t)\mathbf{i}+(-4t)\mathbf{j}+(-10+8t)\mathbf{k}$$

Therefore $$\begin{aligned}l^2 &= (5-t)^2+(-4t)^2+(-10+8t)^2\\ &= 81t^2-170t+125\end{aligned}$$

Now for collision there must be a value of t for which $l = 6$.

So that
$$36 = 81t^2 - 170t + 125 \qquad [1]$$
$$0 = 81t^2 - 170t + 89$$
$$0 = (81t - 89)(t - 1)$$

Hence
$$t = 1 \quad \text{or} \quad t = \tfrac{89}{81}$$

But the spheres will collide only once and that will be after the shorter time interval.
So the spheres *do* collide after 1 unit of time.

Note: Had the roots of equation [1] been complex the conclusion would have been that A and B never collide.

7) Three forces $\mathbf{F}_1 = 2\mathbf{i} + 4\mathbf{j} - \mathbf{k}$, $\mathbf{F}_2 = 3\mathbf{i} - 7\mathbf{j} + 6\mathbf{k}$ and $\mathbf{F}_3 = \mathbf{i} + 5\mathbf{j} - 4\mathbf{k}$ are acting on a particle which undergoes a displacement $12\mathbf{i} + 4\mathbf{j} + 2\mathbf{k}$.
Find the work done by each of the forces.
Find also the work done by the resultant force.

If
$$\mathbf{d} = 12\mathbf{i} + 4\mathbf{j} + 2\mathbf{k}$$

Then the work done by $\mathbf{F}_1$ is $\mathbf{F}_1 . \mathbf{d}$
$$= (2\mathbf{i} + 4\mathbf{j} - \mathbf{k}) . (12\mathbf{i} + 4\mathbf{j} + 2\mathbf{k})$$
$$= 24 + 16 - 2$$
$$= 38 \text{ units}$$

The work done by $\mathbf{F}_2$ is $\mathbf{F}_2 . \mathbf{d}$
$$= (3\mathbf{i} - 7\mathbf{j} + 6\mathbf{k}) . (12\mathbf{i} + 4\mathbf{j} + 2\mathbf{k})$$
$$= 20 \text{ units}$$

The work done by $\mathbf{F}_3$ is $(\mathbf{i} + 5\mathbf{j} - 4\mathbf{k}) . (12\mathbf{i} + 4\mathbf{j} + 2\mathbf{k})$
$$= 24 \text{ units}$$

The resultant force $\mathbf{F}$ is $\mathbf{F}_1 + \mathbf{F}_2 + \mathbf{F}_3$
$$= 6\mathbf{i} + 2\mathbf{j} + \mathbf{k}$$

So the work done by $\mathbf{F}$ is $(6\mathbf{i} + 2\mathbf{j} + \mathbf{k}) . (12\mathbf{i} + 4\mathbf{j} + 2\mathbf{k})$
$$= 72 + 8 + 2$$
$$= 82 \text{ units}$$

Note: It can be seen that, in the previous example, the sum of the individual work done by $\mathbf{F}_1$, $\mathbf{F}_2$ and $\mathbf{F}_3$ is equal to the work done by the resultant force $\mathbf{F}$. This relationship applies to any set of forces acting on a particle and its validity can be proved as follows.

Suppose that a set of forces $\mathbf{F}_1, \mathbf{F}_2, \mathbf{F}_3, \ldots, \mathbf{F}_n$, act on a particle, causing it to be moved through a displacement $\mathbf{d}$.
The total work done by all the forces in the set is given by

$$\mathbf{F}_1.\mathbf{d}+\mathbf{F}_2.\mathbf{d}+\mathbf{F}_3.\mathbf{d}+\ldots+\mathbf{F}_n.\mathbf{d}$$

$$= \mathbf{d}.(\mathbf{F}_1+\mathbf{F}_2+\mathbf{F}_3+\ldots+\mathbf{F}_n)$$

But the resultant force $\mathbf{F} = \mathbf{F}_1+\mathbf{F}_2+\mathbf{F}_3+\ldots+\mathbf{F}_n$

Hence the work done by $\mathbf{F}$ is given by

$$\mathbf{F}.\mathbf{d} = (\mathbf{F}_1+\mathbf{F}_2+\mathbf{F}_3+\ldots+\mathbf{F}_n).\mathbf{d}$$

i.e. when a set of forces act on a particle and displace it, the total work done by the individual forces is equal to the work done by the resultant force.

EXERCISE 1a

1) Two forces $\mathbf{F}_1$ and $\mathbf{F}_2$ act on a particle at a point A with position vector $3\mathbf{i}+2\mathbf{j}+\mathbf{k}$.
$\mathbf{F}_1$ is of magnitude 14N and is in the direction of the vector $6\mathbf{i}+3\mathbf{j}-2\mathbf{k}$,
$\mathbf{F}_2 = 4\mathbf{i}+7\mathbf{j}+6\mathbf{k}$.
Find a vector equation for the line of action of the resultant of $\mathbf{F}_1$ and $\mathbf{F}_2$.
A third force $\mathbf{F}_3$ has a line of action with equation

$$\mathbf{r} = \mathbf{i}-3\mathbf{j}+4\mathbf{k}+\lambda(2\mathbf{i}+5\mathbf{j}-3\mathbf{k})$$

Show that $\mathbf{F}_3$ also acts on the particle at A.

2) Find the equation of the line of action of the resultant of forces $\mathbf{F}_1$ and $\mathbf{F}_2$ if the equations of the lines of action of $\mathbf{F}_1$ and $\mathbf{F}_2$ are respectively

$$\mathbf{r}_1 = 2\mathbf{i}-7\mathbf{j}+\mathbf{k}+\lambda(\mathbf{i}+4\mathbf{j}+2\mathbf{k}) \quad \text{and} \quad \mathbf{r}_2 = 3\mathbf{i}-4\mathbf{j}+\mu(\mathbf{i}+5\mathbf{j}+5\mathbf{k}).$$

and the magnitudes of $\mathbf{F}_1$ and $\mathbf{F}_2$ are $\sqrt{21}$ and $2\sqrt{51}$ respectively.

3) Two forces $\mathbf{F}_1$ and $\mathbf{F}_2$ act through points with position vectors $\mathbf{a}_1$ and $\mathbf{a}_2$ respectively:

$$\mathbf{F}_1 = 2\mathbf{i}-2\mathbf{k}, \quad \mathbf{F}_2 = \mathbf{i}-2\mathbf{j}+\mathbf{k}, \quad \mathbf{a}_1 = \mathbf{i}+2\mathbf{j}-3\mathbf{k} \quad \text{and} \quad \mathbf{a}_2 = 4\mathbf{i}-p\mathbf{k}$$

Find the value of p if the lines of action of $\mathbf{F}_1$ and $\mathbf{F}_2$ intersect.
Find a vector equation of the line of action of the resultant of $\mathbf{F}_1$ and $\mathbf{F}_2$.

4) Three forces $\mathbf{F}_1$, $\mathbf{F}_2$ and $\mathbf{F}_3$ act on a particle.
$\mathbf{F}_1$ has magnitude 14 N and acts in a direction parallel to the line with vector equation $\mathbf{r} = \mathbf{i} - 3\mathbf{j} + 8\mathbf{k} + \lambda(6\mathbf{i} - 2\mathbf{j} + 3\mathbf{k})$.

$\mathbf{F}_2 = a\mathbf{i} + b\mathbf{j} + c\mathbf{k}$.

$\mathbf{F}_3$ is represented in magnitude and direction by the line from A to B where A and B have position vectors $7\mathbf{i} - 2\mathbf{j} + 5\mathbf{k}$ and $3\mathbf{i} + \mathbf{j} + \mathbf{k}$ respectively. Find the values of a, b and c if the particle does not accelerate.

5) The position vectors of the vertices Q and R of a triangle PQR are respectively $8\mathbf{i} + 3\mathbf{j} + 5\mathbf{k}$ and $6\mathbf{i} + 4\mathbf{j} + 9\mathbf{k}$. Two forces $6\mathbf{i} + 4\mathbf{j} + 2\mathbf{k}$ and $8\mathbf{i} + 10\mathbf{j} + 12\mathbf{k}$ act along PQ and PR respectively. A third force $\mathbf{F}$ acts through P.
If the system of forces is in equilibrium, find

(a) the magnitude of the force $\mathbf{F}$,

(b) the position vector of P,

(c) the equation of the line of action of $\mathbf{F}$ in vector form.

6) P and Q are particles moving with constant velocities $\mathbf{v}_P$ and $\mathbf{v}_Q$. Determine whether they collide and if they do, find the position vector of the point where collision occurs, given that

(a) $\mathbf{v}_P = 3\mathbf{i} - \mathbf{j} - 2\mathbf{k}$ and $\mathbf{v}_Q = -7\mathbf{i} + 2\mathbf{j} + \mathbf{k}$. When $t = 0$, P is at the point $(0, 7, 0)$ and Q is at the point $(18, 1, 2)$.

(b) $\mathbf{v}_P = 3\mathbf{i} - 2\mathbf{j} - \mathbf{k}$ and $\mathbf{v}_Q = 3\mathbf{i} + 4\mathbf{j} + 5\mathbf{k}$. P is at a point with position vector $-8\mathbf{i} + 4\mathbf{j} + 4\mathbf{k}$ when $t = 0$ and one second later Q passes through the point with position vector $-5\mathbf{i} - 10\mathbf{j} - 9\mathbf{k}$.

(c) $\mathbf{v}_P = \mathbf{i} - \mathbf{j} - \mathbf{k}$ and $\mathbf{v}_Q = 2\mathbf{i} + 5\mathbf{j} - 3\mathbf{k}$. P and Q are at points $(1, 4, 3)$ and $(7, 6, 2)$ when $t = 0$.

7) Two particles P and Q are observed simultaneously when they are at points with position vectors $\mathbf{p}$ and $\mathbf{q}$. If they are travelling with constant velocity vectors $\mathbf{v}_P$ and $\mathbf{v}_Q$ respectively, find their distance apart after time t and hence find the least distance between them, given that

(a) $\mathbf{p} = -\mathbf{i} + 8\mathbf{j} - 4\mathbf{k}$ $\quad \mathbf{q} = 4\mathbf{i} + 6\mathbf{j} - 15\mathbf{k}$
$\mathbf{v}_P = 5\mathbf{i} + \mathbf{j} - 6\mathbf{k}$ $\quad \mathbf{v}_Q = -3\mathbf{i} + \mathbf{j} - \mathbf{k}$

(b) $\mathbf{p} = -3\mathbf{i} - 3\mathbf{j} - 3\mathbf{k}$ $\quad \mathbf{q} = -7\mathbf{i} - 4\mathbf{j} - 2\mathbf{k}$
$\mathbf{v}_P = \mathbf{i} + \mathbf{j} + \mathbf{k}$ $\quad \mathbf{v}_Q = 3\mathbf{i} + 2\mathbf{j} + \mathbf{k}$

8) Three particles A, B and C, moving with constant velocities $\mathbf{v}_A$, $\mathbf{v}_B$ and $\mathbf{v}_C$ are observed at position vectors **a**, **b** and **c** at the times indicated below. Two of the three will collide. Find which two are on collision courses and determine the time and position of the collision.

$\mathbf{v}_A = 2\mathbf{i}+4\mathbf{j}-7\mathbf{k}$;	$\mathbf{a} = -5\mathbf{i}-10\mathbf{j}+24\mathbf{k}$	at a certain time,
$\mathbf{v}_B = \mathbf{i}-\mathbf{j}+2\mathbf{k}$;	$\mathbf{b} = 3\mathbf{i}+4\mathbf{j}-\mathbf{k}$	one second later,
$\mathbf{v}_C = -5\mathbf{i}+\mathbf{j}+6\mathbf{k}$;	$\mathbf{c} = 6\mathbf{i}+\mathbf{j}-3\mathbf{k}$	two seconds later.

9) A particle A moving with a constant velocity $\mathbf{i}+\mathbf{j}+\mathbf{k}$ passes through a point with position vector $3\mathbf{i}-7\mathbf{j}-4\mathbf{k}$ at the same instant as a particle B passes through a point with position vector $-\mathbf{i}+\mathbf{j}+p\mathbf{k}$. B has a constant velocity vector $2\mathbf{i}-\mathbf{j}-5\mathbf{k}$.

(a) Find the velocity of B relative to A.

(b) Find the value of p if A and B collide.

(c) If $p=-\frac{1}{2}$, find the shortest distance between A and B in the subsequent motion.

10) Three particles A, B and C are moving with constant velocity vectors $\mathbf{v}_A$, $\mathbf{v}_B$ and $\mathbf{v}_C$. If $\mathbf{v}_C=-\mathbf{i}+\mathbf{j}+3\mathbf{k}$ and the velocities of A and B relative to C are $-3\mathbf{i}-8\mathbf{k}$ and $3\mathbf{i}-3\mathbf{j}-10\mathbf{k}$ respectively, find $\mathbf{v}_A$ and $\mathbf{v}_B$. At the same instant A and B are at points with position vectors $11\mathbf{i}-2\mathbf{j}+16\mathbf{k}$ and $-7\mathbf{i}+7\mathbf{j}+22\mathbf{k}$ respectively. Show that A and B collide.

11) Forces of magnitudes 6, 7 and 9 N act in the directions $2\mathbf{i}+2\mathbf{j}+\mathbf{k}$, $6\mathbf{i}-3\mathbf{j}+2\mathbf{k}$ and $7\mathbf{i}+4\mathbf{j}-4\mathbf{k}$ respectively. These forces act on a particle causing a displacement $34\mathbf{i}+10\mathbf{j}$ (metres). Find the work done by each force and show that the total work done by the three forces is equal to the work done by the resultant force.

12) Two constant forces $\mathbf{F}_1$ and $\mathbf{F}_2$ are the only forces acting on a particle P, of mass 4 kg, which is initially at rest at a point with position vector $\mathbf{i}+2\mathbf{j}-\mathbf{k}$. Four seconds later P is at the point with position vector $9\mathbf{i}-2\mathbf{j}+11\mathbf{k}$. If $\mathbf{F}_1=3\mathbf{i}-\mathbf{j}+4\mathbf{k}$ find

(a) the work done by $\mathbf{F}_1$ during the four seconds,

(b) the force vector $\mathbf{F}_2$,

(c) the total work done during the four seconds.

13) Forces $\mathbf{F}_1$, $\mathbf{F}_2$ and $\mathbf{F}_3$ act for a time t on a particle of mass 6 kg which undergoes a displacement from rest of $2\mathbf{i}+7\mathbf{j}-\mathbf{k}$ (metres). If $\mathbf{F}_1=\mathbf{i}+3\mathbf{k}$, $\mathbf{F}_2=4\mathbf{j}-\mathbf{k}$, $\mathbf{F}_3=2\mathbf{i}-3\mathbf{j}-\mathbf{k}$ (newtons) find the total work done by the three forces.

If no other work is done, use the principle of work and energy to find the speed of the particle at time t.

FURTHER APPLICATIONS OF THE SCALAR PRODUCT

Resolving a Vector in a Specified Direction

If a vector $\mathbf{v}$ is inclined at an angle θ to a direction vector $\mathbf{d}$, the component of $\mathbf{v}$ in the direction $\mathbf{d}$ is of magnitude $|\mathbf{v}|\cos\theta$.

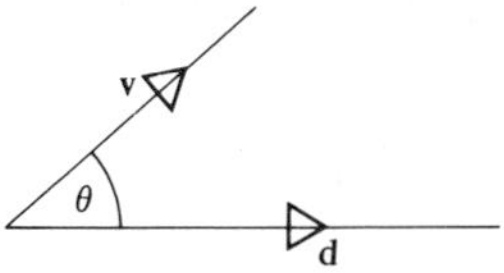

But $$|\mathbf{v}||\mathbf{d}|\cos\theta = \mathbf{v}.\mathbf{d}$$

Hence $$|\mathbf{v}|\cos\theta = \frac{\mathbf{v}.\mathbf{d}}{|\mathbf{d}|} = \mathbf{v}.\hat{\mathbf{d}}$$

This property is very useful in those mechanics problems where a vector has to be resolved in a pair of perpendicular directions in a plane.

Consider, for instance, resolving the vector $2\mathbf{i} + 5\mathbf{j}$, parallel and perpendicular to the direction vector $3\mathbf{i}+4\mathbf{j}$.

If $\mathbf{d}_1 = 3\mathbf{i}+4\mathbf{j}$ then a vector $\mathbf{d}_2$, perpendicular to $\mathbf{d}_1$ can be taken as $4\mathbf{i}-3\mathbf{j}$.

i.e. $$\hat{\mathbf{d}}_1 = \tfrac{1}{5}(3\mathbf{i}+4\mathbf{j}) \quad \text{and} \quad \hat{\mathbf{d}}_2 = \tfrac{1}{5}(4\mathbf{i}-3\mathbf{j})$$

We can now use $\mathbf{v}.\hat{\mathbf{d}}$ to calculate the magnitudes of the components of the vector $2\mathbf{i}+5\mathbf{j}$ in the directions $\mathbf{d}_1$ and $\mathbf{d}_2$, giving

$$(2\mathbf{i}+5\mathbf{j}).\tfrac{1}{5}(3\mathbf{i}+4\mathbf{j}) = \tfrac{26}{5} \quad \text{in the direction } \mathbf{d}_1$$

and $$(2\mathbf{i}+5\mathbf{j}).\tfrac{1}{5}(4\mathbf{i}-3\mathbf{j}) = -\tfrac{7}{5} \quad \text{in the direction } \mathbf{d}_2$$

The negative value, $-\frac{7}{5}$, indicates a component in the direction $-(4\mathbf{i}-3\mathbf{j})$.

Note that we could equally well have used $-4\mathbf{i}+3\mathbf{j}$ as $\mathbf{d}_2$, since this vector also is perpendicular to $\mathbf{d}_1$. If we had chosen $-4\mathbf{i}+3\mathbf{j}$, the second component would have been $+\frac{7}{5}$.

The Power of a Variable Force

Consider a force $\mathbf{F}$ which, at time t, is applied at a point with position vector $\mathbf{r}$.
If, in time δt, the point of application moves through a displacement $\delta\mathbf{r}$, the work done by $\mathbf{F}$ in the time δt is given by

$$\delta\omega \simeq \mathbf{F}.\delta\mathbf{r}$$

The power of the force $\mathbf{F}$ at time t is given approximately by $\dfrac{\delta\omega}{\delta t}$ where

$$\frac{\delta\omega}{\delta t} \simeq \frac{\mathbf{F}.\delta\mathbf{r}}{\delta t} = \mathbf{F}.\frac{\delta\mathbf{r}}{\delta t}$$

$\Rightarrow$ Power $= \mathbf{F}.\mathbf{v}$ where $\mathbf{v}$ is the velocity of the point of application of $\mathbf{F}$ at time t.

The Work Done by a Variable Force

The power, P, of a force is the rate at which it is doing work.

Therefore the work done is given by $\int P\,\mathrm{d}t$

i.e. $$\text{work done} = \int \mathbf{F}.\mathbf{v}\,\mathrm{d}t$$

EXAMPLES 1b

1) Relative to a fixed point O, the position vector of a moving particle P at any time t $(t \geqslant 0)$, is given by $\mathbf{r} = (10-t^2)\mathbf{i}+3t^2\mathbf{j}-4t\mathbf{k}$.
Show that there are only two instants when the direction of motion of P is perpendicular to OP. Find also the value of t when the acceleration of P is perpendicular to OP.

For P, $$\mathbf{r} = (10-t^2)\mathbf{i}+3t^2\mathbf{j}-4t\mathbf{k}$$

$\Rightarrow$ $$\mathbf{v} = -2t\mathbf{i}+6t\mathbf{j}-4\mathbf{k}$$

and $$\mathbf{a} = -2\mathbf{i}+6\mathbf{j}$$

The direction of motion $(\mathbf{v})$ is perpendicular to OP when

$$\mathbf{r}.\mathbf{v} = 0$$

i.e. when $(10-t^2)(-2t)+(3t^2)(6t)+(-4t)(-4) = 0$

$\Rightarrow$ $$-4t+20t^3 = 0$$

$\Rightarrow$ $t = 0$ or $\frac{1}{5}\sqrt{5}$ $\qquad (t \geqslant 0$ so $t \neq -\frac{1}{5}\sqrt{5})$

Therefore there are only two values of t when $\mathbf{v}$ is perpendicular to $\mathbf{r}$.

When the acceleration of P is perpendicular to OP,

$$\mathbf{r}.\mathbf{a} = 0 \Rightarrow (10-t^2)(-2)+3t^2(6) = 0$$

$$\Rightarrow t^2 = 1$$

$$\Rightarrow t = 1 \qquad (t \neq -1)$$

2) Find the magnitude of the component of a vector $\mathbf{v} = 4\mathbf{i} - 3\mathbf{j} + \mathbf{k}$ in the direction (a) $\mathbf{i} - 2\mathbf{j}$ (b) $2\mathbf{i} - \mathbf{j} - 12\mathbf{k}$ (c) $\mathbf{i} + 2\mathbf{j} + 2\mathbf{k}$.
What conclusions can you draw in parts (b) and (c)?

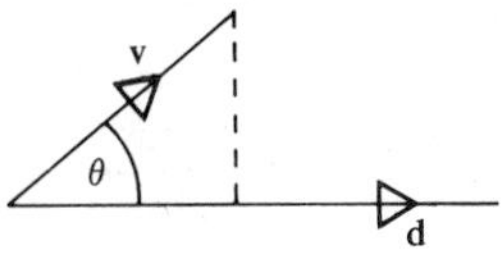

The magnitude of the component of $\mathbf{v}$ in the direction $\mathbf{d}$ is $|\mathbf{v}|\cos\theta = \mathbf{v}.\hat{\mathbf{d}}$

(a) When $\mathbf{d} = \mathbf{i} - 2\mathbf{j}, \quad \hat{\mathbf{d}} = (\mathbf{i} - 2\mathbf{j})/\sqrt{5}$

Hence the required component of $\mathbf{v}$ is of magnitude

$$(4\mathbf{i} - 3\mathbf{j} + \mathbf{k}).(\mathbf{i} - 2\mathbf{j})/\sqrt{5} = (4 + 6 + 0)/\sqrt{5} = 2\sqrt{5}$$

(b) When $\mathbf{d} = 2\mathbf{i} - \mathbf{j} - 12\mathbf{k}, \quad \hat{\mathbf{d}} = (2\mathbf{i} - \mathbf{j} - 12\mathbf{k})/\sqrt{149}$

The required component of $\mathbf{v}$ is therefore of magnitude

$$(4\mathbf{i} - 3\mathbf{j} + \mathbf{k}).(2\mathbf{i} - \mathbf{j} - 12\mathbf{k})/\sqrt{149} = -1/\sqrt{149}$$

(c) When $\mathbf{d} = \mathbf{i} + 2\mathbf{j} + 2\mathbf{k}$ the magnitude of the required component of $\mathbf{v}$ is

$$(4\mathbf{i} - 3\mathbf{j} + \mathbf{k}).(\mathbf{i} + 2\mathbf{j} + 2\mathbf{k})/3 = (4 - 6 + 2)/3 = 0$$

In part (b) the sign of the component is negative showing that the actual component of $\mathbf{v}$ is in the direction $-(2\mathbf{i} - \mathbf{j} - 12\mathbf{k})$.

In part (c), since the component of $\mathbf{v}$ in the direction of $\mathbf{d}$ is zero we conclude that $\mathbf{v}$ is perpendicular to $\mathbf{d}$.

3) The position vector $\mathbf{r}$ of a moving particle of unit mass at time t is given by

$$\mathbf{r} = (4 + 10t)\mathbf{i} + (15 + 20t - 5t^2)\mathbf{j} + (10t^3 - 20)\mathbf{k}$$

Find the work done by the resultant force $\mathbf{F}$ acting on the particle in the time interval from $t = 0$ to $t = 1$. Find also the power of the force when $t = 2$. (Units are kg, m and s.)

The resultant force $\mathbf{F}$ is given by

$$\mathbf{F} = m\mathbf{a} \quad \text{where} \quad m = 1$$

If $\quad \mathbf{r} = (4 + 10t)\mathbf{i} + (15 + 20t - 5t^2)\mathbf{j} + (10t^3 - 20)\mathbf{k}$

then $\quad \mathbf{v} = 10\mathbf{i} + (20 - 10t)\mathbf{j} + 30t^2\mathbf{k}$

and $\quad \mathbf{a} = -10\mathbf{j} + 60t\mathbf{k}$

$\Rightarrow \quad \mathbf{F} = -10\mathbf{j} + 60t\mathbf{k}$

The work done by **F** from $t = 0$ to $t = 1$ is given by

$$W = \int_0^1 \mathbf{F}.\mathbf{v}\,dt$$

$$= \int_0^1 (-10\mathbf{j} + 60t\mathbf{k}).\left\{10\mathbf{i} + (20 - 10t)\mathbf{j} + 30t^2\mathbf{k}\right\} dt$$

$$= \int_0^1 \left\{-10(20 - 10t) + 1800t^3\right\} dt$$

$$= \left[450t^4 + 50t^2 - 200t\right]_0^1$$

$$= 300$$

Therefore the work done is 300 J.

When $t = 2$

$$\mathbf{F} = -10\mathbf{j} + 120\mathbf{k}$$

$$\mathbf{v} = 10\mathbf{i} + 120\mathbf{k}$$

Power is given by

$$\mathbf{F}.\mathbf{v} = (-10\mathbf{j} + 120\mathbf{k}).(10\mathbf{i} + 120\mathbf{k})$$

$$= 14\,400$$

Therefore the power of the force when $t = 2$ is 14.4 kW.

EXERCISE 1b

1) Resolve a vector **v** parallel and perpendicular to a direction vector **d** if:
(a) $\mathbf{v} = 2\mathbf{i} + 3\mathbf{j}$, $\mathbf{d} = \mathbf{i} - \mathbf{j}$,
(b) $\mathbf{v} = 4\mathbf{i} - 3\mathbf{j}$, $\mathbf{d} = 2\mathbf{i} + \mathbf{j}$,
(c) $\mathbf{v} = \mathbf{i} + \mathbf{j}$, $\mathbf{d} = 3\mathbf{i} + 4\mathbf{j}$.
In each case, sketch **v** and the two components in the appropriate directions.

2) Resolve a velocity $\mathbf{v} = 3\mathbf{i} + 2\mathbf{j}$ parallel and perpendicular to the line whose vector equation is $\mathbf{r} = \mathbf{i} - 4\mathbf{j} + \lambda(4\mathbf{i} - \mathbf{j})$.

3) At any time t, the position vector of a particle P is given by

$$\mathbf{r} = \cos t\,\mathbf{i} + 2\sin t\,\mathbf{j} + \sin 2t\,\mathbf{k}$$

where **i**, **j** and **k** are mutually perpendicular vectors each of magnitude 1 m. One of the forces acting on P is given by

$$\mathbf{F} = 2\cos t\,\mathbf{i} + \cos 2t\,\mathbf{j}$$

Find the rate at which **F** is working at time t, and hence find the work done by **F** from $t = 0$ to $t = \frac{1}{2}\pi$.

4) A particle of mass 5 kg moves so that its position vector $\mathbf{r}$ at time t is given by $\mathbf{r} = t^2\mathbf{i} + \sin t\mathbf{j} + \cos t\mathbf{k}$

Find:
(a) the resultant force acting on the particle,
(b) the rate at which the resultant force is working at any time t,
(c) the work done by the resultant force in the time interval $t = 0$ to $t = 4$.
Show that, initially, the acceleration of the particle is perpendicular to its direction of motion.

5) A force $\mathbf{F}$ acts on a particle P of mass m. Find the power of the force at any time t, the work done by the force between $t = 0$ and $t = 2$, and the position vector of P at time t if $\mathbf{F} = 3\mathbf{i} - 4t\mathbf{j} + 6t^2\mathbf{k}$, $m = 1$ and P is initially at rest at the origin.
Deduce the kinetic energy of the particle when $t = 2$.

6) A force given by $\mathbf{F} = 4\mathbf{i} + 16\mathbf{j} + 12t\mathbf{k}$ acts on a particle of mass 2 kg initially at the origin with velocity $2\mathbf{i} - \mathbf{k}$.
Find the power of the force at any time t. Find also the work done by the force in the time interval $t = 0$ to $t = 1$.
Find the velocity and the position vector of the particle when $t = 1$.
Calculate the kinetic energy of the particle when $t = 1$ and verify your answer by considering the relationship between work and energy. (Use kg, m and s.)

MISCELLANEOUS EXERCISE 1

1) The position vector of a particle P at time t is given by

$$\mathbf{r} = t\sin t\mathbf{i} + t\cos t\mathbf{j}$$

where $\mathbf{i}$ and $\mathbf{j}$ are constant orthogonal unit vectors. Calculate the velocity and acceleration vectors of P, and show that its speed is $\sqrt{(1 + t^2)}$. (WJEC)

2) The vectors $\mathbf{F}$ and $\mathbf{u}$ are $(\mathbf{i} - 2\mathbf{j} + 3\mathbf{k})$ and $(8\mathbf{i} + 9\mathbf{j} + 12\mathbf{k})$ respectively. The vector $\mathbf{F}$ is resolved into two components, one in the direction of $\mathbf{u}$ and the other perpendicular to $\mathbf{u}$. Find the magnitude of the component of $\mathbf{F}$ in the direction of $\mathbf{u}$. (U of L)

3) The acceleration $\mathbf{a}$ of a particle of mass m, moving in three dimensional space, is given by the equation

$$\mathbf{a} = (\cos t)\mathbf{i} + (\sin t)\mathbf{j}$$

Find the velocity vector and the position vector of the particle at time t if the particle starts with a velocity $-\mathbf{j} + \mathbf{k}$ at the origin at time $t = 0$. Find also the kinetic energy of the particle and the magnitude of the force acting on the particle at time t. (U of L)

4) *Gravity may be ignored in this question.*
A particle of mass 4 units moves so that its position vector at time t is

$$\mathbf{r} = \mathbf{i}\sin 2t + \mathbf{j}\cos 2t + \mathbf{k}(t^2 - 2t).$$

Find
(a) the kinetic energy of the particle at time t,
(b) the work done by the resultant force in the time interval $t = 0$ to $t = 2$,
(c) the resultant force acting on the particle,
(d) the time when the acceleration of the particle is perpendicular to its direction of motion. (AEB)

5) The position vector $\mathbf{r}$ of a moving particle at time t after the start of the motion is given by

$$\mathbf{r} = (5 + 20t)\mathbf{i} + (95 + 10t - 5t^2)\mathbf{j}$$

Find the initial velocity of the particle.
At time $t = T$ the particle is moving at right angles to its initial direction of motion. Find the value of T and the distance of the particle from its initial position at this time. (JMB)

6) Show that the vectors $\mathbf{r}_1 = 4\mathbf{i} + 3\mathbf{j}$ and $\mathbf{r}_2 = 3\mathbf{i} - 4\mathbf{j}$ are perpendicular.
A billiard table PQRS has one edge PQ in the direction $4\mathbf{i} + 3\mathbf{j}$. Two billiard balls A and B are moving across the table with constant velocity vectors

$$\mathbf{v}_A = 2\mathbf{i} + \mathbf{j} \quad \text{and} \quad \mathbf{v}_B = \mathbf{i} - 5\mathbf{j}.$$

Find the magnitudes of the velocity components of A and B parallel to PQ and QR and find the velocity vector of A relative to B in the direction $\overrightarrow{PQ}$.

7) A bead of mass 2 kg is constrained to move along a smooth straight horizontal wire whose equation is $\mathbf{r} = \mathbf{i} + \lambda(3\mathbf{i} + 2\mathbf{j} + \mathbf{k})$. The bead moves from rest at the point P(4, 2, 1) to the point Q(16, 10, 5) under the action of a force $\mathbf{F} = 7\mathbf{i} + 2\mathbf{k}$. Using the newton and the metre as units for force and distance find:
(a) the work done by $\mathbf{F}$,
(b) the speed of the bead at Q.

8) If the points P_1, P_2 have position vectors $2\mathbf{i} - 5\mathbf{j} + \mathbf{k}$, $-8\mathbf{i} - \mathbf{j} + 4\mathbf{k}$ respectively, and the points Q_1, Q_2 have position vectors $-13\mathbf{j} + 5\mathbf{k}$, $4\mathbf{i} + 3\mathbf{j} - 3\mathbf{k}$ respectively, prove that the lines P_1P_2 and Q_1Q_2 intersect at right angles. Find the position vector of their point of intersection.
A force of magnitude F acting in the direction P_1Q_2 moves a particle from P_2 to Q_2. Determine the work done by the force. (U of L)

9) Two particles A and B are moving with constant velocity vectors $\mathbf{v}_1 = 5\mathbf{i}+3\mathbf{j}-\mathbf{k}$ and $\mathbf{v}_2 = 3\mathbf{i}+4\mathbf{j}-3\mathbf{k}$ respectively.
Find the velocity vector of A relative to B. At time $t = 0$ the particle A is at the point whose position vector is $-4\mathbf{i}+7\mathbf{j}-6\mathbf{k}$.
If A collides with B when $t = 5$, find the position vector of B at $t = 0$.
The velocity of A relative to a third moving particle C is in the direction of the vector $2\mathbf{i}+\mathbf{j}-2\mathbf{k}$ and the velocity of B relative to C is in the direction of the vector $2\mathbf{i}+3\mathbf{j}-6\mathbf{k}$. Find the magnitude and direction of the velocity of C.
(U of L)

10) Three points A, B, C have position vectors

$$4\mathbf{i}+3\mathbf{j}, \quad 3\mathbf{i}+\mathbf{j}+2\mathbf{k}, \quad 7\mathbf{i}+4\mathbf{j}+2\mathbf{k},$$

respectively, referred to an origin O, the unit of distance being the metre.
A particle P leaves O and simultaneously a second particle Q leaves B.
Each particle moves with constant velocity, P along OA and Q along BA.
The speed of Q is 6 m/s and the velocity of P relative to Q is parallel to OC.
Find the speed of P.
Find also, in subsequent motion,
(a) the least distance between P and Q,
(b) the least distance between P and the line BC. (U of L)

11) At time $t = 0$ the position vectors of two particles P and Q are $\mathbf{i}+\mathbf{j}+3\mathbf{k}$ and $4\mathbf{i}+5\mathbf{j}+\mathbf{k}$ respectively. The particles have constant velocity vectors $2\mathbf{i}+\mathbf{j}+2\mathbf{k}$ and $-4\mathbf{j}+3\mathbf{k}$ respectively. Find the position vector of Q relative to P when $t = T$. Show that the distance between the two particles is a minimum when $t = \frac{14}{15}$ and find the minimum distance. Also find the position vector of Q relative to P at this instant. (U of L)

12) Particles A and B start simultaneously from points which have position vectors $-11\mathbf{i}+17\mathbf{j}-14\mathbf{k}$ and $-9\mathbf{i}+9\mathbf{j}-32\mathbf{k}$ respectively. The velocities of A and B are constant and represented by $6\mathbf{i}-7\mathbf{j}+8\mathbf{k}$ and $5\mathbf{i}-3\mathbf{j}+17\mathbf{k}$ respectively. Show that A and B will collide.
A third particle C moves so that its velocity relative to A is parallel to the vector $2\mathbf{i}+3\mathbf{j}+4\mathbf{k}$ and its velocity relative to B is parallel to the vector $\mathbf{i}+2\mathbf{j}+3\mathbf{k}$. Find the velocity of C and its initial position if all three particles collide simultaneously.
(U of L)

13) If forces $\lambda\overrightarrow{AB}$ and $\mu\overrightarrow{AC}$ act along AB and AC, show that the resultant is $(\lambda+\mu)\overrightarrow{AD}$, where D is the point in BC such that $BD:DC = \mu:\lambda$. The position vectors of the points A, B and C are $(\mathbf{i}+2\mathbf{j}+3\mathbf{k})$, $(4\mathbf{i}+2\mathbf{j}-\mathbf{k})$ and $7\mathbf{i}$ respectively, the unit of distance being the metre.
Forces of magnitude 10 N and 7 N act along AB and AC respectively. If D is a point in BC, find the resultant of these forces in the form $n\overrightarrow{AD}$, giving the value of n and the position vector of D.
If these forces move a particle from A to D, find the work done by each force (stating the unit). (AEB)

14) The position vector $\mathbf{r}$ of a particle at time t is

$$\mathbf{r} = 2t^2\mathbf{i}+(t^2-4t)\mathbf{j}+(3t-5)\mathbf{k}.$$

Find the velocity and acceleration of the particle at time t. Show that when $t = 2/5$ the velocity and acceleration are perpendicular to each other. The velocity and acceleration are resolved into components along and perpendicular to the vector $\mathbf{i}-3\mathbf{j}+2\mathbf{k}$. Find the velocity and acceleration components parallel to this vector when $t = 2/5$. (JMB)

15) At time t the position vector of a particle is

$$\mathbf{r} = c\cos nt\,\mathbf{i}+d\sin nt\,\mathbf{j}$$

where c, d $(d \neq c)$ and n are constants and $\mathbf{i}$, $\mathbf{j}$ are constant vectors of unit length perpendicular to each other. Find $\mathbf{v}$ and $\mathbf{a}$, the velocity and acceleration vectors respectively, of the particle. Show that

$$\mathbf{v}.\mathbf{v} = n^2(c^2+d^2-\mathbf{r}.\mathbf{r}),$$

and find the times at which the vectors $\mathbf{v}$ and $\mathbf{a}$ are perpendicular to each other. (WJEC)

16) Two particles A and B, with masses 3 and 4 units respectively, move so that at time t they have position vectors $\mathbf{r}_A$ and $\mathbf{r}_B$, where

$$\mathbf{r}_A = (t-1)\mathbf{i}+\sin\pi t\,\mathbf{j}+(t^2+2)\mathbf{k}$$

$$\mathbf{r}_B = \sin\frac{\pi t}{4}\mathbf{i}+(t^3-8)\mathbf{j}+3t\mathbf{k}$$

Find
(a) the total kinetic energy of the particles at time t,
(b) the magnitude of the resultant force acting on A at time t,
(c) the cosine of the angle which the acceleration of the particle A makes with its path at $t = 1$.
Show that A and B eventually collide and find the position vector of the point of collision. (AEB)

17) *In this question the unit of distance is the metre.*
The points A and B have position vectors $4\mathbf{i}-6\mathbf{j}-12\mathbf{k}$ and $2\mathbf{i}+4\mathbf{j}+4\mathbf{k}$ respectively referred to an origin O. Forces of magnitude 7 N, 3 N, $3\sqrt{10}$ N act along OA, BO, CA respectively in the directions indicated by the order of the letters. Express the three forces in vector form. Find the magnitude of the resultant of the three forces and show that this resultant passes through the midpoint of OB.
Find the vector equation of the line of action of this resultant. (U of L)

18) *In this question the units of force, distance and time are the newton, metre and second respectively and the unit vector* **k** *is vertically upwards.*
A particle of mass 2 kg starts with velocity vector $\mathbf{i}+2\mathbf{j}+3\mathbf{k}$ from a point A which has position vector $4\mathbf{i}+3\mathbf{j}+2\mathbf{k}$. The particle moves under the action of its own weight and a constant force $\mathbf{F}=3\mathbf{i}+4\mathbf{j}+8\mathbf{k}$ and travels from A to a point B in 4 seconds. Find the position vector of B and the speed of the particle when it reaches B. Find also the work done by the force **F** as the particle moves from A to B. (Take g as $10\,\text{m/s}^2$.) (U of L)

19) *In this question the units of distance and time are the metre and second respectively.*
At time $t=0$ a particle A is at the origin and a particle B is at the point with position vector $5\mathbf{i}-10\mathbf{j}-12\mathbf{k}$. Particle A moves with constant velocity vector $2\mathbf{i}$ and B moves with constant velocity vector $4\mathbf{i}+4\mathbf{j}+5\mathbf{k}$. Show that the least distance between A and B in the subsequent motion is $\sqrt{89}$ m.
A third particle C is at the point with position vector $-\mathbf{i}+8\mathbf{j}$ at time $t=0$ and travels with constant velocity to strike B at time $t=3$. Find the velocity vector of C. (U of L)

20) At any time, t, the position vector of a particle of mass 2 kg is given by

$$\mathbf{r} = e^{-t}\{\cos t\,\mathbf{i}+\sin t\,\mathbf{j}\}$$

where **i** and **j** are perpendicular vectors of length 1 m.
Show that the position vector and the acceleration are perpendicular for all times t, and find the resultant force on the particle at time t.
Show that, at time $t=0$, the kinetic energy of the particle is 2 joules and find the kinetic energy at time $t=\frac{1}{2}\pi$ seconds. Hence find the work done by the forces acting on the particle in the time interval $t=0$ to $t=\frac{1}{2}\pi$. (AEB)

CHAPTER 2

THE VECTOR EQUATION OF A CURVE

If $P(x, y, z)$ is a general point on a curve, and $\mathbf{r}$ is the position vector of P relative to the origin O, then

$$\mathbf{r} = x\mathbf{i}+y\mathbf{j}+z\mathbf{k}$$

and this is a vector equation of the curve.

For certain curves the coordinates of P can also be expressed in parametric form. In these cases $\mathbf{r}$ can be expressed in terms of the parameter as the only variable.
For example, any point on the parabola $y^2 = 4ax$ can be expressed in the parametric form $x = ap^2$, $y = 2ap$. Hence the vector equation of this parabola can be written

$$\mathbf{r} = ap^2\mathbf{i}+2ap\mathbf{j}$$

Conversely, if the vector equation of a curve is given, and $\mathbf{r}$ is a function of one parameter, the coefficients of $\mathbf{i}$, $\mathbf{j}$ and $\mathbf{k}$ are the parametric coordinates of a general point on the curve. If the parameter can be eliminated from these coordinates, the Cartesian equation of the curve can be found.
For example, if a curve has a vector equation $\mathbf{r} = a\cos\theta\,\mathbf{i}+a\sin\theta\,\mathbf{j}$, the parametric coordinates of any point on the curve are

$$x = a\cos\theta \qquad y = b\sin\theta$$

Eliminating θ, (using $\cos^2\theta + \sin^2\theta = 1$) gives

$$x^2+y^2 = a^2$$

Therefore $\mathbf{r} = a\cos\theta\,\mathbf{i}+a\sin\theta\,\mathbf{j}$ is the equation of a circle with radius a and and centre O.

The vector equations of other simple conic sections can be recognised easily, as the following table shows.

Vector equation	*Cartesian equation*
$\mathbf{r} = a\cos\theta\,\mathbf{i} + a\sin\theta\,\mathbf{j}$	Circle $x^2+y^2 = a^2$
$\mathbf{r} = ap^2\mathbf{i} + 2ap\,\mathbf{j}$	Parabola $y^2 = 4ax$
$\mathbf{r} = a\cos\theta\,\mathbf{i} + b\sin\theta\,\mathbf{j}$	Ellipse $\dfrac{x^2}{a^2}+\dfrac{y^2}{b^2} = 1$
$\mathbf{r} = a\sec\theta\,\mathbf{i} + b\tan\theta\,\mathbf{j}$	Hyperbola $\dfrac{x^2}{a^2}-\dfrac{y^2}{b^2} = 1$
$\mathbf{r} = cp\,\mathbf{i} + \dfrac{c}{p}\mathbf{j}$	Rectangular hyperbola $xy = c^2$

THE HELIX

A helix is a three-dimensional curve which occurs in many everyday forms. It is the locus of a point that rotates about a fixed axis in such a way that its distance from that axis is constant and it either advances or withdraws a fixed distance along that axis for each complete rotation. A thread wound round a cylinder in such a way that the turns are equally spaced is an example of a helix. The distance between consecutive turns is called the *pitch* of the helix.

Let $\mathbf{r}$ be the position vector of any point P on the helix of pitch p, advancing along the axis Oz, and which passes through the point with position vector $a\mathbf{i}$.

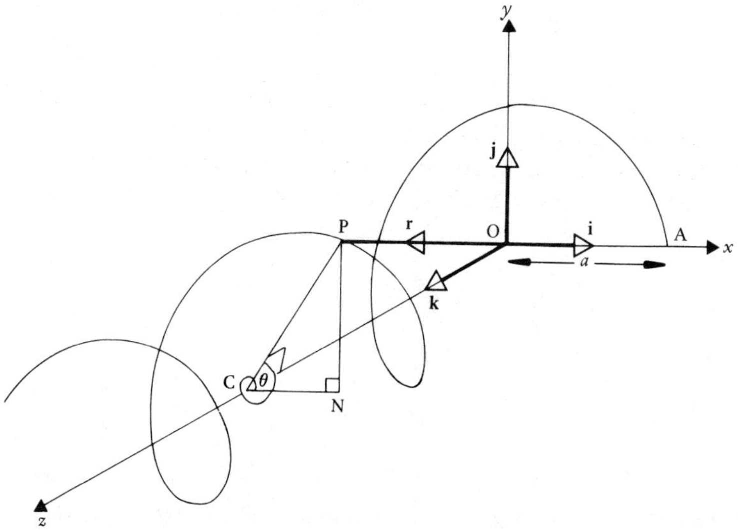

As a point moves from A to P on the curve, let it rotate through an angle θ about Oz. Since the helix passes through the point with position vector $a\mathbf{i}$, every other point on the curve is at a distance a from Oz.

Therefore $\quad CP = a, \quad CN = a\cos\theta, \quad NP = a\sin\theta$

For a turn of 2π the curve advances p along Oz,

so for a turn of θ it advances $\dfrac{p\theta}{2\pi}$, thus $OC = \dfrac{p\theta}{2\pi}$

Therefore $\quad \overrightarrow{CN} = a\cos\theta\,\mathbf{i}, \quad \overrightarrow{NP} = a\sin\theta\,\mathbf{j}, \quad \overrightarrow{OC} = \dfrac{p\theta}{2\pi}\mathbf{k}$

From the diagram, by vector addition,

$$\overrightarrow{OP} = \overrightarrow{CN} + \overrightarrow{NP} + \overrightarrow{OC}$$

hence

$$\mathbf{r} = a\cos\theta\,\mathbf{i} + a\sin\theta\,\mathbf{j} + \frac{p\theta}{2\pi}\mathbf{k}$$

This equation gives the position vector of a general point on the curve and is therefore a vector equation of the given helix.

EXAMPLES 2a

1) Find the vector equation of the circle in the $\mathbf{i}$, $\mathbf{j}$ plane, of radius a and whose centre has position vector $a\mathbf{i} - 3a\mathbf{j}$.

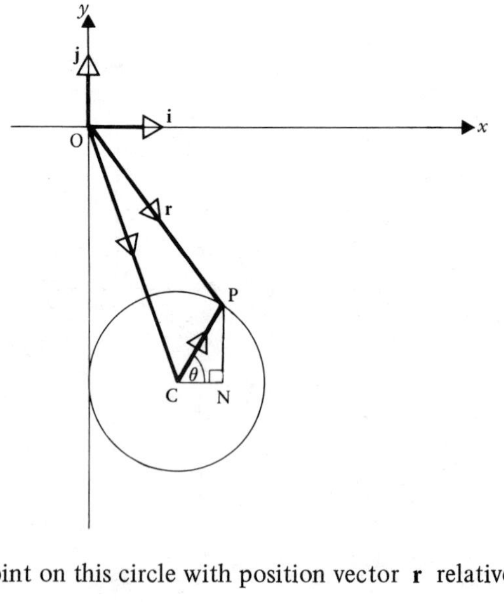

Let P be any point on this circle with position vector $\mathbf{r}$ relative to O.

From the diagram $\quad \mathbf{r} = \overrightarrow{OC} + \overrightarrow{CP}$

$$= \overrightarrow{OC} + \overrightarrow{CN} + \overrightarrow{NP}$$

$$= (a\mathbf{i} - 3a\mathbf{j}) + (a\cos\theta\,\mathbf{i} + a\sin\theta\,\mathbf{j})$$

Therefore $\quad \mathbf{r} = (a + a\cos\theta)\mathbf{i} + (a\sin\theta - 3a)\mathbf{j}$

2) Sketch the curve $\mathbf{r} = p^2\mathbf{i} + 2p\mathbf{j} + 3\mathbf{k}$ and find its Cartesian equations.

From the equation $\quad \mathbf{r} = p^2\mathbf{i} + 2p\mathbf{j} + 3\mathbf{k} \quad$ [1]

we see that the component of $\mathbf{r}$ in the direction of $\mathbf{k}$ is constant.

Therefore the curve must be in the plane $\mathbf{r} = 3\mathbf{k}$.

Rearranging [1] so that the variable components of $\mathbf{r}$ are isolated gives

$$\mathbf{r} - 3\mathbf{k} = p^2\mathbf{i} + 2p\mathbf{j} \quad [2]$$

(Comparing this equation with $\mathbf{r} = ap^2\mathbf{i} + 2ap\mathbf{j}$, we see that it is the equation of a parabola in the plane $\mathbf{r} = 3\mathbf{k}$. This parabola can now be sketched.)

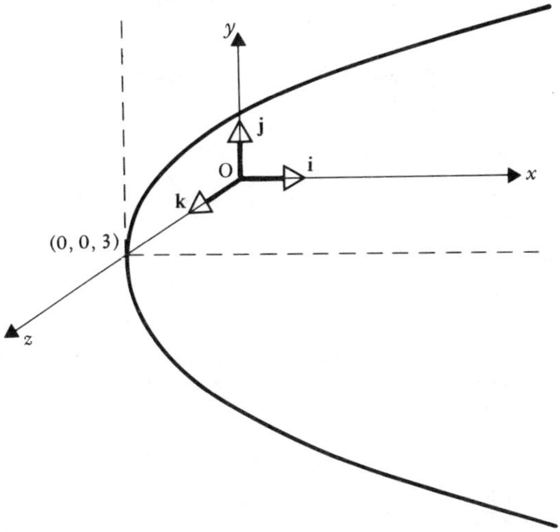

From [1] $\quad x = p^2, \quad y = 2p, \quad z = 3$

Eliminating p from the first two equations gives $y^2 = 4x$

Therefore the Cartesian equations of this curve are

$$y^2 = 4x, \quad z = 3$$

3) Find the position vectors of the points of intersection of the line $\mathbf{r} = (2+s)\mathbf{i}-(2+3s)\mathbf{j}$ with the curve $\mathbf{r} = 4p^2\mathbf{i}-8p^3\mathbf{j}$.

At a point of intersection

$$(2+s)\mathbf{i}-(2+3s)\mathbf{j} = 4p^2\mathbf{i}-8p^3\mathbf{j}$$

Therefore $$2+s = 4p^2 \qquad [1]$$

and $$2+3s = 8p^3 \qquad [2]$$

Eliminating s from [1] and [2] gives

$$2p^3-3p^2+1 = 0$$

$\Rightarrow$ $$(p-1)(p-1)(2p+1) = 0$$

Hence $$p = 1 \quad \text{or} \quad -\tfrac{1}{2}$$

Substituting these values of p into the vector equation of the curve gives

$$\mathbf{r} = 4\mathbf{i}-8\mathbf{j}$$

and $$\mathbf{r} = \mathbf{i}+\mathbf{j}$$

These are the position vectors of the two points of intersection.

Check: when $p=1$, $s=2$ and when $p=-\frac{1}{2}$, $s=-1$.

Substituting these values of s into the vector equation of the *line* also gives

$$\mathbf{r} = 4\mathbf{i}-8\mathbf{j} \quad \text{and} \quad \mathbf{r}=\mathbf{i}+\mathbf{j}$$

EXERCISE 2a

1) Find the Cartesian equations of the following plane curves.
(a) $\mathbf{r} = 2\cos\theta\,\mathbf{i} + 2\sin\theta\mathbf{j}$,
(b) $\mathbf{r} = (2-\sin\theta)\mathbf{i} + (4-\cos\theta)\mathbf{j}$,
(c) $\mathbf{r} = 2\sec\theta\,\mathbf{i} + 2\tan\theta\mathbf{j}$,
(d) $\mathbf{r} = (6-ap^2)\mathbf{i} + (4-2ap)\mathbf{j}$.

2) Find the Cartesian equations of the following curves.
(a) $\mathbf{r} = a\cos\theta\,\mathbf{i} + a\sin\theta\,\mathbf{k}$,
(b) $\mathbf{r} = p\mathbf{j} + p^2\mathbf{k}$,
(c) $\mathbf{r} = 2\cos\phi\mathbf{i} + 3\sin\phi\mathbf{k}$,
(d) $\mathbf{r} = \cos\theta\,\mathbf{i} + \sin\theta\,\mathbf{j} + \mathbf{k}$.

3) Find the position vectors of the points of intersection of
(a) $\mathbf{r} = \cos\theta\,\mathbf{i} + \sin\theta\,\mathbf{j}$
$\mathbf{r} = \lambda\mathbf{i} + 2\lambda\mathbf{j}$,
(b) $\mathbf{r} = 3\cos\theta\,\mathbf{i} + 3\sin\theta\,\mathbf{j}$
$\mathbf{r} = p^2\mathbf{i} + p\mathbf{j}$,
(c) $\mathbf{r} = 3\cos\theta\,\mathbf{i} + 3\sin\theta\,\mathbf{j} + \dfrac{2\theta}{\pi}\mathbf{k}$
$\mathbf{r} = (3-s)\mathbf{i} + s\mathbf{j} + (s-2)\mathbf{k}$.

4) Find the vector equation of the ellipse in the **i**, **j** plane, centre $\mathbf{i}-2\mathbf{j}$, major axis of length $2a$ and minor axis of length $2b$.

5) Find the vector equation of the parabola in the **i**, **j** plane, with axis parallel to **i**, latus rectum of length $4a$ and whose vertex has position vector $\mathbf{i}-2\mathbf{j}$.

6) Find the vector equation of the helix of pitch 4 units which advances along its axis Oz, which has positive rotation and passes through the point with position vector $3\mathbf{i}$.

7) Find the position vector of any point on the circle in the plane parallel to **i** and **k** which has a radius of 3 units and whose centre has position vector $2\mathbf{i}+\mathbf{j}$.

8) Show that $\mathbf{r}=(p^2-3)\mathbf{i}+(5+3p)\mathbf{j}$ is the position vector of a point on a parabola and find the vector equation of its directrix.

MOTION OF A PARTICLE ON A CURVE

For a particle P with position vector **r** at time t, we already know that

$$\mathbf{v} = \frac{d\mathbf{r}}{dt} \qquad \mathbf{a} = \frac{d\mathbf{v}}{dt}$$

and, conversely, that

$$\mathbf{r} = \int \mathbf{v}\, dt \qquad \mathbf{v} = \int \mathbf{a}\, dt$$

These relationships are easy to use when **r** is a function of the time, t.

When P is moving on a curve whose vector equation is given in terms of a parameter other than time, we need adapted forms of the relationships above in order to analyse the motion.

Suppose that **r** is given in terms of a parameter θ, then

$$\mathbf{v} = \frac{d\mathbf{r}}{dt} = \frac{d\mathbf{r}}{d\theta} \times \frac{d\theta}{dt}$$

Hence **v** cannot be found unless $\dfrac{d\theta}{dt}$ is known.

EXAMPLE 2b

A particle P moves on the curve whose vector equation is

$$\mathbf{r} = (2+\cos\theta)\mathbf{i} + (\sin\theta - 1)\mathbf{j}$$

with constant speed u in the sense $\mathbf{i}\to\mathbf{j}$. When the time t is zero, P is at the point with position vector $2\mathbf{i}-2\mathbf{j}$. Show that $\mathbf{a}+u^2\mathbf{r}$ is constant, where **a** is the acceleration of P at time t.

From $\mathbf{r} = (2+\cos\theta)\mathbf{i} + (\sin\theta - 1)\mathbf{j}$ we have

$$x = 2+\cos\theta \qquad y = \sin\theta - 1$$

$$\Rightarrow \qquad (x-2)^2 + (y+1)^2 = 1$$

Therefore P moves on a circle with centre $(2, -1)$ and radius 1.

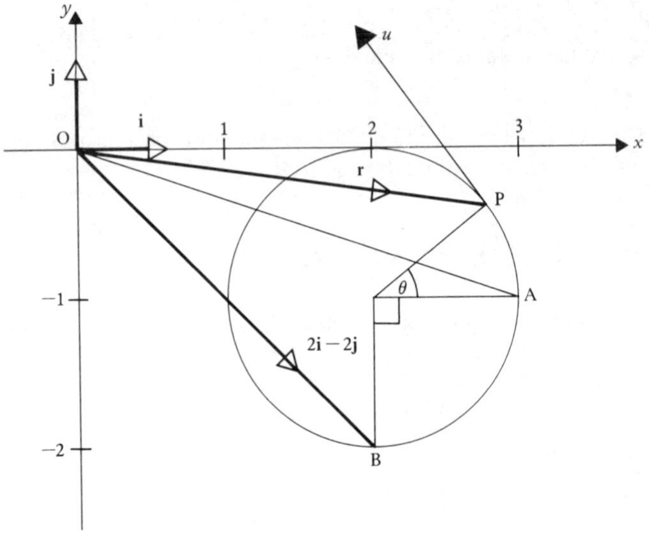

When $t = 0$, $\mathbf{r} = 2\mathbf{i} - 2\mathbf{j} \Rightarrow \cos\theta = 0$ and $\sin\theta = -1$.

Therefore $\theta = -\frac{1}{2}\pi$ when $t = 0$

Now $\frac{d\theta}{dt}$ is the angular velocity of P round the circle

i.e. $$\frac{d\theta}{dt} = \frac{u}{\text{radius of circle}} = u$$

Hence $$\mathbf{v} = \frac{d\mathbf{r}}{dt} = \frac{d\mathbf{r}}{d\theta} \times \frac{d\theta}{dt} = u\frac{d\mathbf{r}}{d\theta}$$

$$\Rightarrow \qquad \mathbf{v} = u[(-\sin\theta)\mathbf{i} + (\cos\theta)\mathbf{j}]$$

Similarly $$\mathbf{a} = \frac{d\mathbf{v}}{d\theta} \times \frac{d\theta}{dt} = u\frac{d\mathbf{v}}{d\theta}$$

$$= u^2[(-\cos\theta)\mathbf{i} + (-\sin\theta)\mathbf{j}]$$

Therefore $u^2\mathbf{r} + \mathbf{a} = 2u^2\mathbf{i} - u^2\mathbf{j}$ which is constant.

EXERCISE 2b

1) The position vector of a particle P is given by $\mathbf{r} = \sin\theta\,\mathbf{i} + \cos\theta\,\mathbf{j} + \theta\,\mathbf{k}$. Find, in terms of θ and ω, the velocity and acceleration of P if $\theta = \omega t$ where ω is a constant.

2) A particle travels, with constant angular velocity ω, round the circle with vector equation $\mathbf{r} = 4\cos\theta\,\mathbf{i} + 4\sin\theta\,\mathbf{j}$. Find the acceleration of the particle, in terms of θ and ω.

3) A particle A moves on the parabola $\mathbf{r} = ap^2\mathbf{i} + 2ap\mathbf{j}$, where p increases at a rate proportional to the square of the time t. If the values of p when $t = 0$ and when $t = 2$ are 0 and 4, find an expression for p in terms of t. Find also the velocity and acceleration of A, giving your answers in terms of p.

4) A particle moves in a plane with constant speed V. If the position vector of the particle at time t is given by $\mathbf{r} = e^p\mathbf{i} - e^{2p}\mathbf{j}$, where p is a parameter, find the velocity of the particle.

DIFFERENTIATION OF A VECTOR OF CONSTANT MAGNITUDE

When we differentiate a vector with variable direction but constant magnitude, we find a special relationship between that vector and its derivative.

Consider a vector $\mathbf{r}$, with constant magnitude r and variable direction $\hat{\mathbf{r}}$. Then $\hat{\mathbf{r}} = l\mathbf{i} + m\mathbf{j} + n\mathbf{k}$ where l, m and n are the direction cosines of $\mathbf{r}$. (l, m and n are variable.)
Differentiating $\hat{\mathbf{r}}$ with respect to any scalar quantity p we have

$$\frac{d\hat{\mathbf{r}}}{dp} = \frac{dl}{dp}\mathbf{i} + \frac{dm}{dp}\mathbf{j} + \frac{dn}{dp}\mathbf{k}$$

Then

$$\hat{\mathbf{r}}\cdot\frac{d\hat{\mathbf{r}}}{dp} = (l\mathbf{i} + m\mathbf{j} + n\mathbf{k})\cdot\left(\frac{dl}{dp}\mathbf{i} + \frac{dm}{dp}\mathbf{j} + \frac{dn}{dp}\mathbf{k}\right)$$

$$= l\frac{dl}{dp} + m\frac{dm}{dp} + n\frac{dn}{dp}$$

$$= \tfrac{1}{2}\frac{d}{dp}(l^2 + m^2 + n^2)$$

But l, m and n are direction cosines, so $l^2 + m^2 + n^2 = 1$.

Therefore

$$\hat{\mathbf{r}}\cdot\frac{d\hat{\mathbf{r}}}{dp} = 0$$

Hence $$r\hat{\mathbf{r}}.r\frac{d\hat{\mathbf{r}}}{dp} = 0 \Rightarrow \mathbf{r}.\frac{d\mathbf{r}}{dp} = 0$$

i.e. a vector of constant magnitude is perpendicular to its derivative with respect to a scalar.

MOTION OF A PARTICLE IN A PLANE USING POLAR COORDINATES

Consider a particle P moving along a curve in the xy plane such that, at time t, P is at the point with polar coordinates (r, θ).

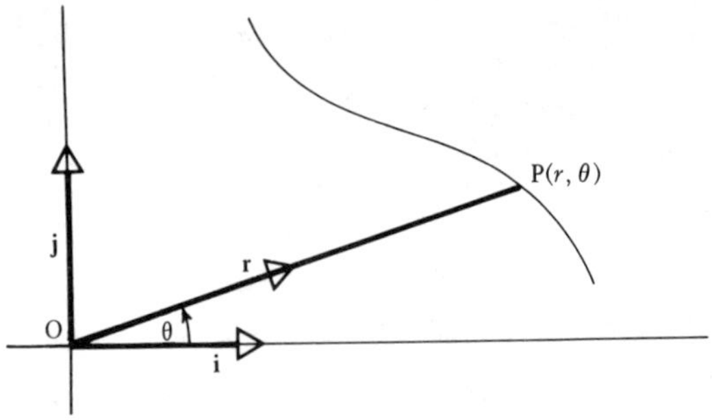

The position vector of P at time t is $\mathbf{r}$, where

$$\mathbf{r} = r\hat{\mathbf{r}} \quad \text{and} \quad \hat{\mathbf{r}} = \cos\theta\,\mathbf{i} + \sin\theta\,\mathbf{j}$$

The velocity of P at time t is $\mathbf{v}$ where

$$\mathbf{v} = \frac{d\mathbf{r}}{dt} = \dot{r}\hat{\mathbf{r}} + r\frac{d\hat{\mathbf{r}}}{dt}$$

But $$\frac{d\hat{\mathbf{r}}}{dt} = \frac{d\hat{\mathbf{r}}}{d\theta}\times\frac{d\theta}{dt} = \dot{\theta}\hat{\mathbf{s}} \quad \text{where} \quad \hat{\mathbf{s}} = -\sin\theta\,\mathbf{i} + \cos\theta\,\mathbf{j}$$

Therefore $$\mathbf{v} = \dot{r}\hat{\mathbf{r}} + r\dot{\theta}\hat{\mathbf{s}}$$

Differentiating again with respect to time gives the acceleration, $\mathbf{a}$, of P at time t.

$$\mathbf{a} = \left(\ddot{r}\hat{\mathbf{r}} + \dot{r}\frac{d\hat{\mathbf{r}}}{dt} + \dot{r}\dot{\theta}\hat{\mathbf{s}} + r\ddot{\theta}\hat{\mathbf{s}} + r\dot{\theta}\frac{d\hat{\mathbf{s}}}{dt}\right)$$

Now $$\frac{d\hat{\mathbf{s}}}{dt} = \frac{d\hat{\mathbf{s}}}{d\theta}\times\frac{d\theta}{dt} = \dot{\theta}(-\cos\theta\,\mathbf{i} - \sin\theta\,\mathbf{j}) = -\dot{\theta}\hat{\mathbf{r}}$$

Therefore $$\mathbf{a} = \ddot{r}\hat{\mathbf{r}} + \dot{r}\dot{\theta}\hat{\mathbf{s}} + \dot{r}\dot{\theta}\hat{\mathbf{s}} + r\ddot{\theta}\hat{\mathbf{s}} - r\dot{\theta}^2\hat{\mathbf{r}}$$
$$= (\ddot{r} - r\dot{\theta}^2)\hat{\mathbf{r}} + (r\ddot{\theta} + 2\dot{r}\dot{\theta})\hat{\mathbf{s}}$$

Now θ is the angle between the radius vector OP and the x axis, at any time t.

Therefore $\dot{\theta}$ is the angular velocity of OP (or of the point P),
and $\ddot{\theta}$ is the angular acceleration of OP (or of P).

Radial and Transverse Components

From the expressions obtained above for $\hat{\mathbf{r}}$ and $\hat{\mathbf{s}}$, we see that $\hat{\mathbf{s}}$ is in a direction given by rotating $\hat{\mathbf{r}}$ through a positive right angle.

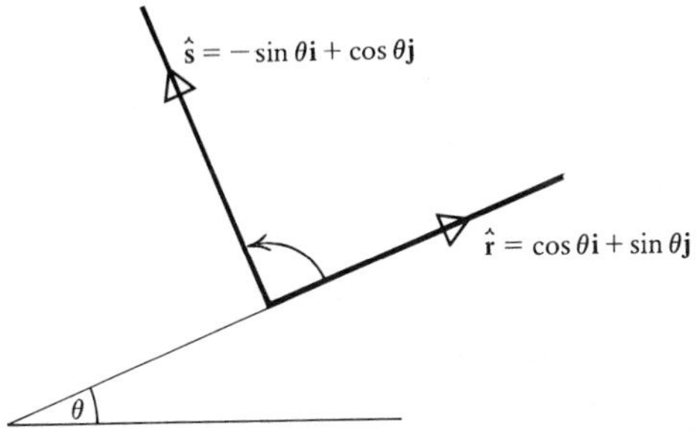

The velocity and acceleration of P, found above, therefore each comprise a pair of perpendicular components.
One of these is along OP and is called the radial component.
The other is perpendicular to OP (in the sense of θ increasing) and is called the transverse component.

Therefore the radial component of velocity is $\dot{r}$
the transverse component of velocity is $r\dot{\theta}$
the radial component of acceleration is $\ddot{r} - r\dot{\theta}^2$
the transverse component of acceleration is $r\ddot{\theta} + 2\dot{r}\dot{\theta}$

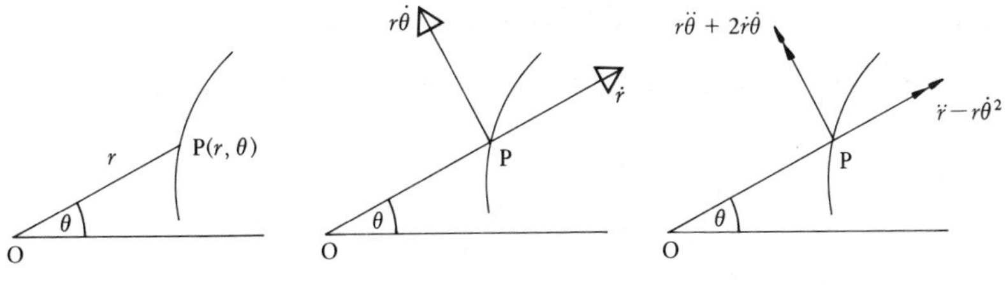

Note. An alternative form for the transverse component of acceleration is

$$\frac{1}{r}\frac{\mathrm{d}}{\mathrm{d}t}(r^2\dot{\theta})$$

In some problems this is a more convenient expression.

EXAMPLES 2c

1) A particle moves round the curve $\mathbf{r} = a(1+\cos\theta)$ with constant angular velocity ω. Find the radial and transverse components of velocity and acceleration at any time t, in terms of ω and θ.

The particle moves round the curve with constant angular velocity ω, therefore

$$\dot{\theta} = \omega \quad \text{and} \quad \ddot{\theta} = 0$$

Now $$r = a(1+\cos\theta)$$

Hence $$\dot{r} = (-a\sin\theta)\dot{\theta} = -a\omega\sin\theta$$

and $$\ddot{r} = (-a\omega\cos\theta)\dot{\theta} = -a\omega^2\cos\theta$$

For velocity the components are

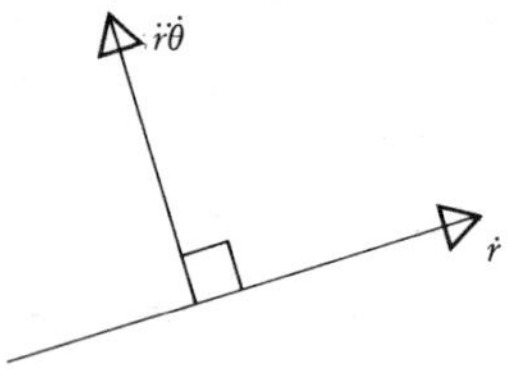

Therefore the radial component is $-a\omega\sin\theta$,
and the transverse component is $a\omega(1+\cos\theta)$.

For acceleration the components are

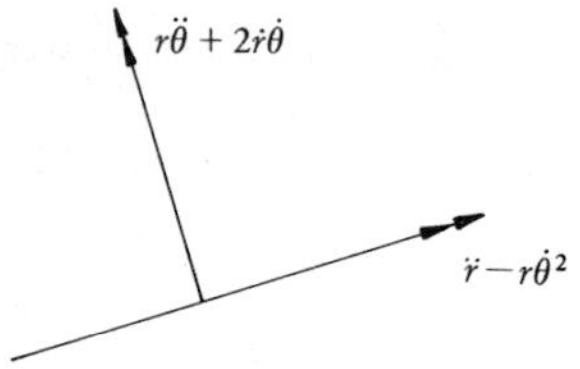

Now $$\ddot{r} - r\dot{\theta}^2 = -a\omega^2\cos\theta - a(1+\cos\theta)\omega^2$$
$$= -a\omega^2(1+2\cos\theta)$$

and $$r\ddot{\theta} + 2\dot{r}\dot{\theta} = 0 + 2(-a\omega\sin\theta)\omega$$

Therefore the radial component is $-a\omega^2(1+2\cos\theta)$,
and the transverse component is $-2a\omega^2\sin\theta$.

2) O is fixed and OA is a straight wire that rotates about O with a constant angular velocity ω. A particle P moves from O along the wire with a constant speed V relative to the wire. Find, in terms of V, ω and t, the radial and transverse components of the acceleration of P at time t after it leaves O. Find also the polar equation of the path of P.

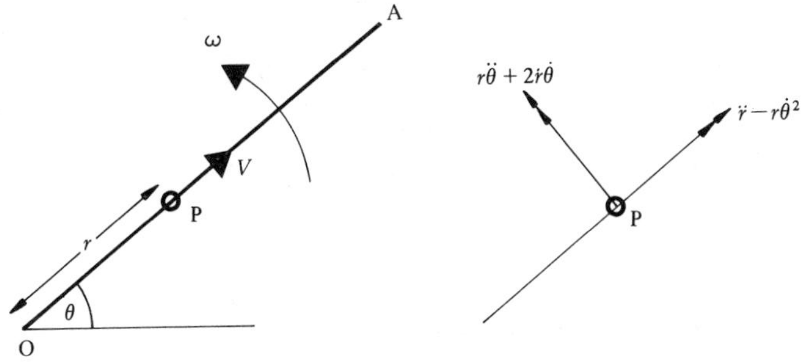

P travels along OA with speed V.

Therefore $$r = Vt \qquad [1]$$

OA rotates with angular speed ω.

Therefore $$\theta = \omega t \qquad [2]$$

Then $$\dot{r} = V, \qquad \ddot{r} = 0$$

$$\dot{\theta} = \omega, \qquad \ddot{\theta} = 0$$

So the components of the acceleration of P are

$$\ddot{r} - r\dot{\theta}^2 = -r\omega^2 \qquad \text{along OA}$$

and $$r\ddot{\theta} + 2\dot{r}\dot{\theta} = 2V\omega \qquad \text{perpendicular to OA.}$$

Eliminating t from equations [1] and [2] gives

$$r = \frac{V\theta}{\omega}$$

and this is the polar equation of the path of P.

3) The polar coordinates, at time t, of a particle of unit mass moving in a plane, are (r, θ). The resultant force acting on the particle is of magnitude $\dfrac{\lambda}{r^2}$ and is always directed towards the pole O. Given that λ and k are constant, show that $r^2\dfrac{d\theta}{dt} = k$ and that $r^3\dfrac{d^2r}{dt^2} = k^2 - \lambda r$.

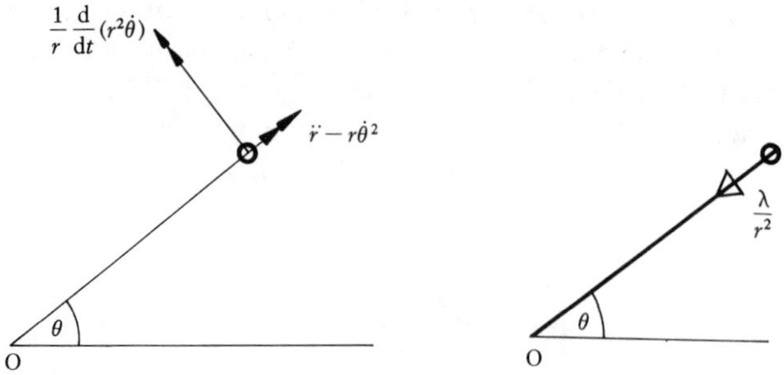

Using Newton's law of motion along the radius vector gives

$$-\frac{\lambda}{r^2} = 1(\ddot{r}-r\dot{\theta}^2) \qquad [1]$$

No force acts perpendicular to the radius vector

therefore $$\frac{1}{r}\frac{d}{dt}(r^2\dot{\theta}) = 0 \qquad [2]$$

From [2] we see that $r^2\dot{\theta}$ is constant

i.e. $$r^2\frac{d\theta}{dt} = k$$

Using $\dot{\theta} = \frac{k}{r^2}$ in equation [1] gives $-\frac{\lambda}{r^2} = \ddot{r}-\frac{k^2}{r^3}$

$$\Rightarrow \qquad r^3\frac{d^2r}{dt^2} = k^2-\lambda r$$

Note that in this example we chose to use the transverse acceleration component in the form $\frac{1}{r}\frac{d}{dt}(r^2\dot{\theta})$.

4) A point P moves round a curve with a constant angular velocity ω. At time t, its polar coordinates are (r, θ) and its radial component of acceleration is $-2r\omega^2$. Initially when $r = a$ and $\theta = 0$, the radial component of velocity is $a\omega$.
Find the equation of the locus of P.

The particle moves with constant angular velocity, therefore

$$\dot{\theta} = \omega \quad \text{and} \quad \ddot{\theta} = 0$$

The radial component of acceleration at time t is $\ddot{r}-r\dot{\theta}^2$ where

$$\ddot{r}-r\dot{\theta}^2 = \ddot{r}-r\omega^2$$

Hence
$$\ddot{r} = -r\omega^2$$

Writing $\ddot{r}$ as $\dot{r}\dfrac{d\dot{r}}{dr}$ gives
$$\dot{r}\frac{d\dot{r}}{dr} = -r\omega^2$$

Therefore
$$\int_{a\omega}^{\dot{r}} \dot{r}\,d\dot{r} = -\int_a^r r\omega^2\,dr$$

$\Rightarrow$
$$\dot{r}^2-a^2\omega^2 = -r^2\omega^2+a^2\omega^2$$

$\Rightarrow$
$$\dot{r} = \omega\sqrt{(2a^2-r^2)}$$

Now
$$\dot{r} = \left(\frac{dr}{d\theta}\right)\left(\frac{d\theta}{dt}\right) = \omega\frac{dr}{d\theta}$$

Therefore
$$\frac{dr}{d\theta} = \sqrt{(2a^2-r^2)}$$

Separating the variables gives

$$\int_a^r \frac{dr}{\sqrt{(2a^2-r^2)}} = \int_0^\theta d\theta$$

i.e.
$$\arcsin\frac{r}{a\sqrt{2}} - \arcsin\frac{1}{\sqrt{2}} = \theta$$

Hence
$$r = a\sqrt{2}\sin(\theta+\pi/4)$$

EXERCISE 2c

1) A particle moves with constant angular velocity ω round the curve with polar equation:
(a) $r = a\theta$,
(b) $r = a(1+\theta)$,
(c) $r = a(1+\sin\theta)$,
(d) $r = a(\cos\theta+\sin\theta)$.
In each case find, in terms of θ and ω, the radial and transverse components of acceleration when the particle is at the point (r, θ).

2) Find the point or points on each of the curves given in question 1, at which the acceleration of the particle is maximum.

3) A particle moves with constant angular velocity ω round the curve $r = a\cos\theta$. Find expressions for the radial and transverse components of acceleration in terms of r and ω when the particle is at the point (r, θ).

4) A particle moves round a curve with constant angular velocity ω. When the particle is at the point (r, θ) it has a radial component of acceleration of $2\omega^2(1-r)$. Initially $r=3$, $\theta=0$, and the radial component of velocity is zero. Find the polar equation of the curve.

5) A particle moves round the curve $r(1+\cos\theta)=a$ with constant angular velocity ω. Find the radial and transverse components of velocity at any instant as functions of θ.

6) A particle P moves with constant angular velocity ω about an origin O. At time t, P is at the point with polar coordinates (r, θ) and its radial component of acceleration is zero. Initially $\theta=0$, $r=1$ and $\dot{r}=\omega$. Find the polar equation of the locus of P.

7) The polar coordinates (r, θ) of a particle P moving in a plane are such that, at time t, $\dot{r}=16\cos\theta$ and $\dot{\theta}=(1-4\sin\theta)^2$
If $r=4$ when $\theta=0$, find the polar equation of the curve and find the radial acceleration of P at the point $(4, 0)$.

8) A particle P is moving on the curve with polar equation $r=4e^{\theta}$. If the radial velocity of P is equal to $\dfrac{3}{r}$, show that the transverse acceleration of P is zero and find the radial acceleration.

MULTIPLE CHOICE EXERCISE 2

(The instructions for answering these questions are given on page ix.)

TYPE I

1) A circle of radius 2 units has its centre at the point $(3, 2)$. The position vector of any point on the circumference of the circle is given by:
(a) $\mathbf{r}=2\cos\theta\,\mathbf{i}+2\sin\theta\,\mathbf{j}$ (b) $\mathbf{r}=2(\cos\theta+3)\mathbf{i}+2(\sin\theta+2)\mathbf{j}$
(c) $\mathbf{r}=(2\cos\theta+3)\mathbf{i}+(2\sin\theta+2)\mathbf{j}$ (d) $\mathbf{r}=3\cos\theta\,\mathbf{i}+2\sin\theta\,\mathbf{j}$

2) The position vector of a particle at time t is $t^2\mathbf{i}-\ln t\,\mathbf{j}+\mathbf{k}$.
The acceleration vector of the particle is:

(a) $2\mathbf{i}-e^t\mathbf{j}$, (b) $2\mathbf{i}+\dfrac{1}{t^2}\mathbf{j}$, (c) $2\mathbf{i}-\dfrac{1}{t^2}\mathbf{j}$, (d) $2t\mathbf{i}-\dfrac{1}{t}\mathbf{j}$, (e) $-\dfrac{1}{2t^3}\mathbf{j}$.

3) A point P moves round the curve $r=a\theta$ such that OP rotates with constant angular velocity ω. The transverse component of acceleration when P is at the point (r, θ) is:
(a) $a\omega^2$ (b) $-r\omega^2$ (c) $-2a\omega^2$ (d) $2a\omega^2$ (e) $r\omega^2$.

4) The vector equation of a curve is $\mathbf{r}=2t\mathbf{i}+t^2\mathbf{j}$. The Cartesian equation of this curve is:
(a) $y^2=4x$ (b) $4y+x^2=0$ (c) $4y=x^2$ (d) $y^2=4ax$ (e) $xy=2$.

5) A particle of unit mass starts from rest at the origin and moves under the action of a force which, t seconds after the particle leaves O, is $6t\mathbf{i}-\mathbf{j}+t^2\mathbf{k}$. The position vector of the particle at time t is:
(a) $t^3\mathbf{i}-\frac{1}{2}t^2\mathbf{j}+\frac{1}{12}t^4\mathbf{k}$ (b) $2\mathbf{k}$ (c) $3t^2\mathbf{i}-t\mathbf{j}+\frac{1}{3}t^3\mathbf{k}$ (d) $6\mathbf{i}+2t\mathbf{k}$.

TYPE II

6) The vector equation of a curve is $\mathbf{r}=3\cos\theta\,\mathbf{i}+2\sin\theta\,\mathbf{j}$.
(a) The curve is an ellipse.
(b) The Cartesian equation of the curve is $\dfrac{x^2}{9}-\dfrac{y^2}{4}=1$.
(c) The centre of the curve is at the point $(3, 2)$.

7) A particle moves with constant angular velocity ω round the curve with polar equation $r=a\cos\theta$.
(a) The radial component of acceleration is $-r\omega^2$.
(b) The radial component of velocity is $-a\omega\sin\theta$.
(c) The transverse component of acceleration is $2\omega\dot{r}$.

8) A particle moves with constant speed round the curve whose vector equation is $\mathbf{r}=(2+\sin\theta)\mathbf{i}+(\cos\theta-3)\mathbf{j}$.
(a) The acceleration of the particle is constant.
(b) The particle is always the same distance from the point $(2, -3)$.
(c) The speed of the particle is equal to 2 units.

9) A particle moves round a curve such that at time t it is at the point with polar coordinates (a, θ), has a radial component of acceleration of $-a\omega^2$ and zero transverse component of acceleration, where a and ω are constants.
(a) $\ddot{r}=0$ (b) $\dot{\theta}\neq 0$ (c) the particle moves with constant speed.

TYPE III

10) (a) At time t, the position vector of a particle is $\mathbf{r}=\cos\omega t\,\mathbf{i}+\sin\omega t\,\mathbf{j}$ where ω is constant.
(b) A particle is moving round a circle with constant speed ω.

11) (a) $\hat{\mathbf{r}}\cdot\dfrac{\mathrm{d}\hat{\mathbf{r}}}{\mathrm{d}t}=0$.

(b) $\mathbf{r}\cdot\dfrac{\mathrm{d}\mathbf{r}}{\mathrm{d}t}=0$.

12) (a) $\mathbf{r}$ is a vector of constant magnitude.

(b) $\dfrac{\mathrm{d}\mathbf{r}}{\mathrm{d}t}$ is perpendicular to $\mathbf{r}$ for all values of t.

TYPE IV

13) A particle moves under the action of a force $\mathbf{F}$. Find the position vector of the particle at time t.
(a) The mass of the particle is m.
(b) $\mathbf{F} = \mathbf{i} + 3\mathbf{j}$.
(c) The initial velocity vector of the particle is $\mathbf{j} - 2\mathbf{k}$.

14) Find, in terms of a, θ and t, the transverse component of acceleration of a particle moving round a curve such that:
(a) the initial speed of the particle is u,
(b) the mass of the particle is m,
(c) the polar equation of the curve is $r(1 + \cos\theta) = a$.

15) A particle moves on a plane curve. Find its position vector when $t = 2$.
(a) The initial velocity vector is $\mathbf{i} + \mathbf{j}$.
(b) The acceleration vector at time t is $3\mathbf{i} - 2t\mathbf{j}$.
(c) When $t = 3$ the particle is at the point $(2, 3)$.

16) Find the vector equation of the ellipse such that:
(a) the centre of the ellipse is at the point $(3, 2)$,
(b) the major axis is of length $2a$,
(c) the minor axis is of length $2b$,
(d) the minor axis is parallel to the x axis.

17) At time t the position vector of a particle A is $\mathbf{r}$ and the velocity vector of a particle B relative to A is $\mathbf{v}$. Find the position vector of B at time t.
(a) $\mathbf{r} = (2\cos 3t)\mathbf{i} + (\sin t)\mathbf{j}$.
(b) $\mathbf{v} = (3\sin 3t)\mathbf{i} + (2\cos t)\mathbf{j}$.
(c) A is initially at the point $2\mathbf{i}$.
(d) B is initially at the origin.

MISCELLANEOUS EXERCISE 2

1) A particle of unit mass moves in the $\mathbf{i}$, $\mathbf{j}$ plane under the action of a variable force $\mathbf{F}$, where $\mathbf{F} = \mathbf{i} + (\sin at)\mathbf{j}$. Initially the particle is at the origin with velocity vector $v\mathbf{i}$. If a and v are constants and t is the time, write down an expression for the particle's acceleration vector and hence find the vector equation of its path.

2) Sketch the circle $\mathbf{r} = \cos\theta\,\mathbf{i} + \sin\theta\,\mathbf{j}$ and the straight line given by $\mathbf{r} = (t+1)\mathbf{i} + (2t+a)\mathbf{j}$ where a is a constant. Find the range of values of a for which the line cuts the circle in two real distinct points.

3) A particle of mass 2 units is moving under the action of a force such that its position vector at time t is

$$\mathbf{r} = 5\cos t\mathbf{i} + 3\sin t\mathbf{j}$$

Sketch the path of the particle and find the force acting on it. Find also the kinetic energy of the particle when $t = \pi/4$.

4) Two particles move so that their position vectors at time t are given by $\mathbf{r}_1 = (3\sin wt)\mathbf{i} + (4\cos wt)\mathbf{j} + \mathbf{k}$ and $\mathbf{r}_2 = t^2\mathbf{i} + 2t\mathbf{j} + \mathbf{k}$. Obtain the Cartesian equations of each path and sketch these paths. Give, in both Cartesian and vector form, the equation of the plane in which the particles move. Show that the paths are perpendicular when $t = 0$ and find the values of t when the accelerations are perpendicular. (AEB)

5) A point moves such that its position vector at time t is $\mathbf{r} = ct\mathbf{i} + \dfrac{c}{t}\mathbf{j}$.

Sketch its path and find the acceleration vector of the point. Show that this acceleration vector is equal to $k\mathbf{j}(\mathbf{r}.\mathbf{j})^3$ where k is a constant, and find k in terms of c.

6) Sketch the curves $\mathbf{r} = (2a\cos\theta)\mathbf{i} + (2a\sin\theta)\mathbf{j}$

$$\mathbf{r} = 3ap^2\mathbf{i} + 3ap\mathbf{j}.$$

Find the position vectors of the points in which they meet.
A particle describes the first curve in such a way that $\theta = \omega t$, where ω is constant and t is the time. Another particle describes the second curve in such a way that $p = kt$, where k is a constant.
Show that the acceleration of each particle is constant in magnitude and find the times at which these accelerations are at right angles. (U of L)

7) During the motion of a particle in a plane its polar coordinates r, θ at time t satisfy the equations

$$\dot{r} = 24\sin\theta \quad \text{and} \quad \dot{\theta} = (5 + 3\cos\theta)^2.$$

Given that $r = 1$ when $\theta = 0$, express r in terms of θ.
Find the value of r at a point where the particle is moving parallel to the line $\theta = 0$. Calculate the radial acceleration at this point. (JMB)

8) The position of a particle P moving in a plane is given by polar coordinates (r, θ) with pole O. The particle moves round the curve

$$r = a(1 + \sin\theta)$$

where a is a positive constant, so that the radius vector OP rotates with constant angular speed ω. Find the speed of the particle in terms of a, ω and θ.
Find also the value of θ when this speed is greatest. (U of L)

9) At time t, the polar coordinates of a particle of unit mass moving in a plane are (r, θ). The only force on the particle is

$$\mathbf{F} = \frac{\mu}{r^3}\hat{\mathbf{r}}$$

where μ is constant and $\hat{\mathbf{r}}$ is a unit vector along the radial direction. Show that

$$\frac{d\theta}{dt} = \frac{c}{r^2} \quad \text{and} \quad \frac{d^2r}{dt^2} = \frac{\mu + c^2}{r^3}$$

where c is constant. (U of L)

10) A particle of mass m moves along the path $r = ae^{\theta}$, where r, θ are polar coordinates and a is a constant, under the action of a force $F(r)$ which is directed towards the pole. Show that

$$F(r) = 2mh^2/r^3$$

where h is a constant, and that the speed of the particle is $(h/r)\sqrt{2}$. Explain the physical significance of the constant h. (U of L)

11) A particle P, of unit mass, moves on a smooth horizontal plane under the action of a force of magnitude $\omega^2 r + \dfrac{\omega^2 a^3}{r^2}$ directed towards a fixed point O on the plane, where ω is a constant and $OP = r$. If P is projected at time $t = 0$ from a point distant a from O with a horizontal speed $4a\omega/\sqrt{3}$ in a direction perpendicular to OP, show that, in the subsequent motion,

$$\dot{r}^2 = -\omega^2(r-a)(r-2a)(3r^2+9ar+8a^2)/(3r^2)$$

and deduce that r lies between a and $2a$. (U of L)

12) Show that the radial and transverse components of the acceleration of a particle moving in a plane are $(\ddot{r} - r\dot{\theta}^2)$ and $(r\ddot{\theta} + 2\dot{r}\dot{\theta})$ respectively.
A particle describes the curve $r = 3e^{\theta}$ so that the radial velocity of the particle when it is at a distance r from the pole O is $2/r$. Show that the acceleration of the particle is $8/r^3$ directed towards O. (U of L)

13) Obtain expressions for the radial and transverse components of acceleration of a point (r, θ) moving along a plane curve. If the angular velocity has the constant value ω, find the equation of the curve, given that the radial component of acceleration is $(a-2r)\omega^2$ and the transverse component is $4a\omega^2 \cos\theta$, where a is a constant. Sketch the curve and find the point on it at which the resultant acceleration is a maximum. (U of L)

14) A point P moves in a path whose polar equation is given by

$$r = \frac{2}{2+\cos\theta}$$

with respect to a pole O and initial line OA. At any time t during the motion $r^2\dot{\theta} = 2$. Write down an expression for $r\dot{\theta}$ in terms of θ and show that

$$\dot{r} = \sin\theta$$

Hence show that the velocity of P is the resultant of two velocities of constant magnitude, one perpendicular to OA and one perpendicular to OP.
Prove that the acceleration of P is directed towards O and that its magnitude is inversely proportional to r^2. (JMB)

15) Show that the radial and transverse components of acceleration of a point (r, θ) moving in a plane curve are $\ddot{r} - r\dot{\theta}^2$ and $2\dot{r}\dot{\theta} + r\ddot{\theta}$ respectively.
The straight line OA rotates with constant angular velocity ω about the fixed point O. The radial acceleration of a point $P(r, \theta)$ on OA, such that initially $r = a$, $\theta = 0$, is always $kr\omega^2$, and the initial radial velocity of P is $a\omega$. Find the equations of the locus of P when $k = 0$, $k = -1$, $k = -2$, and sketch the locus in each case. (U of L)

16) A particle is describing a plane curve and at time t is at the point P whose polar coordinates are (r, θ) referred to pole O. Find expressions for the components of the acceleration of the particle along and perpendicular to the radius vector OP.
A particle P of mass m is travelling along the curve $r(1+\cos\theta) = 2a$ in a horizontal plane so that the radius vector OP is rotating at constant angular speed ω. Prove that the speed of the particle at any instant is $\omega\sqrt{r^3/a}$. When $\theta = \frac{1}{2}\pi$, find the resultant horizontal force acting on the particle. (U of L)

17) Write down expressions for the radial and transverse components of velocity in terms of plane polar coordinates.
A ferry boat B, which travels with constant speed $3u$ in still water, crosses a river, which is flowing with speed u, by always pointing its bows towards its point of destination O, which is directly opposite its starting point A. Given that the river flows in a direction perpendicular to OA, show that if r is the distance OB and θ is the angle AOB, then

$$\cos\theta \frac{dr}{d\theta} = r(\sin\theta - 3)$$ (U of L)

18) The position, at time t, of a particle P moving in a plane is defined by its distance r from a fixed point O and the angle θ that OP makes with a fixed direction Ox. Prove that the components of the acceleration of P along and perpendicular to OP are

$$\frac{d^2r}{dt^2}-r\left(\frac{d\theta}{dt}\right)^2, \quad \frac{1}{r}\frac{d}{dt}\left(r^2\frac{d\theta}{dt}\right)$$

respectively.
A particle Q moves in a horizontal plane about a fixed point O with constant angular velocity ω and with no acceleration along OQ. Denoting by r the distance OQ and given that $r=a$ and $\dfrac{dr}{dt}=3a\omega$ when $t=0$, show that at time t

$$r = a(2e^{\omega t}-e^{-\omega t})$$ (U of L)

19) Derive expressions for the radial and transverse components of acceleration of a particle moving in a plane.
A particle P, of mass 1 kg, moves on a smooth horizontal table and is attached to a fixed point O of the table by a light elastic string, of natural length 1 m and modulus 8/9 N. The position of P on the table is specified by polar coordinates (r,θ), referred to O as pole and a fixed line OA as the initial line, where OP $= r$ m. Initially P is held at a point on OA with the string just taut and is projected at right angles to OA with a speed of 2 m s^{-1}. Show that, t seconds after projection,

(a) $\dfrac{d^2r}{dt^2}=\dfrac{4}{r^3}-\dfrac{8}{9}(r-1)$,

(b) $\left(\dfrac{dr}{dt}\right)^2=4-\dfrac{4}{r^2}-\dfrac{8}{9}(r-1)^2$.

Hence, or otherwise, show that, in the subsequent motion, the length of the string cannot exceed 3 m. (U of L)

20) A particle A, of mass m, is held at rest on a smooth horizontal table. One end of a light inextensible string is attached to A. The string passes through a small smooth hole H in the table, and carries at the other end a particle B, also of mass m, hanging freely. Initially AH $=a$ and the particle A is moving horizontally with speed $\sqrt{(2gh)}$, where $h>a/2$, in a direction perpendicular to the string. If r is the distance AH after time t, show that

$$\dot{r}^2 = gh(1-a^2/r^2)+g(a-r).$$

Show also that if the particle B reaches the table, then the total length of the string cannot exceed

$$[h+\sqrt{(h^2+4ah)}]/2.$$ (U of L)

21) A particle moving on a smooth horizontal plane describes the curve $r = 2a/(2-\sin\theta)$ under the action of a force directed towards the point $r = 0$. When $r = 2a$, the speed of the particle is V. Show that the value of $r^2\,\mathrm{d}\theta/\mathrm{d}t$ is constant and equal to $2aV$.
Show also that
(a) $\mathrm{d}r/\mathrm{d}t = V\cos\theta$,
(b) $\mathrm{d}^2r/\mathrm{d}t^2 = -(2aV^2\sin\theta)/r^2$.
Find the acceleration of the particle when its speed is a maximum. (U of L)

22) A particle P, whose polar coordinates referred to an origin O are (r, θ), moves along a plane curve. Given that $\hat{\mathbf{a}}$ and $\hat{\mathbf{b}}$ are unit vectors parallel to and perpendicular to $\overrightarrow{\mathrm{OP}}$, prove that the velocity $\mathbf{v}$ of P and the acceleration $\mathbf{f}$ can be expressed as

$$\mathbf{v} = \dot{r}\hat{\mathbf{a}} + r\dot{\theta}\hat{\mathbf{b}},$$

$$\mathbf{f} = (\ddot{r} - r\dot{\theta}^2)\hat{\mathbf{a}} + (r\ddot{\theta} + 2\dot{r}\dot{\theta})\hat{\mathbf{b}}.$$

A wheel is rotating with constant angular speed ω about its centre O, and a particle P moves from the centre along a spoke of the wheel with constant speed u relative to the spoke. Find, in terms of u, ω and t, the radial and transverse components of the acceleration of P at time t after it leaves O. Given that P starts at the centre of the wheel, find the magnitudes of the transverse components of velocity and acceleration of P when $t = 2$ s.
(U of L)

CHAPTER 3

THE VECTOR PRODUCT AND ITS APPLICATION

For two vectors **a** and **b**, inclined to each other at an angle θ, the vector product $\mathbf{a} \times \mathbf{b}$ (or $\mathbf{a} \wedge \mathbf{b}$) is defined as

> a vector of magnitude $ab \sin\theta$
> in a direction perpendicular to the plane containing **a** and **b**,
> in the sense of a right-handed screw turned from **a** to **b**.

It follows that the vector product $\mathbf{b} \times \mathbf{a}$ is in the direction of a right-handed screw turned from **b** to **a** which is opposite to the direction of $\mathbf{a} \times \mathbf{b}$.

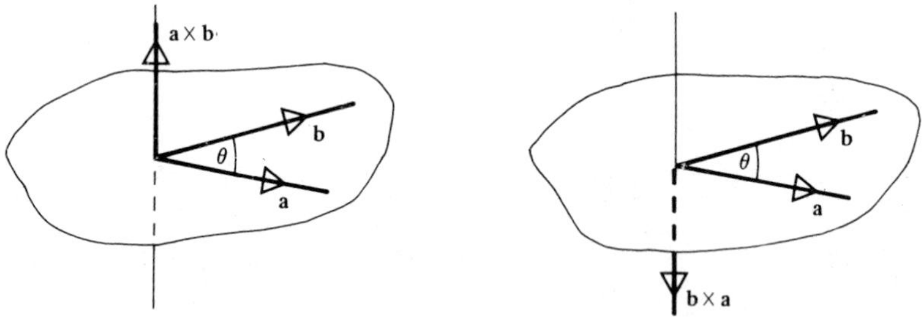

The magnitudes of $\mathbf{a} \times \mathbf{b}$ and $\mathbf{b} \times \mathbf{a}$ are equal however, as $ab \sin\theta = ba \sin\theta$.

Therefore $$\mathbf{a} \times \mathbf{b} = -\mathbf{b} \times \mathbf{a}$$

Thus the vector product is *not* commutative.

The Vector Product of Parallel Vectors

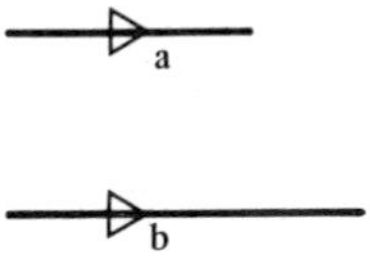

If **a** and **b** are parallel vectors

$$|\mathbf{a}\times\mathbf{b}| = ab\sin\theta$$

but $\sin\theta = 0$

therefore $\mathbf{a}\times\mathbf{b} = \mathbf{0}$

The Vector Product of Perpendicular Vectors

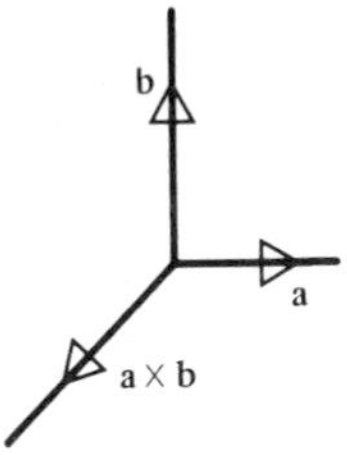

If **a** and **b** are perpendicular vectors,

$\sin\theta = 1$ and $|\mathbf{a}\times\mathbf{b}| = ab$

In this case **a**, **b** and $\mathbf{a}\times\mathbf{b}$ form a right-handed set of three mutually perpendicular vectors as shown in the diagram.

This result is particularly important in the case of the unit vectors **i**, **j** and **k**,

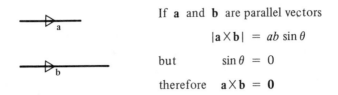

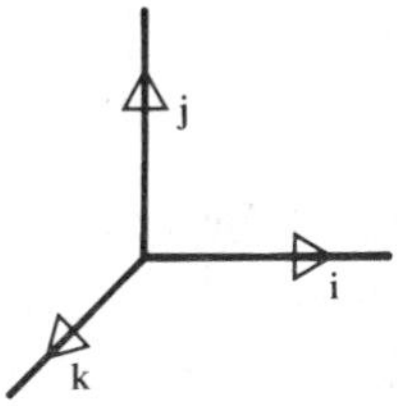

thus $\mathbf{i}\times\mathbf{j} = \mathbf{k}$ and $\mathbf{j}\times\mathbf{i} = -\mathbf{k}$

$\mathbf{j}\times\mathbf{k} = \mathbf{i}$ and $\mathbf{k}\times\mathbf{j} = -\mathbf{i}$

$\mathbf{k}\times\mathbf{i} = \mathbf{j}$ and $\mathbf{i}\times\mathbf{k} = -\mathbf{j}$

also $\mathbf{i}\times\mathbf{i} = \mathbf{j}\times\mathbf{j} = \mathbf{k}\times\mathbf{k} = \mathbf{0}$

Calculation of the Vector Product

When two vectors are each given in Cartesian component form, their vector product can be calculated as follows.

$$(x_1\mathbf{i}+y_1\mathbf{j}+z_1\mathbf{k})\times(x_2\mathbf{i}+y_2\mathbf{j}+z_2\mathbf{k}) = \begin{vmatrix} \mathbf{i} & \mathbf{j} & \mathbf{k} \\ x_1 & y_1 & z_1 \\ x_2 & y_2 & z_2 \end{vmatrix}$$

e.g.

$$(2\mathbf{i}+\mathbf{j}-2\mathbf{k})\times(\mathbf{j}+3\mathbf{k}) = \begin{vmatrix} \mathbf{i} & \mathbf{j} & \mathbf{k} \\ 2 & 1 & -2 \\ 0 & 1 & 3 \end{vmatrix}$$

$$= 5\mathbf{i}-6\mathbf{j}+2\mathbf{k}$$

THE VECTOR MOMENT OF A FORCE ABOUT A POINT

The term 'moment of a force about a point' is meaningless in itself but is a common abbreviation for the turning effect of a force about an axis passing through the point and which is perpendicular to the plane containing the force and the point, i.e. in the diagrams below the moment of **F** about P is the turning effect of **F** about XY.

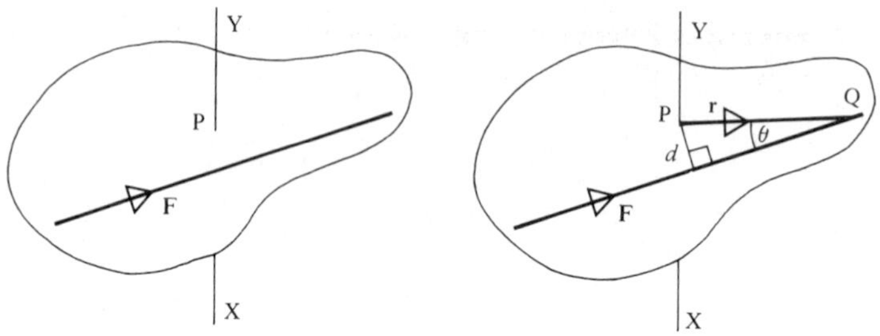

For any point Q on the line of action of **F**, the moment of **F** about P is dF

If **r** is the position vector of Q relative to P then

$$dF = (r\sin\theta)F = |\mathbf{r}\times\mathbf{F}|$$

Now $\mathbf{r}\times\mathbf{F}$ is perpendicular to both **r** and **F**.
The axis of rotation, XY, is also perpendicular to both **r** and **F**.
Therefore $\mathbf{r}\times\mathbf{F}$ is parallel to XY.

Thus the magnitude of the vector $\mathbf{r}\times\mathbf{F}$ is equal to the magnitude of the moment of **F** about the axis XY.

The direction of $\mathbf{r}\times\mathbf{F}$ is along the axis XY and is in the sense of a right-handed screw turned from **r** to **F**.

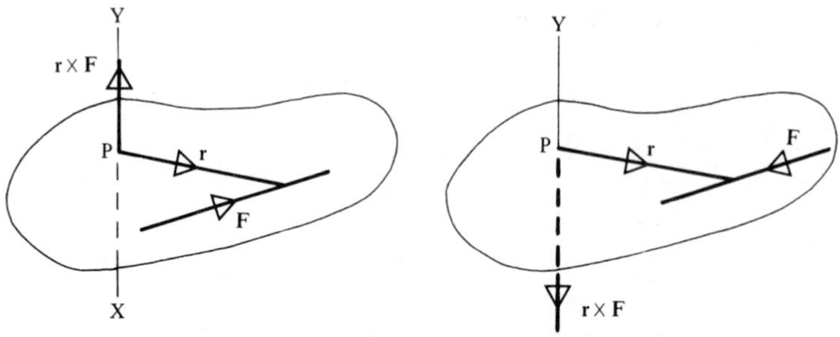

$\mathbf{r}\times\mathbf{F}$ is called the vector moment of **F** about P.

Note. If the vector moment of $\mathbf{F}$ about P is known to be $\mathbf{a}$, then $\mathbf{r}\times\mathbf{F}=\mathbf{a}$. But $\mathbf{r}$ is the position vector of *any* point on the line of action of $\mathbf{F}$, so

$\mathbf{r}\times\mathbf{F} = \mathbf{a}$ is a vector equation of the line of action of $\mathbf{F}$.

EXAMPLES 3a

1) A force $\mathbf{F}$ acts through the point with position vector $\mathbf{r}$. If $\mathbf{F}=\mathbf{i}+2\mathbf{k}$ and $\mathbf{r}=\mathbf{i}+2\mathbf{j}-\mathbf{k}$ find the vector moment of $\mathbf{F}$ about
(a) the origin, (b) the point with position vector $\mathbf{i}-\mathbf{j}$.

(a) The moment of $\mathbf{F}$ about O is $\mathbf{r}\times\mathbf{F}$.

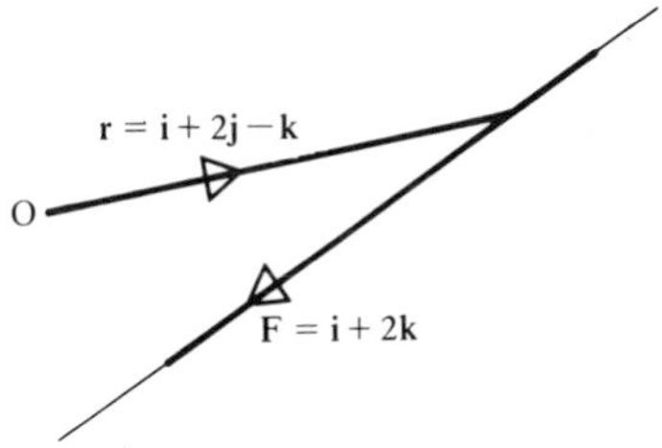

$$\begin{aligned}\mathbf{r}\times\mathbf{F} &= (\mathbf{i}+2\mathbf{j}-\mathbf{k})\times(\mathbf{i}+2\mathbf{k})\\ &= \begin{vmatrix}\mathbf{i} & \mathbf{j} & \mathbf{k}\\ 1 & 2 & -1\\ 1 & 0 & 2\end{vmatrix}\\ &= 4\mathbf{i}-3\mathbf{j}-2\mathbf{k}\end{aligned}$$

(b) The moment of $\mathbf{F}$ about P is $\overrightarrow{PQ}\times\mathbf{F}$.

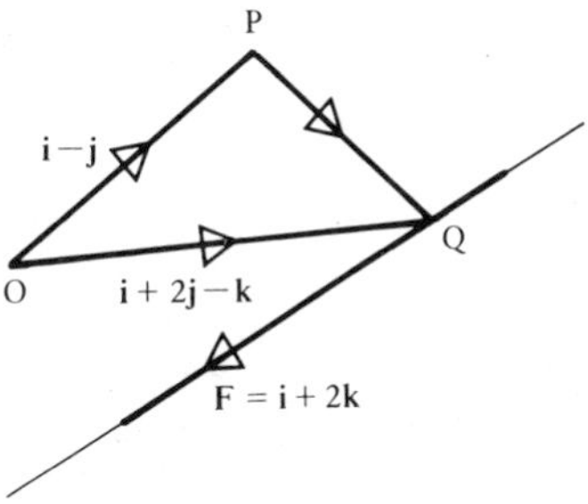

$$\begin{aligned}\overrightarrow{PQ}\times\mathbf{F} &= (\mathbf{r}-\overrightarrow{OP})\times\mathbf{F}\\ &= (3\mathbf{j}-\mathbf{k})\times(\mathbf{i}+2\mathbf{k})\\ &= \begin{vmatrix}\mathbf{i} & \mathbf{j} & \mathbf{k}\\ 0 & 3 & -1\\ 1 & 0 & 2\end{vmatrix}\\ &= 6\mathbf{i}-\mathbf{j}-3\mathbf{k}\end{aligned}$$

2) A force $\mathbf{F}$, where $\mathbf{F}=\mathbf{i}-\mathbf{j}+2\mathbf{k}$, acts through the point with position vector $2\mathbf{i}+\mathbf{k}$. Find the magnitude of the moment of $\mathbf{F}$ about the point A whose position vector is $\mathbf{j}-2\mathbf{k}$. Find also the vector equation of the axis through A.

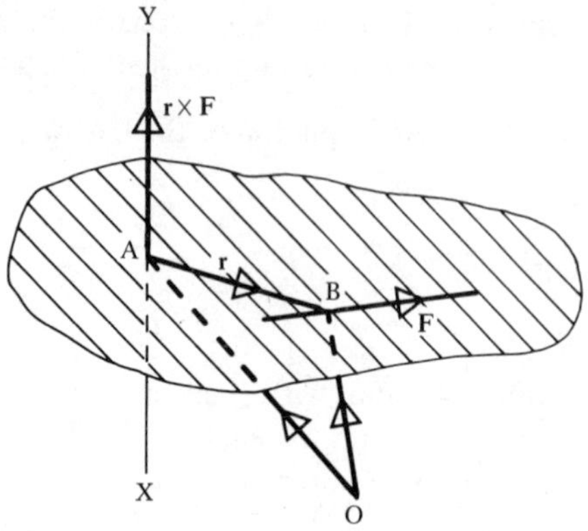

$$\left.\begin{aligned}\overrightarrow{OB} &= 2\mathbf{i}+\mathbf{k}\\ \overrightarrow{OA} &= \mathbf{j}-2\mathbf{k}\end{aligned}\right\} \quad \text{therefore} \quad \overrightarrow{AB} = \overrightarrow{OB}-\overrightarrow{OA} = 2\mathbf{i}-\mathbf{j}+3\mathbf{k}$$

The vector moment of $\mathbf{F}$ about A is $\mathbf{r}\times\mathbf{F}$

$$\begin{aligned} &= \overrightarrow{AB}\times\mathbf{F}\\ &= (2\mathbf{i}-\mathbf{j}+3\mathbf{k})\times(\mathbf{i}-\mathbf{j}+2\mathbf{k})\\ &= \begin{vmatrix}\mathbf{i} & \mathbf{j} & \mathbf{k}\\ 2 & -1 & 3\\ 1 & -1 & 2\end{vmatrix}\\ &= \mathbf{i}-\mathbf{j}-\mathbf{k}\end{aligned}$$

Therefore the magnitude of the moment of $\mathbf{F}$ about A is $\sqrt{3}$.
The direction of $\mathbf{r}\times\mathbf{F}$ is parallel to XY and A is a point on XY, therefore the vector equation of XY is

$$\begin{aligned}\mathbf{r} &= \overrightarrow{OA}+\lambda(\mathbf{r}\times\mathbf{F})\\ &= \mathbf{j}-2\mathbf{k}+\lambda(\mathbf{i}-\mathbf{j}-\mathbf{k})\end{aligned}$$

3) The moment of a force $\mathbf{F}$ about the origin is $3\mathbf{i}-\mathbf{j}+\mathbf{k}$. If $\mathbf{F}=2\mathbf{i}+\mathbf{j}-5\mathbf{k}$ find an equation for the line of action of $\mathbf{F}$ in the form $\mathbf{r}=\mathbf{a}+\lambda\mathbf{b}$.

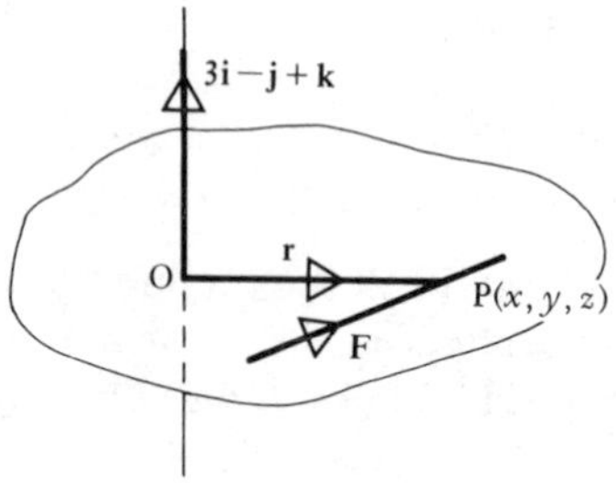

If $\mathbf{r}$ is the position vector of any point $P(x, y, z)$ on the line of action of $\mathbf{F}$ then

$$\mathbf{r} \times \mathbf{F} = 3\mathbf{i} - \mathbf{j} + \mathbf{k}$$

i.e. $$(x\mathbf{i} + y\mathbf{j} + z\mathbf{k}) \times (2\mathbf{i} + \mathbf{j} - 5\mathbf{k}) = 3\mathbf{i} - \mathbf{j} + \mathbf{k}$$

$\Rightarrow$ $$\begin{vmatrix} \mathbf{i} & \mathbf{j} & \mathbf{k} \\ x & y & z \\ 2 & 1 & -5 \end{vmatrix} = 3\mathbf{i} - \mathbf{j} + \mathbf{k}$$

Equating coefficients of $\mathbf{i}$, $\mathbf{j}$, $\mathbf{k}$ gives

$$-5y - z = 3 \qquad [1]$$

$$5x + 2z = -1 \qquad [2]$$

$$x - 2y = 1 \qquad [3]$$

(These three equations are not independent. If they were, unique values for x, y, z could be found, whereas we know that $\mathbf{r}$ is *any* position vector on the line of action of $\mathbf{F}$.)

From [1] and [3] we can find Cartesian equations for the line of action of $\mathbf{F}$.

i.e. $$\frac{x-1}{2} = y = \frac{z+3}{-5} \qquad (= \lambda, \text{ say})$$

Therefore $\mathbf{r} = \mathbf{i} - 3\mathbf{k} + \lambda(2\mathbf{i} + \mathbf{j} - 5\mathbf{k})$ is a vector equation of the line of action of $\mathbf{F}$.

EXERCISE 3a

In questions 1 to 4 a force $\mathbf{F}$ acts through the point with position vector $\mathbf{r}$ relative to the origin O. Find the vector moment of $\mathbf{F}$ about O.

1) $\mathbf{F} = \mathbf{i} + 2\mathbf{j}$, $\mathbf{r} = \mathbf{i} - \mathbf{j}$.

2) $\mathbf{F} = \mathbf{i} + 2\mathbf{k}$, $\mathbf{r} = 2\mathbf{i} - \mathbf{k}$.

3) $\mathbf{F} = \mathbf{i} - \mathbf{j}$, $\mathbf{r} = 3\mathbf{k}$.

4) $\mathbf{F} = \mathbf{i} + \mathbf{j} - 2\mathbf{k}$, $\mathbf{r} = 3\mathbf{i} - \mathbf{j} + \mathbf{k}$.

In questions 5 to 7 a force $\mathbf{F}$ acts through the point with position vector $\mathbf{r}$ relative to O. Find the vector moment of $\mathbf{F}$ about the point P.

5) $\mathbf{F} = 2\mathbf{i} - \mathbf{j}$, $\mathbf{r} = \mathbf{i} + 2\mathbf{j}$, P(0, 1).

6) $\mathbf{F} = \mathbf{i} + \mathbf{j} - \mathbf{k}$, $\mathbf{r} = 3\mathbf{i} - 2\mathbf{j}$, P(0, 1, 0).

7) $\mathbf{F} = \mathbf{i} + 2\mathbf{k}$, $\mathbf{r} = \mathbf{i} + \mathbf{j} - \mathbf{k}$, P(2, 0, 2).

8) A force $\mathbf{F}$ acts through the point with position vector $\mathbf{a}$. Find, in terms of $\mathbf{F}$, $\mathbf{a}$, $\mathbf{b}$, the vector moment of $\mathbf{F}$ about the point with position vector $\mathbf{b}$.

9) Two forces $\mathbf{F}_1$ and $\mathbf{F}_2$ act through the point with position vector $\mathbf{i}+2\mathbf{j}$. Find the vector moment of $\mathbf{F}_1$ and $\mathbf{F}_2$ about the origin and also find the vector moment of the resultant of $\mathbf{F}_1$ and $\mathbf{F}_2$ about the origin, where

$$\mathbf{F}_1 = \mathbf{i}-\mathbf{j}-\mathbf{k} \quad \text{and} \quad \mathbf{F}_2 = 2\mathbf{i}+\mathbf{k}$$

10) Find the magnitude of the moment of the force $\mathbf{F}$ about the origin where $\mathbf{r}$ is the position vector of a point on the line of action of $\mathbf{F}$, where:
(a) $\mathbf{F}=\mathbf{i}+2\mathbf{j}$, $\mathbf{r}=2\mathbf{i}-\mathbf{j}$, (b) $\mathbf{F}=\mathbf{i}+2\mathbf{j}-\mathbf{k}$, $\mathbf{r}=\mathbf{i}-\mathbf{j}+2\mathbf{k}$.

In questions 11 to 13, a force $\mathbf{F}$ acts through a point A. Find the vector equation of the axis through the point B about which the moment of $\mathbf{F}$ is calculated.

11) $\mathbf{F}=\mathbf{i}-2\mathbf{j}$, A(0, 1, 0), B(0, 0, 0).

12) $\mathbf{F}=\mathbf{j}+2\mathbf{k}$, A($\mathbf{i}+\mathbf{k}$), B($\mathbf{i}+2\mathbf{j}-\mathbf{k}$).

13) $\mathbf{F}=2\mathbf{i}+\mathbf{j}-\mathbf{k}$, A(1, 1, 2), B(−1, 2, −1).

In questions 14 to 16 a force $\mathbf{F}$ has a vector moment $\mathbf{a}$ about the origin. Find the vector equation of the line of action of $\mathbf{F}$ if:

14) $\mathbf{F}=(\mathbf{i}+\mathbf{j})$, $\mathbf{a}=4\mathbf{k}$.

15) $\mathbf{F}=(2\mathbf{i}-\mathbf{j})$, $\mathbf{a}=(\mathbf{i}+2\mathbf{j})$.

16) $\mathbf{F}=(\mathbf{i}+\mathbf{j}-\mathbf{k})$, $\mathbf{a}=(3\mathbf{i}-2\mathbf{j}+\mathbf{k})$.

17) The moment of a non-zero force $\mathbf{F}$ about a point A is the same as its moment about another point B. Show that the line of action of the force $\mathbf{F}$ is parallel to AB. (JMB)

18) Find the vector moment of the force $\mathbf{i}+2\mathbf{j}-3\mathbf{k}$ about the point (3, 0, −2), given that the line of action of the force passes through the point (1, 5, 2).

19) A force $\mathbf{F}$ acts at the point with position vector $\mathbf{r}$. Express as a vector product the moment of this force about the point with position vector $\mathbf{a}$.
A force of unit magnitude has equal moments about points with position vectors $\mathbf{j}$ and $\mathbf{i}+2\mathbf{j}-\mathbf{k}$. Find the possible forces. (JMB)

20) The force $\mathbf{F}$ given by $\mathbf{F}=4\mathbf{i}+\mathbf{k}$ is such that its line of action passes through a point A whose distance from the origin O is 3. The moment of $\mathbf{F}$ about O is $2\mathbf{i}+6\mathbf{j}-8\mathbf{k}$. Find the coordinates of A.

VECTOR MOMENT OF A COUPLE

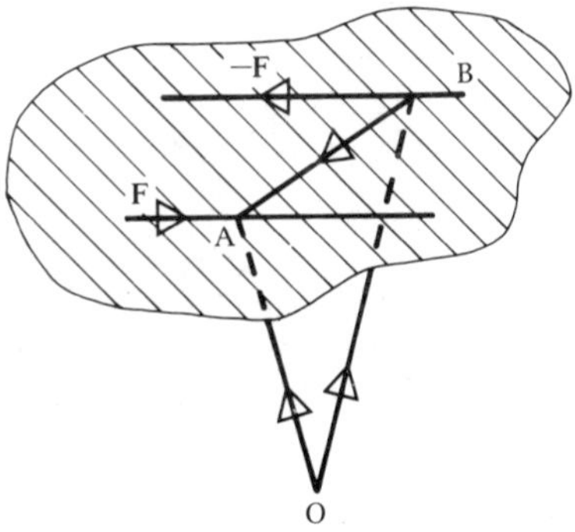

Consider the forces **F** and −**F** acting along parallel lines. Such a pair of forces is a couple and we know that the moment of a couple is the same about any point in the plane of the couple. Now consider the sum of the moments of these forces about a point O, not in the plane of the couple.

If A is a point on the line of action of **F**, and B is a point on the line of action of −**F**, the sum of the moments of the forces about O is

$$\overrightarrow{OA} \times \mathbf{F} + \overrightarrow{OB} \times (-\mathbf{F})$$
$$= (\overrightarrow{OA} - \overrightarrow{OB}) \times \mathbf{F}$$
$$= \overrightarrow{BA} \times \mathbf{F}$$

The expression $\overrightarrow{BA} \times \mathbf{F}$ is called the vector moment of the couple.

We see that the characteristics of the vector moment of the couple are

(a) $\overrightarrow{BA}$ is independent of O.
(b) $\overrightarrow{BA} \times \mathbf{F}$ is perpendicular to the plane of the couple.
(c) $|\overrightarrow{BA} \times \mathbf{F}| = Fd$ which is the magnitude of the moment of couple.

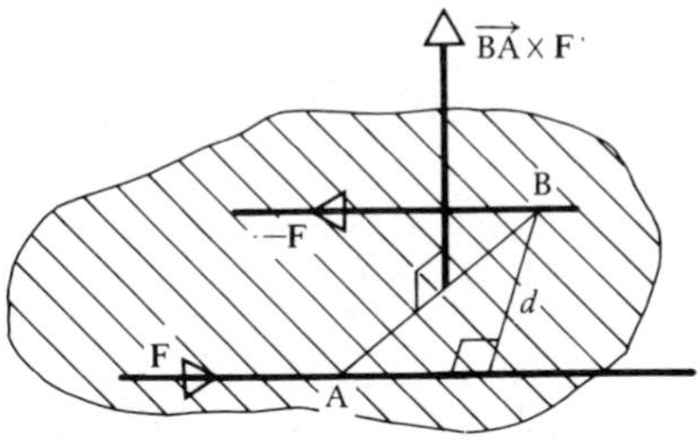

Therefore

> the vector moment of a couple is the same about any point whether or not that point is in the plane of the couple.

When a problem refers to a single vector **G** as representing a couple it must be remembered that **G** is the vector moment of that couple and therefore is a vector perpendicular to the plane containing the couple.

Equivalent Couples

Two couples are equivalent if their vector moments are equal, i.e. the magnitudes and directions of their vector moments must be the same. This can only be so if the couples act in parallel planes.

EXAMPLE 3b

A couple comprises a force $\mathbf{F}$ acting along the line l_1 and a force $-\mathbf{F}$ acting along the line l_2. If $\mathbf{F} = 3(\mathbf{i}-\mathbf{j}+\mathbf{k})$ and the vector equations of l_1, l_2 are $\mathbf{r} = \mathbf{i}+\mathbf{j}+t(\mathbf{i}-\mathbf{j}+\mathbf{k})$, $\mathbf{r} = \mathbf{i}+\mathbf{k}+s(\mathbf{i}-\mathbf{j}+\mathbf{k})$ find the vector moment of the couple and hence a unit vector perpendicular to the plane of the couple.

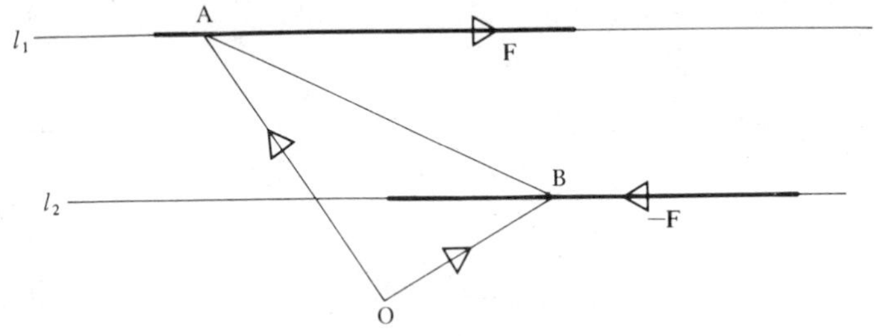

One force $\mathbf{F}$ acts along the line with equation $\mathbf{r} = \mathbf{i}+\mathbf{j}+t(\mathbf{i}-\mathbf{j}+\mathbf{k})$; therefore $\mathbf{i}+\mathbf{j}$ is the position vector of a point A on this line. Similarly $\mathbf{i}+\mathbf{k}$ is the position vector of a point B on the line l_2.

The vector moment of the couple is $\overrightarrow{BA} \times \mathbf{F}$

i.e.
$$(\mathbf{j}-\mathbf{k}) \times 3(\mathbf{i}-\mathbf{j}+\mathbf{k}) = 3\begin{vmatrix} \mathbf{i} & \mathbf{j} & \mathbf{k} \\ 0 & 1 & -1 \\ 1 & -1 & 1 \end{vmatrix}$$
$$= 3(-\mathbf{j}-\mathbf{k})$$

This is perpendicular to the plane of the couple, therefore a unit vector in this direction is $\frac{1}{2}\sqrt{2}\,(-\mathbf{j}-\mathbf{k})$.

EXERCISE 3b

In questions 1 to 4, two forces $\mathbf{F}$ and $-\mathbf{F}$ act through the points with position vectors $\mathbf{r}_1$ and $\mathbf{r}_2$. Find the moment of this couple and hence find the distance between the lines of action of the two forces.

1) $\mathbf{F} = \mathbf{i}-2\mathbf{j}+2\mathbf{k}$, $\mathbf{r}_1 = \mathbf{i}+\mathbf{j}$, $\mathbf{r}_2 = \mathbf{i}-\mathbf{k}$.

2) $\mathbf{F} = 2\mathbf{i}+6\mathbf{j}-3\mathbf{k}$, $\mathbf{r}_1 = 2\mathbf{j}$, $\mathbf{r}_2 = \mathbf{i}+\mathbf{j}+\mathbf{k}$.

3) $\mathbf{F} = 8\mathbf{i} - \mathbf{j} + 4\mathbf{k}$, $\mathbf{r}_1 = \mathbf{i} - \mathbf{j} - \mathbf{k}$, $\mathbf{r}_2 = 3\mathbf{i}$.

4) $\mathbf{F} = \mathbf{i} + \mathbf{j} + \mathbf{k}$, $\mathbf{r}_1 = 2\mathbf{i} + \mathbf{k}$, $\mathbf{r}_2 = \mathbf{j} - \mathbf{k}$.

5) A couple is formed from two forces of magnitude $\sqrt{3}$ units acting along the lines whose vector equations are $\mathbf{r} = \mathbf{i} + 2\mathbf{j} + t(\mathbf{i} - \mathbf{j} + \mathbf{k})$ and $\mathbf{r} = 2\mathbf{i} - \mathbf{k} + s(\mathbf{i} - \mathbf{j} + \mathbf{k})$. Find the magnitude of the moment of this couple. Find also a vector which is perpendicular to the plane containing the couple.

Resultant Vector Moment

The resultant vector moment of a set of forces about a particular point is the sum of the vector moments of the separate forces about that point.

If a set of forces reduces to a single force, the sum of the vector moments of the separate forces about a particular point is equal to the vector moment of the resultant force about that point.

If a set of forces reduces to a couple, the sum of the vector moments of the separate forces about a point is equal to the vector moment of the couple.

If a set of forces is in equilibrium then, about *any* point, the sum of the vector moments of the separate forces is zero.

SYSTEMS OF FORCES IN THREE DIMENSIONS

Two systems of forces are *equivalent* if their effect on a body is the same in all respects.

The *resultant* of a set of forces is the simplest possible equivalent system and the original set of forces is said to *reduce* to its resultant system.

We know that a system of coplanar forces is equivalent either to a single force, or to a couple, or that the system is in equilibrium. But if we consider a set of non-coplanar forces a fourth possibility exists in addition to the three already mentioned.

Consider, for example, two forces whose lines of action are skew. Such a pair of forces cannot be in equilibrium and they cannot reduce either to a single force or to a couple but are equivalent to a combination of a force and couple where the plane of the couple does not contain the force.

Thus a system of *non-coplanar* forces can be
(a) in equilibrium,

or equivalent to

either (b) a single force,
or (c) a couple,
or (d) a combination of a couple and non-coplanar force.

Suppose that S is a system of forces $\mathbf{F}_1,\ \mathbf{F}_2,\ \mathbf{F}_3 \ldots$ acting through points with position vectors $\mathbf{r}_1,\ \mathbf{r}_2,\ \mathbf{r}_3 \ldots$

If S reduces to a *single force* $\mathbf{R}$ acting through a point with position vector $\mathbf{a}$ then

$$\Sigma\mathbf{F} = \mathbf{R} \quad \text{and} \quad \Sigma\mathbf{r}\times\mathbf{F} = \mathbf{a}\times\mathbf{R}$$

If S reduces to a *couple* of moment $\mathbf{G}$ then

$$\Sigma\mathbf{F} = \mathbf{0} \quad \text{and} \quad \Sigma\mathbf{r}\times\mathbf{F} = \mathbf{G}$$

If S is in *equilibrium* then

$$\Sigma\mathbf{F} = \mathbf{0} \quad \text{and} \quad \Sigma\mathbf{r}\times\mathbf{F} = \mathbf{0}$$

Identification of Force Systems

The properties given above can be used to determine the nature of a force system.

To prove that a system of forces is in *equilibrium*, it is necessary to show that

$$\textit{both} \quad \Sigma\mathbf{F} = \mathbf{0} \quad \textit{and} \quad \Sigma\mathbf{r}\times\mathbf{F} = \mathbf{0}$$

(Either condition on its own is not sufficient because $\Sigma\mathbf{F} = \mathbf{0}$ is also true for a system equivalent to a couple; and $\Sigma\mathbf{r}\times\mathbf{F} = \mathbf{0}$ applies also to a system whose resultant is a single force through O.)

To prove that a system of forces reduces to a *couple*, it is necessary to show that

$$\textit{both} \quad \Sigma\mathbf{F} = \mathbf{0} \quad \textit{and} \quad \Sigma\mathbf{r}\times\mathbf{F} \neq \mathbf{0}$$

(Again either condition on its own is insufficient.)

Showing that a system of forces reduces to a single force is not so straightforward because, if $\Sigma\mathbf{F} \neq \mathbf{0}$ the system reduces *either* to a single force *or* to the combination of a couple and a non-coplanar force.

To distinguish between these two possibilities we must investigate the value of $\Sigma\mathbf{r}\times\mathbf{F}$.

If the system reduces to a single force $\mathbf{R}$ then

either the line of action of $\mathbf{R}$ passes through the origin in which case $\Sigma\mathbf{r}\times\mathbf{F} = \mathbf{0}$.

or the line of action of $\mathbf{R}$ does not pass through O in which case the resultant moment of the system about O acts in the plane containing O and $\mathbf{R}$.

This means that the resultant *vector* moment of the system about O is perpendicular to the resultant force, i.e. $\Sigma\mathbf{F}.\Sigma\mathbf{r}\times\mathbf{F}=\mathbf{0}$, therefore, to prove that a system of forces reduces to a *single force* it is necessary to show that

$$\textit{both}\quad \Sigma\mathbf{F}\neq\mathbf{0}\quad\textit{and}\quad\begin{cases}\text{either } \Sigma\mathbf{r}\times\mathbf{F}=\mathbf{0}\\ \text{or}\quad \Sigma\mathbf{F}.\Sigma\mathbf{r}\times\mathbf{F}=\mathbf{0}\end{cases}$$

Note that the condition $\Sigma\mathbf{F}.\Sigma\mathbf{r}\times\mathbf{F}=\mathbf{0}$ includes the possibility that $\Sigma\mathbf{r}\times\mathbf{F}=\mathbf{0}$. However it is sensible to work out $\Sigma\mathbf{r}\times\mathbf{F}$ first, and only if it is not zero to evaluate $\Sigma\mathbf{F}.\Sigma\mathbf{r}\times\mathbf{F}$.

If the system reduces to a couple together with a non-coplanar force then the resultant moment of the system about O is non-zero and acts in a plane that does *not* contain the resultant force. Therefore the resultant vector moment is *not* perpendicular to the resultant force. It follows that,

to prove that a system of forces reduces to *a couple and a non-coplanar force* it is necessary to show that

$$\textit{both}\quad \Sigma\mathbf{F}\neq\mathbf{0}\quad\textit{and}\quad \Sigma\mathbf{F}.\Sigma\mathbf{r}\times\mathbf{F}\neq\mathbf{0}$$

EXAMPLES 3c

1) Prove that the following system of forces is in equilibrium:

$\mathbf{F}_1=3\mathbf{i}-\mathbf{j}+\mathbf{k}$ acting at the point $\mathbf{r}_1=\mathbf{i}+\mathbf{j}$,
$\mathbf{F}_2=-2\mathbf{i}+4\mathbf{j}-5\mathbf{k}$ acting at the point $\mathbf{r}_2=3\mathbf{i}-3\mathbf{j}+5\mathbf{k}$,
$\mathbf{F}_3=-\mathbf{i}-3\mathbf{j}+4\mathbf{k}$ acting at the point $\mathbf{r}_3=-2\mathbf{j}+4\mathbf{k}$.

$$\Sigma\mathbf{F}=(3\mathbf{i}-\mathbf{j}+\mathbf{k})+(-2\mathbf{i}+4\mathbf{j}-5\mathbf{k})+(-\mathbf{i}-3\mathbf{j}+4\mathbf{k})=\mathbf{0}$$

$$\Sigma\mathbf{r}\times\mathbf{F}=(\mathbf{i}+\mathbf{j})\times(3\mathbf{i}-\mathbf{j}+\mathbf{k})+(3\mathbf{i}-3\mathbf{j}+5\mathbf{k})\times(-2\mathbf{i}+4\mathbf{j}-5\mathbf{k})+(-2\mathbf{j}+4\mathbf{k})\times(-\mathbf{i}-3\mathbf{j}+4\mathbf{k})$$

$$=\begin{vmatrix}\mathbf{i}&\mathbf{j}&\mathbf{k}\\1&1&0\\3&-1&1\end{vmatrix}+\begin{vmatrix}\mathbf{i}&\mathbf{j}&\mathbf{k}\\3&-3&5\\-2&4&-5\end{vmatrix}+\begin{vmatrix}\mathbf{i}&\mathbf{j}&\mathbf{k}\\0&-2&4\\-1&-3&4\end{vmatrix}$$

$$=(\mathbf{i}-\mathbf{j}-4\mathbf{k})+(-5\mathbf{i}+5\mathbf{j}+6\mathbf{k})+(4\mathbf{i}-4\mathbf{j}-2\mathbf{k})=\mathbf{0}$$

Both $\Sigma\mathbf{F}=\mathbf{0}$ and $\Sigma\mathbf{r}\times\mathbf{F}=\mathbf{0}$

Therefore the system is in equilibrium.

(An alternative method is to show that $\Sigma\mathbf{F}=\mathbf{0}$ and that the lines of action of the *three* given forces are concurrent.)

2) Forces $\mathbf{F}_1$, $\mathbf{F}_2$ and $\mathbf{F}_3$ act through the points with position vectors $\mathbf{r}_1$, $\mathbf{r}_2$ and $\mathbf{r}_3$ where:

$$\mathbf{F}_1 = \mathbf{i}+\mathbf{j}-\mathbf{k}, \qquad \mathbf{r}_1 = \mathbf{i}+\mathbf{j}+\mathbf{k}$$

$$\mathbf{F}_2 = 2\mathbf{i}-\mathbf{j}+3\mathbf{k}, \qquad \mathbf{r}_2 = -\mathbf{i}-2\mathbf{j}+\mathbf{k}$$

$$\mathbf{F}_3 = -3\mathbf{i}-4\mathbf{j}+\mathbf{k}, \qquad \mathbf{r}_3 = 4\mathbf{i}+5\mathbf{j}$$

When a fourth force $\mathbf{F}_4$ is added it reduces the system to equilibrium. Find $\mathbf{F}_4$ and a vector equation of its line of action.

The system is in equilibrium therefore $\Sigma\mathbf{F} = \mathbf{0}$

i.e. $$\mathbf{F}_1+\mathbf{F}_2+\mathbf{F}_3+\mathbf{F}_4 = \mathbf{0}$$

$\Rightarrow$ $$(\mathbf{i}+\mathbf{j}-\mathbf{k})+(2\mathbf{i}-\mathbf{j}+3\mathbf{k})+(-3\mathbf{i}-4\mathbf{j}+\mathbf{k})+\mathbf{F}_4 = \mathbf{0}$$

Therefore $$\mathbf{F}_4 = 4\mathbf{j}-3\mathbf{k}$$

Let $\mathbf{r}_4$ be the position vector of any point on the line of action of $\mathbf{F}_4$.

The system is in equilibrium therefore $\Sigma\mathbf{r}\times\mathbf{F} = \mathbf{0}$

$\Rightarrow$ $$(\mathbf{i}+\mathbf{j}+\mathbf{k})\times(\mathbf{i}+\mathbf{j}-\mathbf{k})+(-\mathbf{i}-2\mathbf{j}+\mathbf{k})\times(2\mathbf{i}-\mathbf{j}+3\mathbf{k})$$
$$+(4\mathbf{i}+5\mathbf{j})\times(-3\mathbf{i}-4\mathbf{j}+\mathbf{k})+\mathbf{r}_4\times(4\mathbf{j}-3\mathbf{k}) = \mathbf{0}$$

If $\mathbf{r}_4 = x\mathbf{i}+y\mathbf{j}+z\mathbf{k}$ we have

$$\begin{vmatrix}\mathbf{i} & \mathbf{j} & \mathbf{k}\\ 1 & 1 & 1\\ 1 & 1 & -1\end{vmatrix} + \begin{vmatrix}\mathbf{i} & \mathbf{j} & \mathbf{k}\\ -1 & -2 & 1\\ 2 & -1 & 3\end{vmatrix} + \begin{vmatrix}\mathbf{i} & \mathbf{j} & \mathbf{k}\\ 4 & 5 & 0\\ -3 & -4 & 1\end{vmatrix} + \begin{vmatrix}\mathbf{i} & \mathbf{j} & \mathbf{k}\\ x & y & z\\ 0 & 4 & -3\end{vmatrix} = \mathbf{0}$$

$$(-2\mathbf{i}+2\mathbf{j})+(-5\mathbf{i}+5\mathbf{j}+5\mathbf{k})+(5\mathbf{i}-4\mathbf{j}-\mathbf{k})+(-3y-4z)\mathbf{i}+3x\mathbf{j}+4x\mathbf{k} = \mathbf{0}$$

$\Rightarrow$ $$-2\mathbf{i}+3\mathbf{j}+4\mathbf{k}+(-3y-4z)\mathbf{i}+3x\mathbf{j}+4x\mathbf{k} = \mathbf{0}$$

or $$(-2-3y-4z)\mathbf{i}+(3+3x)\mathbf{j}+(4+4x)\mathbf{k} = \mathbf{0}$$

Hence $$x = -1 \quad \text{and} \quad 3y+4z = -2$$

Taking $y = \dfrac{-2-4z}{3} = \lambda$ gives

$$\mathbf{r} = -\mathbf{i}+\lambda\mathbf{j}+\frac{(-2-3\lambda)}{4}\mathbf{k}$$

and this is a vector equation of the line of action of $\mathbf{F}_4$.

3) The points A(0, 1, 1), B(3, −1, 0), C(2, 0, 4), D(1, 3, −1) form a skew quadrilateral ABCD. Forces $\overrightarrow{AB}$, $\overrightarrow{BC}$, $\overrightarrow{CD}$, $\overrightarrow{DA}$ act along the sides AB, BC, CD, DA of this quadrilateral. Prove that this system of forces is equivalent to a couple and find its vector moment.

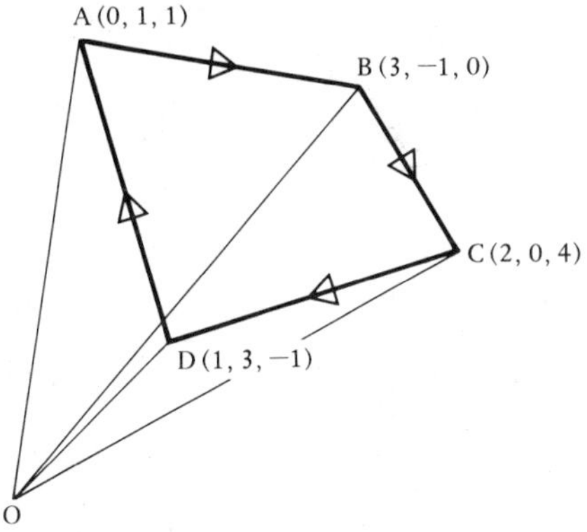

$$\overrightarrow{AB} = \overrightarrow{OB} - \overrightarrow{OA} = 3\mathbf{i} - 2\mathbf{j} - \mathbf{k} = \mathbf{F}_1$$

$$\overrightarrow{BC} = \overrightarrow{OC} - \overrightarrow{OB} = -\mathbf{i} + \mathbf{j} + 4\mathbf{k} = \mathbf{F}_2$$

$$\overrightarrow{CD} = \overrightarrow{OD} - \overrightarrow{OC} = -\mathbf{i} + 3\mathbf{j} - 5\mathbf{k} = \mathbf{F}_3$$

$$\overrightarrow{DA} = \overrightarrow{OA} - \overrightarrow{OD} = -\mathbf{i} - 2\mathbf{j} + 2\mathbf{k} = \mathbf{F}_4$$

$$\mathbf{F}_1 + \mathbf{F}_2 + \mathbf{F}_3 + \mathbf{F}_4 = \mathbf{0}, \quad \text{i.e.} \quad \Sigma\mathbf{F} = \mathbf{0}$$

The sum of the moments of the forces about O is

$$\overrightarrow{OA} \times \mathbf{F}_1 + \overrightarrow{OB} \times \mathbf{F}_2 + \overrightarrow{OC} \times \mathbf{F}_3 + \overrightarrow{OD} \times \mathbf{F}_4$$

$$= \begin{vmatrix} \mathbf{i} & \mathbf{j} & \mathbf{k} \\ 0 & 1 & 1 \\ 3 & -2 & -1 \end{vmatrix} + \begin{vmatrix} \mathbf{i} & \mathbf{j} & \mathbf{k} \\ 3 & -1 & 0 \\ -1 & 1 & 4 \end{vmatrix} + \begin{vmatrix} \mathbf{i} & \mathbf{j} & \mathbf{k} \\ 2 & 0 & 4 \\ -1 & 3 & -5 \end{vmatrix} + \begin{vmatrix} \mathbf{i} & \mathbf{j} & \mathbf{k} \\ 1 & 3 & -1 \\ -1 & -2 & 2 \end{vmatrix}$$

$$= (\mathbf{i} + 3\mathbf{j} - 3\mathbf{k}) + (-4\mathbf{i} - 12\mathbf{j} + 2\mathbf{k}) + (-12\mathbf{i} + 6\mathbf{j} + 6\mathbf{k}) + (4\mathbf{i} - \mathbf{j} + \mathbf{k})$$

$$= -11\mathbf{i} - 4\mathbf{j} + 6\mathbf{k} \neq \mathbf{0}$$

We now have $\Sigma\mathbf{r} \times \mathbf{F} \neq \mathbf{0}$ and $\Sigma\mathbf{F} = \mathbf{0}$.

Therefore the forces reduce to a couple of vector moment $-11\mathbf{i} - 4\mathbf{j} + 6\mathbf{k}$.

4) The vertices of a triangle ABC are at the points A(0, 1, 2), B(3, −1, 1), C(1, 0, 1). Forces $\mathbf{F}_1$, $\mathbf{F}_2$, $\mathbf{F}_3$ of magnitudes $\sqrt{14}$ N, $2\sqrt{5}$ N, $3\sqrt{3}$ N act along the sides AB, BC, CA of the triangle in the sense indicated by the letters. Prove that this system of forces is equivalent to a single force and find the Cartesian equation of its line of action.

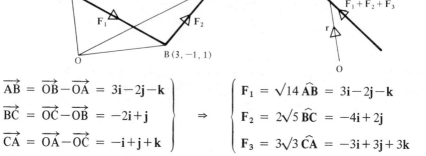

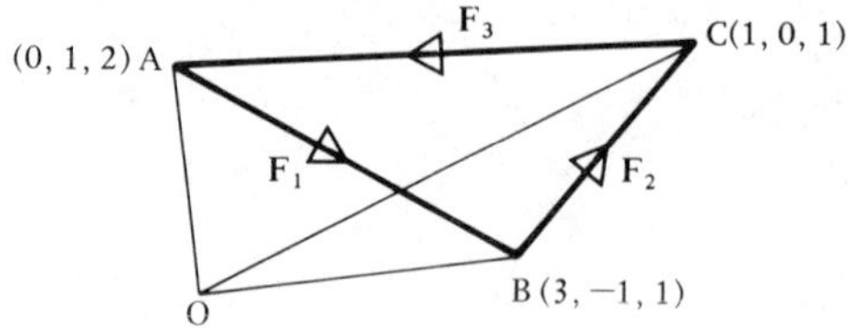

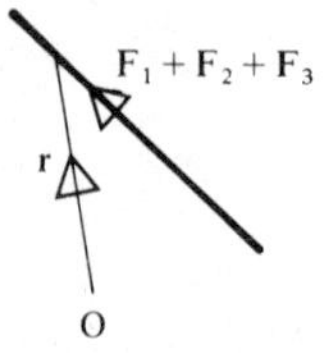

$$\left.\begin{aligned}\overrightarrow{AB} &= \overrightarrow{OB}-\overrightarrow{OA} = 3\mathbf{i}-2\mathbf{j}-\mathbf{k}\\ \overrightarrow{BC} &= \overrightarrow{OC}-\overrightarrow{OB} = -2\mathbf{i}+\mathbf{j}\\ \overrightarrow{CA} &= \overrightarrow{OA}-\overrightarrow{OC} = -\mathbf{i}+\mathbf{j}+\mathbf{k}\end{aligned}\right\} \Rightarrow \left\{\begin{aligned}\mathbf{F}_1 &= \sqrt{14}\,\widehat{\mathbf{AB}} = 3\mathbf{i}-2\mathbf{j}-\mathbf{k}\\ \mathbf{F}_2 &= 2\sqrt{5}\,\widehat{\mathbf{BC}} = -4\mathbf{i}+2\mathbf{j}\\ \mathbf{F}_3 &= 3\sqrt{3}\,\widehat{\mathbf{CA}} = -3\mathbf{i}+3\mathbf{j}+3\mathbf{k}\end{aligned}\right.$$

Therefore $\qquad \Sigma\mathbf{F} = -4\mathbf{i}+3\mathbf{j}+2\mathbf{k} \neq \mathbf{0}$

Therefore the forces cannot be in equilibrium, nor can they be equivalent to a couple.

Also, because the three points, A, B and C are coplanar, $\mathbf{F}_1$, $\mathbf{F}_2$ and $\mathbf{F}_3$ are coplanar; therefore the force system cannot reduce to a force together with non-coplanar couple.

Having eliminated all other possibilities we see that the resultant is the single force $-4\mathbf{i}+3\mathbf{j}+2\mathbf{k}$.

Now, if $\mathbf{r}$ is the position vector of any point on the line of action of the resultant force, then

$$\overrightarrow{OA}\times\mathbf{F}_1+\overrightarrow{OB}\times\mathbf{F}_2+\overrightarrow{OC}\times\mathbf{F}_3 = \mathbf{r}\times\Sigma\mathbf{F}$$

$$\Rightarrow \begin{vmatrix}\mathbf{i} & \mathbf{j} & \mathbf{k}\\ 0 & 1 & 2\\ 3 & -2 & -1\end{vmatrix} + \begin{vmatrix}\mathbf{i} & \mathbf{j} & \mathbf{k}\\ 3 & -1 & 1\\ -4 & 2 & 0\end{vmatrix} + \begin{vmatrix}\mathbf{i} & \mathbf{j} & \mathbf{k}\\ 1 & 0 & 1\\ -3 & 3 & 3\end{vmatrix} = \begin{vmatrix}\mathbf{i} & \mathbf{j} & \mathbf{k}\\ x & y & z\\ -4 & 3 & 2\end{vmatrix}$$

$$\Rightarrow \qquad -2\mathbf{i}-4\mathbf{j}+2\mathbf{k} = (2y-3z)\mathbf{i}+(-2x-4z)\mathbf{j}+(3x+4y)\mathbf{k}$$

Therefore
$$\begin{cases}2y-3z = -2 & [1]\\ x+2z = 2 & [2]\\ 3x+4y = 2 & [3]\end{cases}$$

Equations [1], [2] and [3] are not independent, therefore any pair of equations from [1], [2] and [3] will serve as the Cartesian equations of the line of action of the resultant force.

Using equations [1] and [3] and writing them in standard form gives

$$y = \frac{3z-2}{2} \quad \text{and} \quad y = \frac{2-3x}{4} = \frac{3x-2}{-4}$$

Therefore $$\frac{x-\frac{2}{3}}{-4} = \frac{y}{3} = \frac{z-\frac{2}{3}}{2}$$

These are the Cartesian equations of the line of action of the resultant force.

5) A force **F** acts at the point A, with position vector **i**−**k** and **F** = **i**−**j**+2**k**. Prove that this force is equivalent to a force **F** acting at the point B, with position vector **i**+2**j**+**k**, together with a couple. Find the vector moment of the couple.

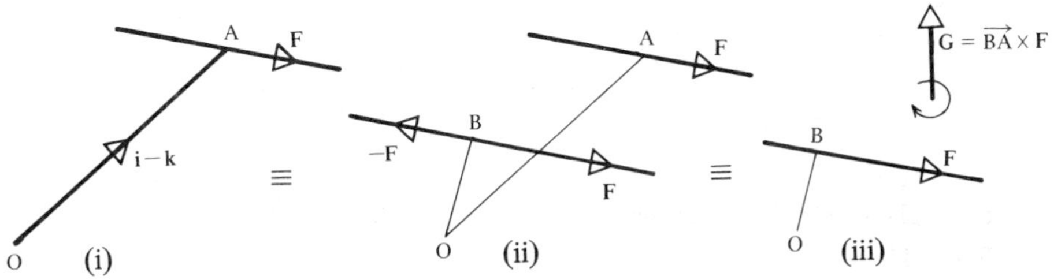

If we introduce forces **F** and −**F** acting at B, these two forces, being in equilibrium, do not alter the effect of the original force.
Therefore the force in (i) is equivalent to the forces in (ii).

If we now consider the combination of the force **F** acting at A and the force −**F** acting at B, we see that they form a couple of moment $\overrightarrow{BA}\times\mathbf{F}$.

$$\overrightarrow{BA}\times\mathbf{F} = (-2\mathbf{j}-2\mathbf{k})\times(\mathbf{i}-\mathbf{j}+2\mathbf{k})$$

$$= \begin{vmatrix} \mathbf{i} & \mathbf{j} & \mathbf{k} \\ 0 & -2 & -2 \\ 1 & -1 & 2 \end{vmatrix} = -6\mathbf{i}-2\mathbf{j}+2\mathbf{k}$$

Therefore the force **F** acting at A is equivalent to the force **F** acting at B together with a couple of moment −6**i**−2**j**+2**k**.

(The general property that any force acting through a given point can be replaced by an equal force acting through another point, together with a couple, is given in *Mechanics and Probability*.)

6) The points A, B, C have position vectors **a**, **b**, **c** respectively. Forces AB, BC and CA act round the sides of the triangle ABC in the sense indicated by the order of the letters. Verify that these forces are equivalent to a couple.

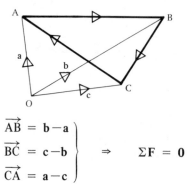

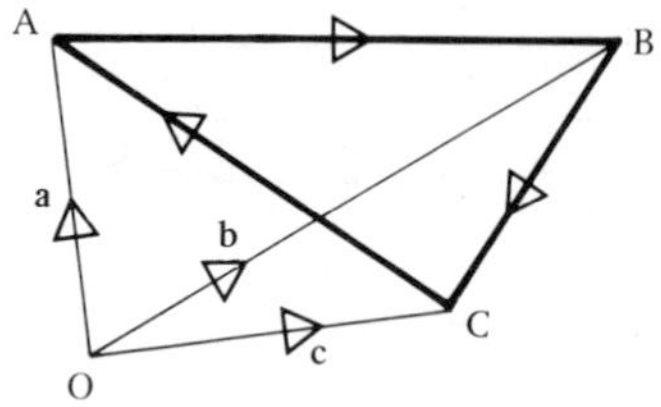

$$\left.\begin{aligned}\overrightarrow{AB} &= \mathbf{b}-\mathbf{a}\\ \overrightarrow{BC} &= \mathbf{c}-\mathbf{b}\\ \overrightarrow{CA} &= \mathbf{a}-\mathbf{c}\end{aligned}\right\} \Rightarrow \quad \Sigma\mathbf{F} = \mathbf{0}$$

Therefore the forces either are in equilibrium or reduce to a couple.

Taking moments about A gives

$$\begin{aligned}&\mathbf{0}\times\overrightarrow{AB} + \overrightarrow{AB}\times\overrightarrow{BC} + \mathbf{0}\times\overrightarrow{AC}\\ &= \overrightarrow{AB}\times\overrightarrow{BC}\end{aligned}$$

But $\frac{1}{2}|\overrightarrow{AB}\times\overrightarrow{BC}| =$ area of ΔABC

i.e. $\overrightarrow{AB}\times\overrightarrow{BC}$ is not zero.

Therefore the forces are equivalent to a couple whose magnitude is equal to twice the area of ΔABC.

7) Forces $\mathbf{F}_1$, $\mathbf{F}_2$ and $\mathbf{F}_3$ act through points whose position vectors are $\mathbf{a}_1$, $\mathbf{a}_2$ and $\mathbf{a}_3$ respectively, where

$$\mathbf{F}_1 = \mathbf{i}+\mathbf{j}-\mathbf{k} \qquad \mathbf{a}_1 = \mathbf{i}-\mathbf{j}$$

$$\mathbf{F}_2 = \mathbf{i}-\mathbf{j} \qquad \mathbf{a}_2 = \mathbf{i}+\mathbf{k}$$

$$\mathbf{F}_3 = 2\mathbf{i}+3\mathbf{j}+\mathbf{k} \qquad \mathbf{a}_3 = \mathbf{j}-\mathbf{k}$$

Find the simplest system of forces that is equivalent to the given system

$$\Sigma\mathbf{F} = 4\mathbf{i}-3\mathbf{j}$$

$$\Sigma\mathbf{r}\times\mathbf{F} = \begin{vmatrix}\mathbf{i} & \mathbf{j} & \mathbf{k}\\ 1 & -1 & 0\\ 1 & 1 & -1\end{vmatrix} + \begin{vmatrix}\mathbf{i} & \mathbf{j} & \mathbf{k}\\ 1 & 0 & 1\\ 1 & -1 & 0\end{vmatrix} + \begin{vmatrix}\mathbf{i} & \mathbf{j} & \mathbf{k}\\ 0 & 1 & -1\\ 2 & 3 & 1\end{vmatrix} = -\mathbf{k}$$

$$\Sigma\mathbf{F}\cdot\Sigma\mathbf{r}\times\mathbf{F} = (4\mathbf{i}-3\mathbf{j})\cdot(-\mathbf{k}) = 0$$

When $\Sigma\mathbf{F} \neq \mathbf{0}$ and $\Sigma\mathbf{F}.\Sigma\mathbf{r}\times\mathbf{F} = \mathbf{0}$, the resultant is a single force. In this case the single force is $4\mathbf{i}-3\mathbf{j}$.

Let the position vector of any point on the line of action of the resultant be $\mathbf{r} = x\mathbf{i}+y\mathbf{j}+z\mathbf{k}$.

Then $\qquad \mathbf{r}\times(4\mathbf{i}-3\mathbf{j}) = -\mathbf{k}$

i.e. $$\begin{vmatrix} \mathbf{i} & \mathbf{j} & \mathbf{k} \\ x & y & z \\ 4 & -3 & 0 \end{vmatrix} = -\mathbf{k} \Rightarrow z = 0, \quad 3x+4y = 1$$

If $x = \lambda$, then $y = \dfrac{1-3\lambda}{4} \Rightarrow \mathbf{r} = \lambda\mathbf{i}+\frac{1}{4}(1-3\lambda)\mathbf{j}$

Therefore the system is equivalent to the single force, $4\mathbf{i}-3\mathbf{j}$, acting along the line whose vector equation is $\mathbf{r} = \lambda\mathbf{i}+\frac{1}{4}(1-3\lambda)\mathbf{j}$.

EXERCISE 3c

1) Prove that the following system of forces is in equilibrium:
$\mathbf{F}_1 = \mathbf{i}-\mathbf{j}$ acting at the point $\mathbf{r}_1 = \mathbf{i}+\mathbf{k}$,
$\mathbf{F}_2 = \mathbf{i}-\mathbf{k}$ acting at the point $\mathbf{r}_2 = 2\mathbf{i}$,
$\mathbf{F}_3 = 2\mathbf{j}+\mathbf{k}$ acting at the point $\mathbf{r}_3 = \mathbf{i}-2\mathbf{j}$,
$\mathbf{F}_4 = -2\mathbf{i}-\mathbf{j}$ acting at the point $\mathbf{r}_4 = 3\mathbf{i}+\mathbf{j}+\mathbf{k}$.

2) Prove that the following system of forces reduces to a couple and find the vector moment of the couple.
$\mathbf{F}_1 = \mathbf{i}+2\mathbf{j}-3\mathbf{k}$ acting at the point $\mathbf{r}_1 = \mathbf{i}-\mathbf{j}+2\mathbf{k}$,
$\mathbf{F}_2 = \mathbf{i}-\mathbf{j}+2\mathbf{k}$ acting at the point $\mathbf{r}_2 = 2\mathbf{i}+\mathbf{j}-\mathbf{k}$,
$\mathbf{F}_3 = -3\mathbf{i}+\mathbf{j}-3\mathbf{k}$ acting at the point $\mathbf{r}_3 = 3\mathbf{i}+\mathbf{k}$,
$\mathbf{F}_4 = \mathbf{i}-2\mathbf{j}+4\mathbf{k}$ acting at the point $\mathbf{r}_4 = \mathbf{j}-2\mathbf{k}$.

3) Three forces $\mathbf{F}_1$, $\mathbf{F}_2$ and $\mathbf{F}_3$ act through points with position vectors $2\mathbf{i}-\mathbf{j}$, $2\mathbf{i}+\mathbf{k}$ and $8\mathbf{i}+\mathbf{j}-2\mathbf{k}$ respectively.
$\mathbf{F}_1 = \mathbf{i}-2\mathbf{j}$; $\mathbf{F}_2 = \mathbf{j}+\mathbf{k}$; $\mathbf{F}_3$ is of magnitude $\sqrt{11}$ and acts along the line whose equation is $\mathbf{r} = 5\mathbf{i}-\mathbf{k}+\lambda(3\mathbf{i}+\mathbf{j}-\mathbf{k})$.
Prove that the three forces are equivalent to a single force and find a vector equation for its line of action.

4) The points A(0, 1, 1), B(1, −1, 0), C(1, 1, 0), D(0, 0, 1), E(−1, 0, 1) form a skew pentagon ABCDE. Forces **AB**, **BC**, **CD**, **DE**, **EA** act round the sides AB, BC, CD, DE, EA of the pentagon. Prove that this system of forces is equivalent to a couple and find the magnitude of the moment of this couple.

5) The line of action of a force $3\mathbf{i}+2\mathbf{j}-\mathbf{k}$ passes through the origin and the line of action of a second force $-3\mathbf{i}-\mathbf{j}+2\mathbf{k}$ passes through the point (3, 1, 2). Reduce the two forces to a single force acting at the origin together with a couple.

6) Forces $4\mathbf{j}$ and $3\mathbf{k}$ act through the points with position vectors $(\mathbf{i}+\mathbf{j})$ and $(\mathbf{j}+\mathbf{k})$ respectively. A third force acts through the point with position vector $(\mathbf{i}+\mathbf{k})$ and is such that the three forces are equivalent to a couple. Find the vector moment and the magnitude of this couple.

7) The forces $(q\mathbf{j}+r\mathbf{k})$, $(r\mathbf{k}+p\mathbf{i})$ and $(p\mathbf{i}+q\mathbf{j})$ act respectively through the three points with position vectors $p\mathbf{i}$, $q\mathbf{j}$ and $r\mathbf{k}$, where p, q and r are constants. Show that the force system is equivalent to a single force through the origin and find its magnitude.

8) In each of the following cases find the simplest system of forces which is equivalent to the system given:

(a) $\mathbf{F}_1 = 2\mathbf{i}-\mathbf{j}$ acting at the point $\mathbf{r}_1 = \mathbf{i}+2\mathbf{j}$,
$\mathbf{F}_2 = 3\mathbf{i}+\mathbf{j}$ acting at the point $\mathbf{r}_2 = 2\mathbf{i}+3\mathbf{j}$,
$\mathbf{F}_3 = -\mathbf{i}+\mathbf{j}$ acting at the point $\mathbf{r}_3 = -2\mathbf{j}$.

(b) $\overrightarrow{AB}$, $\overrightarrow{BC}$, $\overrightarrow{CA}$ where the points A, B, C have position vectors $\mathbf{a}$, $\mathbf{b}$, and $\mathbf{c}$ respectively.

(c) $\mathbf{F} = \mathbf{i}+3\mathbf{j}$ acting at the origin and a couple of vector moment $3\mathbf{k}$.

(d) $\mathbf{F}_1 = \mathbf{i}-\mathbf{j}+\mathbf{k}$ acting at the point $\mathbf{r}_1 = \mathbf{i}-2\mathbf{k}$,
$\mathbf{F}_2 = -2\mathbf{i}+3\mathbf{j}+\mathbf{k}$ acting at the point $\mathbf{r}_2 = -\mathbf{i}+2\mathbf{j}-\mathbf{k}$,
$\mathbf{F}_3 = -2\mathbf{j}-\mathbf{k}$ acting at the point $\mathbf{r}_3 = 3\mathbf{j}$.

(e) $\overrightarrow{AB}$, $\overrightarrow{BC}$, $\overrightarrow{CA}$ and $\overrightarrow{AD}$ where ABCD is a regular tetrahedron.

9) Forces $\mathbf{F}_1$, $\mathbf{F}_2$, $\mathbf{F}_3$ act at points $\mathbf{r}_1$, $\mathbf{r}_2$, $\mathbf{r}_3$ where
$\mathbf{F}_1 = 3\mathbf{i}-\mathbf{j}+2\mathbf{k}$ $\mathbf{r}_1 = 3\mathbf{i}-\mathbf{k}$,
$\mathbf{F}_2 = -\mathbf{i}-4\mathbf{j}+\mathbf{k}$ $\mathbf{r}_2 = 2\mathbf{i}-4\mathbf{j}$,
$\mathbf{F}_3 = \mathbf{i}+\mathbf{j}-2\mathbf{k}$ $\mathbf{r}_3 = -3\mathbf{j}+5\mathbf{k}$.
When a fourth force $\mathbf{F}_4$ is added the system is in equilibrium. Find $\mathbf{F}_4$ and a vector equation of its line of action.

10) A force $\mathbf{i}-\mathbf{j}+2\mathbf{k}$ acts at the point with position vector $-\mathbf{i}-\mathbf{j}+\mathbf{k}$. Show that this force is equivalent to an equal force acting through the origin together with a couple. Find the vector moment of the couple.

11) A force $\mathbf{i}+\mathbf{j}-\mathbf{k}$ acts through the origin together with a couple of moment $\mathbf{i}+\mathbf{j}+2\mathbf{k}$. Prove that the vector moment of this couple is perpendicular to the force and hence show that the couple and force are coplanar. Show that the couple and force are equivalent to a single force and find a vector equation of its line of action.

12) Prove that the following system of forces is equivalent to a couple together with a non-coplanar force:
$\mathbf{F}_1 = \mathbf{i}+\mathbf{j}$ acting through the point $\mathbf{a}_1 = 3\mathbf{i}+\mathbf{j}+\mathbf{k}$,
$\mathbf{F}_2 = \mathbf{i}+\mathbf{k}$ acting through the point $\mathbf{a}_2 = \mathbf{i}$,
$\mathbf{F}_3 = 2\mathbf{j}-\mathbf{k}$ acting through the point $\mathbf{a}_3 = 2\mathbf{j}$.

13) A force $\mathbf{i}+2\mathbf{j}-\mathbf{k}$ acts through a point with position vector $2\mathbf{i}$ together with a couple of vector moment $2\mathbf{i}-\mathbf{j}$. Prove that this system of forces cannot be reduced any further and find an equivalent system where the resultant force acts through the origin.

MULTIPLE CHOICE EXERCISE 3

(The instructions for answering these questions are given on page ix.)

TYPE I

1) A force $\mathbf{F}$ acts through the point A. If the position vector of A is $\mathbf{a}$ and the position vector of another point B is $\mathbf{b}$, the vector moment of $\mathbf{F}$ about B is:
(a) $\mathbf{F}\times\mathbf{a}$ (b) $\mathbf{b}\times\mathbf{F}$ (c) $\mathbf{0}$ (d) $(\mathbf{a}-\mathbf{b})\times\mathbf{F}$ (e) $\mathbf{a}\times\mathbf{F}$.

2) If $\mathbf{a}=\mathbf{i}+\mathbf{j}$, $\mathbf{b}=2\mathbf{i}-\mathbf{j}$ then $\mathbf{a}\times\mathbf{b}$ is:
(a) $2\mathbf{i}-\mathbf{j}$ (b) $-3\mathbf{k}$ (c) $\mathbf{0}$ (d) $\mathbf{i}$ (e) $\sqrt{10}$.

3) The vertices A, B, C of ΔABC have position vectors $\mathbf{a}$, $\mathbf{b}$, $\mathbf{c}$ respectively. The area of ΔABC is:
(a) $\mathbf{a}\times\mathbf{b}$ (b) $\frac{1}{2}|\mathbf{b}\times\mathbf{c}|$ (c) $\frac{1}{2}|(\mathbf{a}-\mathbf{b})\times(\mathbf{b}-\mathbf{c})|$
(d) $(\mathbf{b}-\mathbf{a})\times(\mathbf{a}-\mathbf{c})$.

4) Forces $\mathbf{i}-2\mathbf{j}$ and $-\mathbf{i}+2\mathbf{j}$ act through the points whose position vectors are $\mathbf{k}$ and $3\mathbf{k}$. The vector sum of the moments of these forces about O is:
(a) $2(-2\mathbf{i}-\mathbf{j})$ (b) $\mathbf{0}$ (c) $2(\mathbf{i}-2\mathbf{j})$ (d) $4\mathbf{k}$ (e) $2\mathbf{k}$.

5) A force $\mathbf{F}$ acts along the line whose vector equation is $\mathbf{r}=\mathbf{a}+\lambda\mathbf{b}$. The moment of $\mathbf{F}$ about O is:
(a) $\mathbf{b}\times\mathbf{F}$ (b) $\mathbf{a}\times\mathbf{F}$ (c) $\mathbf{F}\times\mathbf{a}$ (d) $\mathbf{a}.\mathbf{F}$ (e) $\mathbf{F}\times\mathbf{b}$.

6) Forces $\mathbf{i}-2\mathbf{j}+\mathbf{k}$, $2\mathbf{i}+\mathbf{j}-\mathbf{k}$, $3\mathbf{i}$ act through the origin. The resultant of these forces is:
(a) a couple (b) a single force (c) equilibrium (d) none of these.

TYPE II

7) ABC is a triangle.
(a) The area of $\Delta\text{ABC}=\frac{1}{2}|\mathbf{AB}\times\mathbf{BC}|$.
(b) Forces $\mathbf{AB}$, $\mathbf{BC}$, $\mathbf{CA}$ acting round the sides of the triangle are in equilibrium.
(c) The vector $\mathbf{BC}\times\mathbf{CA}$ is perpendicular to the plane containing ΔABC.

8) A system of forces $\mathbf{F}_1$, $\mathbf{F}_2$ and $\mathbf{F}_3$ act through the points with position vectors $\mathbf{r}_1$, $\mathbf{r}_2$, $\mathbf{r}_3$. The system is in equilibrium.
(a) $\Sigma \mathbf{r} \times \mathbf{F} = \mathbf{0}$.
(b) The lines of action of the forces are concurrent.
(c) $\mathbf{F}_1$, $\mathbf{F}_2$ and $\mathbf{F}_3$ are coplanar.

9) $\mathbf{a}$ and $\mathbf{b}$ are perpendicular vectors.
(a) $\mathbf{a} \times \mathbf{b} = \mathbf{0}$.
(b) $\mathbf{a} \times \mathbf{b}$, $\mathbf{a}$, $\mathbf{b}$ are mutually perpendicular.
(c) $|\mathbf{a} \times \mathbf{b}| = ab$.

10) A force $\mathbf{F}$ acts along the line whose vector equation is $\mathbf{r} = \mathbf{a} + \lambda \mathbf{b}$. The moment of $\mathbf{F}$ about O is:
(a) $\mathbf{a} \times \mathbf{F}$ (b) $(\mathbf{a} + \mathbf{b}) \times \mathbf{F}$ (c) $\mathbf{b} \times \mathbf{F}$.

TYPE III

11) (a) Two couples are equivalent.
(b) Two couples act in the same plane.

12) (a) $\mathbf{a} = \lambda \mathbf{b}$.
(b) $\mathbf{a} \times \mathbf{b} = \mathbf{0}$.

13) (a) A system of forces $\mathbf{F}_1$, $\mathbf{F}_2$, $\mathbf{F}_3, \ldots$ is in equilibrium.
(b) $\Sigma \mathbf{F} = \mathbf{0}$.

14) (a) $\mathbf{a} \times \mathbf{b} = ab\hat{\mathbf{n}}$.
(b) $\mathbf{a} . \mathbf{b} = 0$.

15) (a) $\mathbf{a}$ and $\mathbf{b}$ are non-parallel vectors and $\mathbf{a} \neq \mathbf{0}$, $\mathbf{b} \neq \mathbf{0}$.
(b) $\mathbf{a} \times \mathbf{b} = \mathbf{b} \times \mathbf{a}$.

TYPE IV

16) Forces $\mathbf{F}_1$, $\mathbf{F}_2$, $\mathbf{F}_3$, $\mathbf{F}_4$ act through points whose position vectors are $\mathbf{r}_1$, $\mathbf{r}_2$, $\mathbf{r}_3$, $\mathbf{r}_4$. Find the resultant of this system.
(a) $\Sigma \mathbf{F} = \mathbf{0}$ (b) $\Sigma \mathbf{r} = \mathbf{0}$ (c) $\Sigma \mathbf{r} \times \mathbf{F} = \mathbf{0}$.

17) A force $\mathbf{F}$ acting through the point with position vector $\mathbf{a}$ is equivalent to a force $\mathbf{F}$ acting through the point with position vector $\mathbf{b}$ together with a couple. Find the magnitude of the moment of this couple.
(a) $\mathbf{F} = \mathbf{i} + \mathbf{j}$ (b) $\mathbf{a} = 2\mathbf{k}$ (c) $\mathbf{b} = \mathbf{i} - \mathbf{j}$.

18) A particle moves under the action of a constant force $\mathbf{F}$. Find the Cartesian equation of the path of the particle.
(a) The particle moves in a straight line.
(b) The acceleration of the particle is $\mathbf{i} - \mathbf{j}$.
(c) The moment of $\mathbf{F}$ about O is $\mathbf{i} + \mathbf{j}$.

19) ABC is a triangle. Find the direction cosines of a normal to the plane of the triangle.
(a) $\overrightarrow{AB} = \mathbf{i}+\mathbf{j}$.
(b) The position vector of B is $2\mathbf{i}-\mathbf{j}+\mathbf{k}$.
(c) $\overrightarrow{AC} = 2\mathbf{i}-\mathbf{k}$.

TYPE V

20) Any three forces whose lines of action are concurrent are in equilibrium.

21) Three non zero vectors $\mathbf{a}$, $\mathbf{b}$ and $\mathbf{c}$ are such that $\mathbf{a}\times\mathbf{b} = \mathbf{a}\times\mathbf{c}$, therefore $\mathbf{b} = \mathbf{c}$ or $\mathbf{a} = \lambda(\mathbf{b}-\mathbf{c})$.

22) ABCD is a parallelogram. The unit vector perpendicular to the plane of ABCD is $\mathbf{AB}\times\mathbf{BC}$ divided by the area of ABCD.

23) A system of forces is such that $\Sigma\,\mathbf{r}\times\mathbf{F} = \mathbf{0}$, therefore the forces must be in equilibrium.

MISCELLANEOUS EXERCISE 3

1) A particle P moves so that its position vector $\mathbf{r}$ at time t is given by

$$\mathbf{r} = \mathbf{a}\cos\omega t + \mathbf{b}\sin\omega t,$$

where $\mathbf{a}$, $\mathbf{b}$ are constant vectors and ω is a constant. Show that

$$\mathbf{r}\times\frac{d\mathbf{r}}{dt} \text{ is independent of } t.$$

2) A force $\mathbf{F}$ acts at the point with position vector $\mathbf{r}$. Express as a vector product the moment of this force about the point with position vector $\mathbf{a}$.
A force of unit magnitude has equal moments about points with position vectors $(0,1,0)$ and $(1,2,-1)$. Find the possible values of the components of the force.

3) Three forces $\mathbf{F}_1 = 4\mathbf{i}+\mathbf{j}+2\mathbf{k}$, $\mathbf{F}_2 = \mathbf{i}-2\mathbf{j}+\mathbf{k}$ and $\mathbf{F}_3 = -5\mathbf{i}+\mathbf{j}-3\mathbf{k}$ act at the points with position vectors $6\mathbf{i}+4\mathbf{j}+\mathbf{k}$, $\mathbf{i}+5\mathbf{j}-2\mathbf{k}$ and $\mathbf{i}+\mathbf{j}+\mathbf{k}$ respectively. Show that the forces reduce to a couple and calculate its magnitude.
The force $\mathbf{F}_3$ is now removed from the system and replaced by the force $\mathbf{F}_4$ such that the forces $\mathbf{F}_1$, $\mathbf{F}_2$ and $\mathbf{F}_4$ are in equilibrium. Find
(a) the magnitude of $\mathbf{F}_4$,
(b) a vector equation for the line along which $\mathbf{F}_4$ acts.

4) *In this question distances are measured in metres, and forces in newtons.* The vertices A, B, C of a triangle ABC have position vectors $3\mathbf{i}+\mathbf{j}+2\mathbf{k}$, $\mathbf{i}+5\mathbf{j}+6\mathbf{k}$ and $4\mathbf{i}-\mathbf{j}+4\mathbf{k}$ respectively, relative to an origin O.
Forces $\mathbf{F}_1$, $\mathbf{F}_2$ and $\mathbf{F}_3$ act along the sides $\overrightarrow{AB}$, $\overrightarrow{BC}$ and $\overrightarrow{CA}$ respectively. Given that $|\mathbf{F}_1| = 18\text{ N}$, $|\mathbf{F}_2| = 14\text{ N}$ and $|\mathbf{F}_3| = 6\text{ N}$, find $\mathbf{F}_1$, $\mathbf{F}_2$ and $\mathbf{F}_3$ in terms of $\mathbf{i}$, $\mathbf{j}$ and $\mathbf{k}$. Show that the resultant of these forces is $-2\mathbf{i}+4\mathbf{j}+4\mathbf{k}$ and find the total moment of the forces about the origin O.
A force $\mathbf{F}_4 = \lambda(-\mathbf{i}+3\mathbf{j}+2\mathbf{k})$, where λ is a constant, is now added to the system. Given that $\mathbf{F}_4$ acts through the point with position vector $7\mathbf{j}$ relative to O, find the value of λ such that the total moment of all the forces about O is zero and find the resultant of all the forces in this case. (AEB)

5) (a) Define the moment of a force $\mathbf{F}$ about a point A as a vector product. Hence, or otherwise, prove that the moment of the couple formed by the forces $\mathbf{F}$ and $-\mathbf{F}$ acting in different lines is the same about all points.
(b) A force of unit magnitude acts through the origin O of a rectangular system of axes Ox, Oy and Oz and has equal moments about each of the points $(1, 1, 0)$ and $(2, 0, 1)$. Find the possible values of the components of the force in the directions Ox, Oy and Oz. (JMB)

6) Two forces $\mathbf{F}_1 = -6\mathbf{i}-\mathbf{j}+2\mathbf{k}$ and $\mathbf{F}_2 = 6\mathbf{i}+3\mathbf{j}-5\mathbf{k}$ act through the points A and B which have position vectors $\mathbf{r}_1 = 5\mathbf{j}+\mathbf{k}$ and $\mathbf{r}_2 = -2\mathbf{i}+3\mathbf{j}+2\mathbf{k}$ respectively. Find the sum of the vector moments of $\mathbf{F}_1$ and $\mathbf{F}_2$ about the origin and show that there is no point about which this sum is zero.
A third force $\mathbf{F}_3$ which acts parallel to $\overrightarrow{AB}$ is added to the system; the three forces are equivalent to a force which acts in the $\mathbf{i}$-$\mathbf{j}$ plane, together with a couple. Find $\mathbf{F}_3$. (AEB)

7) (a) Express as a vector product the moment of a force $\mathbf{F}$, acting at a point with position vector $\mathbf{r}$, about a point A, whose position vector is $\mathbf{r}_0$. Hence, using vector methods, prove that the moment of a couple is the same about all points.
(b) A force of unit magnitude has equal moments about the points with position vectors $(0, 0, 1)$ and $(1, -1, 2)$. Find the possible values of the components of the force. (AEB)

8) *In this question the unit of force is the newton and the unit of length is the metre.*
Forces $\mathbf{F}_1 = \mathbf{i}+2\mathbf{j}+3\mathbf{k}$ and $\mathbf{F}_2 = 2\mathbf{i}+\mathbf{k}$ act at points with position vectors $2\mathbf{i}+5\mathbf{j}+c\mathbf{k}$ and $5\mathbf{i}+c\mathbf{j}+2\mathbf{k}$ respectively. Given that these forces meet at a point, find the value of c and determine a vector equation of the line of action of the resultant of forces $\mathbf{F}_1$ and $\mathbf{F}_2$.
Show that the sum of the moments of these forces about the origin is $12\mathbf{i}-4\mathbf{j}-7\mathbf{k}$. (U of L)

9) Three forces of magnitudes 26, $4\sqrt{41}$ and 15 newtons act respectively along the sides $\overrightarrow{OA}$, $\overrightarrow{AB}$ and $\overrightarrow{BO}$ of the triangle AOB. The position vectors of A and B relative to O are $5\mathbf{i}+12\mathbf{j}$ and $3\mathbf{i}+4\mathbf{k}$ respectively, where the unit of distance is the metre. Show that the resultant is $-3\mathbf{i}-4\mathbf{k}$ and find its magnitude.
Find also the magnitude of the moment of the resultant about O. (AEB)

10) *In this question, the unit of force is the newton and the unit of length is the metre.*
Forces $\mathbf{F}_1 = 2\mathbf{i}-\mathbf{j}+3\mathbf{k}$, $\mathbf{F}_2 = 4\mathbf{i}+\mathbf{j}+5\mathbf{k}$, $\mathbf{F}_3 = -6\mathbf{i}-8\mathbf{k}$ act respectively at points with position vectors

$$\mathbf{i}-2\mathbf{j}-\mathbf{k} \qquad 7\mathbf{i}-2\mathbf{j}+7\mathbf{k} \qquad 3\mathbf{i}-3\mathbf{j}+\mathbf{k}$$

Show that the lines of action of $\mathbf{F}_1$ and $\mathbf{F}_2$ meet at the point with position vector $3\mathbf{i}-3\mathbf{j}+2\mathbf{k}$.
Show that this system of forces is equivalent to a couple, and find the magnitude of the couple.
If the force $\mathbf{F}_3$ is now replaced by a force $\mathbf{F}_4$ such that $\mathbf{F}_1$, $\mathbf{F}_2$ and $\mathbf{F}_4$ are in equilibrium, find a vector equation of the line of action of $\mathbf{F}_4$. (U of L)

11) A force $2\mathbf{i}+\mathbf{j}+\mathbf{k}$ acts at the point $4\mathbf{i}-8\mathbf{j}+8\mathbf{k}$ and another force $\mathbf{i}-2\mathbf{j}+2\mathbf{k}$ acts at the point $a\mathbf{i}-4\mathbf{j}$, the units of force and distance being the newton and the metre respectively. Show that, if the lines of action of the forces intersect, then $a = -3$.
Find
(a) the magnitude of the resultant,
(b) the vector equation of the line of action of the resultant,
(c) the moments of the resultant about the origin and about the point $-\mathbf{i}-10\mathbf{j}+6\mathbf{k}$, expressing your answers in vector form. (AEB)

12) Forces $-9\mathbf{i}+\mathbf{j}-2\mathbf{k}$, $3\mathbf{i}+2\mathbf{j}-3\mathbf{k}$ and $6\mathbf{i}-3\mathbf{j}+5\mathbf{k}$ act through points with position vectors $-11\mathbf{i}+2\mathbf{j}-5\mathbf{k}$, $\mathbf{i}-4\mathbf{j}+5\mathbf{k}$, and $-8\mathbf{i}+4\mathbf{j}-8\mathbf{k}$ respectively. Prove that these forces are equivalent to a couple, and find the moment of this couple. (U of L)

13) Show that each of the following systems of forces is in equilibrium:
(a) forces $3\overrightarrow{AB}$ and $4\overrightarrow{AC}$, where A, B and C have position vectors $(4\mathbf{i}-\mathbf{k})$, $(4\mathbf{j}+3\mathbf{k})$ and $(7\mathbf{i}-3\mathbf{j}-4\mathbf{k})$ respectively:
(b) forces $\mathbf{F}_1 = 3\mathbf{i}+4\mathbf{j}+5\mathbf{k}$ acting at the point $L(7\mathbf{i}+9\mathbf{j}+11\mathbf{k})$,
$\mathbf{F}_2 = \mathbf{i}+\mathbf{j}+\mathbf{k}$ acting at the point $M(4\mathbf{i}+4\mathbf{j}+4\mathbf{k})$,
$\mathbf{F}_3 = -4\mathbf{i}-5\mathbf{j}-6\mathbf{k}$ acting at the point $N(5\mathbf{i}+6\mathbf{j}+7\mathbf{k})$.
Find the cosine of the angle between the lines of action of the forces $\mathbf{F}_1$ and $\mathbf{F}_2$. (AEB)

14) The vertices of a tetrahedron ABCD have position vectors $\mathbf{a}$, $\mathbf{b}$, $\mathbf{c}$, $\mathbf{d}$ respectively, where

$$\mathbf{a} = 3\mathbf{i}-4\mathbf{j}+\mathbf{k} \qquad \mathbf{b} = 4\mathbf{i}+4\mathbf{j}-2\mathbf{k}$$
$$\mathbf{c} = 4\mathbf{i}+\mathbf{k} \qquad \mathbf{d} = \mathbf{i}-2\mathbf{j}+\mathbf{k}$$

Forces of magnitude 30 and $3\sqrt{13}$ units act along CB and CD respectively. A third force acts at A. If the system reduces to a couple, find the magnitude of this couple and the force at A. Find also a unit vector along the axis of the couple. (U of L)

The units in the following questions are the metre and the newton.

15) Two forces $\mathbf{i}-\mathbf{j}+\mathbf{k}$ and $2\mathbf{i}+3\mathbf{j}-\mathbf{k}$ act at points whose position vectors are $3\mathbf{i}+5\mathbf{l}$ and $-\mathbf{i}-\mathbf{j}+4\mathbf{k}$ respectively. Show that the lines of action of these two forces intersect and find $\mathbf{a}$, the position vector of the point of intersection. Find

(a) the magnitude of the resultant of these forces,
(b) a vectoı equation of its line of action.

Find also the vector moment of this resultant about

(c) the origin,
(d) the point with position vector $\mathbf{i}+\mathbf{j}+3\mathbf{k}$.

A third force $\lambda(\mathbf{i}-\mathbf{j}+6\mathbf{k})$, $\lambda > 0$, acting through the point with position vector $\mathbf{a}$, is added to the system. If the magnitude of the resultant of the three forces is 13 N, find λ. (AEB)

16) Two forces $(3\mathbf{i}+2\mathbf{j}+\mathbf{k})$ and $(\mathbf{i}+2\mathbf{j}+3\mathbf{k})$ act at the points B and C, whose position vectors are $(4\mathbf{i}-\mathbf{j}+\mathbf{k})$ and $(3\mathbf{i}+\mathbf{j}+6\mathbf{k})$ respectively. Find the force through the origin O and the couple which are together equivalent to these forces. Find the magnitude of the couple. Show that the lines of action of the forces through B and C meet and find the position vector of the point of intersection of these lines. (U of L)

17) The line of action of a force $\mathbf{F}$ passes through a point A and the position vector of A relative to an origin O is denoted by $\mathbf{r}$. Define the vector moment of $\mathbf{F}$ about the point O and show that it is independent of the position of the point A on the line of action of $\mathbf{F}$.
A system of five forces consists of the three forces $\mathbf{F}_1 = (\mathbf{i}-2\mathbf{j}+2\mathbf{k})$, $\mathbf{F}_2 = (2\mathbf{i}+6\mathbf{k})$ and $\mathbf{F}_3 = (\mathbf{i}-2\mathbf{j}-4\mathbf{k})$ all acting through the origin O, together with a force $\mathbf{F}_4 = (\mathbf{i}-2\mathbf{j}-\mathbf{k})$ acting through the point whose position vector is $(-2\mathbf{i}+4\mathbf{j}+2\mathbf{k})$ and a force $\mathbf{F}_5 = (-\mathbf{i}-2\mathbf{j}-7\mathbf{k})$ acting through the point whose position vector is $(\mathbf{i}-2\mathbf{j}-\mathbf{k})$. Reduce the system to a force $\mathbf{R}$ acting at O together with a couple $\mathbf{G}$.
Hence, or otherwise, verify that the system is equivalent to a single force $\mathbf{F} = (4\mathbf{i}-8\mathbf{j}-4\mathbf{k})$ acting through the point with position vector $(\mathbf{i}-\mathbf{j}+\mathbf{k})$. (U of L)

18) Forces $(\mathbf{i}+2\mathbf{j}+\mathbf{k})$ and $(2\mathbf{i}+2\mathbf{j}+3\mathbf{k})$ act at points B and C, whose position vectors are $(4\mathbf{i}+4\mathbf{j}+6\mathbf{k})$ and $(\mathbf{i}+2\mathbf{j}+\mathbf{k})$ respectively. Find the force through the origin O and the couple which are together equivalent to these forces. Find also the magnitude of the couple.
Show that the lines of action of the forces through B and C meet and find the position vector of the point of intersection of these lines. (U of L)

19) Forces, $\mathbf{F}_1$, $\mathbf{F}_2$, $\mathbf{F}_3$, where

$$\mathbf{F}_1 = (3\mathbf{i}-\mathbf{j}+2\mathbf{k})$$
$$\mathbf{F}_2 = (-\mathbf{i}-4\mathbf{j}+\mathbf{k})$$
$$\mathbf{F}_3 = (\mathbf{i}+\mathbf{j}-2\mathbf{k})$$

act at points with position vectors $\mathbf{r}_1$, $\mathbf{r}_2$, $\mathbf{r}_3$, where

$$\mathbf{r}_1 = (6\mathbf{i}-\mathbf{j}+\mathbf{k})$$
$$\mathbf{r}_2 = (\mathbf{i}-8\mathbf{j}+\mathbf{k})$$
$$\mathbf{r}_3 = (\mathbf{i}-2\mathbf{j}+3\mathbf{k})$$

When a fourth force $\mathbf{F}_4$ is added to these three forces, the system is in equilibrium. Find $\mathbf{F}_4$ and a vector equation of its line of action.
Find also the moment of $\mathbf{F}_4$ about the origin O. (U of L)

20) Let $\mathbf{i}$, $\mathbf{j}$ be orthogonal unit vectors in a plane and let

$$\mathbf{F} = X\mathbf{i}+Y\mathbf{j}$$

be a force acting at a point P in the plane whose position vector relative to the origin O is

$$\mathbf{r} = x\mathbf{i}+(-y\mathbf{j})$$

Prove that the force is equivalent to an equal force at O together with a couple.
Three variable forces $\mathbf{F}_1$, $\mathbf{F}_2$, $\mathbf{F}_3$ act at points with position vectors $\mathbf{0}$, $\mathbf{i}+\mathbf{j}$, $-3\mathbf{i}+2\mathbf{j}$ respectively and, at time t,

$$\mathbf{F}_1 = 2\cos t\mathbf{i}, \quad \mathbf{F}_2 = \cos t\mathbf{i}+2\sin t\mathbf{j}, \quad \mathbf{F}_3 = 3\sin t\mathbf{i}+\cos t\mathbf{j}$$

If the system is reduced to a single force $\mathbf{F}$ at O and a couple G, find the values of $\mathbf{F}$ and G. Deduce the equation of the line of action of the resultant and show that this line passes through a fixed point which is independent of t.
(O)

21) A force system consists of three forces

$$\mathbf{F}_1 = \mathbf{i}-\mathbf{j}+2\mathbf{k}$$
$$\mathbf{F}_2 = \mathbf{i}+3\mathbf{j}-\mathbf{k}$$
$$\mathbf{F}_3 = s\mathbf{i}+t\mathbf{j}+2\mathbf{k}$$

where s and t are constants. They act respectively through the points whose position vectors are

$$3\mathbf{i}-\mathbf{j}+\mathbf{k}, \quad \mathbf{j}+2\mathbf{k}, \quad \mathbf{k}$$

Obtain, in terms of s and t, the equivalent force system consisting of a single force $\mathbf{F}$ acting through the origin and a couple of moment $\mathbf{G}$.
Determine s and t so that $\mathbf{G}$ is parallel to $\mathbf{F}$. (U of L)

22) (a) A particle P moves with constant velocity $\mathbf{v}$ and, at time $t = 0$, the position vector of P referred to a fixed origin O is $\mathbf{r}_0$. Write down the vector equation of the path of P and show that P is (or was) closest to O at time

$$t = -\frac{\mathbf{v}\cdot\mathbf{r}_0}{\mathbf{v}\cdot\mathbf{v}}$$

(b) A system of three forces consists of a force $\mathbf{F}_1 = \mathbf{i}-\mathbf{j}+\mathbf{k}$ acting through the point whose position vector is $2\mathbf{i}+2\mathbf{k}$, a force $\mathbf{F}_2 = \mathbf{j}+2\mathbf{k}$ acting through the point whose position vector is $\mathbf{i}-\mathbf{k}$, and a force $\mathbf{F}_3$.

(i) If $\mathbf{F}_3$ acts through the origin and the system reduces to a couple, find $\mathbf{F}_3$ and the magnitude of the couple.

(ii) If the system is in equilibrium and $\mathbf{F}_3$ acts through the point $a\mathbf{i}+\mathbf{j}+4\mathbf{k}$, find the value of a. (U of L)

23) A force $\mathbf{F}_1$, of magnitude 26 N, acts along the direction of the vector $(4\mathbf{i}-3\mathbf{j}+12\mathbf{k})$. Given that the line of action of $\mathbf{F}_1$ passes through the point which has position vector $(2\mathbf{i}+\mathbf{j}-\mathbf{k})$, find the moment of $\mathbf{F}_1$ about the origin O.
A bead moves along a smooth straight wire from the point A to the point B, where

$$\overrightarrow{OA} = (3\mathbf{i}-2\mathbf{j}+\mathbf{k}), \quad \overrightarrow{OB} = (5\mathbf{i}-22\mathbf{j}+2\mathbf{k}),$$

under the influence of $\mathbf{F}_1$ and the reaction of the wire only.
Find the work done by $\mathbf{F}_1$ in this motion. (U of L)p

24) A car in a fairground ride moves along a continuous track. At time t the position vector of the car relative to an origin O at horizontal ground level is

$$\mathbf{r}(t) = 40\sin 2\omega t\,\mathbf{i} + 40\cos\omega t\,\mathbf{j} + 10(1-\cos\omega t)\,\mathbf{k}$$

where $\omega = 1/10\ \text{rad s}^{-1}$, and $\mathbf{k}$ points vertically upwards.

(a) How long does each lap take?

(b) Find the maximum height of the car above ground level and determine its velocity and acceleration at this point.

(c) Calculate the vector $\dot{\mathbf{r}}(0)\times\dot{\mathbf{r}}(t)$ and hence show that the car's velocity is always perpendicular to $\mathbf{j}+4\mathbf{k}$.

(d) Deduce that the motion takes place in a plane and given that $\mathbf{i}$, $\mathbf{j}$, $\mathbf{k}$ are unit vectors along axes Ox, Oy, Oz respectively, find the Cartesian equation of this plane. (U of L)

CHAPTER 4

IMPACT

DIRECT IMPACT

The work on impact in *Mechanics and Probability* is limited to examples of direct impact, i.e. impact in which the directions of motion of the colliding objects are along the line of action of the impulses, acting at the instant of impact.

The mechanical principles used to solve these problems are as follows.

(a) At the moment of impact, two equal and opposite impulses act, one on each of the colliding objects.

(b) When an impulse acts on an object, it causes an increase in momentum equal in magnitude and direction to the impulse.

(c) Unless the colliding objects coalesce on impact, Newton's law of restitution can be used, i.e. for two colliding objects

$$\text{separation speed} = e \times \text{approach speed}$$

where e is the coefficient of restitution for the two particular objects.

In general $$0 \leqslant e \leqslant 1$$

If $e = 1$ the collision is perfectly elastic.
If $e = 0$ the collision is inelastic and there is no bounce.

(d) If both colliding objects are free to move, the equal and opposite impulses cause equal and opposite changes in momentum; therefore the total momentum in the direction of the impulses is unchanged by the impact.

OBLIQUE IMPACT

At the instant of *any* impact, two equal and opposite impulses act in a direction perpendicular to the colliding surfaces. If, just before impact, at least one of the colliding objects was moving in a direction *different* from the line of action of the impulses, the impact is called *oblique* or *indirect*.
The principles used for direct impact apply equally well to oblique impact but the following extra facts are also needed.

> In a direction in which an object experiences no impulse, no change in momentum occurs.

> The law of restitution applies *only* along the line of impulses (this is an experimental result).

These extra considerations make it clear that, when solving problems on oblique impact, different approaches are needed to deal with the motion parallel to, and perpendicular to, the line of action of the impulses.

Impact with a Fixed Surface

Consider a sphere A, of mass m, moving with speed u on a smooth horizontal plane in a direction which is at $30°$ to a smooth vertical wall with which it collides and then bounces off.
At the moment of impact, an impulse perpendicular to the wall acts on the sphere. (An equal and opposite impulse acts on the wall.)

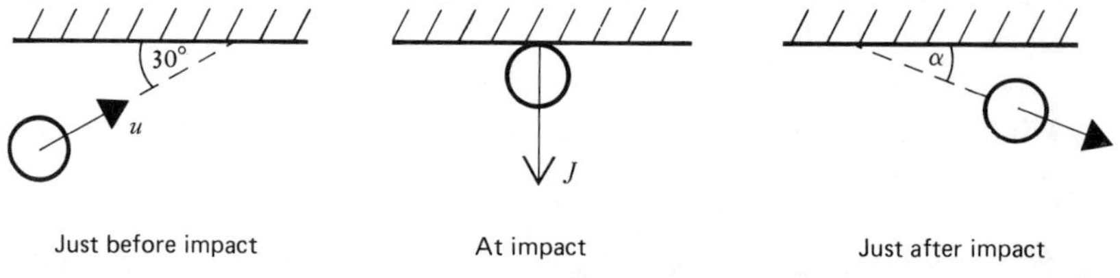

The impulse J causes a change in momentum perpendicular to the wall. Parallel to the wall, where there is no impulse, there is no change in momentum and therefore no change in the velocity component. This property is incorporated in the following diagrams.

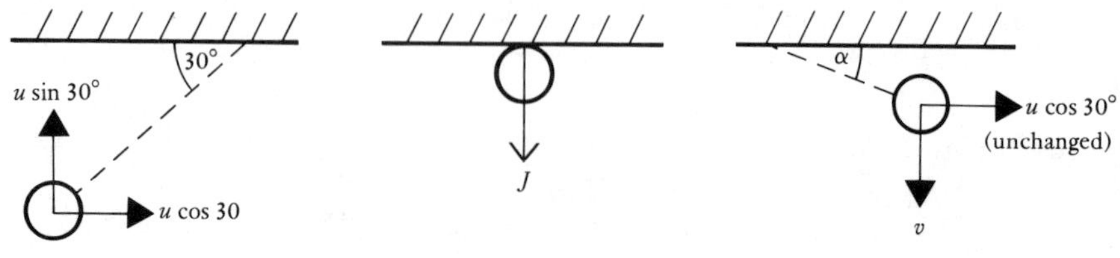

Along the line of the impulse we use:

(a) Impulse = increase in momentum

$$\Rightarrow \quad J = mv - m(-u\sin 30^\circ)$$

$$\Rightarrow \quad J = m(v + \tfrac{1}{2}u) \qquad [1]$$

(b) Law of restitution

$$\Rightarrow \quad v = eu\sin 30^\circ$$

$$\Rightarrow \quad v = \tfrac{1}{2}eu \qquad [2]$$

These two equations, together with the unchanged velocity component parallel to the wall, provide solutions to most problems of this type.

Note that the first set of three diagrams (opposite), which were drawn to illustrate the given information, were not suitable as 'working diagrams' for the solution. The second set of three are the working diagrams. It is unwise to attempt any oblique impact problem without such a set of clear diagrams showing the appropriate velocity components.

EXAMPLES 4a

1) A smooth sphere is projected along a horizontal plane and collides obliquely with a vertical wall. Just before impact the sphere is moving with a speed of $10\,\text{m s}^{-1}$ at an angle of 60° to the wall.
If $e = \frac{1}{4}$ find the speed of the sphere after impact and the angle, α, which its direction of motion then makes with the wall.

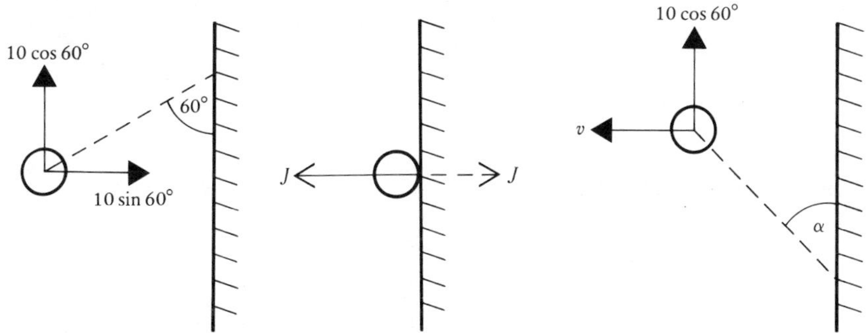

(Parallel to the wall there is no impulse, therefore no change in speed. This is shown on the diagram.)

Using the law of restitution perpendicular to the wall we have,

$$v = \tfrac{1}{4}(10\sin 60^\circ)$$

$$= \tfrac{5}{4}\sqrt{3}$$

The speed after impact is given by

$$[(10\cos 60^\circ)^2 + v^2]^{1/2} = [5^2 + (\tfrac{5}{4}\sqrt{3})^2]^{1/2}$$

i.e. the speed after impact is $\frac{5}{4}\sqrt{19}\ \mathrm{m\,s^{-1}}$

Also $$\tan\alpha = \frac{v}{10\cos 60^\circ}$$

$$= \frac{5\sqrt{3}}{(4)(5)}$$

Hence $$\alpha = \arctan(\tfrac{1}{4}\sqrt{3})$$

Note. In this problem we did not use impulse = change in momentum because the value of J was not required.

2) The coefficient of restitution between a snooker ball and the side cushion is $\frac{1}{3}$. If the ball hits the cushion and then rebounds at right angles to its original direction, show that the angles made with the side cushion by the directions of motion before and after impact are 60° and 30° respectively.

Let the original speed be u, in a direction making an angle θ with the side cushion.

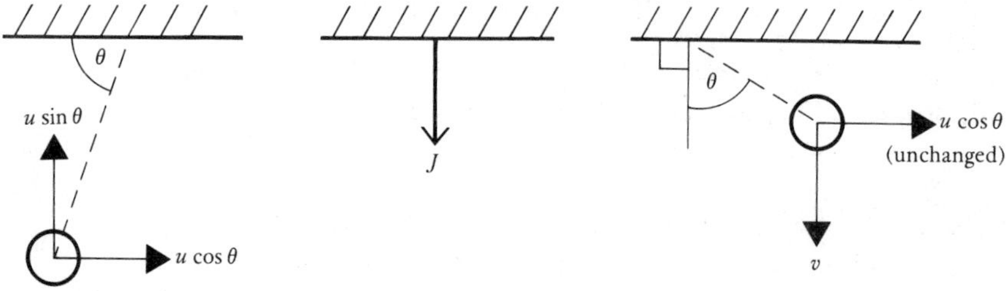

Using the law of restitution gives

$$v = \tfrac{1}{3}(u\sin\theta)$$

After impact, $$\tan\theta = \frac{u\cos\theta}{v} = \frac{3\cos\theta}{\sin\theta}$$

$\Rightarrow$ $$\tan^2\theta = 3$$

$\Rightarrow$ $$\tan\theta = \sqrt{3} \qquad (\theta \text{ is acute})$$

$\Rightarrow$ $$\theta = 60^\circ$$

Therefore the directions of motion before and after impact are at 60° and 30° to the cushion.

3) A smooth sphere of mass m is moving on a horizontal plane with a velocity $3\mathbf{i}+\mathbf{j}$ when it collides with a vertical wall which is parallel to the vector $\mathbf{j}$. If the coefficient of restitution between the sphere and the wall is $\frac{1}{2}$, find
(a) the velocity of the sphere after impact,
(b) the loss in kinetic energy caused by the impact,
(c) the impulse $\mathbf{J}$ that acts on the sphere.

Using impulse = increase in momentum gives

$$\mathbf{J} = m(\mathbf{v}) - m(3\mathbf{i}+\mathbf{j})$$

where $\mathbf{v}$ is the velocity of the sphere after impact.

To find $\mathbf{v}$ we must separate the velocity components parallel and perpendicular to the wall.

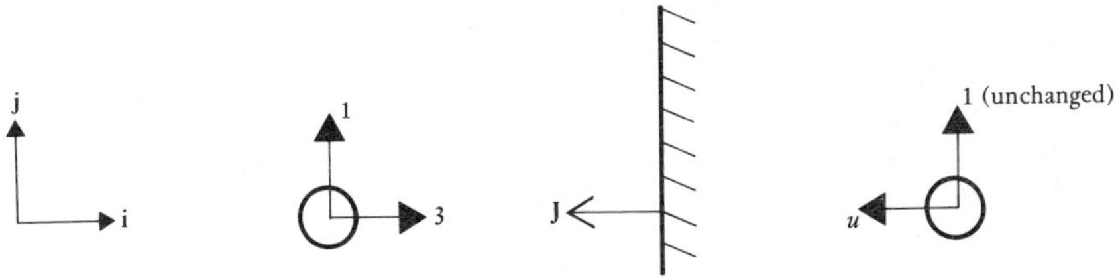

Using the law of restitution gives

$$u = (\tfrac{1}{2})(3) \quad \Rightarrow \quad \mathbf{v} = -\tfrac{3}{2}\mathbf{i}+\mathbf{j}$$

(a) Therefore the velocity of the sphere after impact is $-\frac{3}{2}\mathbf{i}+\mathbf{j}$

(b) The loss in K.E. $= \frac{1}{2}m(3^2+1^2) - \frac{1}{2}m(\{\frac{3}{2}\}^2+1^2)$
$= \frac{27}{8}m$

(c) $\mathbf{J} = m(-\frac{3}{2}\mathbf{i}+\mathbf{j}) - m(3\mathbf{i}+\mathbf{j}) = -\frac{9}{2}m\mathbf{i}$

Note that this problem was fairly easy to solve because the impulse was parallel to $\mathbf{i}$ and the velocity components parallel and perpendicular to the impulse were therefore the coefficients of $\mathbf{i}$ and $\mathbf{j}$, in the velocity vector. If, however, the direction of the impulse is not parallel either to $\mathbf{i}$ or to $\mathbf{j}$, the necessary velocity components are not immediately obvious. In this case we can calculate the components using the fact that the magnitude of a vector $\mathbf{v}$ in a direction $\mathbf{d}$ is given by $\mathbf{v}\cdot\hat{\mathbf{d}}$.

4) A sphere of mass m is moving with a velocity $4\mathbf{i}-\mathbf{j}$ when it hits a wall and rebounds with velocity $\mathbf{i}+3\mathbf{j}$. Find the impulse it receives. Find also the coefficient of restitution between the sphere and the wall.

Using impulse = increase in momentum gives

$$\mathbf{J} = m(\mathbf{i}+3\mathbf{j}) - m(4\mathbf{i}-\mathbf{j})$$
$$= m(-3\mathbf{i}+4\mathbf{j})$$

(To find the coefficient of restitution we require the velocity components, before and after impact, in the direction of **J**, i.e. in the direction $-3\mathbf{i}+4\mathbf{j}$.)

The unit vector $\hat{\mathbf{d}}_1$ in the direction of **J** is $\frac{1}{5}(-3\mathbf{i}+4\mathbf{j})$.

The magnitudes of the velocity components in this direction are:

$$(4\mathbf{i}-\mathbf{j}).\tfrac{1}{5}(-3\mathbf{i}+4\mathbf{j}) = -\tfrac{16}{5} \quad \text{before impact}$$

$$(\mathbf{i}+3\mathbf{j}).\tfrac{1}{5}(-3\mathbf{i}+4\mathbf{j}) = \tfrac{9}{5} \quad \text{after impact}$$

(The significance of the negative sign in $-\frac{16}{5}$ is simply an indication that this component is in a direction opposite to that of **J**; the *speed* of approach to the wall is therefore $\frac{16}{5}$ and the speed of separation is $\frac{9}{5}$.)

Newton's law of restitution gives

$$\tfrac{9}{5} = (e)(\tfrac{16}{5})$$

$$\Rightarrow \quad e = \tfrac{9}{16}$$

Note. The speeds perpendicular to **J** are not required in this problem. When they are required, they can be found using $\mathbf{v}.\hat{\mathbf{d}}_2$ where $\mathbf{d}_2$ is perpendicular to $\mathbf{d}_1$. In this problem $\mathbf{d}_2$ could be taken as $4\mathbf{i}+3\mathbf{j}$. This is parallel to the wall.

EXERCISE 4a

1) A sphere whose velocity is $2\mathbf{i}-5\mathbf{j}$ collides with a smooth plane which is perpendicular to **j**.
If the coefficient of restitution between the sphere and the plane is $\frac{2}{3}$, find the velocity of the sphere after impact.

2) A smooth sphere, travelling horizontally with a speed of $2\ \text{m s}^{-1}$, collides with a vertical wall at an angle of $45°$. If the coefficient of restitution between the sphere and the wall is $\frac{1}{4}$ find the speed of the sphere after impact. Find also the angle between the wall and the direction of motion of the sphere after impact.

3) A smooth sphere, moving with speed u, collides obliquely with a plane and bounces off at an angle of $30°$ to the plane. The coefficient of restitution between the sphere and the plane is $\frac{1}{2}$. Find the angle between the plane and the direction of motion of the sphere as it approaches the plane, and its speed after impact.

4) A smooth sphere is projected along horizontal ground and collides obliquely with a vertical wall. It hits the wall when moving at $3\ \text{m s}^{-1}$ at an angle of $30°$ to the wall. Find the velocity of the sphere just after impact with the wall if
(a) $e=\frac{1}{2}$, (b) $e=1$, (c) $e=0$.

5) A smooth sphere travelling on horizontal ground impinges obliquely on a vertical wall and rebounds at right angles to its original direction of motion. If the sphere is moving at $60°$ to the wall before impact, find the value of e.

6) A smooth sphere of unit mass is moving with speed $5\ \text{m s}^{-1}$ on a horizontal plane, when it collides, at an angle of $45°$ with a vertical wall. It bounces off in a direction making $30°$ with the wall. Find the coefficient of restitution between the sphere and the wall. Find also the impulse given to the sphere at the moment of impact.

7) A smooth sphere of mass m has a velocity $3\mathbf{i}-2\mathbf{j}$ when it collides with a rail in the direction $12\mathbf{i}+5\mathbf{j}$.
The velocity of the sphere after impact is $\frac{3}{2}\mathbf{i}+\frac{8}{5}\mathbf{j}$. Find:
(a) the coefficient of restitution between the sphere and the rail,
(b) the impulse exerted by the sphere on the rail,
(c) the kinetic energy lost at impact.

IMPACT BETWEEN TWO OBJECTS EACH FREE TO MOVE

When there is a collision between objects that are free to move, the impulses that occur act in equal and opposite pairs so

the total momentum in any direction is unchanged by the impact.

When two smooth spheres collide, the impulses act along the line joining their centres.
The solution of problems of this type is simplified by

(a) resolving all velocities parallel and perpendicular to the line of impulses.
(b) drawing at least two diagrams, one showing the velocity components just before impact, another showing the velocity components just after impact. An extra diagram showing the impulses acting at the moment of collision is often helpful.
(c) remembering that the momentum of *each* object is unchanged in a direction in which no impulse acts.
(d) appreciating that when a stationary object is struck, it begins to move in the direction of the impulse it receives.

EXAMPLES 4b

1) Two smooth spheres A and B, of equal radius but of masses m and M, are free to move on a horizontal table. A is projected with speed u towards B which is at rest. On impact, the line joining their centres is inclined at an angle θ to the velocity of A before impact. If e is the coefficient of restitution between the spheres, find the speed with which B begins to move. If A's path after impact is perpendicular to its path before impact, show that

$$\tan^2\theta = \frac{eM-m}{M+m}$$

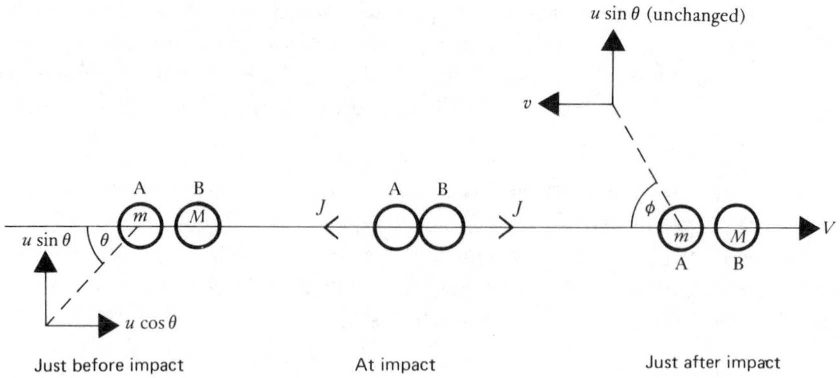

(When B is struck by the impulse J it begins to move in the direction of J as shown in the diagram.)

Along the line of centres we apply

(a) conservation of linear momentum

i.e. $$mu\cos\theta = MV - mv \qquad [1]$$

(b) law of restitution

i.e. $$eu\cos\theta = V + v \qquad [2]$$

From [1] and [2] $\times M$ we get

$$(1+e)mu\cos\theta = (M+m)V$$

i.e. B begins to move with a speed $\dfrac{(1+e)mu\cos\theta}{M+m}$.

Also from [1] and [2] $\times M$ we get

$$(eM-m)u\cos\theta = (M+m)v$$

$$\Rightarrow \qquad v = \frac{(eM-m)u\cos\theta}{(M+m)}$$

Hence $$\tan\phi = \frac{u\sin\theta}{v} = \frac{(M+m)\tan\theta}{(eM-m)}$$

But the paths of A before and after impact are at right angles, therefore $\cot\phi = \tan\theta$.

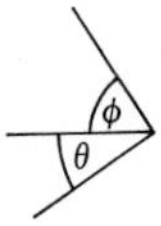

Hence
$$\frac{(eM-m)}{(M+m)\tan\theta} = \tan\theta$$

$\Rightarrow$
$$\tan^2\theta = \frac{eM-m}{M+m}$$

2) Two smooth spheres, A and B, of equal radius, lie on a horizontal table. A is of mass m and B is of mass $3m$. The spheres are projected towards each other with velocity vectors $5\mathbf{i}+2\mathbf{j}$ and $2\mathbf{i}-\mathbf{j}$ respectively and when they collide the line joining their centres is parallel to the vector $\mathbf{i}$.

If the coefficient of restitution between A and B is $\frac{1}{3}$, find their velocities after impact and the loss in kinetic energy caused by the collision. Find also the magnitude of the impulses that act at the instant of impact.

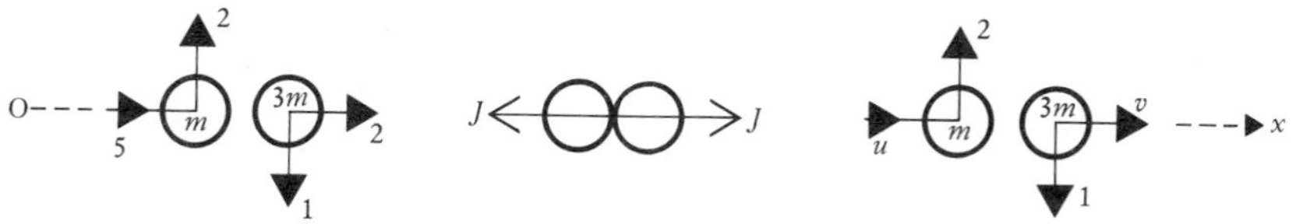

(Because the line of centres at impact, Ox, is parallel to the vector $\mathbf{i}$, the velocity components of A and B parallel and perpendicular to the line of centres are the coefficients of $\mathbf{i}$ and $\mathbf{j}$ in the velocity vectors. Components perpendicular to Ox are unchanged by the impact.)

Applying conservation of linear momentum and the law of restititution along Ox gives

$$5m+(3m)(2) = mu+3mv$$

and
$$\tfrac{1}{3}(5-2) = v-u$$

Hence $u=2$ and $v=3$.

The velocities of A and B after impact are therefore
$2\mathbf{i}+2\mathbf{j}$ and $3\mathbf{i}-\mathbf{j}$ respectively.

Before impact the kinetic energy of A is $\frac{1}{2}m(5^2+2^2) = \frac{29}{2}m$
and of B is $\frac{1}{2}(3m)(2^2+1^2) = \frac{15}{2}m$

After impact the kinetic energy of A is $\frac{1}{2}m(2^2+2^2) = 4m$
and of B is $\frac{1}{2}(3m)(3^2+1^2) = 15m$

Therefore the loss in K.E. at impact is

$$\tfrac{29}{2}m+\tfrac{15}{2}m-4m-15m = 3m$$

To find the value of J we consider the change in momentum along Ox, for one sphere only.

For B,

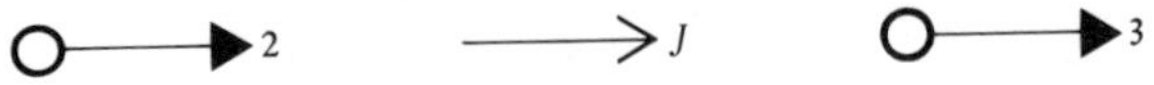

$$\Rightarrow \qquad J = 3m(3-2)$$

$$\Rightarrow \qquad J = 3m$$

3) A sphere A is of mass m and another sphere B of identical size but of mass $2m$, move towards each other with velocities $\mathbf{i}+2\mathbf{j}$ and $-\mathbf{i}+3\mathbf{j}$ respectively. They collide when their line of centres is parallel to $\mathbf{i}-\mathbf{j}$. If $e=\frac{1}{2}$ find the velocities of A and B after impact.

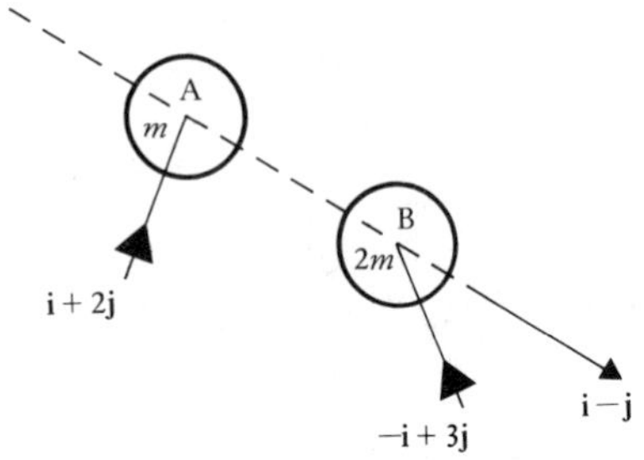

Before impact

(Before we can draw working diagrams, the velocity components, before and after impact, parallel and perpendicular to the line of centres must be found. To do this we need unit vectors in these two directions.)

Let the unit vector along the line of centres be $\hat{\mathbf{a}}$ where

$$\hat{\mathbf{a}} = \frac{1}{\sqrt{2}}(\mathbf{i}-\mathbf{j})$$

and a unit vector perpendicular to the line of centres is $\hat{\mathbf{b}}$ where

$$\hat{\mathbf{b}} = \frac{1}{\sqrt{2}}(\mathbf{i}+\mathbf{j})$$

The velocity components in these directions can now be found.

Sphere	Magnitude of Component	
	parallel to $\hat{\mathbf{a}}$	parallel to $\hat{\mathbf{b}}$
A	$(\mathbf{i}+2\mathbf{j})\cdot\frac{1}{\sqrt{2}}(\mathbf{i}-\mathbf{j}) = -\frac{1}{2}\sqrt{2}$	$(\mathbf{i}+2\mathbf{j})\cdot\frac{1}{\sqrt{2}}(\mathbf{i}+\mathbf{j}) = \frac{3}{2}\sqrt{2}$
B	$(-\mathbf{i}+3\mathbf{j})\cdot\frac{1}{\sqrt{2}}(\mathbf{i}-\mathbf{j}) = -2\sqrt{2}$	$(-\mathbf{i}+3\mathbf{j})\cdot\frac{1}{\sqrt{2}}(\mathbf{i}+\mathbf{j}) = \sqrt{2}$

Hence

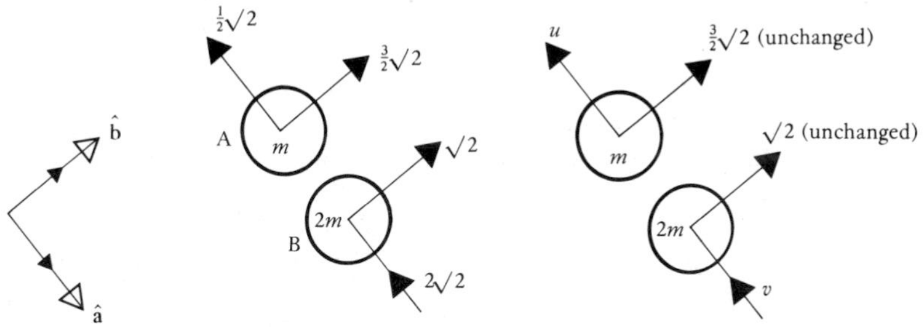

Using conservation of linear momentum and the law of restitution along the line of centres gives

$$m(\tfrac{1}{2}\sqrt{2})+2m(2\sqrt{2}) = mu+2mv$$

and

$$\tfrac{1}{2}(2\sqrt{2}-\tfrac{1}{2}\sqrt{2}) = u-v$$

Hence $u = 2\sqrt{2}$ and $v = \frac{5}{4}\sqrt{2}$.

The velocity of A after impact is

$$-u\hat{\mathbf{a}}+\tfrac{3}{2}\sqrt{2}\hat{\mathbf{b}} = (-2\sqrt{2})\frac{1}{\sqrt{2}}(\mathbf{i}-\mathbf{j})+(\tfrac{3}{2}\sqrt{2})\frac{1}{\sqrt{2}}(\mathbf{i}+\mathbf{j})$$

$$= \tfrac{1}{2}(-\mathbf{i}+7\mathbf{j})$$

The velocity of B after impact is

$$-v\hat{\mathbf{a}}+\sqrt{2}\hat{\mathbf{b}} = (-\tfrac{5}{4}\sqrt{2})\frac{1}{\sqrt{2}}(\mathbf{i}-\mathbf{j})+(\sqrt{2})\frac{1}{\sqrt{2}}(\mathbf{i}+\mathbf{j})$$

$$= \tfrac{1}{4}(-\mathbf{i}+9\mathbf{j})$$

EXERCISE 4b

1) A uniform smooth sphere moving with speed V on a horizontal surface strikes a stationary identical sphere, the direction of motion being at $60°$ to the line of centres at impact. If the coefficient of restitution between the spheres is $\frac{1}{2}$, find the speed of the second sphere immediately after impact.

2) A smooth sphere A collides with a stationary identical sphere B. Just before impact the direction of motion of A is inclined at $45°$ to the line of centres of A and B at impact. Show that just after impact the direction of motion of A is inclined at an angle θ to the line of centres where

$$(1-e)\tan\theta = 2$$

(e is the coefficient of restitution between A and B.)

3) A smooth sphere A, of mass m, has velocity vector $5\mathbf{i}+12\mathbf{j}$. A second smooth sphere B, of mass $3m$ has velocity $-3\mathbf{i}+4\mathbf{j}$.
The spheres collide and on impact their line of centres is in the direction $\mathbf{j}$. If sphere A moves in the direction $\mathbf{i}+\mathbf{j}$ immediately after impact, find the coefficient of restitution.

4) Two smooth spheres A and B of equal radius and mass are moving on a horizontal table with velocity vectors $\mathbf{i}+2\mathbf{j}$, $-3\mathbf{i}+\mathbf{j}$ respectively and collide when the line joining their centres is parallel to $\mathbf{i}$. Find the velocity vectors of A and B after the impact if (a) $e=\frac{1}{2}$, (b) $e=1$, (c) the collision is inelastic.

5) Two smooth spheres A and B of equal radius and mass lie on a horizontal surface. B is at rest and A is projected towards B with velocity vector $4\mathbf{i}+3\mathbf{j}$ and they collide when their line of centres is parallel to the vector $\mathbf{i}$. If B moves off with speed 3 units, find the value of e and the velocity vector of A after impact.

6) Two smooth spheres X and Y of equal radius and mass lie on a horizontal billiard table ABCD. X lies at rest and Y is projected towards X with speed u. The spheres collide when their line of centres is parallel to the edge AB. After the impact Y moves directly towards the cushion edge AB and after collision with the cushions both spheres are moving with equal speed. If e, the coefficient of restitution, is the same for all impacts, find e and the direction in which Y was moving before the first collision.

7) A smooth sphere X of mass m lies at rest at the centre of a billiard table ABCD. A second smooth sphere Y of equal radius but mass $2m$ is at rest at the midpoint of the edge BC. If Y is to be projected towards X so that after collision X moves towards the pocket at A, what must be the direction of the line of centres at impact? If $e=\frac{1}{5}$ and Y is projected with speed u and X moves off with speed $u/10$, find the direction in which Y was projected.

8) Two identical smooth spheres are moving on a horizontal table with velocity vectors $3\mathbf{i}+4\mathbf{j}$ and $-\mathbf{i}+\mathbf{j}$ and collide when the line joining their centres is parallel to the vector $\mathbf{i}$. If the coefficient of restitution between the spheres is $\frac{1}{2}$, find the velocity vectors of the spheres after impact. Find also the ratio of the magnitudes of the velocities, before and after impact, of the spheres relative to each other. If, at this instant of impact, the centres of the spheres are 2 units of distance apart, find the distance between their centres 1 unit of time later.

9) A smooth sphere A of mass $4m$, moving with velocity $9\mathbf{j}$ is struck so that it moves with velocity $(8\mathbf{i}-6\mathbf{j})$. Find the impulse given to A.
The sphere A then collides with a second smooth sphere B, of equal radius but of mass $2m$. Before the collision B is moving with velocity $(2\mathbf{i}+4\mathbf{j})$ and, at the instant of collision, the line of centres of A and B is parallel to $\mathbf{j}$. The coefficient of restitution between A and B is $\frac{1}{4}$. Calculate the velocities of A and B after the collision.

MULTIPLE CHOICE QUESTIONS

These are included in the exercise at the end of Chapter 5 on p. 99.

MISCELLANEOUS EXERCISE 4

1) Two equal uniform, smooth spheres of mass m are at rest in contact on a smooth horizontal table. A third uniform smooth sphere, of the same radius but of mass $3m$, moves on the table with speed u in a direction perpendicular to the line of centres of the other two spheres and strikes them simultaneously. The coefficient of restitution between any two spheres is e. Find
(a) the speeds of the three spheres after impact,
(b) the magnitude and direction of the impulse received by each sphere.
(AEB)

2) A red ball is stationary on a rectangular billiard table OABC. It is then struck by a white ball of equal mass and equal radius with velocity $u(-2\mathbf{i}+11\mathbf{j})$ where $\mathbf{i}$ and $\mathbf{j}$ are unit vectors along OA and OC respectively. After impact the red and white balls have velocities parallel to the vectors $-3\mathbf{i}+4\mathbf{j}$, $2\mathbf{i}+4\mathbf{j}$ respectively. Prove that the coefficient of restitution between the two balls is $\frac{1}{2}$.
(U of L)

3) Show that the vectors $\mathbf{p}_1 = 3\mathbf{i}+4\mathbf{j}$ and $\mathbf{p}_2 = 4\mathbf{i}-3\mathbf{j}$ are at right angles. If $\mathbf{n}$ and $\mathbf{t}$ are unit vectors in the directions $\mathbf{p}_1$ and $\mathbf{p}_2$ respectively, express each of the vectors $\mathbf{v}_1 = 3\mathbf{i}$ and $\mathbf{v}_2 = \mathbf{i}+\mathbf{j}$ in the form $\mathbf{v} = a\mathbf{n}+b\mathbf{t}$ where a and b are scalar constants and state the values of these constants in each case.
Two billiard balls A_1 and A_2 of equal mass collide; at the instant of collision A_1 and A_2 have velocity vectors $\mathbf{v}_1$ and $\mathbf{v}_2$ respectively, as given above, and $\mathbf{n}$ is the unit vector along the line of centres. If the coefficient of restitution between the billiard balls is $\frac{1}{2}$, find their velocity vectors after impact in terms of $\mathbf{n}$ and $\mathbf{t}$. Hence find their velocity vectors after impact in terms of $\mathbf{i}$ and $\mathbf{j}$.
(AEB)

4) A rectangular billiard table is a long and b wide $(a > b)$. The coefficient of restitution when a particle hits any side cushion is e. If a smooth particle hits a cushion and comes off in a direction at right angles to its original direction, show that the angles made by the two directions with the cushion are $\operatorname{arccot}\sqrt{e}$ and $\arctan\sqrt{e}$.
A smooth particle is projected from a point A on a shorter side. It moves on a rectangular path, hitting each side cushion in turn, and returns to A. Show that A divides the shorter side in the ratio $(a\sqrt{e}-be):(b-a\sqrt{e})$. (C)

5) A smooth uniform sphere A, moving with speed u, impinges on an identical sphere B at rest, the direction of motion just before impact being inclined at an angle α to the line of centres. Find the magnitude and direction of the velocities of A and B after the impact in terms of u, α and the coefficient of restitution e.
Given that $\tan^2\alpha = \frac{8}{27}$ and $e = \frac{2}{3}$, show that
(a) the speed of A is halved by the impact,
(b) the direction of motion of A is turned through the angle $\tan^{-1}(\frac{2}{5}\sqrt{6})$
(JMB)

6) Show that the vectors $\mathbf{p}_1 = 5\mathbf{i}+12\mathbf{j}$ and $\mathbf{p}_2 = 12\mathbf{i}-5\mathbf{j}$ are at right angles.
If $\mathbf{n}$ and $\mathbf{t}$ are unit vectors in the directions $\mathbf{p}_1$ and $\mathbf{p}_2$ respectively, show that the vectors $13\mathbf{i}+13\mathbf{j}$ and $26\mathbf{i}$ can be expressed as $17\mathbf{n}+7\mathbf{t}$ and $10\mathbf{n}+24\mathbf{t}$ respectively.
Two perfectly elastic smooth spheres, S_1 and S_2, of equal masses and equal radii, collide. At the instant of collision S_1 and S_2 have velocity vectors $\mathbf{v}_1 = 13\mathbf{i}+13\mathbf{j}$ and $\mathbf{v}_2 = 26\mathbf{i}$ respectively and $\mathbf{n}$ is the unit vector in the direction of the line of centres. Find the velocity vectors of the spheres after impact in terms of $\mathbf{n}$ and $\mathbf{t}$ and hence find the velocity vectors after impact in terms of $\mathbf{i}$ and $\mathbf{j}$. (U of L)

7) A smooth sphere A impinges obliquely on a stationary sphere B of equal mass. The directions of motion of sphere A before impact and after impact make angles α and β respectively with the line of centres at the instant of impact. Show that

$$\cot\beta = \tfrac{1}{2}(1-e)\cot\alpha$$

where e is the coefficient of restitution.
Find, in terms of α and e, the tangent of the angle through which the sphere A is deflected, and show that, if α is varied, this angle is a maximum when

$$\tan^2\alpha = \tfrac{1}{2}(1-e)$$

(JMB)

8) A smooth sphere A of mass $3M$ is at rest on a smooth horizontal table when it is struck obliquely by another smooth sphere B of equal radius and mass M. Before impact the sphere B moves along the surface of the table with a velocity u in a direction which, at the moment of impact of the two spheres, makes an acute angle α with their common line of centres. After impact the sphere B moves in a direction at right angles to its original direction of motion. Prove that the coefficient of restitution e between the two spheres must be greater than $\frac{1}{3}$ and show that

$$-\frac{1}{\sqrt{2}} \leqslant \tan\alpha \leqslant \frac{1}{\sqrt{2}}$$

Given that $\alpha = \frac{\pi}{6}$, find e and find, in terms of u, the speed of A after impact. (AEB)

9) A smooth uniform ball travelling along a smooth horizontal table collides with a second smooth uniform ball of the same mass and radius which is at rest on the table. At the moment of impact the line of centres makes an angle of 30° with the direction in which the first ball is moving. If the coefficient of restitution between the balls is e, show that the first ball is deflected by the impact through an angle θ, where

$$\tan\theta = \frac{(1+e)\sqrt{3}}{5-3e}$$

(O)

10) A smooth sphere S of mass km, where $k > 1$, is at rest on a smooth horizontal table. A second smooth sphere T of mass m and the same radius as S is moving towards S and impinges obliquely on it. If, after impact, the directions of motion of the two spheres are perpendicular, prove that the coefficient of restitution is $1/k$.
Given that $k = 2$, and that the kinetic energy lost due to impact is one quarter of the original kinetic energy, determine the inclination of the initial direction of motion to the line of centres. (U of L)

11) If $16\mathbf{i} + 13\mathbf{j} = \alpha(3\mathbf{i} + 4\mathbf{j}) + \beta(4\mathbf{i} - 3\mathbf{j})$, find α and β.
A smooth uniform sphere P is stationary on a rectangular horizontal table OABC. Sphere P is struck by an identical sphere Q which is moving on the table with a velocity $(16\mathbf{i} + 13\mathbf{j})u$, where $\mathbf{i}$ and $\mathbf{j}$ are unit vectors along OA and OC respectively. After impact spheres P and Q have velocities in the directions of the vectors $3\mathbf{i} + 4\mathbf{j}$ and $7\mathbf{i} + \mathbf{j}$ respectively. Prove that the coefficient of restitution between the spheres is $\frac{1}{2}$. (AEB)

12) A smooth sphere A, of mass m and moving with speed u, collides obliquely with a stationary sphere B, of mass $2m$, the coefficient of restitution between the spheres being $\frac{1}{3}$. At the instant of impact the velocity of A makes an angle θ with the line of centres of the spheres. Immediately after impact the velocity of A makes an angle ϕ with the line of centres. Find the speed with which B starts to move and show that $\tan\phi = 9\tan\theta$.
Show that the angle through which A is deflected by the collision is greatest when $\tan\theta = \frac{1}{3}$. Find the tangent of this greatest angle of deflection.

(U of L)

13) Two smooth spheres A and B, of equal radii but of masses m and $2m$ respectively, are sliding on a horizontal table. The velocity vectors of A and B are $\lambda(\mathbf{i}+\mathbf{j})$, $(\lambda > 0)$, and $\mathbf{j}$ respectively. They collide when their line of centres is parallel to the vector $\mathbf{i}$. The coefficient of restitution between them is $e(\leqslant 1)$. If, after collision, sphere B is moving in the direction $\mathbf{i}+2\mathbf{j}$, show that $\frac{3}{4} \leqslant \lambda \leqslant \frac{3}{2}$.
Given that $\lambda = \frac{5}{4}$, find the value of e and the kinetic energy lost in the impact.

(AEB)

CHAPTER 5

PROJECTILES ON INCLINED PLANES

In *Mechanics and Probability*, standard equations for the motion of a projectile were derived. These equations were based on horizontal and vertical components of acceleration, velocity and displacement and they are valid for any projectile during its flight in a resistance-free medium.

It sometimes happens that a particle is projected, not along a horizontal surface but up or down an inclined plane. In such cases the standard projectile equations can be used for the motion of the particle up to the moment of its impact with the inclined plane. At that instant the displacement components of the particle also satisfy the equation of the plane. This extra property provides one method for dealing with such problems.

Note. Care must be taken to check how the angle of projection is given. Sometimes the angle is measured from the horizontal; in other cases it is taken from the inclined plane.

EXAMPLES 5a

1) A particle is projected with an initial velocity of $20\,\text{m}\,\text{s}^{-1}$ at $60°$ to the horizontal from a point O at the bottom of a plane inclined at $45°$ to the horizontal. The particle hits the plane again at a point A, where OA is a line of greatest slope.

Find (a) the time of flight,
(b) the range on the plane.

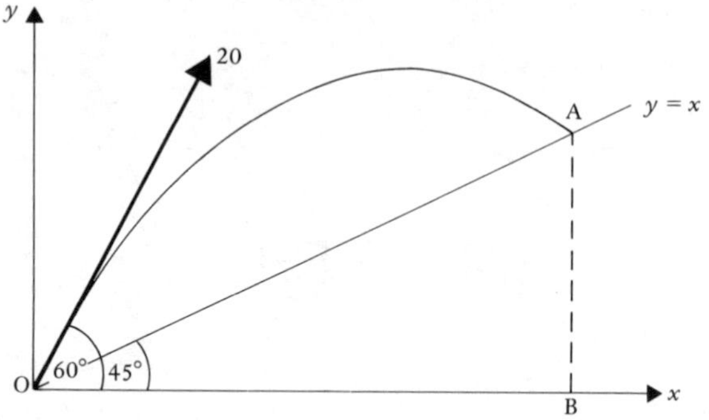

For the projectile, at any time t,

$$x = 20t\cos 60°$$

$$y = 20t\sin 60° - (\tfrac{1}{2})(10)t^2$$

The point A is on the plane so, when the particle reaches A, $y = x$

Therefore $$20t(\tfrac{1}{2}\sqrt{3}) - 5t^2 = 20t(\tfrac{1}{2})$$

$\Rightarrow$ $$t = 2(\sqrt{3}-1) \qquad (t = 0 \text{ at O})$$

(a) The time of flight is $2(\sqrt{3}-1)$ seconds.

(b) The range on the plane is OA and the x coordinate of the projectile when $t = 2(\sqrt{3}-1)$ is OB,

i.e. $$OB = 40(\sqrt{3}-1)\cos 60° = 20(\sqrt{3}-1)$$

Now $$OA = OB\sec 45°$$

$$= 20(\sqrt{3}-1)\sqrt{2}$$

Therefore the range on the plane is $20(\sqrt{6}-\sqrt{2})$m.

2) A particle is projected from a point O on a plane inclined at an angle α to the horizontal. The particle hits the plane at right angles at A, where OA is a line of greatest slope, O being lower than A. If the particle is projected at an angle θ to the horizontal, find the relationship between θ and α.

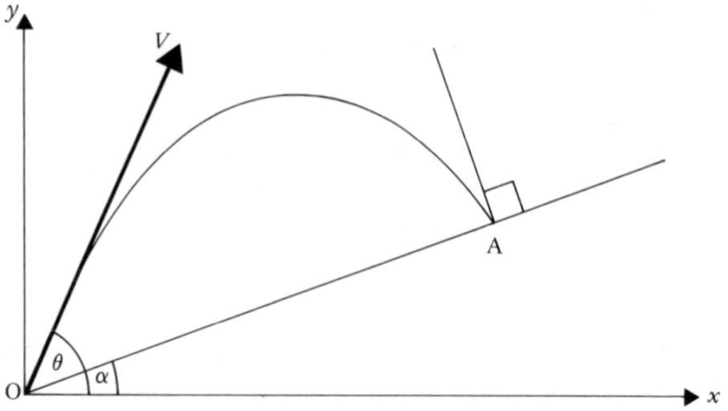

The coordinates of A satisfy both the equation of the projectile and the equation of the line of greatest slope of the plane. Therefore, at A,

$$y = x\tan\theta - \frac{gx^2}{2V^2\cos^2\theta}$$

and $$y = x\tan\alpha$$

$$\Rightarrow \qquad \tan\alpha = \tan\theta - \frac{gx}{2V^2\cos^2\theta} \qquad [1]$$

The direction of flight at A is inclined at $(\alpha+\frac{1}{2}\pi)$ to Ox, therefore

$$\frac{dy}{dx} = \tan(\alpha+\tfrac{1}{2}\pi) = -\cot\alpha$$

$$\Rightarrow \qquad \tan\theta - \frac{gx}{V^2\cos^2\theta} = -\cot\alpha \qquad [2]$$

Eliminating $\dfrac{gx}{V^2\cos^2\theta}$ from [1] and [2] gives

$$2\tan\theta - 2\tan\alpha = \tan\theta + \cot\alpha$$

$$\Rightarrow \qquad \tan\theta = \cot\alpha + 2\tan\alpha$$

3) A particle is projected from a point A, whose position vector is $2\mathbf{i}+3\mathbf{j}$, with initial velocity vector $20\mathbf{i}+30\mathbf{j}+20\mathbf{k}$, where $\mathbf{i}$ and $\mathbf{k}$ are horizontal and $\mathbf{j}$ is the unit vector in the direction of the upward vertical. Find a vector equation for the path of the particle.

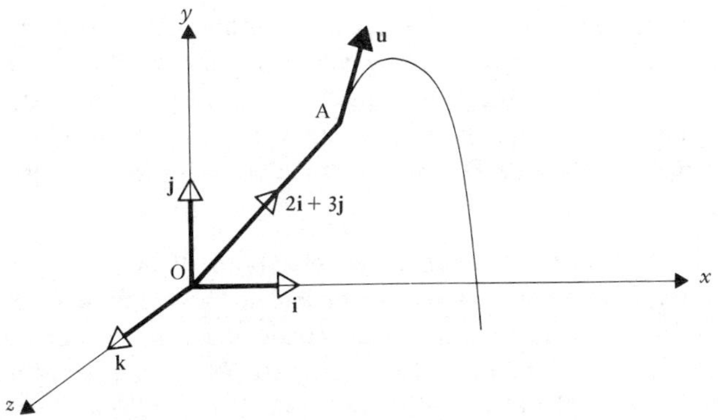

While the particle is in flight it has a constant acceleration g vertically downward.

Therefore the acceleration vector at any time t is $\mathbf{a}$ where

$$\mathbf{a} = -g\mathbf{j}$$

and the velocity vector (by integration) is $\mathbf{v}$ where

$$\mathbf{v} = -gt\mathbf{j}+\mathbf{c}$$

When $t = 0$, $\mathbf{v} = 20\mathbf{i}+30\mathbf{j}+20\mathbf{k}$

therefore $$\mathbf{v} = 20\mathbf{i}+(30-gt)\mathbf{j}+20\mathbf{k}$$

The position vector at any time t is given by

$$\mathbf{r} = 20t\mathbf{i}+(30t-\tfrac{1}{2}gt^2)\mathbf{j}+20t\mathbf{k}+\mathbf{c}_1$$

When $t = 0$, $\mathbf{r} = 2\mathbf{i}+3\mathbf{j}$

therefore $$\mathbf{r} = (20t+2)\mathbf{i}+(30t-\tfrac{1}{2}gt^2+3)\mathbf{j}+20t\mathbf{k}$$

This is a vector equation of the path of the projectile.

EXERCISE 5a

(Take $g = 10\,\text{m s}^{-2}$.)

1) A particle is projected with an initial velocity of $30\,\text{m s}^{-1}$ at $45°$ to the horizontal from a point O on an inclined plane. The particle hits the plane again at a point A where OA is inclined at $30°$ to the horizontal. Find the time for which the particle is in the air and the distance OA if:
(a) A is higher than O, (b) A is lower than O.

2) A particle is projected at an angle θ to the horizontal with an initial speed of $100\,\text{m s}^{-1}$ from a point O on a plane inclined at $30°$ to the horizontal. The plane containing the path passes through a line of greatest slope of the inclined plane. Find, in terms of θ, the range of the particle up the plane. Deduce the maximum range up the plane.

3) A particle is projected from the origin O with initial velocity vector $20\mathbf{i}+20\mathbf{j}$ where $\mathbf{i}$ is horizontal and $\mathbf{j}$ vertically upward. The path of the particle crosses the line with vector equation $\mathbf{r} = \lambda(3\mathbf{i}+\mathbf{j})$ at the point A. Write down the vector equation of the path of the projectile and hence or otherwise find the distance OA and the time taken for the particle to reach A.

4) A particle is projected with an initial velocity of $30\,\text{m s}^{-1}$ at $\arctan\frac{4}{3}$ to the horizontal from a point O on a plane inclined at $\arctan\frac{1}{2}$ to the horizontal. The path of the particle lies in the vertical plane through a line of greatest slope of the inclined plane. Find the range of the particle on the plane if the particle is projected (a) up the plane, (b) down the plane.

5) A plane is inclined at an angle α to the horizontal. A particle P is projected from a point O on the plane at an angle θ *to the plane* and directly up the plane. When the particle strikes the plane it is travelling horizontally. Show that $\tan(\theta + \alpha) = 2 \tan \alpha$.

6) A particle is projected with an initial velocity vector $50\mathbf{i} + 30\mathbf{j} + 50\mathbf{k}$ from the point with position vector $5\mathbf{i} - 3\mathbf{k}$, where $\mathbf{i}$ and $\mathbf{k}$ are horizontal and $\mathbf{j}$ is vertically upward. Find vector expressions for the velocity and position of the particle at any time t. If the xz plane is ground level find the time of flight of the particle. Find also the vector equation of the plane that contains the path of the particle.

7)

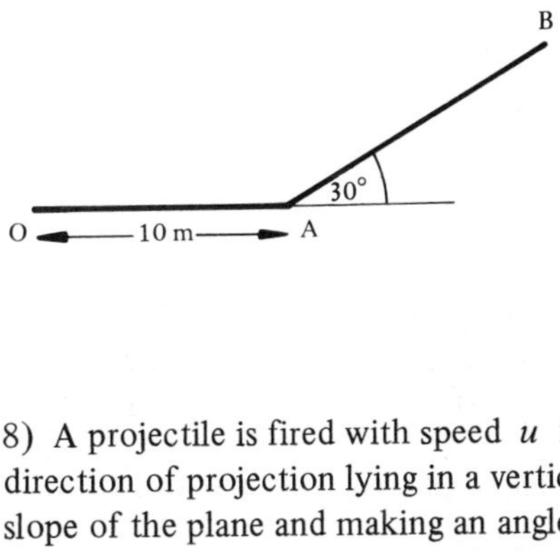

The diagram shows a sloping embankment AB where OA is level ground. A particle is projected towards the slope from O with initial velocity $30\,\text{m s}^{-1}$ at $45°$ to the horizontal. Find the time of flight of the particle and the distance from A of the point at which it hits the slope AB.

8) A projectile is fired with speed u from a point on an inclined plane, the direction of projection lying in a vertical plane containing a line of greatest slope of the plane and making an angle θ with this line of greatest slope. The plane is inclined at angle α to the horizontal where $\theta + \alpha < \frac{1}{2}\pi$. Find the time that elapses before the projectile strikes the plane.

9) A particle is projected from a point O of a plane inclined at an angle α to the *vertical*. The angle of projection is θ with the vertical and the plane of motion contains a line of greatest slope of the plane. When the particle hits the plane (at a point higher than O) it is moving horizontally. Prove that $\tan \alpha = 2 \tan \theta$.

10) A particle is projected from a point O on level ground towards a smooth vertical wall 50 m from O and hits the wall. The initial velocity of the particle is $30\,\text{m s}^{-1}$ at $45°$ to the horizontal and the coefficient of restitution between the particle and the wall is e. Find the distance from O of the point at which the particle hits the ground again if:

(a) $e = 0$, (b) $e = 1$, (c) $e = \frac{1}{2}$.

INCLINED AXES

As we saw in the preceding section, some problems involving projection on an inclined plane can be solved using equations for the motion of the projectile referred to horizontal and vertical axes. This method, however, sometimes results in an awkward solution and an alternative approach is therefore needed. This refers the motion of the projectile to axes that are parallel and perpendicular to the inclined plane.
With this frame of reference a completely new set of equations is required.

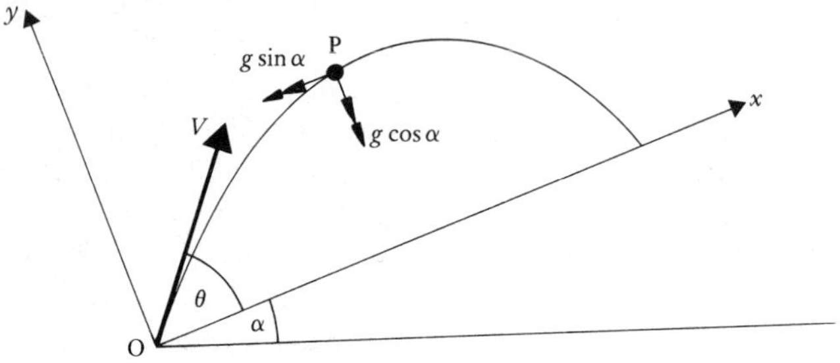

Consider a particle P, projected with speed V at an angle θ to a plane which is inclined at an angle α to the horizontal.

Resolving the gravitational acceleration in the direction of the axes shown we have

$$\ddot{x} = -g \sin\alpha \qquad \text{and} \qquad \ddot{y} = -g\cos\alpha$$

Integrating with respect to time, and using initial velocity components $V\cos\theta$ and $V\sin\theta$, gives

$$\dot{x} = V\cos\theta - gt\sin\alpha \qquad \text{and} \qquad \dot{y} = V\sin\theta - gt\cos\alpha$$

Integrating again and using $x = 0 = y$ when $t = 0$ gives

$$x = Vt\cos\theta - \tfrac{1}{2}gt^2\sin\alpha \quad \text{and} \quad y = Vt\sin\theta - \tfrac{1}{2}gt^2\cos\alpha$$

These equations can be used in a variety of problems but it is unwise to regard them as quotable. It is better to derive the required equations each time, starting from the acceleration components.

Note that the equation of the trajectory is not easy to find when inclined axes are used.

SPECIAL PROPERTIES

Consider again a particle P, projected with speed V up a plane inclined at an angle α to the horizontal, where the initial velocity of P is inclined at an angle θ to a line of greatest slope.

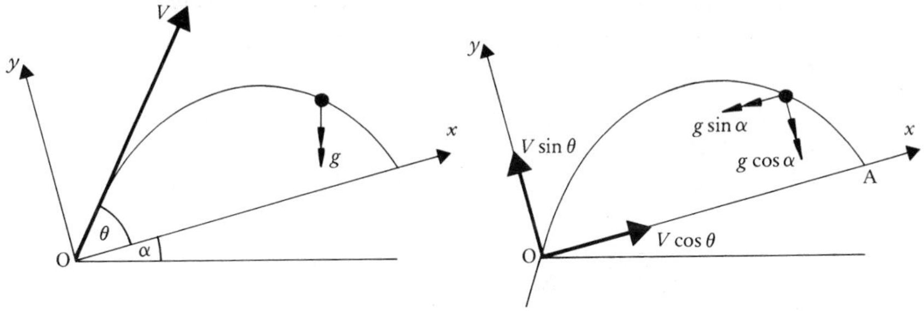

As before we have:

$$\ddot{x} = -g \sin\alpha \qquad \ddot{y} = -g\cos\alpha$$

$$\dot{x} = V\cos\theta - gt\sin\alpha \qquad \dot{y} = V\sin\theta - gt\cos\alpha$$

$$x = Vt\cos\theta - \tfrac{1}{2}gt^2\sin\alpha \qquad y = Vt\sin\theta - \tfrac{1}{2}gt^2\cos\alpha$$

These basic equations can be used to find a number of results of special interest.

The Time of Flight

The time of flight is the time taken for P to travel from O to A.

At O and A, $y = 0$ i.e. $Vt\sin\theta - \frac{1}{2}gt^2\cos\alpha = 0$

$$\Rightarrow \qquad t = 0 \quad \text{or} \quad \frac{2V\sin\theta}{g\cos\alpha}$$

But $t = 0$ at O, therefore

> the time of flight is $\dfrac{2V\sin\theta}{g\cos\alpha}$

The Range up the Plane

The distance OA is the range, R, up the plane and this is the value of x when $t = \dfrac{2V\sin\theta}{g\cos\alpha}$.

Therefore
$$R = (V\cos\theta)\left(\frac{2V\sin\theta}{g\cos\alpha}\right) - (\tfrac{1}{2}g\sin\alpha)\left(\frac{2V\sin\theta}{g\cos\alpha}\right)^2$$
$$= \left(\frac{2V^2\sin\theta}{g\cos^2\alpha}\right)(\cos\theta\cos\alpha - \sin\theta\sin\alpha)$$

i.e. the range up the plane is $\dfrac{2V^2\sin\theta\cos(\theta+\alpha)}{g\cos^2\alpha}$.

The Maximum Range up the Plane

Rearranging the expression for the general range up the plane we have
$$R = \frac{V^2}{g\cos^2\alpha}\{\sin(2\theta+\alpha) - \sin\alpha\}$$

Now V and α are constant so R is greatest when $\sin(2\theta+\alpha)$ is greatest.

This is when $\sin(2\theta+\alpha) = 1$ therefore
$$R_{\text{maximum}} = \frac{V^2(1-\sin\alpha)}{g\cos^2\alpha}$$

When
$$\sin(2\theta+\alpha) = 1$$
$$2\theta + \alpha = 90°$$
$$\Rightarrow \quad \theta = 45° - \tfrac{1}{2}\alpha$$

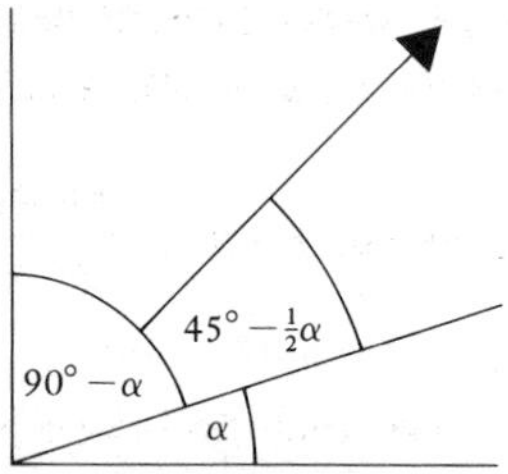

i.e. the range up the plane is maximum when the direction of projection bisects the angle between the upward slope of the plane and the vertical.

Note. The expression for maximum range can be given in another form, as follows.
$$R_{\text{maximum}} = \frac{V^2(1-\sin\alpha)}{g\cos^2\alpha} = \frac{V^2(1-\sin\alpha)}{g(1-\sin^2\alpha)} = \frac{V^2}{g(1+\sin\alpha)}$$

PROJECTION DOWN AN INCLINED PLANE

When a particle is projected down an inclined plane it is more convenient to take the x axis down the plane as shown.

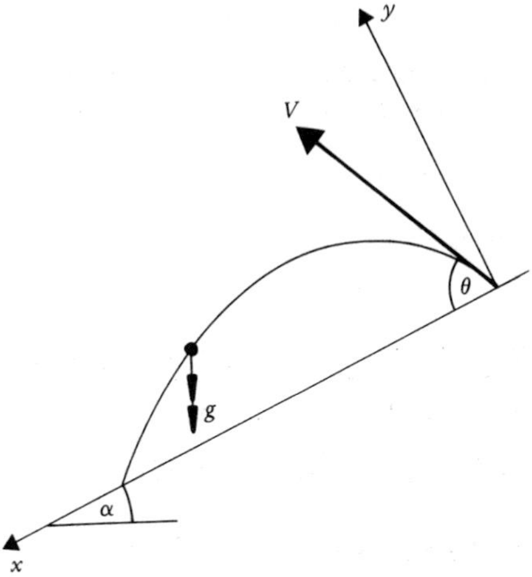

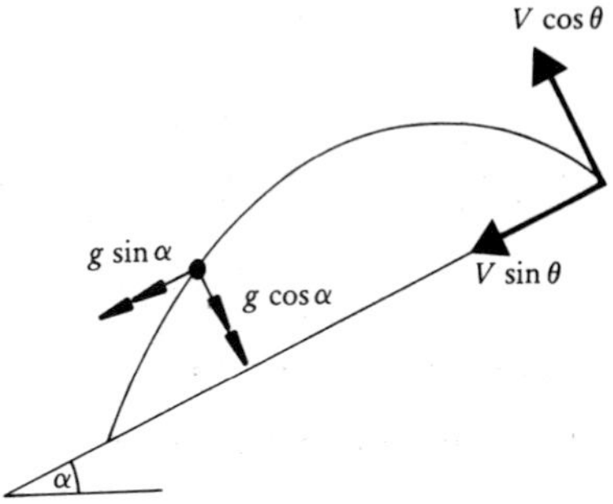

Then we have

$$\ddot{x} = g \sin\alpha \qquad \ddot{y} = -g\cos\alpha$$

$$\dot{x} = V\cos\theta + gt\sin\alpha \qquad \dot{y} = V\sin\theta - gt\cos\alpha$$

$$x = Vt\cos\theta + \tfrac{1}{2}gt^2\sin\alpha \qquad y = Vt\sin\theta - \tfrac{1}{2}gt^2\cos\alpha$$

These equations can be used, in the same way as before, to derive any required result, including the special properties of the trajectory, i.e.

the time of flight is $\dfrac{2V\sin\theta}{g\cos\alpha}$

the range *down* the plane is $\dfrac{2V^2\sin\theta\cos(\theta-\alpha)}{g\cos^2\alpha}$

the maximum range down the plane is $\dfrac{V^2(1+\sin\alpha)}{g\cos^2\alpha}$, or $\dfrac{V^2}{g(1-\sin\alpha)}$,

and it occurs when the direction of projection bisects the angle between the vertical and the downward slope of the plane.

Note. The *results* given above should not generally be regarded as quotable. Most problems should be dealt with from first principles using the *methods* explained above.

EXAMPLES 5b

1) A particle P is projected with an initial velocity of $20\,\mathrm{m\,s^{-1}}$ at $75°$ to the horizontal from a point O at the bottom of a plane inclined at $45°$ to the horizontal. The particle hits the plane again at a point A, where OA is a line of greatest slope.

Find (a) the time of flight,

(b) the range on the plane,

(c) the greatest height above the plane reached by the particle.

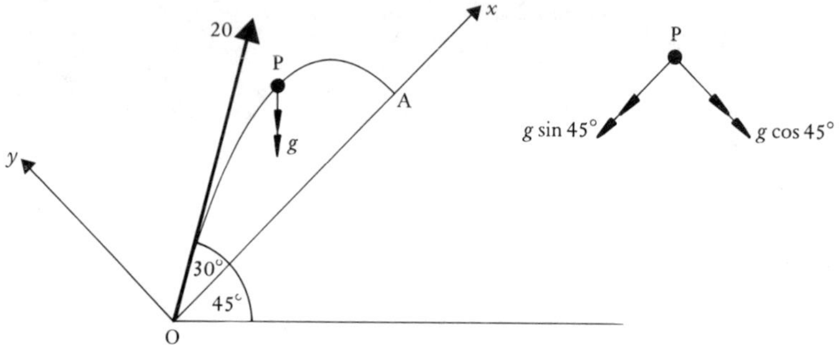

Taking axes as shown,

$$\ddot{x} = -\tfrac{1}{2}g\sqrt{2} \qquad \ddot{y} = -\tfrac{1}{2}g\sqrt{2}$$

Integrating with respect to t and using initial velocity components $20\cos 30°$ and $20\sin 30°$ gives

$$\dot{x} = 10\sqrt{3}-\tfrac{1}{2}gt\sqrt{2} \qquad \dot{y} = 10-\tfrac{1}{2}gt\sqrt{2}$$

and

$$x = 10t\sqrt{3}-\tfrac{1}{4}gt^2\sqrt{2} \qquad y = 10t-\tfrac{1}{4}gt^2\sqrt{2}$$

(a) At A, $y = 0 \Rightarrow 10t-\tfrac{1}{4}gt^2\sqrt{2} = 0$

$\Rightarrow \qquad t = 0$ or $t = 2\sqrt{2} \qquad (g = 10\,\mathrm{m\,s^{-2}})$

Therefore the time of flight is $2\sqrt{2}$ seconds.

(b) The range on the plane is OA and this is the value of x when $t = 2\sqrt{2}$,

i.e.

$$x = (10\sqrt{3})(2\sqrt{2})-\tfrac{1}{4}(10)(2\sqrt{2})^2\sqrt{2}$$
$$= 20\sqrt{6}-20\sqrt{2}$$

Therefore the range is $20(\sqrt{6}-\sqrt{2})\,\mathrm{m}$.

(c)

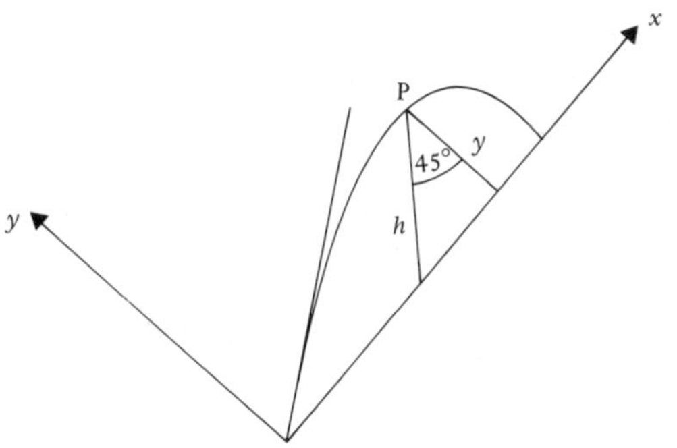

At any time t, the height h of P above the plane is given by

$$h = y \sec 45°$$
$$= (10t - \tfrac{1}{4}gt^2\sqrt{2})\sqrt{2}$$
$$= 10t\sqrt{2} - 5t^2$$
$$= -5(t^2 - 2t\sqrt{2})$$
$$= -5\{(t-\sqrt{2})^2 - 2\}$$

Therefore the maximum height of P above the plane is 10 m and this occurs when $t = \sqrt{2}$.

Note that the maximum height above the plane and the maximum distance from the plane, i.e. h and y, occur at the same time, which is after half the time of flight.

The reader can compare this example with Example 5a, No 1. Parts (a) and (b) of this example could have been done using horizontal and vertical axes, but part (c) is better done using inclined axes. This is because neither h nor y can be obtained *directly* from horizontal and vertical axes.

2) A particle is projected from a point O on a smooth plane inclined to the horizontal at $\arctan \frac{1}{2}$. The particle is projected at $\arctan \frac{3}{4}$ to the plane, hits the plane at a higher point A and rebounds. OA is a line of greatest slope and e is the coefficient of restitution between P and the plane. If P continues to move up the plane after impact find the possible range of values of e.

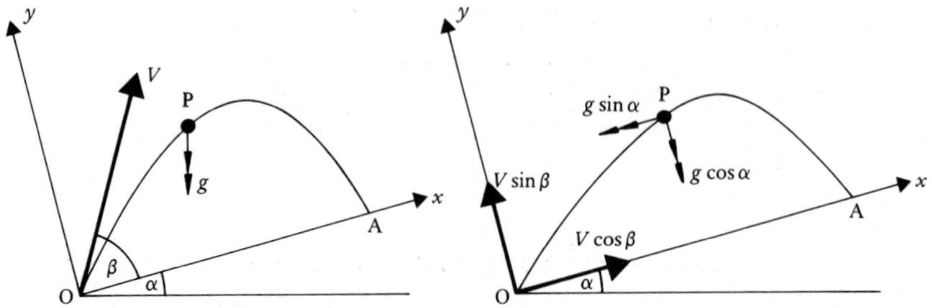

In the diagrams, $\tan\alpha = \frac{1}{2} \quad \Rightarrow$ 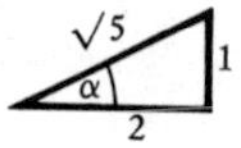

and $\tan\beta = \frac{3}{4} \quad \Rightarrow$ 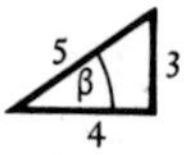

Taking inclined axes as shown,

$$\ddot{x} = -\tfrac{1}{5}g\sqrt{5} \qquad \ddot{y} = -\tfrac{2}{5}g\sqrt{5}$$

$$\dot{x} = \tfrac{4}{5}V - \tfrac{1}{5}gt\sqrt{5} \qquad \dot{y} = \tfrac{3}{5}V - \tfrac{2}{5}gt\sqrt{5}$$

$$x = \tfrac{4}{5}Vt - \tfrac{1}{10}gt^2\sqrt{5} \qquad y = \tfrac{3}{5}Vt - \tfrac{1}{5}gt^2\sqrt{5}$$

When the particle hits the plane at A, $y = 0$,

i.e. $$\tfrac{3}{5}Vt - \tfrac{1}{5}gt^2\sqrt{5} = 0 \quad \Rightarrow \quad t = (3V\sqrt{5}/5g)$$

Therefore, at A, $$\dot{x} = \tfrac{4}{5}V - \tfrac{1}{5}g\sqrt{5}(3V\sqrt{5}/5g) = \tfrac{1}{5}V$$

$$\dot{y} = \tfrac{3}{5}V - \tfrac{2}{5}g\sqrt{5}(3V\sqrt{5}/5g) = -\tfrac{3}{5}V$$

We can now investigate the effect of the impact.

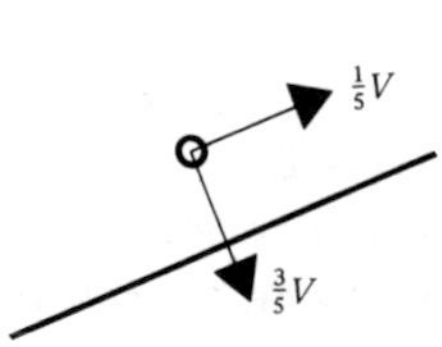

Just before impact

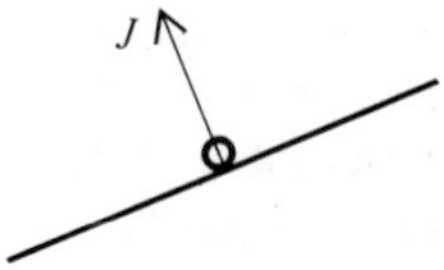

At impact

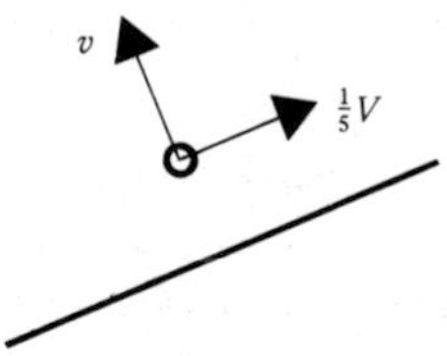

Just after impact

Using the law of restitution perpendicular to the plane gives

$$v = e(\tfrac{3}{5}V)$$

Parallel to the plane the velocity component is unchanged.

Just after impact, the resultant velocity of P is inclined to the plane at an angle θ, where

$$\tan\theta = v \div (\tfrac{1}{5}V) = (\tfrac{3}{5}eV) \div (\tfrac{1}{5}V) = 3e$$

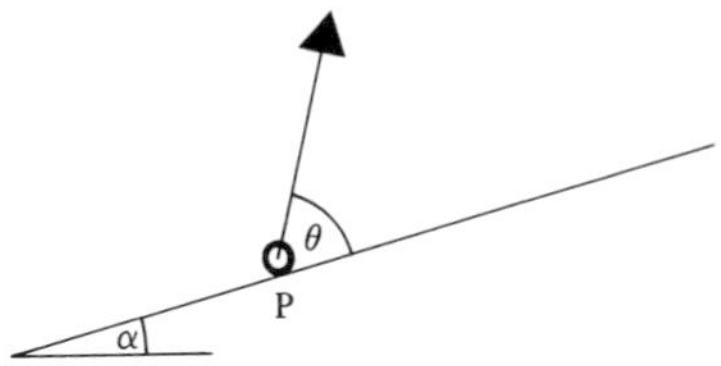

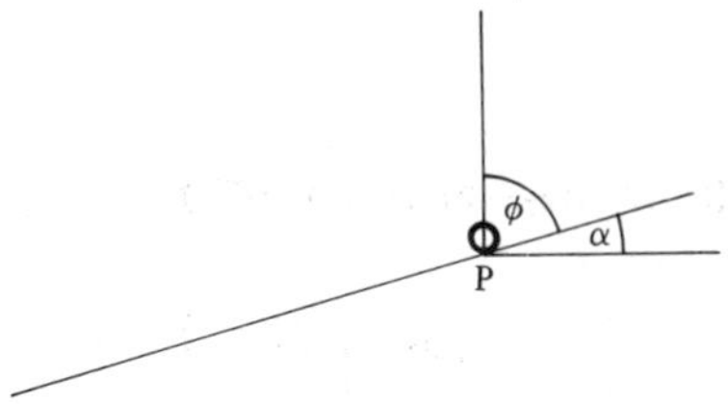

If P continues to move up the plane then

$$\theta < \phi \quad \Rightarrow \quad \tan\theta < \tan\phi$$

Now $\phi = \tfrac{1}{2}\pi - \alpha \quad \Rightarrow \quad \tan\phi = \cot\alpha.$

Therefore $\tan\theta < \cot\alpha$

i.e. $3e < 2$

$\Rightarrow$ $e < \tfrac{2}{3}$

We also know that $e \geqslant 0$, therefore the range of values of e for which P continues to move up the plane after impact is

$$0 \leqslant e < \tfrac{2}{3}$$

EXERCISE 5b

1) A particle is projected, with speed u, from a point O on a plane inclined at $30°$ to the horizontal and strikes the plane at right angles at a point A where AO is a line of greatest slope of the plane and O is lower than A. If the coefficient of restitution between the particle and the plane is $\tfrac{1}{2}$, find the speed of the particle after the first impact.

2) A particle is projected from a point O on a smooth plane, which is inclined at $\arctan\tfrac{1}{3}$ to the horizontal, and strikes the plane again at A where OA is a line of greatest slope and A is higher than O. The initial velocity of the particle is 40 m s^{-1} at $\arctan 3$ to the horizontal. Determine in which direction the particle moves after striking the plane at A if:
(a) the impact is inelastic, (b) the impact is perfectly elastic.

3) A particle is projected with speed u from a point on a plane which is inclined at an angle β to the horizontal. The particle is projected at an angle α to the horizontal in a vertical plane through a line of greatest slope of the plane. Show that the range up the plane is

$$\frac{2u^2 \sin(\alpha-\beta)\cos\alpha}{g\cos^2\beta}$$

Deduce that the range up the plane is a maximum when $\alpha = \frac{1}{4}\pi + \frac{1}{2}\beta$.

4) A particle projected from a point A on a smooth plane of inclination α strikes the plane at B, where AB is a line of greatest slope with B higher than A. The velocity of projection is V at an angle θ with the plane. Prove that the time of flight to B is $(2V\sin\theta)/(g\cos\alpha)$.
The coefficient of restitution between the particle and the plane is e and when the particle strikes the plane at B it rebounds in a vertical direction. Show that $e = \cot\theta\cot\alpha - 2$.

5) An inclined plane makes an angle α with the horizontal. A particle is projected up the slope from a point on the plane with a velocity V at an angle θ to the plane, the direction of projection being in the vertical plane containing a line of greatest slope. Show that the time of flight is $2V\sin\theta/(g\cos\alpha)$ and find the range of the particle along the plane.
If the particle is travelling horizontally when it strikes the plane, show that

$$\tan\theta = \frac{\sin\alpha\cos\alpha}{1+\sin^2\alpha}$$

6) A plane is inclined at an angle α to the horizontal and OA is a line of greatest slope. A particle P is projected at $20\ \mathrm{m\,s^{-1}}$ from O, in the vertical plane containing OA, at an angle β to the inclined plane. If $\tan\alpha = \frac{5}{12}$ and $\tan\beta = \frac{3}{4}$ find the greatest height of P above the inclined plane during its flight when
(a) P is projected up the plane, (b) P is projected down the plane.

7) A particle P is projected under gravity from a point A on a plane inclined at an angle α to the horizontal. The velocity of projection is V at $\arctan\frac{1}{2}$ to an upward line of greatest slope of the plane and the motion takes place in the vertical plane through a line of greatest slope. When P hits the plane at a point B, it is travelling at an angle α to the horizontal. Find the time taken for the particle to travel from A to B in terms of V, g and a. Find also the possible values of $\tan\alpha$.

8) From the top of a vertical cliff 30 m high a particle P is projected with speed $10\ \mathrm{m\,s^{-1}}$. Show that the maximum distance from the foot of the cliff that P can reach is $10\sqrt{7}$ m, assuming that the ground at the foot of the cliff is horizontal.

SUMMARY

There are several considerations that should be borne in mind when solving problems concerning projectiles, some of which are covered in *Mechanics and Probability*.

A summary of the main points is set out below.

1. While a projectile is in flight it has a constant acceleration g vertically downward.
2. When a particle is projected from a point on a plane, the maximum range on that plane is achieved when the angle of projection bisects the angle between the slope of the plane and the vertical. In the special case of a horizontal plane this angle of projection becomes 45°.
3. The choice between horizontal and vertical axes, or axes parallel and perpendicular to an inclined plane is helped by remembering that inclined axes are usually better for problems involving impact or distance above the plane.
4. Velocity and displacement components in the chosen directions should be derived afresh for each problem, starting with the appropriate acceleration components.

MULTIPLE CHOICE EXERCISE 5

(The instructions for answering these questions are given on page ix.)

TYPE I

1) A particle is projected from a point on a plane which is inclined at 20° to the horizontal. The vertical plane containing the path of the projectile also contains a line of greatest slope of the plane. For the maximum range down the plane the angle of projection (measured from the horizontal) is:
(a) 60° (b) 80° (c) 55° (d) 35° (e) 45°.

2) A smooth sphere is moving on a horizontal surface with velocity vector $3\mathbf{i}+\mathbf{j}$ immediately before it hits a vertical wall. The wall is parallel to the vector $\mathbf{j}$ and the coefficient of restitution between the wall and sphere is $\frac{1}{3}$. The velocity vector of the sphere after it hits the wall is:
(a) $\mathbf{i}+\mathbf{j}$ (b) $3\mathbf{i}-\frac{1}{3}\mathbf{j}$ (c) $-\mathbf{i}+\mathbf{j}$ (d) $\mathbf{i}-\mathbf{j}$ (e) $-\mathbf{i}-\frac{1}{3}\mathbf{j}$.

3) A particle which is projected from a point on an inclined plane strikes the plane and rebounds. After impact the particle retraces its original path. The coefficient of restitution between the particle and the plane is:
(a) 0 (b) -1 (c) $\frac{1}{2}$ (d) 1 (e) $\frac{1}{3}$.

4) 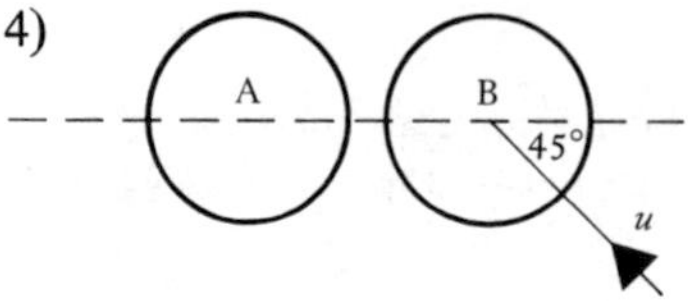

The diagram shows the velocities just before collision of two smooth spheres of equal radius and mass. The impact is perfectly elastic. The velocities just after impact are:

(a)

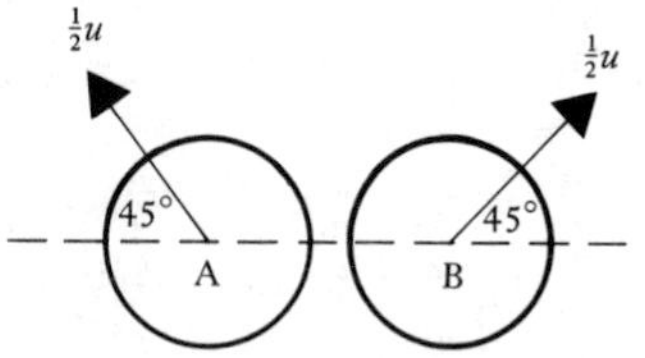

(b)

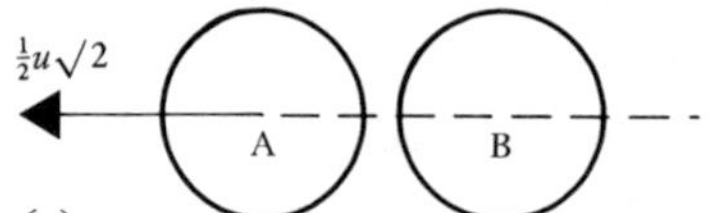

(c)

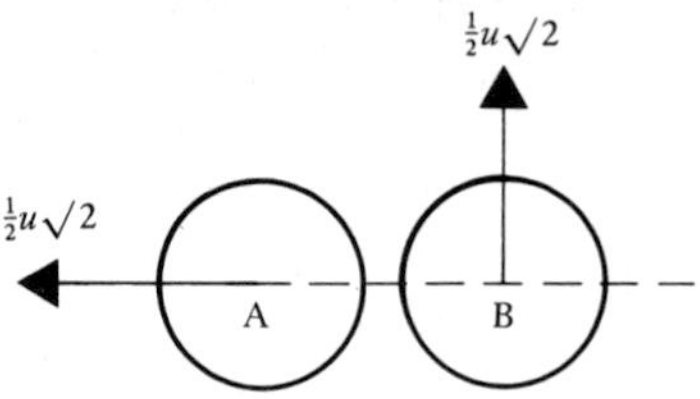

(d)

TYPE II

5) A smooth sphere moving on a horizontal surface collides indirectly with a vertical wall. The coefficient of restitution between the sphere and the wall is $e\ (<1)$.

(a) The momentum of the sphere perpendicular to the wall is unchanged by the impact.
(b) The component of velocity parallel to the wall before impact is equal to the component of velocity parallel to the wall after impact.
(c) There is no loss in the kinetic energy of the sphere due to the impact.
(d) If u is the component of velocity perpendicular to the wall before impact and v is the component of velocity in the same direction after impact then $eu + v = 0$.

6) The diagram shows the vertical plane containing the path of a particle which is projected from a point O on an inclined plane.

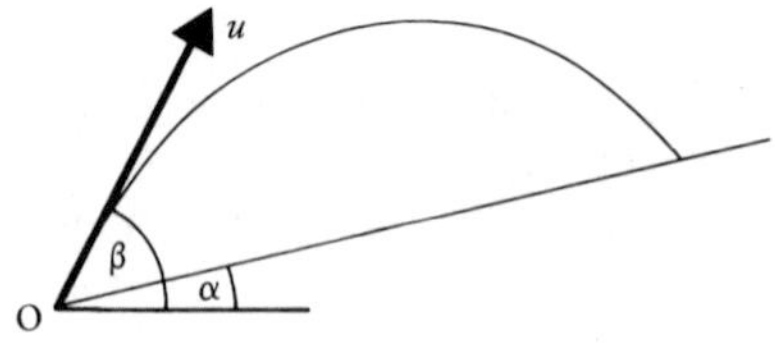

(a) The path of the projectile is a parabola.
(b) If T is the time of flight, the particle is at its greatest height above the plane after an interval $\frac{1}{2}T$.
(c) If the range on the plane is maximum then $\beta = \frac{1}{4}\pi + \frac{1}{2}\alpha$.
(d) The kinetic energy of the particle is maximum as it leaves O.

7) Two spheres A and B of equal mass are free to move on a smooth horizontal surface. A and B move towards each other with velocity vectors $a\mathbf{i}+b\mathbf{j}$ and $c\mathbf{i}+d\mathbf{j}$ respectively and collide when the line joining their centres is parallel to $\mathbf{i}$. After impact A and B have velocity vectors $p\mathbf{i}+q\mathbf{j}$ and $r\mathbf{i}+s\mathbf{j}$ respectively. The coefficient of restitution between the spheres is e (<1).
(a) $b=q$ (b) $c=r$ (c) $a+c=p+r$ (d) $ea=p$.

8) From a point O on a plane inclined at α to the horizontal a particle is projected down the plane with initial velocity u at an angle β to the horizontal. Taking axes as shown:

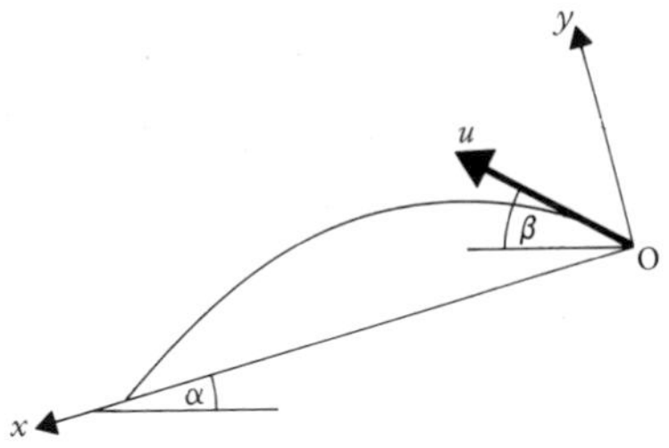

(a) $\ddot{x}=-g\sin\alpha$ (b) $\ddot{y}=-g\cos\alpha$ (c) $\ddot{x}=g\cos\alpha$ (d) $\ddot{y}=g\sin\alpha$.

TYPE III

9) (a) A smooth sphere moving on a horizontal surface strikes a vertical wall indirectly.
(b) A smooth sphere moving on a horizontal surface strikes a vertical wall and rebounds in a direction perpendicular to the wall.

10) Two smooth spheres A and B move towards each other on a horizontal surface and collide.
(a) A and B move directly towards each other.
(b) A is brought to rest by the impact.

11)

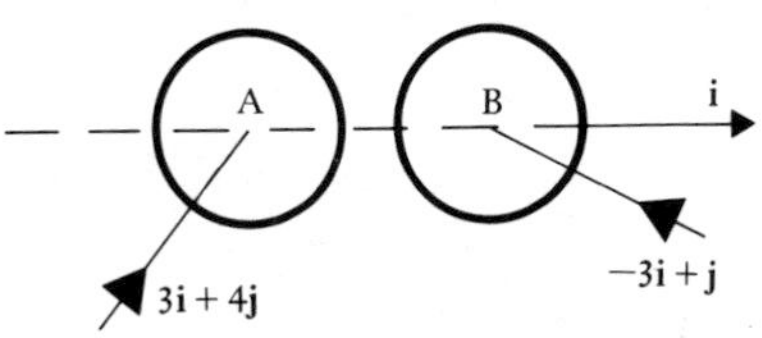

Two spheres A and B of equal radius and mass move towards each other with velocities as shown and collide when the line joining their centres is parallel to $\mathbf{i}$.

(a) $e=\frac{1}{2}$.
(b) The velocities of A and B after impact are $-\frac{3}{2}\mathbf{i}+4\mathbf{j}$ and $\frac{3}{2}\mathbf{i}+\mathbf{j}$ respectively.

TYPE IV

12) A particle is projected from a point O on an inclined plane. Find the maximum range of the particle up the plane.
(a) The plane is inclined at 20° to the horizontal and a line of greatest slope lies in the vertical plane containing the path of the projectile.
(b) The initial speed of the particle is $30\ \text{m s}^{-1}$.
(c) The mass of the particle is 0.2 kg.

13) Two smooth spheres A and B of equal radius move towards each other on a horizontal table and collide. Find the velocity of each sphere after impact.
(a) The initial velocities of A and B are $3\mathbf{i}+2\mathbf{j}$ and $2\mathbf{i}-3\mathbf{j}$ respectively.
(b) When A and B collide the line joining their centres is parallel to $\mathbf{i}+\mathbf{j}$.
(c) The coefficient of restitution between A and B is $\frac{1}{2}$.

14) A particle is projected from a point O on a smooth inclined plane and hits the plane again at a point A and rebounds. Find the distance from O of the point at which the particle hits the plane for the third time.
(a) OA is inclined at 15° to the horizontal with A higher than O.
(b) The initial velocity of the particle is $40\ \text{m s}^{-1}$ at 40° to the horizontal.
(c) Immediately after hitting the plane at A the velocity of the particle is inclined at an angle of 20° to the horizontal.

15) Two smooth spheres are moving on a horizontal table and collide. Find the coefficient of restitution between the spheres.
(a) The spheres are of equal mass and each has a radius a.
(b) Before impact the centres of the spheres are moving along parallel paths distant $\frac{1}{2}a$ apart.
(c) After impact the angle between their paths is 30°.

16) A particle is projected from a point O on an inclined plane and hits the plane again at a point A. Find the angle of projection of the particle.
(a) OA is inclined at α to the horizontal with A lower than O.
(b) The distance OA is l.
(c) The speed of the particle as it hits the plane at A is u.

TYPE V

17) When a perfectly elastic impact occurs between a moving object and a fixed object there is no loss in kinetic energy.

18) When two objects, both of which are free to move, collide and coalesce their total momentum is unchanged by the collision.

19) When two objects which are both free to move, collide, their total mechanical energy is always unchanged by the collision.

MISCELLANEOUS EXERCISE 5

1) A particle is launched at time $t = 0$ so that it follows the path

$$\mathbf{r} = (-15+5t)\mathbf{i} + (70+30t-5t^2)\mathbf{k}$$

(Distances are measured in metres and time in seconds.)
Find
(a) the position vector of the point of projection and the velocity vector of the particle at time $t = 0$.
(b) the speed of the particle at time t.
The particle meets the line through the point with position vector $5\mathbf{i}$ and gradient $\mathbf{i}+19\mathbf{k}$ at the point A. Show that this occurs when $t = 5$. Find the position vector of the point A and its distance from the point of projection. Find also the time at which the particle is moving at right angles to its initial direction of motion. (AEB)

2) Two particles of equal mass are projected at the same instant and with the same speed $\sqrt{(gl)}$ from points A and B, distant l apart, on a line of greatest slope of a plane inclined at $30°$ to the horizontal. A is at a level higher than B. The particle at A is projected horizontally towards B and the particle at B is projected at an angle of $60°$ above the horizontal towards A. Prove that the particles will collide and that if they coalesce, the combined mass will begin to move in a direction inclined at $30°$ below the horizontal. (U of L)

3) A particle is projected with speed u from a point on a plane which is inclined at an angle β to the horizontal. The particle is projected at an angle α to the horizontal in a vertical plane through a line of greatest slope of the plane. Show that the range up the plane is

$$\frac{2u^2 \sin(\alpha-\beta)\cos\alpha}{g\cos^2\beta}$$

Deduce that the range up the plane is a maximum when $\alpha = \frac{1}{4}\pi + \frac{1}{2}\beta$.
If the *maximum* range down the plane is twice the maximum range up the plane, find the angle β. (U of L)

4) A particle is projected with velocity V, at an angle of elevation α to the horizontal, from a point on a plane inclined at angle $\beta\,(<\alpha)$ to the horizontal. The path of the particle is in a vertical plane through a line of greatest slope of the plane. If R_1 and R_2 are the respective maximum ranges when the particle is fired up the plane and down the plane, show that

(a) $R_1 = \dfrac{V^2}{g(1+\sin\beta)}$, (b) $\sin\beta = \dfrac{R_2-R_1}{R_2+R_1}$. (AEB)

5) A perfectly elastic particle is projected from a point O on a fixed smooth plane which is inclined at an angle α to the horizontal. The velocity of projection is in the vertical plane through the line of greater slope through O and makes an angle θ with the upwards line of slope. By using coordinate axes along the perpendicular to this line, or otherwise, prove that, if $\tan\theta = \frac{1}{2}\cot\alpha$, the particle will retrace its path to O after bouncing on the plane. (O)

6) A particle, projected with speed u at an angle α to the horizontal, moves freely under gravity. Find $\dot{x}$, $\dot{y}$, the horizontal and vertical components of its velocity, and x, y, its horizontal and vertical displacements respectively, at any time $t \geqslant 0$ during the motion.

Show that $2\dfrac{y}{x} - \dfrac{\dot{y}}{\dot{x}} = \tan\alpha$ throughout the motion.

A particle is projected at an angle α to the horizontal from the foot of an inclined plane so as to strike the inclined plane. It moves in the vertical plane containing the line of greatest slope, which is inclined at $\tan^{-1}(\frac{1}{2})$ to the horizontal.

(a) Find α if the particle strikes the inclined plane at right angles.
(b) If $\alpha = \tan^{-1} 2$, at what angle to the horizontal would it be travelling just before impact? (WJEC)

7) A ball is projected under gravity from a point A on an inclined plane which makes an angle α to the horizontal. The velocity of projection is V at an angle θ to an upward line of greatest slope of the plane. The motion takes place in the vertical plane through a line of greatest slope and the particle hits the plane at a point which is further up the plane than A. The coefficient of restitution between the plane and the ball is e. Show that the ball first hits the plane after a time $\dfrac{2V\sin\theta}{g\cos\alpha}$ and find the time taken between the first and second bounce.

Prove that the point of second impact of the ball and the plane will be higher than the first impact if $\tan\alpha\tan\theta < \dfrac{1}{2+e}$.

If on the second bounce the ball is moving perpendicular to the plane, find e in terms of α and θ. (AEB)

8) A particle is projected at an angle α to the horizontal from a point A on a plane inclined at an angle $\arctan(1/2)$ to the horizontal. The particle moves in a vertical plane through a line of greatest slope and strikes the plane again at a point B higher up the plane. Find $\tan\alpha$

(a) if the particle is moving horizontally at B,
(b) if the particle strikes the plane normally at B. (U of L)

9) A particle P is projected horizontally, with speed V, from a point O on a plane which is inclined at an angle β to the horizontal. The particle hits the plane at a point A which is on the line of greatest slope through O. Show that the time of flight is

$$\frac{2V}{g}\tan\beta.$$

Find the tangent of the acute angle between the horizontal and the direction of motion of P when P reaches A.
A second particle Q is projected from O, with speed V, in a direction perpendicular to the plane. Find the time taken for Q to return to the plane and show that Q hits the plane at A. (JMB)

10) A vehicle is moving with constant acceleration kg up a slope of inclination α, the floor of the vehicle being parallel to the slope. Show that if a particle is falling freely inside the vehicle its acceleration relative to the vehicle makes an angle β with the floor, where

$$\cot\beta = k\sec\alpha + \tan\alpha$$

A particle is projected inside the vehicle from a point A on the floor, its initial velocity relative to the vehicle being V at an angle $\theta\ (>\beta)$ with the floor, as shown in the diagram. It strikes the ceiling at B, where AB is perpendicular to the floor and $AB = h$. By considering motion parallel to the slope, show that the time of flight is

$$\frac{2V\cos\theta}{g\cos\alpha\cot\beta}$$

and deduce that $$h = \frac{2V^2\cos\theta\tan\beta\sin(\theta-\beta)}{g\cos\alpha\cos\beta}$$

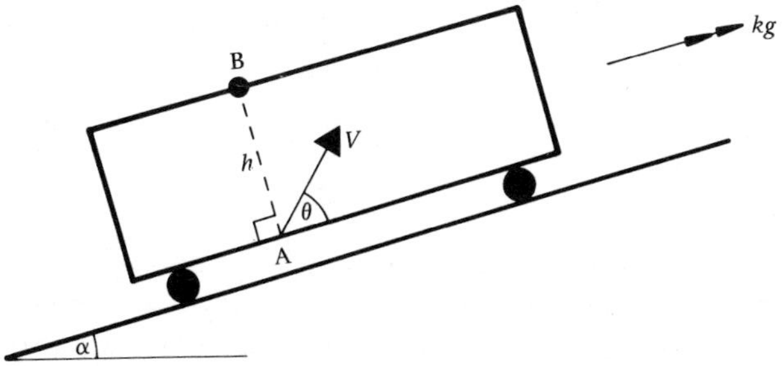

(C)

11) A particle is projected under gravity from a point O up a plane which is inclined at an angle β to the horizontal. The velocity of projection V makes an angle α with the plane, and the motion takes place in a vertical plane through a line of greatest slope. Given that axes Ox and Oy are taken along and perpendicular to a line of greatest slope respectively, find the coordinates of the position of the particle at time t. Deduce the time of flight of the particle, and show that its range up the inclined plane can be expressed as

$$\frac{V^2 [\sin(2\alpha+\beta)-\sin\beta]}{g\cos^2\beta}$$

Hence find the maximum value of the range as α varies.
When the range of the particle up the plane is $(\sqrt{3}-1)$ times its maximum range and $\beta=\frac{1}{6}\pi$, find the two values of α with which this can be achieved.

(AEB)

12) A particle is projected with speed V from a point on a plane which is inclined to the horizontal at an angle α. The direction of projection is in a vertical plane containing a line of greatest slope of the inclined plane, and makes an angle θ with that line.

(a) Prove that, as θ varies, the range up the slope is a maximum when

$$\theta = \tfrac{1}{4}\pi-\tfrac{1}{2}\alpha$$

(b) Prove that, if the particle strikes the plane at right angles above the point of projection,

$$\tan\theta = \tfrac{1}{2}\cot\alpha$$

(U of L)

13) An elastic ball is projected with speed V from a point O of a smooth plane inclined at an angle θ to the horizontal. The initial direction of projection is up the plane, makes an angle $\theta+\phi$ with the horizontal and lies in a vertical plane containing a line of greatest slope of the plane. If the ball ceases to bounce at the instant when it returns to O, show that

$$\tan\theta\tan\phi = 1-e$$

(U of L)

14) A smooth plane is inclined at an angle β to the horizontal, where $\beta<\pi/4$. A particle is projected with velocity $\sqrt{(ga)}$ at an angle 2β with the horizontal from a point O of the plane. The vertical plane containing the path of the particle also contains a line of greatest slope of the inclined plane and the particle moves towards the upper part of the plane. Show that the particle strikes the plane at a distance $(2a\cos 2\beta\sin\beta)/\cos^2\beta$ from O. (U of L)

15) A particle is projected with speed V at an angle θ to the horizontal from the point $x = 0$, $y = 0$. If the x and y axes are horizontal and vertically upwards respectively, show from first principles that the equation of the path of the particle is

$$y = x \tan\theta - \frac{gx^2}{2V^2}(1 + \tan^2\theta)$$

The point of projection is the lowest point on the inner surface of a bowl formed by rotating the curve $x^2 = 4ay$, where a is a positive constant, about the y axis. Show that the particle strikes the bowl at a horizontal distance

$$x = \frac{4aV^2 \tan\theta}{V^2 + 2ag + 2ag\tan^2\theta}$$

from the origin.
Find the greatest horizontal range of the particle in the bowl for a constant value of V as θ varies. (AEB)

CHAPTER 6

MOTION IN A STRAIGHT LINE. HARMONIC MOTION. MOTION OF A BODY WITH VARIABLE MASS

The way in which a particle moves depends on several factors, viz. the nature of the forces acting on the particle, the initial conditions (i.e. initial velocity, position, etc.) and the mass of the particle, which may or may not be constant. The relationship between the applied forces and the motion of the body is expressed in Newton's second law of motion.
For a body of constant mass this law is used in the form $\mathbf{F} = m\mathbf{a}$.

The acceleration $\mathbf{a}$ can be expressed in a variety of ways.
For instance, if a particle is moving in a straight line Ox,

a can be expressed as $\ddot{x}, \ \dfrac{d\dot{x}}{dt}, \ \dfrac{dv}{dt}$ or $v\dfrac{dv}{dx}$

THE BASIC EQUATION OF MOTION

When we apply the appropriate form of Newton's second law to analyse the motion of a particle under the action of known forces, we derive a differential equation which is called the *basic equation of motion* of the particle.

MOTION OF A PARTICLE IN A STRAIGHT LINE

Consider a particle of constant mass m moving along a straight line Ox in the direction Ox under the action of a constant force mk directed towards O and a resisting force mkv^2 where v is the speed of the particle.

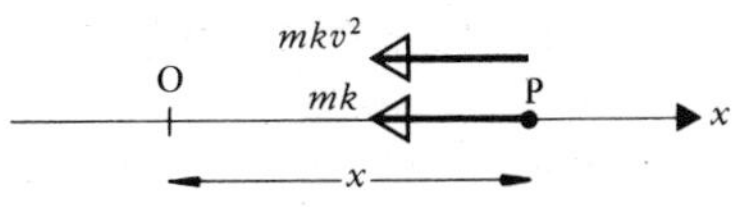

Forces acting

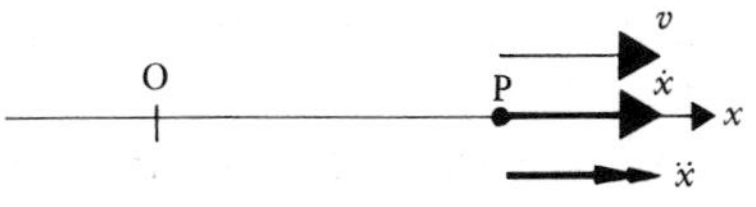

Velocity and acceleration

When the particle is at P, where its displacement from O is x, the resultant force acting on it in the direction Ox is

$$-(mk + mk\dot{x}^2) \qquad (v = \dot{x})$$

Using Newton's second law of motion gives

$$-(mk + mk\dot{x}^2) = m\ddot{x}$$

or

$$\ddot{x} + k\dot{x}^2 + k = 0 \qquad [1]$$

This is the basic equation of motion for the particle, and it is a second order differential equation in x and t as it has a term in $\frac{d^2x}{dt^2}$.

Before the motion can be analysed further, a complete solution of this equation must be found which involves two stages of integration and therefore two constants of integration. These constants can be evaluated if two facts are known about the conditions of the particle at some given time (e.g. the values of x and $\dot{x}$ when $t = 0$).

Note that this equation of motion is valid only when the particle is moving away from O in the sense Ox,
e.g. when the particle is moving leftwards towards O, $v = -\dot{x}$ and the resisting force mkv^2 acts away from O.

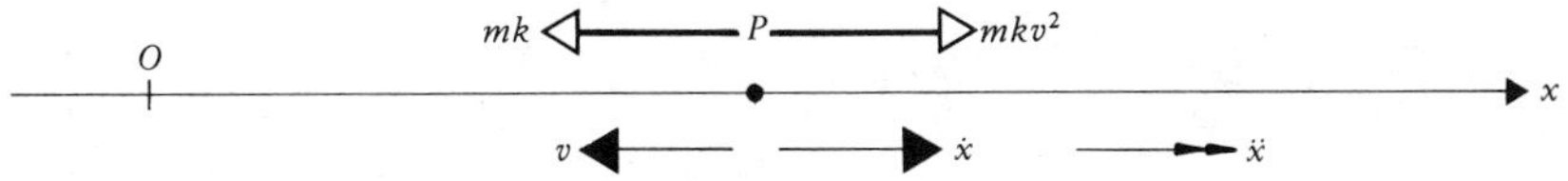

In this case the basic equation of motion is

$$mk(-\dot{x})^2 - mk = m\ddot{x} \quad \Rightarrow \quad \ddot{x} - k\dot{x}^2 + k = 0$$

which is different from equation [1] above.

When an equation of motion is derived for a particular direction, it should not be used for the opposite direction (or the opposite side of O) without checking its validity. Sometimes the equation remains valid but in other cases, as in the example above, different equations are required.

In many problems where a particle is moving in a straight line under the action of known forces, the basic equation of motion is a differential equation. The solution of these equations will not always be shown in full in the text since the methods available for solving such differential equations can be found in standard text books on Pure Mathematics.

Note. Motion in a straight line is sometimes called rectilinear motion. Care must be taken not to confuse the term 'linear motion' with motion in a straight line. Linear motion means motion in a line, which may or may not be straight.

EXAMPLES 6a

1) A particle moves in a straight line Ox under the action of a force which, acting alone, would give it a constant acceleration of $5\ \mathrm{m\,s^{-2}}$ away from O. The particle actually moves in a medium which retards it at the rate of $2v$ when its speed is v. Derive the equation of motion for this particle.

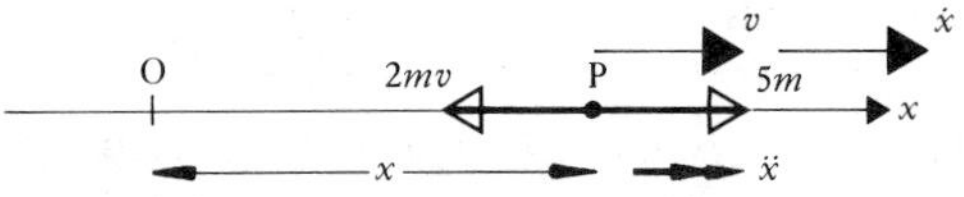

Let the mass of the particle be m

Hence the accelerating force $= 5m$

and the retarding force $= -2m\dot{x}$

Therefore the resultant force in the direction Ox is

$$5m - 2m\dot{x}$$

and $5m - 2m\dot{x} = m\ddot{x}$ $\qquad (\mathbf{F} = m\mathbf{a})$

so the equation of motion is $\ddot{x} + 2\dot{x} - 5 = 0$

Note that this equation is valid when the particle is moving in either sense along Ox. This is because, when the direction of motion changes the direction of the retarding force $2mv$ changes, but simultaneously v becomes $-\dot{x}$. Thus the retarding force remains as $-2m\dot{x}$ in the direction Ox.

2) An engine of mass M is accelerating on the level with a constant power H, against resistances which are proportional to the square of its speed. Find the equation of motion of the engine.

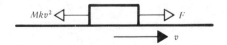

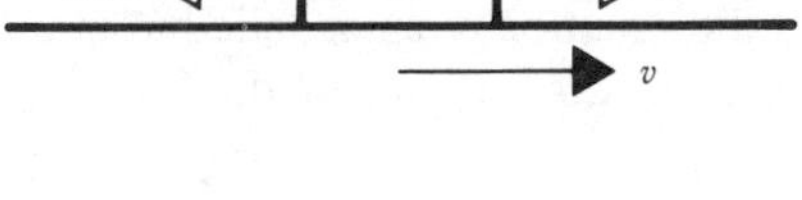

When the speed of the engine is v, the driving force F is $\dfrac{H}{v}$

Therefore the resultant force on the engine is $\dfrac{H}{v} - Mkv^2$

Using Newton's second law of motion gives $\dfrac{H}{v} - Mkv^2 = M\dfrac{dv}{dt}$

This is the equation of motion of the engine.

3) A particle moves along a straight line Ox such that its displacement from O at time t is x where

$$\frac{d^2x}{dt^2} + 3\frac{dx}{dt} + 6 = 0$$

Describe a possible physical situation which would give rise to such an equation of motion.

If, when $t = 1$, $\dfrac{dx}{dt} = 2$ and $x = 0$ find the speed when $t = 0$.

Sketch the graph of x against t and hence give a brief description of the motion of the particle.

Rearranging the equation of motion gives $\ddot{x} = -3\dot{x} - 6$

or $-3m\dot{x} - 6m = m\ddot{x}$

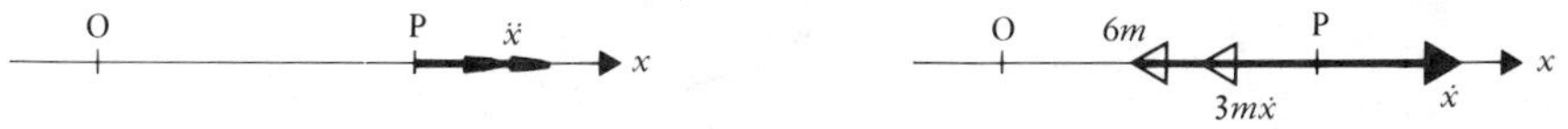

Thus, from Newton's second law of motion, the resultant force acting on the particle is $-3m\dot{x} - 6m$. This is made up of a constant force in the direction xO and a force directly proportional to the speed of the particle in the direction opposite to the direction of motion.

One physical situation to which this equation of motion applies could be: a particle which, with an initial velocity in the direction Ox, is acted on by a constant braking force and is moving in a medium which offers a resistance proportional to its speed. (No doubt the reader can think of many other alternatives which would give rise to the given equation of motion.)

Writing $\dot{x}$ as v and $\ddot{x}$ as $\dfrac{dv}{dt}$, the equation of motion becomes

$$\frac{dv}{dt} + 3v + 6 = 0$$

Separating the variables gives

$$\int_2^v \frac{dv}{v+2} = \int_1^t -3\,dt$$

Therefore $$\ln\left(\frac{v+2}{4}\right) = -3t+3$$

or $$v = 4e^{3-3t}-2$$

Therefore when $t = 0$, $$v = 4e^3-2$$

Replacing v by $\frac{dx}{dt}$ in [1] gives

$$\frac{dx}{dt} = 4e^{3-3t}-2$$

Integrating again with respect to time we have,

$$\Big[x\Big]_0^x = \Big[-\tfrac{4}{3}e^{3-3t} - 2t\Big]_1^t$$

$\Rightarrow$ $$x = \tfrac{10}{3} - \tfrac{4}{3}e^{3-3t} - 2t$$

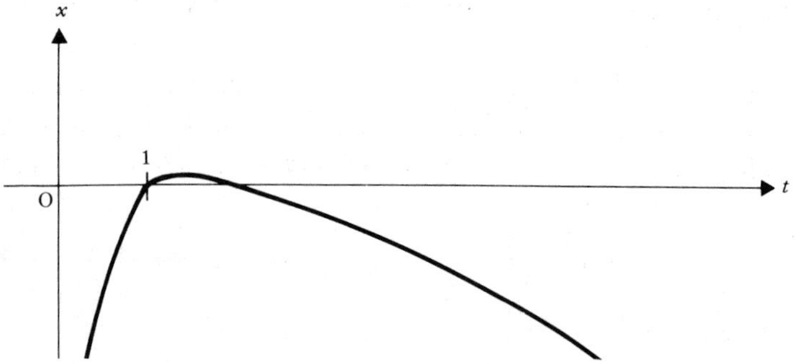

From the graph we see that the particle initially moves in the positive sense along Ox, passing through O when $t = 1$ and returns to O, continuing thereafter to move in the negative sense along Ox.

EXERCISE 6a

(Take $g = 10\,\mathrm{m\,s^{-2}}$ when necessary.)

In questions 1 to 4 derive the equation of motion from the physical situation described but do not solve the equation.

1) A particle of mass m is moving along a straight line Ox in the direction Ox under the action of two forces, one proportional to its displacement from O and directed towards O and the other a resistance which is constant.

2) A particle of mass m is projected vertically upward under gravity in a medium which offers a resistance kv when the speed is v.

3) A particle of mass m is attached to one end of a light elastic string of natural length a and modulus mg, the other end of the string is attached to a fixed point O on a smooth horizontal surface. The particle is released from rest at a horizontal distance $a+l$ from O, the subsequent motion taking place in a medium which offers a resistance mkv when the speed of the particle is v.

4) A vehicle of mass M is moving on a level road with constant power H against a resistance of Mkv^2 when the speed is v.

In questions 5 to 7 suggest a possible physical situation which would result in the following equations of motion.

5) $\dfrac{d^2x}{dt^2}+k\dfrac{dx}{dt} = 0$

6) $\dfrac{d^2x}{dt^2}+n^2x = 0$

7) $\dfrac{d^2x}{dt^2}+kv+g = 0$

8) A particle of mass m moves along a straight line Ox under the action of a constant force $4m$ towards O and a retarding force of $3mv$ when the speed is v. If when $t=0$, $x=0$ and $\dot{x}=3$ find the displacement from O when the particle first comes to rest.

9) A particle P moves in a straight line on a smooth horizontal plane. The initial velocity of P is V and the resisting force is kvm where m is the mass of P and v is its speed. Find v at any subsequent time t.

10) The equation of motion of a particle moving on the x axis is

$$\ddot{x}+2k\dot{x}+n^2x = 0$$

If k and n are positive constants, with $k<n$, find the time that elapses between two successive maximum values of $|x|$ and show that this time is constant.

11) A particle of unit mass moves along the line Ox under the action of a force $(9+v^2)$ in the direction Ox where v is the speed of the particle at time t. If $v=0$ when $t=0$, find v when $t=\frac{1}{12}\pi$.

12) A particle P moves with speed v in a straight line on a smooth horizontal table against a resistance of kv^2 per unit mass, where k is a constant. After travelling a distance s the initial speed V is reduced to $\frac{1}{2}V$. Find k in terms of s.

13) A particle of mass m slides from rest down a straight groove inclined at 30° to the horizontal against resistances which are equal to $5mv^2$ when the speed of the particle is v. Derive the equation of motion of this particle, taking x as the displacement from the initial position. Find the speed of the particle after it has been moving for $\frac{1}{5}$ s and show that the particle has a terminal velocity of $1\ \text{m s}^{-1}$.

14) A car of mass 1000 kg is accelerating on a level road with constant power of 25 kW against resistances of magnitude ten times the speed of the car. Derive the equation of motion and find the distance covered when the car accelerates from rest to a speed of $25\ \text{m s}^{-1}$.

HARMONIC MOTION

There are many forms of oscillatory motion, the most straightforward being simple harmonic motion. This type of motion is analysed in *Mechanics and Probability* but a reminder of its properties is given here.

(a) A particle whose acceleration is directed towards a fixed origin and is proportional to the distance from that origin, moves with S.H.M.

(b) The basic equation of S.H.M. is

$$\ddot{x} = -n^2x \quad \text{or} \quad \ddot{\theta} = -n^2\theta$$

(c) Further relationships, which can usually be quoted, are:

$$v^2 = n^2(a^2 - x^2) \quad \text{where } a \text{ is the amplitude}$$

$$x = a\cos nt$$

$$T = \frac{2\pi}{n} \quad \text{where } T \text{ is the periodic time.}$$

(d) The motion of a particle moving in a resistance-free medium on a stretched elastic string can be shown to be S.H.M. during the time that the string is taut. Attached to an oscillating spring however a particle always performs S.H.M. since the spring never becomes slack.

(e) Remember that some problems about particles moving on elastic strings or springs can be solved without using the equations of S.H.M. When considering only velocity and position, the principle of conservation of mechanical energy provides the best solution when there is no resisting force.

The reader is recommended to revise this work before progressing further.

DAMPED HARMONIC MOTION

Simple harmonic motion is much less likely to occur naturally than a more familiar type of oscillatory motion is that performed by, for example, the end of a tuning fork after it has been struck. The amplitude of this motion gets progressively smaller and the oscillations eventually cease. This is an example of a form of damped oscillations.

Forces Causing Damped Oscillations

When the resultant force acting on a particle is proportional to the displacement from a fixed point and is directed towards that fixed point, the particle performs S.H.M. If, in addition, there is a force resisting the motion of the particle, the oscillations may be damped or the nature of the motion may be completely changed. This resisting force can take many forms, but we shall confine our analysis to resisting forces that are directly proportional to the speed of the particle.

The Equation of Motion

Consider a particle of mass m moving along a straight line Ox. The forces acting on the particle when its displacement from O is x are

$$\begin{cases} \text{a force } mn^2x \text{ directed towards O,} \\ \text{a resisting force } 2mkv, \end{cases}$$

where v is the speed of the particle, and k and n are constants.

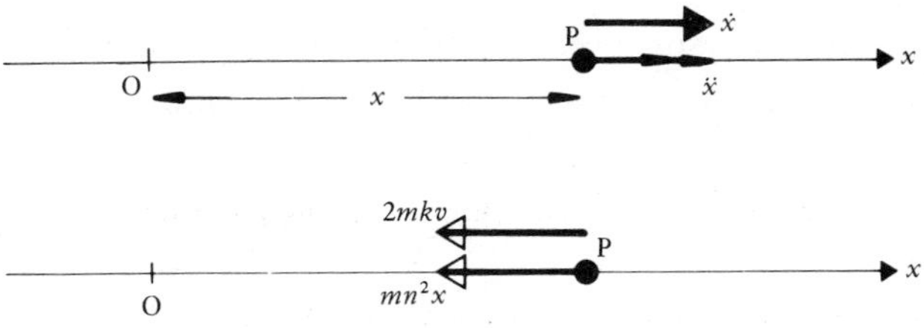

The resultant force in the direction Ox is $-mn^2x - 2mk\dot{x}$

Newton's second law of motion gives

$$m\ddot{x} = -mn^2x - 2mk\dot{x}$$

i.e.

$$\frac{d^2x}{dt^2} + 2k\frac{dx}{dt} + n^2x = 0 \qquad [1]$$

This is the basic equation of motion for any particle moving in a straight line under the action of the forces described above.

To analyse the motion more fully, equation [1] must be integrated. It is a second order linear differential equation and its solution depends on the nature of the roots of the equation

$$p^2 + 2kp + n^2 = 0 \quad \Rightarrow \quad p = -k \pm \sqrt{k^2 - n^2} \qquad [2]$$

(1) *Real distinct roots* i.e. $k^2 > n^2$.
The complete solution of [1] is

$$x = Ae^{-kt + t\sqrt{(k^2 - n^2)}} + Be^{-kt - t\sqrt{(k^2 - n^2)}} \qquad [3]$$

(2) *Equal roots* i.e. $k^2 = n^2$.
In this case the complete solution of [1] has the form

$$x = e^{-kt}(A + Bt) \qquad [4]$$

(3) *Complex roots* i.e. $k^2 < n^2$.
In this case the complete solution of [1] can be expressed in one of two forms,

either $$x = Ce^{-kt}\cos\left[t\sqrt{(n^2 - k^2)} + \epsilon\right] \qquad [5]$$

or $$x = e^{-kt}\left[A\cos t\sqrt{n^2 - k^2} + B\sin t\sqrt{n^2 - k^2}\right]$$

where A, B, C, ϵ are constants of integration.

(In this context, the first of these forms is more useful.)

(For a full analysis of the solution of second order linear differential equations refer to a standard text book on Pure Mathematics.)

As equations [3], [4] and [5] represent the three possible complete solutions to equation [1] we can now determine what form the motion takes in each of these three cases.

CASE 1. $x = Ae^{-kt + t\sqrt{(k^2 - n^2)}} + Be^{-kt - t\sqrt{(k^2 - n^2)}}$

There is not more than one value of t for which $x = 0$ and not more than one value for t for which $\dot{x} = 0$, therefore this motion is not oscillatory.

CASE 2. $x = e^{-kt}(A + Bt)$

Again this motion is not oscillatory as there is only one finite value of t for which $x = 0$, and only one finite value of t for which $\dot{x} = 0$.

CASE 3. $x = Ce^{-kt}\cos[t\sqrt{(n^2-k^2)}+\epsilon]$.

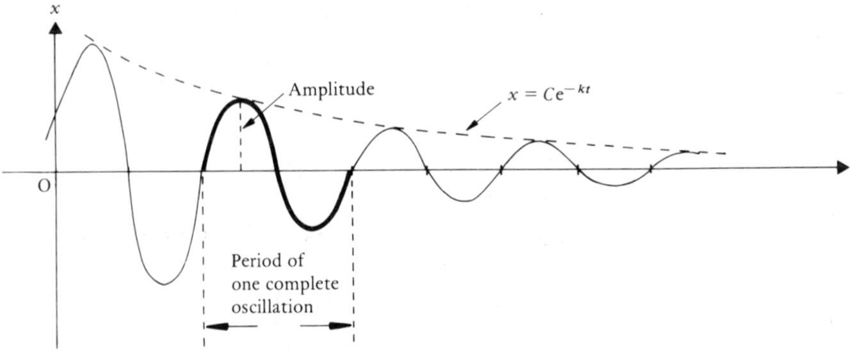

This motion is oscillatory and is called *damped harmonic motion*.
The oscillations die away as t increases. The equation $x = Ce^{-kt}$ forms an upper boundary for the curve and e^{-kt} is called the damping factor.

One complete oscillation is the path travelled by the particle in the interval of time between passing in the same direction through O on successive occasions.

The *period of an oscillation* is the time taken to complete one oscillation.

The *amplitude of an oscillation* is the maximum value of x attained in that oscillation.

Properties of Damped Harmonic Motion

In order to investigate some of the properties of D.H.M. we will consider a case where the constants c, k and n are given numerical values.
Suppose that a particle moves along a straight line Ox such that its displacement x from O at time t is given by

$$x = 2e^{-t}\cos(3t+\tfrac{1}{3}\pi) \qquad [1]$$

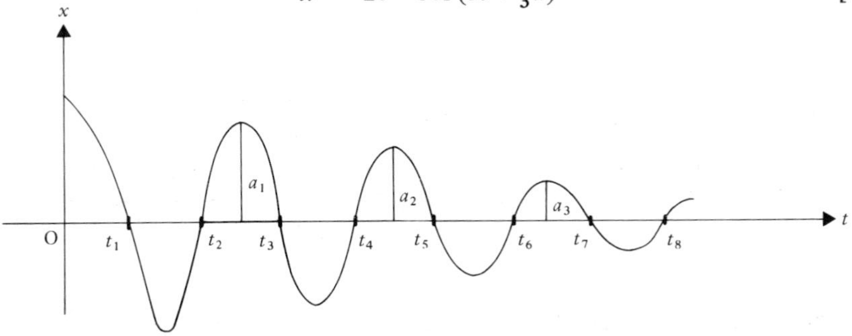

We will take complete oscillations as the paths travelled by the particle in the intervals of time from t_2 to t_4, t_4 to t_6 etc. (We could equally well use the intervals of time t_1 to t_3 etc.)

Period of the Oscillations

When $x = 0$, $\cos(3t+\frac{1}{3}\pi) = 0$

i.e. $$3t = (2N-1)\tfrac{1}{2}\pi - \tfrac{1}{3}\pi \qquad (N = 1, 2, 3, \ldots)$$

Thus the particle passes through O at times

$$\tfrac{1}{3}(\tfrac{1}{2}\pi-\tfrac{1}{3}\pi),\quad \tfrac{1}{3}(\tfrac{3}{2}\pi-\tfrac{1}{3}\pi),\quad \tfrac{1}{3}(\tfrac{5}{2}\pi-\tfrac{1}{3}\pi),\quad \ldots$$

Hence the successive times when the particle passes through O are separated by *equal intervals* of $\frac{1}{3}\pi$ seconds.

Therefore the *period of these oscillations is constant* and equal to $\frac{2}{3}\pi$ seconds.

Amplitude

The stationary values of x occur when $\dot{x} = 0$ and, from equation [1],

$$\dot{x} = -2e^{-t}\cos(3t+\tfrac{1}{3}\pi) - 6e^{-t}\sin(3t+\tfrac{1}{3}\pi)$$

Therefore, when $\dot{x} = 0$ $\tan(3t+\frac{1}{3}\pi) = -\frac{1}{3}$

$$\Rightarrow \qquad \cos(3t+\tfrac{1}{3}\pi) = \pm\frac{1}{\sqrt{10}}$$

Hence $$3t+\tfrac{1}{3}\pi = N\pi - \arctan\tfrac{1}{3}$$

$$\Rightarrow \qquad 3t = N\pi-\alpha \quad \text{where} \quad \alpha = \tfrac{1}{3}\pi+\arctan\tfrac{1}{3}$$

Taking $N = 1, 2, 3, 4, 5, \ldots$ gives the values of t for alternating maximum and minimum values of x.

When $N = 1, 3, 5, \ldots$ is substituted in equation [1],

$\cos(3t+\frac{1}{3}\pi) = -\dfrac{1}{\sqrt{10}}$ so the value of x is negative.

For $N = 2, 4, 6, \ldots$ $\cos(3t+\frac{1}{3}\pi) = +\dfrac{1}{\sqrt{10}}$ so x has positive, and therefore maximum, values of $\dfrac{2}{\sqrt{10}}e^{-t}$.

Therefore successive values of the amplitude are

$$a_1 = \frac{6}{\sqrt{10}}e^{-(2\pi-\alpha)/3},\quad a_2 = \frac{6}{\sqrt{10}}e^{-(4\pi-\alpha)/3},\quad \ldots,\quad a_n = \frac{6}{\sqrt{10}}e^{-(2n\pi-\alpha)/3}$$

From this we see that $$\frac{a_{n+1}}{a_n} = e^{-2/3\pi}$$

i.e. the ratio of the amplitudes of successive oscillations is constant.

Therefore the *amplitude of successive oscillations decreases with time in geometric progression* of common ratio e^{-T} where T is the period of oscillations.

The properties which this example demonstrates are, in fact, general properties of damped harmonic motion. This can be proved from the general equation for this type of motion, $x = Ce^{-t} \cos [t\sqrt{(n^2 - k^2)} + \epsilon]$, giving the results in the summary on P.

FORCED HARMONIC MOTION

A further type of oscillatory motion results from the action of two forces where

one force would, alone, produce S.H.M.,
the other force acts at all times in the direction of motion.

Consider a particle P of mass m, moving in a straight line Ox under the action of two forces such that, when its displacement from O is x, one force is mn^2x towards O and the other force is $2kmv$ in the direction of motion.

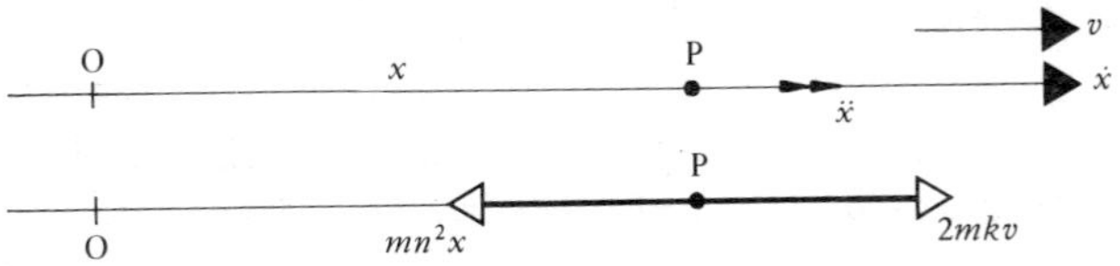

When the particle is moving in the direction Ox, the equation of its motion is

$$2mk\dot{x} - mn^2x = m\ddot{x} \quad \Rightarrow \quad \ddot{x} - 2k\dot{x} + n^2x = 0$$

(A check will show that this equation is valid for motion in either direction.)

When this differential equation is solved we find that if $k^2 \geqslant n^2$ the motion is not oscillatory (see p. 115).
If $k^2 < n^2$ however, the solution is

$$x = A\,e^{kt} \cos [t\sqrt{(n^2 - k^2)} + \epsilon]$$

where A and ϵ are constants.

This motion is oscillatory but the amplitude of successive oscillations *increases* with time.

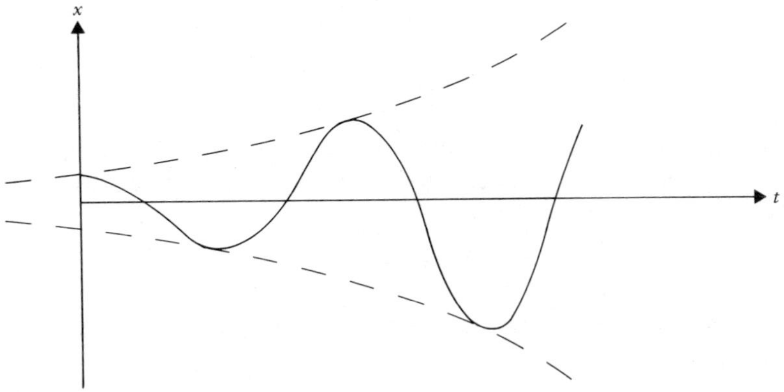

These oscillations of increasing amplitude are known as *forced vibrations* and they occur in *forced harmonic motion*.

In practice the aim is usually to avoid setting up forced vibrations, as an object oscillating further and further from the centre position is likely to destroy itself. For instance, it is popularly supposed that when troops are crossing a suspension bridge they are ordered to stop marching and to walk 'out of step', to avoid setting up forced vibrations in the bridge.

Mathematically, forced harmonic motion involves the same methods as were used for damped harmonic motion. The solutions of the two equations of motion differ only in the sign of the exponential index.

SUMMARY

When a particle of mass m moves in a straight line Ox under the action of a force mn^2x directed *towards* O and a resisting force $2mkv$, the equation of motion is

$$\frac{d^2x}{dt^2} + 2k\frac{dx}{dt} + n^2x = 0$$

Provided that $k^2 < n^2$, the complete solution of this equation can be given in the form

$$x = Ce^{-kt}\cos[t\sqrt{(n^2-k^2)}+\epsilon]$$

where C and ϵ are constants of integration.
This equation represents damped harmonic oscillations, and e^{-kt} is called the damping factor. The period of these oscillations is constant and equal to

$$\frac{2\pi}{\sqrt{(n^2-k^2)}}$$

The amplitude of the oscillations *decreases* with time in geometric progression.

If $k^2 \geqslant n^2$, the motion is not oscillatory.

If the force $2mkv$ is not a resisting force but acts instead in the direction of motion, forced harmonic motion results when $k^2 < n^2$, and we have oscillatory motion for which

$$x = Ce^{kt}\cos[t\sqrt{(n^2-k^2)}+\epsilon]$$

These oscillations are of constant period but the amplitude *increases* in geometric progression.

EXAMPLES 6b

1) A particle moving in a straight line Ox has a displacement x from O at time t where x satisfies the equation,

$$\frac{d^2x}{dt^2}+4\frac{dx}{dt}+13x = 0$$

Show that the motion of the particle is oscillatory. If, when $t=0$, $x=3$ and $\dot{x}=-6$, show that the particle oscillates about O with a constant period.

Comparing $$\frac{d^2x}{dt^2}+4\frac{dx}{dt}+13x = 0 \qquad [1]$$

with $$\frac{d^2x}{dt^2}+2k\frac{dx}{dt}+n^2x = 0$$

we see that $k=2$ and $n^2=13$, i.e. $k^2<n^2$.

Therefore the complete solution of equation [1] is

$$x = Ce^{-2t}\cos(3t+\epsilon) \qquad \text{where } C \text{ and } \epsilon \text{ are constants.} \qquad [2]$$

Therefore equation [1] is the equation of motion of an oscillating particle.

Now, when $t=0$, $x=3$ therefore $3 = C\cos\epsilon$ [3]

Also $\dot{x} = -2Ce^{-2t}\cos(3t+\epsilon)-3Ce^{-2t}\sin(3t+\epsilon)$

When $t=0$, $\dot{x}=-6$, so $-6 = -2C\cos\epsilon-3C\sin\epsilon$ [4]

[4] ÷ [3] gives $-2 = -2-3\tan\epsilon \Rightarrow 3\tan\epsilon = 0$

Hence $\epsilon=0$ and $C=3$

Therefore equation [2] becomes

$$x = 3e^{-2t}\cos 3t$$

When $x=0$,

$$3e^{-2t}\cos 3t = 0$$

$$\Rightarrow \qquad \cos 3t = 0$$

$$\Rightarrow \qquad 3t = (2N+1)\tfrac{1}{2}\pi$$

Hence $$t = (2N+1)\tfrac{1}{6}\pi, \quad N = 1, 2, 3, \ldots$$

This shows that the particle passes through O at successive times separated by equal intervals of $\frac{1}{3}\pi$ s, therefore the particle oscillates about O with a constant period of $\frac{2}{3}\pi$ s.

2) A particle of mass m is attached to one end of a light elastic string of natural length l and modulus $2mg$. The other end of the string is attached to a fixed point O. With the string vertical, the particle is pulled down a distance l below the equilibrium position and then released. The subsequent motion takes place in a medium which offers a resistance $mv\sqrt{(g/l)}$ when the speed is v. Show that the particle performs damped harmonic motion about the equilibrium position without the string going slack.

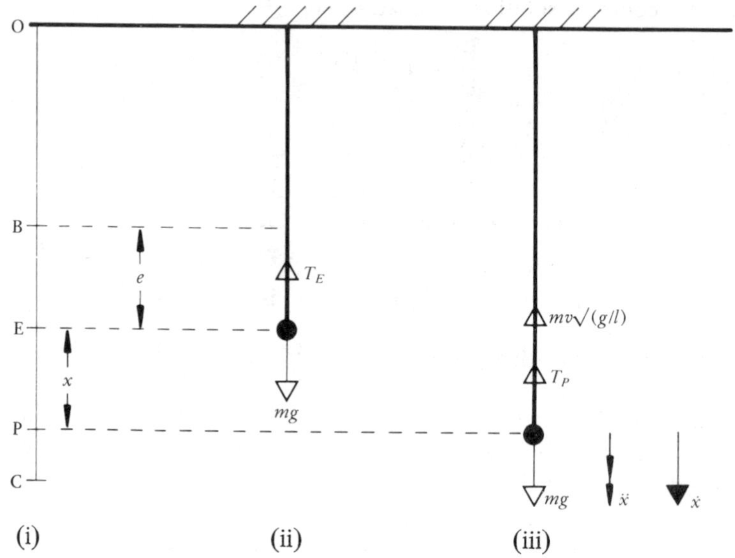

In diagram (i) OB is the natural length of the string,
E is the equilibrium position of the particle,
P is a general position with displacement x from E,
C is the initial position of the particle.

In diagram (ii) $$T_E = \frac{2mge}{l} \qquad \text{(Hooke's law)}$$

As the particle is in equilibrium, $T_E = mg$ and $e = \frac{1}{2}l$

In diagram (iii), where the extension is $\frac{1}{2}l + x$,

$$T_P = \frac{2mg}{l}(\tfrac{1}{2}l + x)$$

The resultant downward force on the particle is

$$mg - \frac{2mg}{l}(\tfrac{1}{2}l + x) - m\sqrt{\frac{g}{l}}\,\dot{x}$$

Applying Newton's law in this direction gives

$$mg - \frac{2mg}{l}(\tfrac{1}{2}l + x) - m\sqrt{\frac{g}{l}}\,\dot{x} = m\ddot{x}$$

$$\Rightarrow \qquad \ddot{x} + \sqrt{\frac{g}{l}}\,\dot{x} + \frac{2g}{l}x = 0 \qquad [1]$$

Comparing with $\ddot{x} + 2k\dot{x} + n^2x = 0$ we see that $k^2 < n^2$.
Therefore the complete solution of equation [1] is

$$x = Ce^{-t\surd(g/4l)}\cos[t\surd(7g/4l) + \epsilon] \qquad [2]$$

where C and ϵ are constants of integration.
The motion of this particle is therefore damped harmonic motion and, provided that the string does not go slack, the period is $2\pi\sqrt{\dfrac{4l}{7g}}$.

Now the particle is initially released from rest at C, where $t = 0$ and $x = l$.

Using these values in equation [2] gives $l = C\cos\epsilon$.

We also know that the particle is at its lowest position when it is about to be released from C; therefore the first maximum value of x is l, and this occurs when $t = 0$.

(If the string is to remain taut in the subsequent motion, the greatest height of the particle above E must not exceed $\frac{1}{2}l$; i.e. the magnitude of the next stationary value of x must not exceed $\frac{1}{2}l$.)

The next stationary value of x occurs when half the periodic time has elapsed, i.e. when $t = \pi\sqrt{\dfrac{4l}{7g}}$.

At this time,

$$x = -Ce^{-\pi/\sqrt{7}}\cos\epsilon$$
$$= -(e^{-\pi/\sqrt{7}})(l) \qquad (C\cos\epsilon = l)$$

Therefore the maximum height above E to which the particle rises is h where

$$h = \frac{l}{e^{\pi/\sqrt{7}}} = \frac{l}{3.27}$$

i.e.

$$h < \tfrac{1}{2}l$$

Therefore the string does not go slack in the subsequent motion.

3) A particle P, of mass 1 kg, is attached to the midpoint of an elastic string of natural length 2 m and modulus of elasticity 20 N. The ends of the string are attached to two points A and B, distant 6 m apart on a smooth horizontal plane. Initially P is released from rest at a point between A and B, and distant 2 m from A. Immediately a force, equal in magnitude to four times the speed, begins to act in the direction of motion. If the string breaks when the tension in it exceeds 90 N, show that the string breaks before P comes to instantaneous rest.

Consider the forces acting when P is distant x from O, the midpoint of AB.

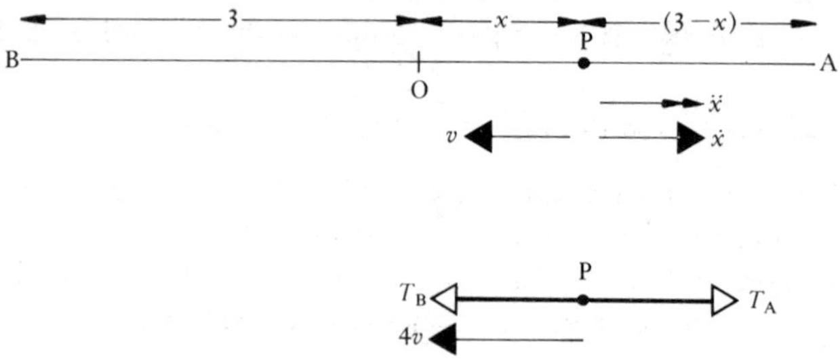

While P moves from A to B, $v = -\dot{x}$

For the half-string BP, $T_B = \frac{20}{1}(2+x)$

and for the half-string AP, $T_A = \frac{20}{1}(2-x)$

Newton's law of motion gives

$$T_A - T_B - 4v = \ddot{x} \qquad (m = 1)$$

$$\Rightarrow \quad 20(2-x) - 20(2+x) + 4\dot{x} = \ddot{x}$$

$$\Rightarrow \quad \ddot{x} - 4\dot{x} + 40x = 0$$

The solution of this equation is

$$x = Ae^{2t}\cos\left[t\sqrt{40-4} + \epsilon\right]$$

and this is the equation of forced harmonic motion.

Now $x = 1$ when $t = 0$ so $A\cos\epsilon = 1$.

Also $\dot{x} = 2Ae^{2t}\cos[6t+\epsilon] - 6Ae^{2t}\sin[6t+\epsilon]$

and $\dot{x} = 0$ when $t = 0$ giving $\tan\epsilon = \frac{1}{3}$.

Hence $\cos\epsilon = \dfrac{3}{\sqrt{10}}$ and $A = \frac{1}{3}\sqrt{10}$.

(P will continue to move towards B until either $\dot{x}=0$ or the string snaps.)

If $\dot{x}=0$, $\tan(6t+\epsilon) = \frac{1}{3}$

$\Rightarrow$ $6t+\epsilon = n\pi+\arctan\frac{1}{3}$

$\Rightarrow$ $6t = n\pi$ $(\epsilon = \arctan\frac{1}{3})$

The first value of t, after $t=0$, to satisfy this equation is $t=\frac{1}{6}\pi$.

For this value of t, $\cos(6t+\epsilon) = -\dfrac{3}{\sqrt{10}}$

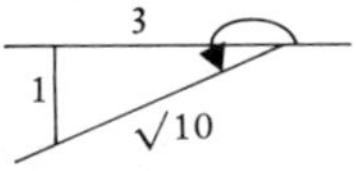

therefore $x = (\frac{1}{3}\sqrt{10})(e^{\pi/3})\left(-\dfrac{3}{\sqrt{10}}\right) = -e^{\pi/3} = -2.85$

i.e. the particle has passed through O and is 2.85 m from O.

If the particle should reach this position, the length of the half-string AP would be 5.85 m and the tension in it would be $\frac{20}{1}\times 4.85$ N.
This tension is greater than the breaking tension of 90 N therefore, before P could reach its first position of instantaneous rest, the string AP would break.

EXERCISE 6b

1) A particle moves along a straight line Ox such that its displacement x from O at time t is given by:

(a) $\ddot{x}+3\dot{x}+4x=0$

(b) $\dfrac{d^2x}{dt^2}+4\dfrac{dx}{dt}+4=0$

(c) $\ddot{x}+2\sqrt{\dfrac{g}{l}}\dot{x}+\dfrac{3g}{l}x=0$

(d) $\ddot{x}+9x=0$

(e) $\ddot{x}-3\dot{x}+x=0$

(f) $\dfrac{d^2x}{dt^2}-\dfrac{dx}{dt}+3x=0$

Determine in each case whether the motion is harmonic; if it is, state whether it is simple, damped, or forced harmonic motion and write down the period of oscillation.

2) A particle moves along a straight line Ox such that at time t its displacement from O is x where

$$\ddot{x} + 2\dot{x} + 3x = 0$$

Show that the particle performs damped harmonic motion. If when $t = 0$, $x = 0$ and $\dot{x} = 3$, find the value of x when $t = 2$.

3) The equation of motion for a particle P moving in a straight line Ox is

$$\ddot{x} - 4\dot{x} + 9x = 0$$

where x is the displacement of P from O at time t.
Show that P performs forced harmonic motion.
If $x = 1$ and $\dot{x} = 1$ when $t = 0$ find the value of x when $t = 1$.

4) A particle moves along a straight line Ox such that its displacement x from O at time t is given by

$$x = 2e^{-t}\cos 3t$$

Prove that the particle performs oscillations about O with a constant period and that the amplitude of the motion decreases in geometric progression. Find the common ratio of this geometric progression.

5) A light elastic string of natural length $2a$ and modulus mg has a particle of mass m attached to its midpoint. One end of the string is attached to a point A and the other end to a point B, both on a smooth horizontal table where $AB = 4a$. The particle is held at a point C where ACB is a straight line and $AC = a$, and released. The subsequent motion takes place in a medium which offers a resistance $mv\sqrt{g/2a}$ when the speed of the particle is v. Derive the equation of motion of this particle and determine whether the motion is damped harmonic.

6) A light elastic string of natural length $2a$ and modulus $2mg$ has a particle of mass m attached to its midpoint. One end of the string is fixed to a point A, the other end is fixed to a point B, a distance $4a$ vertically below A. Find the equilibrium position of the particle.
The particle is held at a distance $3a$ below A and released from rest in a medium which offers a resistance of $mv\sqrt{g/a}$ when the speed of the particle is v.
Show that the particle performs damped harmonic motion about the equilbrium position and find the maximum height to which the particle rises above the equilibrium position.

7) One end of a light elastic string of natural length a and modulus $4mn^2a$ is attached to a fixed point A on a smooth horizontal plane. The other end is attached to a particle of mass m placed on the plane at a distance a from A. The particle is suddenly given a velocity V directly away from A and thereafter is subject to an air resistance of $2mnv$, where v is the speed. Show that, before the string slackens, its extension at any time t after the motion begins is $\dfrac{u}{k\sqrt{3}}\,\mathrm{e}^{-kt}\sin k\sqrt{3}t$ and find the greatest extension.

8) One end of a light inextensible string of length l is fixed and a particle of mass m is attached to the other end. This simple pendulum performs small oscillations in a fixed vertical plane and the air resistance is $2mk$ times the speed of the particle, where k is a positive constant. Show that, if powers of θ higher than the first are neglected, the equation of motion of the particle is

$$\frac{\mathrm{d}^2\theta}{\mathrm{d}t^2} + 2k\frac{\mathrm{d}\theta}{\mathrm{d}t} + \frac{g\theta}{l} = 0$$

where θ is the inclination of the string to the downward vertical at time t.

MOTION OF A PARTICLE WITH VARYING MASS

Newton's second law of motion states that, when an external force is applied to a body, the rate of increase of momentum produced is directly proportional to the applied force.

In the S.I. system of units, the constant of proportion is unity so the rate of increase in momentum is equal to the force which produces it,

i.e. $$\mathbf{F} = \frac{\mathrm{d}}{\mathrm{d}t}(m\mathbf{v})$$

So far this law has been used in the form $\mathbf{F} = m\mathbf{a}$ but this applies only to a body of constant mass.

Some objects however have variable mass. A rocket, for example, loses mass by ejecting burnt fuel; a particle in a saturated vapour can gain mass as moisture condenses on it. A more general form of Newton's law is required in such cases.

Consider a body moving so that, at time t, its mass is m, its velocity is $\mathbf{v}$ and the resultant force acting on the body is $\mathbf{F}$.
Suppose that, at this instant, a particle of mass δm and velocity $\mathbf{u}$ coalesces with the body so that, at time δt, the body has mass $m+\delta m$ and velocity $\mathbf{v}+\delta\mathbf{v}$.

Just before the particle coalesces with the body,

the momentum of the body is $m\mathbf{v}$

and the momentum of the particle is $\delta m\mathbf{u}$.

Immediately afterwards,

the momentum of the augmented body is $(m+\delta m)(\mathbf{v}+\delta\mathbf{v})$.

Thus the increase in momentum in the time interval δt is

$$(m+\delta m)(\mathbf{v}+\delta\mathbf{v}) - m\mathbf{v} - \delta m\mathbf{u}$$

The impulse of the force acting on the body in the same time interval is approximately $\mathbf{F}\delta t$.

Therefore $$\mathbf{F}\delta t \simeq m\delta\mathbf{v} + (\mathbf{v}-\mathbf{u})\delta m + \delta m\,\delta\mathbf{v}$$

As $\delta t \to 0$ this equation becomes

$$\mathbf{F} = m\frac{\mathrm{d}\mathbf{v}}{\mathrm{d}t} + (\mathbf{v}-\mathbf{u})\frac{\mathrm{d}m}{\mathrm{d}t}$$

If the mass increment has no velocity of its own before becoming part of the body, i.e. $\mathbf{u}=\mathbf{0}$, the equation becomes $\mathbf{F} = m\frac{\mathrm{d}\mathbf{v}}{\mathrm{d}t} + \mathbf{v}\frac{\mathrm{d}m}{\mathrm{d}t}$.

This equation is also obtained directly from $\mathbf{F} = \frac{\mathrm{d}}{\mathrm{d}t}(m\,\mathbf{v})$.

Note that δm can be negative if a body loses mass by ejecting matter from itself, e.g. a rocket burning fuel.

EXAMPLES 6c

1) A body with initial mass M is projected vertically upwards with speed g/k where k is a constant. At any subsequent time t, its speed is v, and its mass is Me^{kt}. The particles of added mass were all at rest when picked up by the body. Use Newton's second law of motion for a body of varying mass to show that,

$$\frac{\mathrm{d}}{\mathrm{d}t}(Mve^{kt}) = -Mge^{kt}$$

Hence find the mass of the body at its highest point.

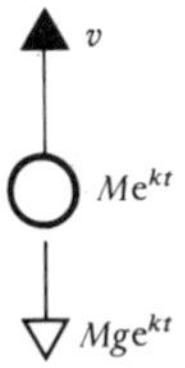

The momentum of the body at time t is $(Me^{kt})(v)$, and the force acting on the body is Mge^{kt} downwards.

Newton's law, $\mathbf{F} = \frac{d}{dt}(m\mathbf{v})$, gives

$$-Mge^{kt} = \frac{d}{dt}(Mve^{kt}) \qquad [1]$$

The initial momentum is Mg/k and the momentum at the highest point is zero, therefore if $t = T$ at the highest point, integrating equation [1] gives

$$-\int_0^T Mge^{kt}\,dt = \left[Mve^{kt}\right]^0_{\text{momentum} = Mg/k}$$

$$\Rightarrow \qquad \frac{Mg}{k}(1-e^{kT}) = 0-\frac{Mg}{k} \qquad \Rightarrow \qquad e^{kT} = 2$$

The mass is then $Me^{kT} = 2M$.

2) A rocket, with an initial mass of 1000 kg, is launched vertically upwards from rest under gravity. The rocket burns fuel at the rate of 10 kg per second. The burnt matter is ejected vertically downwards with a speed of $2000\,\text{m s}^{-1}$ relative to the rocket. If burning ceases after one minute find the maximum velocity of the rocket.
(Take g as constant at $10\,\text{m s}^{-2}$.)

At time t, taking the upward direction as positive,

the mass of the rocket is $(1000-10t)$

the force acting on the rocket is $-(1000-10t)g$

the velocity of the rocket is v

the velocity of the fuel ejected is $-(2000-v)$

the rate of increase of mass is -10.

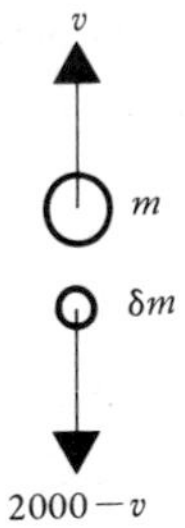

Using $F = m\frac{dv}{dt} + (v-u)\frac{dm}{dt}$ gives

$$-(1000-10t)g = (1000-10t)\frac{dv}{dt} + (2000)(-10)$$

$$\Rightarrow \qquad -(1000-10t) = (100-t)\frac{dv}{dt} - 2000 \qquad (g = 10)$$

Initially $t = 0$ and $v = 0$; after 60 seconds $v = v_{max}$

therefore $$\int_0^{60} \frac{1000+10t}{100-t}\,dt = \int_0^{v_{max}} dv$$

$$\Rightarrow \qquad v_{max} = 10\int_0^{60} \frac{-100+t+200}{100-t}\,dt$$

$$= 10\Big[-t-200\ln|100-t|\Big]_0^{60}$$

$$= -600-2000\ln\tfrac{40}{100}$$

$$= 2000\ln 2.5-600$$

The maximum velocity of the rocket is $200(10\ln 2.5-3)\,\text{m s}^{-1}$.

EXERCISE 6c

1) A rocket is moving vertically upward against gravity. Its mass at time t is $(M-mt)$ and it expels burnt fuel at a speed u vertically downward relative to the rocket. Derive the equation of motion of the rocket but do not solve it.

2) A rocket of initial mass M has a mass $M(1-\frac{1}{3}t)$ at time t. The rocket is launched from rest vertically upwards under gravity and expels burnt fuel at a speed u relative to the rocket vertically downward. Find the speed and height above the launching pad when $t = 1$.

3) A small object P, of mass m_0, is projected vertically upward with speed $\sqrt{2gh}$ and, while ascending, P picks up moisture from the atmosphere; the drops of moisture are at rest when picked up. When P is at a height x above the point of projection its mass is $m_0(1+\lambda x)$ when λ is a constant. Prove that H, the greatest height reached, is given by

$$1+\lambda H = (1+3\lambda h)^{1/3}$$

MISCELLANEOUS EXERCISE 6

1) A particle of mass m moves along a horizontal straight line under the action only of a resisting force of magnitude mv^2/a, where v is its speed and a is a positive constant. Given that the particle is projected from a point O at time $t = 0$ with speed u, show that, when it is at a distance x from O, its speed is $ue^{-x/a}$. (U of L)

2) A particle is projected vertically upwards with speed U in a medium in which the resistance to motion is proportional to the square of the speed. Given that U is also the speed for which the resistance offered by the medium is equal to the weight of the particle, show that the time of ascent is $\pi U/(4g)$ and that the distance ascended is $[U^2/(2g)]\ln 2$. (U of L)

3) A particle P of mass m moves in a medium which produces a resistance of magnitude mkv, where v is the speed of P and k is a constant. The particle P is projected vertically upwards in this medium with speed g/k. Show that P comes momentarily to rest after time $(\ln 2)/k$.
Find, in terms of k and g, the greatest height above the point of projection reached by P. (U of L)

4) A ball of mass m is projected vertically downwards under a constant gravitational force of magnitude g per unit mass. The air resistance produces a force of magnitude kv per unit mass opposing this motion, where k is a positive constant and v is the speed. Given that the initial speed is u, obtain the differential equation satisfied by v.
Show that, whatever the value of u, the speed of the ball tends to the limiting value g/k.
Find the distance travelled by the ball in time T. (U of L)

5) A particle of mass m moves under gravity and against a resistance $\mathbf{R}$, where $\mathbf{R} = -mk\mathbf{v}$, $\mathbf{v}$ being the velocity of the particle and k a positive constant. At time t the particle is at the point (x, y) with respect to axes Ox. Oy drawn horizontally and vertically upwards respectively from O. Show that x satisfies the differential equation

$$\frac{d^2x}{dt^2} = -k\frac{dx}{dt}$$

and find the corresponding equation for y.
Given that at time $t = 0$ the particle is at O and moving with speed V at an angle α to the horizontal, find x and y in terms of g, k, V, α and t. Given that $V\sin\alpha > 0$, show that the particle reaches its maximum height after a time

$$\frac{1}{k}\ln\left(1 + \frac{kV\sin\alpha}{g}\right)$$ (JMB)

6) A particle moves in a straight line under a constant force kV^2 per unit mass in the direction of motion and a resistive force kv^2 per unit mass, where V, k are positive constants and v is the speed after time t. The particle starts from rest at O at time $t = 0$. Show that the distance of the particle from O at time t is $(1/k)\ln[\cosh(kVt)]$.
Show further that the speed v is always less than V. (U of L)

7) A particle of mass m is projected vertically upwards under gravity in a medium which exerts a resisting force of magnitude $mg(v/a)^2$, where v is the speed of the particle and a is a constant. If U is the speed of projection, show that the greatest height of the particle above the point of projection is

$$\frac{a^2}{2g}\ln\left(\frac{a^2+U^2}{a^2}\right)$$

If V is the speed of the particle on returning to the point of projection, show that

$$\frac{1}{V^2}-\frac{1}{U^2} = \frac{1}{a^2}$$ (U of L)

8) A particle is projected vertically upwards with speed u from a point O. The air resistance is proportional to the fourth power of the velocity and is initially equal to the weight of the particle. If v is the speed of the particle when it is at a height z above O, show that

$$\frac{d}{dz}(\tfrac{1}{2}v^2)+g\left(1+\frac{v^4}{u^4}\right) = 0$$

Solve this equation by writing w for v^2 and so prove that the particle is instantaneously at rest when its height above O is $\pi u^2/8g$. (WJEC)

9) A ship of mass M moves from rest under a constant propelling force Mf and against a resistance of magnitude Mkv^2, where v is the speed of the ship and k is a constant. Show that, when the ship has travelled a distance b, its speed is given by

$$kv^2 = f(1-e^{-2kb})$$

The engines are then reversed, thus producing a retarding force Mf in addition to the resistance Mkv^2. Show that the additional distance c travelled before the ship comes to rest is given by

$$e^{2kc}+e^{-2kb} = 2$$ (U of L)

10) A ship, of mass m, is propelled in a straight line through the water by a propeller which develops a constant force of magnitude F. When the speed of the ship is v, the water causes a drag, of magnitude kv, where k is a constant, to act on the ship. The ship starts from rest at time $t = 0$. Show that the ship reaches half its theoretical maximum speed of F/k when $t = (m\ln 2)/k$. When the ship is moving with speed $F/(2k)$, an emergency occurs and the captain reverses the engines so that the propeller force, which remains of magnitude F, acts backwards. Show that the ship covers a further distance

$$\frac{mF}{k^2}\left[\frac{1}{2}-\ln\left(\frac{3}{2}\right)\right]$$

on its original course, which may be assumed to remain unchanged, before being brought to rest. (U of L)

11) A particle moves under gravity in a medium which offers a resistance of magnitude kw per unit mass, where w is the speed of the particle and k is a positive constant. The particle is projected from the origin O at time $t = 0$ with horizontal and vertically-upward components of velocity u_0 and v_0, respectively. Show that the horizontal and vertical displacements of the particle from O at time t are given respectively by

$$x = \frac{u_0}{k}(1-e^{-kt})$$

$$y = \left(\frac{g}{k^2}+\frac{v_0}{k}\right)(1-e^{-kt}) - \frac{gt}{k}$$

Find the Cartesian equation of the trajectory of the particle. (JMB)

12) A particle of mass m is projected vertically upwards with speed u in a medium which exerts a resisting force of magnitude mkv, where v is the speed of the particle and k is a positive constant. Find the time taken to reach the highest point and show that the greatest height attained above the point of projection is

$$\frac{1}{k^2}\left[uk - g\ln\left(1+\frac{ku}{g}\right)\right]$$

Find, in terms of k, g and T, the speed of the particle at time T after it has reached its greatest height, and hence, or otherwise, show that this speed tends to a finite limit as T increases indefinitely. (U of L)

13) A particle falls from rest under the action of gravity from a height h above a horizontal plane. The resistance to the motion of the particle is kv^2 per unit mass, where v is its speed and k is a positive constant. The coefficient of restitution between the particle and the ground is λ. Show that the maximum height H reached on the first rebound is given by

$$2kH = \ln\left[1+\lambda^2(1-e^{-2kh})\right]$$

Show also that, if kh is small, $\lambda \approx \sqrt{(H/h)}$. (U of L)

14) The equation of motion of a particle moving on the x axis is

$$\ddot{x} + 2k\dot{x} + n^2x = 0$$

Given that k and n are positive constants, with $k < n$, show that the time between two successive maxima of $|x|$ is constant. (U of L)

15) A particle of unit mass is moving in a horizontal straight line so that its displacement x from a fixed point of the line at time t satisfies the differential equation

$$4\frac{d^2x}{dt^2} + 12\frac{dx}{dt} + 13x = 0$$

Describe a physical situation which could give rise to such an equation of motion. Solve the equation given that $x = 4$ and $dx/dt = 6$ when $t = 0$. Show from the form of the solution that the motion is oscillatory with a constant period, and that the amplitudes of successive oscillations decrease in geometric progression. (U of L)

16) A particle of mass m is suspended from a fixed point by a spring of natural length l and modulus $5mn^2l$. When in motion it is resisted by a force of magnitude $2mn$ times its speed. Initially the particle is hanging in equilibrium and it is then projected vertically downwards with speed V. If x is the displacement downwards at time t from the equilibrium position, show that

$$\frac{d^2x}{dt^2} + 2n\frac{dx}{dt} + 5n^2x = 0$$

Find x as a function of t and sketch the graph of this function. Show that the particle is instantaneously at rest when $nt = \frac{1}{2}k\pi + \alpha$, where k is a non-negative integer and α is the acute angle such that $\tan 2\alpha = 2$. (General solutions of differential equations may be quoted.) (JMB)

17) A particle of unit mass is tied to one end of a light elastic string, of natural length a and modulus $2an^2$, and the other end of the string is attached to a fixed point O. The particle is released from rest at a point a distance $2a$ vertically below O. When the particle is moving with velocity v, the air resistance is $2nv$. Prove that, when the extension of the string is x,

$$\frac{d^2x}{dt^2} + 2n\frac{dx}{dt} + 2n^2x = g$$

Prove also that the particle first comes to rest before the string becomes slack provided that

$$2an^2 < g(e^{\pi} + 1)$$

(O)

18) A particle moves in a vertical plane under gravity and a resistance whose magnitude per unit mass is k times the speed of the particle, where k is a positive constant. The coordinates of the particle at time t, with respect to a horizontal axis Ox and a vertically upward axis Oy, are (x, y).
Show that

$$\frac{d^2x}{dt^2} = -k\frac{dx}{dt}, \qquad \frac{d^2y}{dt^2} = -k\frac{dy}{dt} - g$$

Given that the particle is projected when $t = 0$ from O with speed V at an angle α above the horizontal axis Ox, find expressions for x and y in terms of t. (JMB)

19) A particle of mass m is attached to one end of an elastic string of modulus mg and natural length l. The other end of the string is attached to a fixed point O. The particle is held below O so that the string is vertical and of length $3l$, and is then released. The subsequent motion takes place in a medium which offers a resistance equal to $(3g/l)^{1/2}mv$, where v is the speed of the particle. Show that the extension x of the string at time t satisfies the differential equation

$$l\frac{d^2x}{dt^2} + (3gl)^{1/2}\frac{dx}{dt} + gx = gl$$

Prove that the string does not become slack, and that the particle is next at rest when its depth below O is $l(2 - e^{-\sqrt{3}\pi})$. (U of L)

20) A particle of mass m is attached to the end A of a light spring AB of natural length l and modulus of elasticity $8mn^2l$ which is lying at rest, straight and unstretched, on a smooth horizontal table.
From time $t = 0$ onwards the end B is made to move with constant speed V in the direction AB and in the subsequent motion the particle is subject to a resistance of magnitude $4mn$ times its speed.
Show that the extension x of the spring satisfies the differential equation

$$\frac{d^2x}{dt^2} + 4n\frac{dx}{dt} + 8n^2x = 4nV$$

Find x in terms of t and show that the extension of the spring is approximately constant for large t and give the value of this constant. (JMB)

21) A particle P of mass m is suspended from a fixed point by a spring of natural length l and modulus $2mn^2l$. The particle is projected vertically downwards with speed V from its equilibrium position. The motion of the particle is resisted by a force of magnitude $2mn$ times its speed acting in a direction opposite to its motion. Given that x is the displacement of P downwards from the equilbrium position at time t, show that

$$\frac{d^2x}{dt^2}+2n\frac{dx}{dt}+2n^2x = 0$$

Find x in terms of t and sketch the graph of x against t.
Show that P is instantaneously at rest when $nt = (k+\frac{1}{4})\pi$, where $k \in \mathbb{N}$.
(U of L)

22) A smooth rod AB of length $2a$ is fixed at an angle of $30°$ to the horizontal with the end A uppermost. A bead of mass m can move on the rod and is connected to the point A by a light spring of natural length a and modulus of elasticity $10mn^2a$, where n is a positive constant.
The motion of the bead is opposed by a force acting along the rod and of magnitude $2mn$ times the speed of the bead. The bead is released from rest at B and at time t $(t>0)$ after its release it is at a distance $x<2a$ down the rod from A. Show that

$$\frac{d^2x}{dt^2}+2n\frac{dx}{dt}+10n^2x = 10n^2a+\frac{1}{2}g$$

and that n^2 must be greater than $g/(20a)$.

Find x at time t and show that $\dfrac{dx}{dt}=0$ when $t=\dfrac{k\pi}{3n}$, where k is an integer.
(JMB)

23) Show that the roots of the equation

$$\lambda^2+k\lambda+\omega^2 = 0$$

are distinct and both negative when $k>2\omega>0$. A particle moves along the x axis under the action of a force $\omega^2|x|$ per unit mass directed towards the origin and a resisting force kv per unit mass, where k and ω are positive constants and x, v are the displacement from the origin and speed respectively after time t. Show that the differential equation satisfied by x and t is

$$\frac{d^2x}{dt^2}+k\frac{dx}{dt}+\omega^2x = 0$$

The particle starts from rest when $x=a$. Show that, if $k>2\omega>0$,

$$(\lambda_2-\lambda_1)x = a(\lambda_2 e^{-\lambda_1 t}-\lambda_1 e^{-\lambda_2 t})$$

where $-\lambda_1, -\lambda_2$ are the roots of the quadratic equation

$$\lambda^2+k\lambda+\omega^2 = 0$$

Further, show that the particle does not pass through the origin. (U of L)

24) A particle moves on the Ox axis and its displacement x at time t is governed by the equation

$$\frac{d^2x}{dt^2} + 2k\frac{dx}{dt} + n^2x = 0$$

where k and n are positive constants.

Given that $x = a$ and $\dfrac{dx}{dt} = 0$ when $t = 0$, find x in each of the two cases

(a) $k^2 = 2n^2$, (b) $k^2 = \frac{3}{4}n^2$.

Show that in case (b) the time interval between successive stationary values of x is constant. (U of L)

25) Initially the total mass of a rocket is M, of which kM is the mass of the fuel. Starting from rest, the rocket gives itself a constant vertical acceleration of magnitude g by ejecting fuel with constant speed u relative to itself. If m denotes its remaining mass at time t, show that the rate of decrease of m with respect to t is $2mg/u$, and deduce that

$$m = Me^{-2gt/u}$$

Find in terms of M, u and k an expression for the kinetic energy of the rocket when the fuel is exhausted. Find the value of k for which this energy is a maximum.
(Assume that the height reached is sufficiently small for g to be considered constant.) (U of L)

26) A rocket vehicle uses fuel at a rate μ, so that

$$dm/dt = -\mu$$

where m is the vehicle's mass at time t. The fuel used is expelled backwards from the vehicle with a constant velocity V relative to the vehicle. The vehicle starts from rest at time $t = 0$ with a total mass m_0 and reaches a final speed v_1 at a time t_1, the mass at that time being m_1. If no external forces act on the vehicle and v denotes the speed at time t, use the conservation of linear momentum to show that

$$m\,dv/dt = \mu V$$

and deduce that

$$v_1 = V\log_e(m_0/m_1)$$

If, for a particular vehicle, $m_0 = 10\,000$ kg, $V = 2500\text{ m s}^{-1}$, $v_1 = 5000\text{ m s}^{-1}$, $t_1 = 200$ s and μ is constant, find numerical values for μ, m_1 and the greatest and least acceleration of the vehicle during the period of powered flight. (WJEC)

27) A particle whose initial mass is m is projected vertically upwards at time $t = 0$ with speed gT, where T is a constant. At time t its speed is u and its mass has increased to $m\mathrm{e}^{t/T}$. If the added mass is at rest when it is acquired, show that

$$\frac{\mathrm{d}}{\mathrm{d}t}(mu\,\mathrm{e}^{t/T}) = -mg\,\mathrm{e}^{t/T}$$

Deduce that the mass of the particle at its highest point is $2m$.
If, instead, the added mass is falling with constant speed gT when it is acquired, find the mass of the particle at its highest point. (U of L)

CHAPTER 7

CENTRE OF MASS. STABILITY OF EQUILIBRIUM

CENTRE OF MASS

Consider a set of n particles whose masses are $m_1, m_2, m_3, \ldots, m_p, \ldots, m_n$, and whose position vectors relative to an origin O are $\mathbf{r}_1, \mathbf{r}_2, \mathbf{r}_3, \ldots, \mathbf{r}_p, \ldots, \mathbf{r}_n$.
The centre of mass of the set of particles is defined as the point with position vector $\mathbf{r}_M$ where

$$\mathbf{r}_M = \left(\sum_{P=1}^{n} m_P \mathbf{r}_P\right) \div \left(\sum_{P=1}^{n} m_P\right)$$

EXAMPLE 7a

Find the position vector of the centre of mass of three particles A, B and C where the mass and position vector of each particle is given in the following table.

Particle	Mass	Position vector
A	2 units	$\mathbf{i}+4\mathbf{j}-7\mathbf{k}$
B	5 units	$3\mathbf{i}-2\mathbf{j}+\mathbf{k}$
C	3 units	$\mathbf{i}-6\mathbf{j}+13\mathbf{k}$

The position vector $\mathbf{r}_M$ of the centre of mass of A, B and C is given by

$$\mathbf{r}_M = \frac{2(\mathbf{i}+4\mathbf{j}-7\mathbf{k})+5(3\mathbf{i}-2\mathbf{j}+\mathbf{k})+3(\mathbf{i}-6\mathbf{j}+13\mathbf{k})}{2+5+3}$$

$$= \frac{(2+15+3)\mathbf{i}+(8-10-18)\mathbf{j}+(-14+5+39)\mathbf{k}}{10}$$

Hence $\mathbf{r}_M = 2\mathbf{i}-2\mathbf{j}+3\mathbf{k}$

We shall now show that the position of the centre of mass of a system of particles does not depend on the choice of origin.
Suppose that, relative to an origin O, the centre of mass of a set of particles is at a point C with position vector $\mathbf{r}_M$ and that, relative to an origin O_1, the centre of mass is at a point C_1 with position vector $\mathbf{s}_M$. The position vector of O_1 relative to O is $\mathbf{d}$.

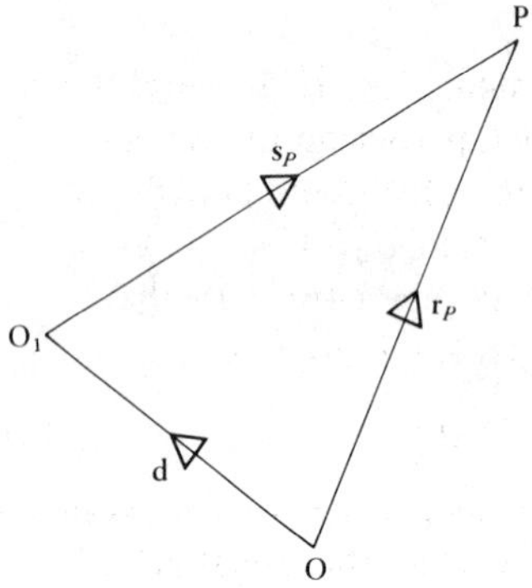

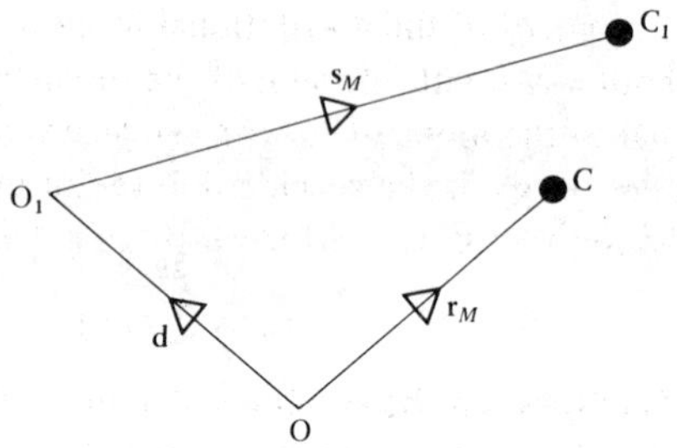

For each particle $\qquad \mathbf{r}_P = \mathbf{d} + \mathbf{s}_P$

For the set of particles

$$\mathbf{r}_M = \frac{\Sigma\, m_P \mathbf{r}_P}{\Sigma\, m_P}$$

$$= \frac{\Sigma\, m_P (\mathbf{d} + \mathbf{s}_P)}{\Sigma\, m_P}$$

$$= \frac{\mathbf{d}\, \Sigma\, m_P + \Sigma\, m_P \mathbf{s}_P}{\Sigma\, m_P}$$

$$= \mathbf{d} + \frac{\Sigma\, m_P \mathbf{s}_P}{\Sigma\, m_P}$$

But, by definition, the position vector of the centre of mass relative to O_1 is given by

$$\mathbf{s}_M = \frac{\Sigma\, m_P \mathbf{s}_P}{\Sigma\, m_P}$$

Hence $\qquad \mathbf{r}_M = \mathbf{d} + \mathbf{s}_M$

Thus C and C_1 coincide

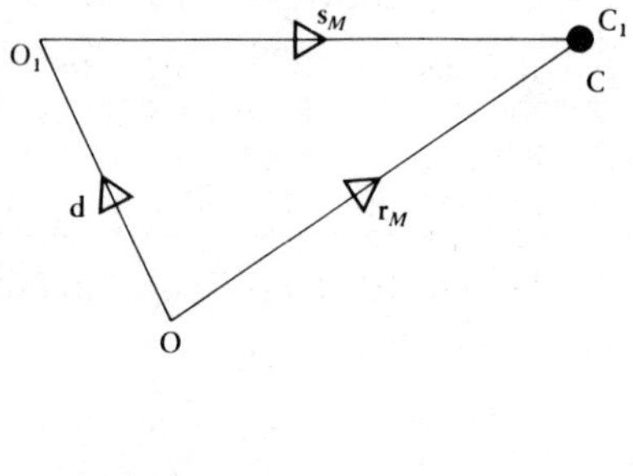

DISTINCTION BETWEEN CENTRE OF MASS AND CENTRE OF GRAVITY

The position of the centre of mass of a system depends only upon the mass and position of each constituent particle,

i.e. $$\mathbf{r}_M = \frac{\Sigma\, m_p \mathbf{r}_p}{\Sigma\, m_p} \qquad [1]$$

The location of G, the centre of gravity of the system, depends however upon the moment of the gravitational force acting on each particle in the system (about any point, the sum of the moments for all the constituent particles is equal to the moment for the whole system concentrated at G).
Hence, if $\mathbf{g}_P$ is the acceleration vector due to gravity of a particle P, the position vector $\mathbf{r}_G$ of the centre of gravity of the system is given by:

$$\mathbf{r}_G \times \sum m_P \mathbf{g}_P = \sum (\mathbf{r}_P \times m_P \mathbf{g}_P) \qquad [2]$$

It is only when the system is in a *uniform gravitational field*, where the acceleration vector due to gravity $(\mathbf{g})$ is the same for all particles, that equation [2]

becomes $$\mathbf{r}_G = \frac{\Sigma\, \mathbf{r}_P m_P}{\Sigma\, m_P} = \mathbf{r}_M$$

In this case, therefore, the centre of gravity and the centre of mass coincide.

If, however, the gravitational field is *not* uniform and $\mathbf{g}_P$ is not constant then, in general, equation [2] cannot be simplified and $\mathbf{r}_G \neq \mathbf{r}_M$.

Thus, for a system of particles in a uniform gravitational field, the centre of mass and the centre of gravity are identical points *but* in a variable gravitational field, the centre of mass and the centre of gravity are, in general, two distinct points.

The calculation of the centre of gravity of a system of particles in a non-uniform gravitational field is beyond the scope of this book, but it is important that the distinction between centre of mass and centre of gravity is understood.
In this book it will be assumed, unless otherwise stated, that all analysis is carried out in a uniform gravitational field.

EXERCISE 7a

1) Four particles of masses 4, 1, 2 and 5 kg are placed at points with position vectors $\mathbf{i}+\mathbf{j}+3\mathbf{k}$, $4\mathbf{i}-3\mathbf{j}+\mathbf{k}$, $5\mathbf{i}-\mathbf{j}$ and $2\mathbf{j}-\mathbf{k}$ respectively. Find the position vector of their centre of mass.

2) Particles of masses $3m$, $2m$, $5m$ and $2m$ are placed at points with position vectors $\mathbf{a}$, $\mathbf{b}$, $3\mathbf{a}$ and $4\mathbf{b}$ respectively. Find the position vector of their centre of mass.

3) Three particles at points A, B and C have a centre of mass at the point with coordinates $(1, 2, 3)$. The mass at $A(4, -1, 2)$ is 2 kg and the mass at $B(-6, 1, -5)$ is 1 kg. If C is the point $(p, q, 8)$, calculate the values of p and q and the mass at C.

4) Find the centre of gravity of three particles of masses 2, 2 and 3 kg if they are at points with coordinates $(1, 0, 2)$, $(3, 1, -4)$ and $(2, -3, -1)$ respectively.

5) OABC is a parallelogram. The position vectors relative to O of A and C are $\mathbf{a}$ and $\mathbf{c}$ respectively. Find the centre of mass of particles placed at O, A, B and C whose masses are 1, 2, 3 and 4 kg respectively.

6) Particles of masses 3, 3 and 1 unit are placed at points with position vectors $a\mathbf{i}+2\mathbf{j}+2\mathbf{k}$, $6\mathbf{i}+b\mathbf{k}$ and $c\mathbf{j}-3\mathbf{k}$ respectively. If their centre of mass is at the point with position vector $3\mathbf{i}+\mathbf{j}-3\mathbf{k}$, evaluate a, b and c. If the particle of mass 1 unit is then replaced by one of mass 3 units, where is the new centre of mass?

PROPERTIES OF THE MOTION OF THE CENTRE OF MASS

In this section of the work we shall analyse the motion of the centre of mass of a set of particles.

Consider a typical particle P of mass m_P which, at time t has a position vector $\mathbf{r}_P$, a velocity vector $\mathbf{v}_P$, an acceleration vector $\mathbf{a}_P$ and is under the action of a single force $\mathbf{F}_P$. Initially the position and velocity vectors of P are $\mathbf{R}_P$ and $\mathbf{V}_P$. Similarly $\mathbf{r}_M$, $\mathbf{v}_M$ and $\mathbf{a}_M$ are the position, velocity and acceleration vectors of the centre of mass at time t, its initial position and velocity vectors being $\mathbf{R}_M$ and $\mathbf{V}_M$.

Acceleration

By definition
$$\mathbf{r}_M = \frac{\Sigma\, m_P \mathbf{r}_P}{\Sigma\, m_P}$$

or
$$\mathbf{r}_M \sum m_P = \sum m_P \mathbf{r}_P$$

Differentiating twice with respect to time we have

$$\frac{\mathrm{d}^2\mathbf{r}_M}{\mathrm{d}t^2} \sum m_P = \sum m_P \frac{\mathrm{d}^2\mathbf{r}_P}{\mathrm{d}t^2}$$

or
$$\mathbf{a}_M \sum m_P = \sum m_P \mathbf{a}_P \qquad [1]$$

Now applying Newton's law to the motion of one particle,

$$\mathbf{F}_P = m_P \mathbf{a}_P$$

So for the whole set of particles

$$\sum \mathbf{F}_P = \sum m_P \mathbf{a}_P \qquad [2]$$

From [1] and [2] we see that

$$\sum \mathbf{F}_P = \left(\sum m_P\right) \mathbf{a}_M$$

But $\Sigma\, \mathbf{F}_P$ is the resultant force acting on the system, $\Sigma\, m_P$ is the total mass of the system and $\mathbf{a}_M$ is the acceleration of the centre of mass.

Thus the acceleration of the centre of mass of a system is the same as that of a particle whose mass is the total mass of the system, acted upon by the resultant of the forces acting on the system.

EXAMPLES 7b

1) Find the acceleration vector of the centre of mass of particles A, B, C and D whose masses are 1, 2, 3 and 4 units respectively and which are moving under the action of force vectors $\mathbf{i}-\mathbf{j}+3\mathbf{k}$, $4\mathbf{i}-3\mathbf{j}$, $2\mathbf{j}+\mathbf{k}$ and $5\mathbf{i}-4\mathbf{k}$ respectively.

The resultant force $\mathbf{F}$ is given by

$$\begin{aligned}\mathbf{F} &= (\mathbf{i}-\mathbf{j}+3\mathbf{k})+(4\mathbf{i}-3\mathbf{j})+(2\mathbf{j}+\mathbf{k})+(5\mathbf{i}-4\mathbf{k})\\ &= 10\mathbf{i}-2\mathbf{j}\end{aligned}$$

The total mass of the system is 10 units. Hence the acceleration vector $\mathbf{a}_M$ of the centre of mass is given by

$$\begin{aligned}\mathbf{a}_M &= \text{resultant force/total mass}\\ &= (10\mathbf{i}-2\mathbf{j})/10\end{aligned}$$

Note. Since in this problem the constituent particles are moving under the action of constant forces, the acceleration of the centre of mass also is constant.

Momentum

The impulse applied to each particle in the system is equal to the change in its momentum, thus

$$\mathbf{I}_P = m_P \mathbf{v}_P - m_P \mathbf{V}_P$$

Hence for the whole system,

$$\sum \mathbf{I}_P = \sum m_P \mathbf{v}_P - \sum m_P \mathbf{V}_P \qquad [3]$$

Now for the centre of mass,

$$\mathbf{r}_M \sum m_P = \sum m_P \mathbf{r}_P$$

Differentiating with respect to time gives

$$\frac{d\mathbf{r}_M}{dt} \sum m_P = \sum m_P \frac{d\mathbf{r}_P}{dt}$$

But at time t, $\quad \dfrac{d\mathbf{r}_M}{dt} = \mathbf{v}_M \quad$ and $\quad \dfrac{d\mathbf{r}_P}{dt} = \mathbf{v}_P$

and when $t = 0$, $\quad \dfrac{d\mathbf{r}_M}{dt} = \mathbf{V}_M \quad$ and $\quad \dfrac{d\mathbf{r}_P}{dt} = \mathbf{V}_P$

Therefore at time t $\quad \mathbf{v}_M \sum m_P = \sum m_P \mathbf{v}_P$

and initially $\quad \mathbf{V}_M \sum m_P = \sum m_P \mathbf{V}_P$

Subtracting $\quad (\mathbf{v}_M - \mathbf{V}_M) \sum m_P = \sum m_P \mathbf{v}_P - \sum m_P \mathbf{V}_P \quad$ [4]

Equation [3] shows that the resultant impulse acting on the system is equal to the change in the resultant momentum of the set of particles.
From equation [4] we see that

> the change in resultant momentum of the system is the same as the change in momentum of a particle, of mass equal to the total mass of the system, placed at the centre of mass.

EXAMPLES 7b (continued)

2) Force vectors $2\mathbf{i}+5\mathbf{j}-\mathbf{k}$, $\mathbf{i}-\mathbf{j}+3\mathbf{k}$ and $3\mathbf{i}+\mathbf{j}$ act respectively on three particles of masses 4, 2 and 3 units which are initially at rest. Find the resultant momentum of the system after 4 seconds. Find the impulse of each force and show that the vector sum of these impulses is equal to the change in the momentum of a particle of mass 9 units at the centre of mass.

The resultant force $\quad \mathbf{F} = (2\mathbf{i}+5\mathbf{j}-\mathbf{k}) + (\mathbf{i}-\mathbf{j}+3\mathbf{k}) + (3\mathbf{i}+\mathbf{j})$

$$= 6\mathbf{i}+5\mathbf{j}+2\mathbf{k}$$

The total mass is $4+2+3 = 9$ units.

Now the motion of the centre of mass is the same as the motion of a particle of mass 9 units subjected to the force $\mathbf{F}$.
Hence, applying Newton's law,

$$\mathbf{F} = 9\mathbf{a}_M$$

Integrating with respect to time gives

$$(6\mathbf{i}+5\mathbf{j}+2\mathbf{k})t = 9\mathbf{v}_M + \mathbf{V}$$

When $t = 0,\quad \mathbf{v}_M = \mathbf{0}\quad \text{so}\quad \mathbf{V} = \mathbf{0}$

Therefore $(6\mathbf{i}+5\mathbf{j}+2\mathbf{k})t = 9\mathbf{v}_M$

So, when $t = 4$, $\mathbf{v}_M = \frac{4}{9}(6\mathbf{i}+5\mathbf{j}+2\mathbf{k})$

The resultant momentum of the system is equal to the momentum of a particle of mass 9 units at the centre of mass, i.e. $9\mathbf{v}_M$.

Thus the resultant momentum is $24\mathbf{i}+20\mathbf{j}+8\mathbf{k}$.

Using $\mathbf{I} = \int \mathbf{F}\,\mathrm{d}t$ for each particle, where $\mathbf{I}$ is the impulse vector, we have,

$$\mathbf{I}_1 = \int_0^4 (2\mathbf{i}+5\mathbf{j}-\mathbf{k})\,\mathrm{d}t = 8\mathbf{i}+20\mathbf{j}-4\mathbf{k}$$

$$\mathbf{I}_2 = \int_0^4 (\mathbf{i}-\mathbf{j}+3\mathbf{k})\,\mathrm{d}t = 4\mathbf{i}-4\mathbf{j}+12\mathbf{k}$$

$$\mathbf{I}_3 = \int_0^4 (3\mathbf{i}+\mathbf{j})\,\mathrm{d}t = 12\mathbf{i}+4\mathbf{j}$$

$$\begin{aligned}\text{Resultant impulse} &= (8\mathbf{i}+20\mathbf{j}-4\mathbf{k}) + (4\mathbf{i}-4\mathbf{j}+12\mathbf{k}) + (12\mathbf{i}+4\mathbf{j}) \\ &= 24\mathbf{i}+20\mathbf{j}+8\mathbf{k} \\ &= 9\mathbf{v}_M\end{aligned}$$

Therefore the resultant impulse is equal to the change in momentum of a mass of 9 units at the centre of mass.

Work Done and Change in Kinetic Energy

Consider a set of particles moving under the action of a set of constant forces. When a force $\mathbf{F}_P$ is applied to one particle P of mass m_P and moves it from its initial position vector $\mathbf{R}_P$ to a general position vector $\mathbf{r}_P$, the work done by $\mathbf{F}_P$ is $\mathbf{F}_P\,.\,(\mathbf{r}_P-\mathbf{R}_P)$.
The total work done by all the forces acting on the system of particles is therefore given by

$$\sum \mathbf{F}_P\,.\,(\mathbf{r}_P-\mathbf{R}_P) = \sum \mathbf{F}_P\,.\,\mathbf{r}_P - \sum \mathbf{F}_P\,.\,\mathbf{R}_P$$

Now consider the action of a force $\Sigma\, \mathbf{F}_P$ on a particle of mass $\Sigma\, m_P$ at the centre of mass, causing a displacement from the initial position vector $\mathbf{R}_M$ to a general position vector $\mathbf{r}_M$.

The work done by the force $\Sigma\, \mathbf{F}_P$ is given by

$$\left(\sum \mathbf{F}_P\right).(\mathbf{r}_M - \mathbf{R}_M) = \left(\sum \mathbf{F}_P\right).\left(\frac{\Sigma\, m_P \mathbf{r}_P}{\Sigma\, m_P} - \frac{\Sigma\, m_P \mathbf{R}_P}{\Sigma\, m_P}\right)$$

In general this is *not* equal to $\Sigma\, \mathbf{F}_P . \mathbf{r}_P - \Sigma\, \mathbf{F}_P . \mathbf{R}_P$

So the total work done by the individual forces displacing the individual particles is *not* in general, equal to the work done by the resultant force displacing a particle of mass $\Sigma\, m_P$ at the centre of mass of the system.

If $\mathbf{F}_P$ is the resultant force acting on the particle P, the change in kinetic energy of P is equal to the work done by $\mathbf{F}_P$.
It then follows immediately from the last section that

the total change in kinetic energy of all the individual particles is *not* in general, equal to the change in kinetic energy of a particle of mass $\Sigma\, m_P$ at the centre of mass of the system.

Clearly the two conclusions just reached are 'negative' results, i.e. they make no constructive contribution to the solution of those problems which require the calculation of the work done by a number of forces acting on a system of particles, or the change in kinetic energy so caused.
Nevertheless they are important results, since they show that it is invalid to use the motion of the centre of mass of a system of *independent* particles subjected to *random* forces, in evaluating the total work done or the change in kinetic energy of the system as a whole.

EXAMPLES 7b (continued)

3) Two forces $\mathbf{F}_1$ and $\mathbf{F}_2$ act respectively on particles A and B which are initially at rest.
If $\mathbf{F}_1 = 2\mathbf{i}+\mathbf{j}-3\mathbf{k}$, $\mathbf{F}_2 = 4\mathbf{i}+\mathbf{k}$, A is of mass 3 kg and B is of mass 5 kg, find the change in kinetic energy of A and B in the first 12 seconds of motion.
Find also the change in kinetic energy of a particle of mass 8 kg, initially at rest and subjected to a force $\mathbf{F}_1 + \mathbf{F}_2$.

For particle A, the acceleration vector $\mathbf{a}_1$ is given by

$$\mathbf{a}_1 = \tfrac{1}{3}\mathbf{F}_1$$

Hence $$\mathbf{v}_1 = \tfrac{1}{3}\int \mathbf{F}_1 \, \mathrm{d}t = \tfrac{1}{3}(2\mathbf{i}+\mathbf{j}-3\mathbf{k})t + \mathbf{V}_1$$

But $\mathbf{v}_1 = \mathbf{0}$ when $t = 0$, so $\mathbf{V}_1 = \mathbf{0}$.

Therefore, after 12 seconds,

$$\mathbf{v}_1 = \tfrac{12}{3}(2\mathbf{i}+\mathbf{j}-3\mathbf{k})$$

$$|\mathbf{v}_1|^2 = 16[2^2+1^2+(-3)^2] = 16 \times 14$$

$$(\text{K.E.})_A = \tfrac{1}{2}(3)(16 \times 14)\,\text{J} = 336\,\text{J}$$

Similarly for particle B,

$$\mathbf{a}_2 = \tfrac{1}{5}\mathbf{F}_2$$

$$\mathbf{v}_2 = \int \mathbf{a}_2 \, \mathrm{d}t = \tfrac{1}{5}(4\mathbf{i}+\mathbf{k})t + \mathbf{0}$$

Hence, after 12 seconds,

$$\mathbf{v}_2 = \tfrac{12}{5}(4\mathbf{i}+\mathbf{k})$$

$$|\mathbf{v}_2|^2 = (\tfrac{12}{5})^2[4^2+1^2] = \tfrac{144}{25} \times 17$$

$$(\text{K.E.})_B = \tfrac{1}{2}(5)(\tfrac{144}{25} \times 17)\,\text{J} = 244.8\,\text{J}$$

The change in kinetic energy of A and B together is therefore 580.8 J. Considering now a force $\mathbf{F}_1 + \mathbf{F}_2$ acting on a mass of $(3+5)$ kg, causing an acceleration vector $\mathbf{a}$ we have,

$$\mathbf{a} = \tfrac{1}{8}(\mathbf{F}_1 + \mathbf{F}_2) = \tfrac{1}{8}(6\mathbf{i}+\mathbf{j}-2\mathbf{k})$$

$$\Big[\mathbf{v}\Big]_{\mathbf{0}}^{\mathbf{v}} = \int_0^t \mathbf{a} \, \mathrm{d}t = \tfrac{1}{8}t(6\mathbf{i}+\mathbf{j}-2\mathbf{k}) + \mathbf{0}$$

After 12 seconds $$\mathbf{v} = \tfrac{12}{8}(6\mathbf{i}+\mathbf{j}-2\mathbf{k})$$

$$|\mathbf{v}|^2 = (\tfrac{3}{2})^2[6^2+1^2+(-2)^2] = \tfrac{9}{4} \times 41$$

$$\text{K.E.} = \tfrac{1}{2}(8)(\tfrac{9}{4} \times 41)\,\text{J} = 369\,\text{J}$$

Note. This example verifies that the total change in kinetic energy of the particles is not equal to the change in kinetic energy of the total mass subjected to the resultant force.

MOTION OF THE CENTRE OF MASS IN A UNIFORM GRAVITATIONAL FIELD

If the only forces acting on a set of particles are the weights of those particles then (using the notation introduced earlier in this chapter)

$$\mathbf{F}_P = m_P \mathbf{g}$$

Thus the total work done by all the forces in a specified time is given by

$$\begin{aligned}\sum \mathbf{F}_P \cdot (\mathbf{r}_P - \mathbf{R}_P) &= \sum m_P \mathbf{g} \cdot (\mathbf{r}_P - \mathbf{R}_P) \\ &= \mathbf{g} \cdot \left(\sum m_P \mathbf{r}_P - \sum m_P \mathbf{R}_P\right)\end{aligned}$$

Now the work done when a force $\Sigma\, \mathbf{F}_P$ acts on a mass $\Sigma\, m_P$ at the centre of mass is given by

$$\begin{aligned}\left(\sum \mathbf{F}_P\right) \cdot (\mathbf{r}_M - \mathbf{R}_M) &= \left(\sum m_P \mathbf{g}\right) \cdot \left(\frac{\Sigma\, m_P \mathbf{r}_P - \Sigma\, m_P \mathbf{R}_P}{\Sigma\, m_P}\right) \\ &= \mathbf{g} \cdot \left(\sum m_P \mathbf{r}_P - \sum m_P \mathbf{R}_P\right)\end{aligned}$$

In this case we see that the total work done by the individual forces *is* equal to the work done by the resultant force applied to the total mass concentrated at the centre of mass.

Note. In this case all the forces act in the same direction so the scalar sum of the work they all do is consistent with the work done by the resultant force (a vector sum) in that direction.

Now in a uniform gravitation field, the centre of mass and centre of gravity of a system coincide.

Therefore the properties of the linear motion of a system of particles moving under gravity in a uniform gravitational field correspond in all respects to the motion produced when the resultant force acts on the total mass concentrated at the centre of gravity.

CENTROID

The centroid of an object is a geometric centre. For instance the centroid of a circle is the point of intersection of two diameters; the centroid of a parallelogram is the point of intersection of its diagonals; the centroid of a triangle is the point of concurrence of its medians.
In certain circumstances the centre of mass or the centre of gravity of the object happens to lie at the centroid, but this is not always the case. If the mass of the body is uniformly distributed, the centre of mass and the centroid coincide. But when the mass is unevenly distributed throughout the body it is most unlikely that the centroid and the centre of mass will be the same point.
For example, the centroid of a triangle PQR is at C as shown in diagram (i).

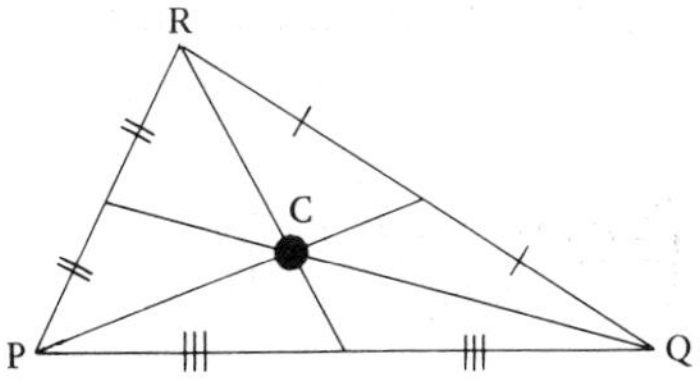

(i)

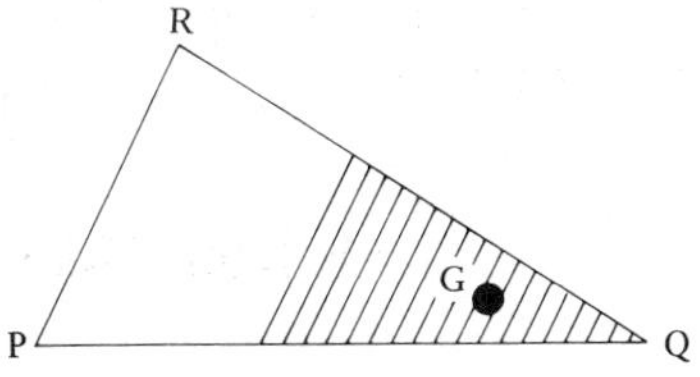

(ii)

But if the mass of a triangular lamina PQR is heavily concentrated in the shaded section as shown in diagram (ii) the centre of mass will be approximately at point G.
It is therefore misleading to use the term centroid when referring to centre of mass or to centre of gravity. Unfortunately this is a fairly common practice and the student must decide with care what is really meant by centroid whenever this term is encountered. Amongst the variety of expressions which include the word *centroid*, the two that follow are fairly frequently used and again they require thoughtful interpretation.

(a) The centre of mass of a set of particles of various masses may be called the *weighted centroid*.

(b) If each particle of a set has an *associated number* (n, say) then, using the usual notation, the point with position vector $\dfrac{\Sigma\, \mathbf{r}_p n_p}{\Sigma\, n_p}$ may be called the *centroid of a set of particles with associated numbers*.

SUMMARY

1. The position vector of the centre of mass of a set of particles is given by:

$$\mathbf{r}_M = \frac{\Sigma\, m_P \mathbf{r}_P}{\Sigma\, m_P}$$

2. The acceleration of the centre of mass is the same as the acceleration of a particle of mass $\Sigma\, m_P$ whose position vector is $\mathbf{r}_M$ subjected to a force $\Sigma\, \mathbf{F}_P$.

3. The change in the resultant momentum of the system is the same as the change in momentum of a particle of mass $\Sigma\, m_P$ at the centre of mass, under the action of a force $\Sigma\, \mathbf{F}_P$.

4. In general the total work done by the individual forces in displacing the individual particles is *not* equal to the work done by the resultant force acting on the concentrated mass at the centre of mass.
Further, the total change in kinetic energy of all the particles is *not*, in general, equal to the change in kinetic energy of the concentrated mass subjected to the resultant force.

EXERCISE 7b

1) Forces $\mathbf{F}_1$, $\mathbf{F}_2$ and $\mathbf{F}_3$ act on particles A, B and C whose masses are m_1, m_2 and m_3 respectively. Find the acceleration of the centre of mass of the particles if:

(a) $\mathbf{F}_1 = 3\mathbf{i}+7\mathbf{j}-\mathbf{k}$, $\mathbf{F}_2 = 2\mathbf{i}-\mathbf{j}+6\mathbf{k}$, $\mathbf{F}_3 = \mathbf{i}+4\mathbf{j}+2\mathbf{k}$,
$m_1 = 2$ units, $m_2 = 5$ units, $m_3 = 3$ units.

(b) $\mathbf{F}_1 = \mathbf{i}-2\mathbf{k}$, $\mathbf{F}_2 = 3\mathbf{i}+4\mathbf{j}$, $\mathbf{F}_3 = 2\mathbf{i}-\mathbf{j}-\mathbf{k}$,
$m_1 = m_2 = m_3 = m$.

2) Forces $\mathbf{F}_1 = 4\mathbf{i}+3\mathbf{j}-\mathbf{k}$ and $\mathbf{F}_2 = 3\mathbf{i}-5\mathbf{j}+8\mathbf{k}$ act on two particles of masses 2 kg and 1 kg respectively. The particles are initially at rest at points with position vectors $\mathbf{i}+\mathbf{j}+\mathbf{k}$ and $2\mathbf{j}-7\mathbf{k}$ respectively. Find:

(a) the initial position vector of the centre of mass,

(b) the acceleration of the centre of mass,

(c) the total momentum of the particles after 2 seconds.

3) ABCD is a square framework of rigid light rods which rests on a smooth horizontal plane. Particles of masses m, $2m$, $3m$ and $4m$ are attached at A, B, C and D. A force of magnitude $3F$ is applied at A towards B and a force of magnitude $4F$ is applied at B towards C. Find the linear acceleration of the centre of mass. If the mass at D is removed what effect does this have on (a) the magnitude (b) the direction of the acceleration of the centre of mass?

4) Two particles of mass 1 kg lie at rest at points P, Q on a smooth horizontal surface. A force of magnitude 2 N is applied to the particle at P in a direction inclined at 60° to PQ. Find the acceleration of the centre of mass of the particles. Find also the total momentum and the total kinetic energy of the particles after 5 seconds. If the force of magnitude 2 N is now removed and replaced by two forces each of magnitude 1 N in the same direction as the original force, one applied to P and one applied to Q, in what way does this affect the motion of the centre of mass of the particles?

5) At time t, two particles of equal mass 3 kg have velocity vectors

$$(2+3t)\mathbf{i}+(-1+t)\mathbf{j} \quad \text{and} \quad (3-4t)\mathbf{i}+(2+5t)\mathbf{j}$$

Determine the acceleration vector of their centre of mass and the total work done on the particles in the first two seconds.

6) A, B and C are points with coordinates (4, 0), (4, 3) and (0, 3) respectively relative to an origin O. Forces of magnitudes 6, 10 and 4 N begin to act respectively on particles of equal mass at A, B and C, each force being directed towards O. Determine the initial position vector of the centre of mass of the particles and state the direction in which the centre of mass begins to move.

7) Three particles of masses 3, 3 and 4 units are at rest on a smooth horizontal plane at points A, B and C with position vectors $\mathbf{i}+4\mathbf{j}$, $5\mathbf{i}+10\mathbf{j}$ and $3\mathbf{i}+2\mathbf{j}$ respectively.
Write down the position vector of the centre of mass of the particles.
If constant forces of magnitudes $3\sqrt{17}$, $10\sqrt{5}$ and $\sqrt{13}$ units respectively now act on the particles in the directions $\overrightarrow{OA}$, $\overrightarrow{OB}$ and $\overrightarrow{OC}$, find the magnitude of the initial acceleration of the centre of mass. Find the total work done by the forces at the instant when the total momentum vector of the particles is $48\mathbf{i}+102\mathbf{j}$.

STABILITY OF EQUILIBRIUM

Consider first a smooth bead threaded on to a wire whose shape is shown in the diagram and which is fixed in a vertical plane.

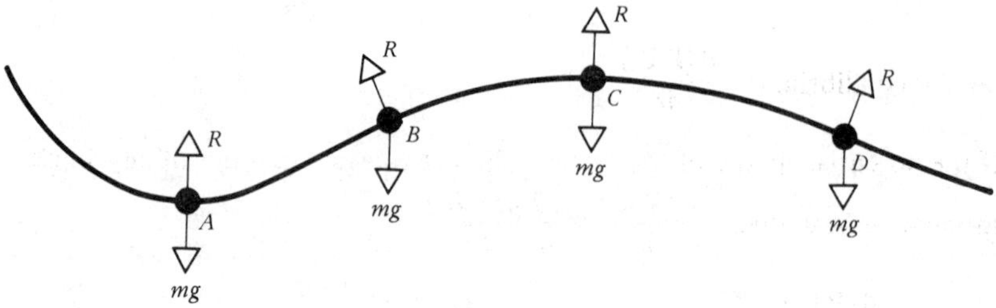

If the bead is placed at A or at C it will rest in equilibrium (mg and R are equal and opposite in these positions).
When placed at points such as B or D however the bead is not in equilibrium. When the bead is slightly displaced from A, its natural tendency is to return to A which is called a *position of stable equilibrium*. When the bead is slightly displaced from C however, it will continue to move further from C which is called a *position of unstable equilibrium*.
It is interesting to notice that:

(a) the total mechanical energy of the bead remains constant as it moves along the wire, since the normal reaction R between the bead and the wire is always perpendicular to the direction of motion of the bead and hence R does no work.

(b) the potential energy of the bead has a minimum value at A and a maximum value at C.

The above example illustrates the concept of stable and unstable equilibrium. We will now investigate a more general case.
Consider a system whose total mechanical energy remains constant (i.e. a conservative system).

Then $$\text{P.E.}+\text{K.E.} = \lambda \qquad \text{(a constant)}$$

where the term P.E. includes all the potential energy of the system whether gravitational or elastic.

But $$\text{K.E.} = \tfrac{1}{2}mv^2$$

So $$\text{P.E.} = \lambda - \tfrac{1}{2}mv^2$$

Differentiating with respect to time gives

$$\frac{\mathrm{d}\,(\text{P.E.})}{\mathrm{d}t} = 0 - mv\frac{\mathrm{d}v}{\mathrm{d}t}$$

Now in a position of equilibrium the resultant force acting on the system is zero.

Therefore the resultant acceleration is also zero, i.e. $\dfrac{\mathrm{d}v}{\mathrm{d}t} = 0$

So, for equilibrium, $\dfrac{\mathrm{d}\,(\text{P.E.})}{\mathrm{d}t} = 0$

If the potential energy of the system can be expressed in terms of any single variable, θ say, then $\dfrac{\mathrm{d}\,(\text{P.E.})}{\mathrm{d}t} = \dfrac{\mathrm{d}\theta}{\mathrm{d}t}\dfrac{\mathrm{d}\,(\text{P.E.})}{\mathrm{d}\theta}$

Thus if $\dfrac{\mathrm{d}\,(\text{P.E.})}{\mathrm{d}\theta} = 0$ if follows that $\dfrac{\mathrm{d}\,(\text{P.E.})}{\mathrm{d}t} = 0$

and the system is in equilibrium.

Hence for a conservative system whose potential energy can be expressed in terms of a single variable θ,

equilibrium positions are given by $\dfrac{\mathrm{d}\,(\text{P.E.})}{\mathrm{d}\theta} = 0.$

Stable Equilibrium

When a body is slightly displaced in any direction from a position of equilibrium E, it will tend to return to E if E is below the displaced position. If this condition is satisfied, the potential energy at E has a minimum value.

So an equilibrium position where $\theta = \alpha$ is stable if $\dfrac{\mathrm{d}^2\,(\text{P.E.})}{\mathrm{d}\theta^2} > 0$

(**Note.** The condition given above is the conclusion of an intuitive argument rather than a proof, which is too difficult at this stage. However the condition so derived is correct and can be used with confidence.)
So the conditions for stable equilibrium are

$$\frac{\mathrm{d}\,(\text{P.E.})}{\mathrm{d}\theta} = 0 \quad\text{and}\quad \frac{\mathrm{d}^2\,(\text{P.E.})}{\mathrm{d}\theta^2} > 0$$

Conversely a position for which

$$\frac{\mathrm{d}\,(\text{P.E.})}{\mathrm{d}\theta} = 0 \quad\text{and}\quad \frac{\mathrm{d}^2\,(\text{P.E.})}{\mathrm{d}\theta^2} < 0$$

is a position of unstable equilibrium.

EXAMPLES 7c

1) Two uniform smooth rods AB and BC, each of length $2a$ and mass m, are smoothly jointed at B. They rest symmetrically in a vertical plane over a beam of width $2b$. If the rods are not horizontal, show that there is a position of equilibrium only if $b \leqslant a$.

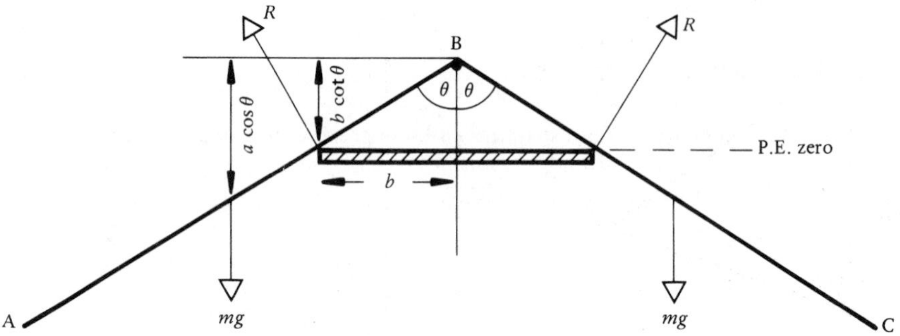

Let the rods be equally inclined to the vertical at a general angle θ.
(As θ varies, the two reactions, R, do no work so the system is conservative and we can use the potential energy method to investigate the equilibrium of the rod.)
In general the potential energy of the rod is given by

$$\text{P.E.} = -2mg(a\cos\theta - b\cot\theta)$$

Hence
$$\frac{\mathrm{d}(\text{P.E.})}{\mathrm{d}\theta} = -2mg(-a\sin\theta + b\operatorname{cosec}^2\theta)$$

$$= 2mg(a\sin\theta - b\operatorname{cosec}^2\theta)$$

There are equilibrium positions where $\dfrac{\mathrm{d}(\text{P.E.})}{\mathrm{d}\theta} = 0$

i.e. where
$$a\sin\theta - b\operatorname{cosec}^2\theta = 0$$

or
$$\sin^3\theta = \frac{b}{a}$$

This equation has a real solution only if $\dfrac{b}{a} \leqslant 1$.

In this case $\dfrac{\mathrm{d}(\text{P.E.})}{\mathrm{d}\theta} = 0$ when $\theta = \arcsin\left(\dfrac{b}{a}\right)^{1/3}$

Therefore the rod is in equilibrium when $\theta = \arcsin\left(\dfrac{b}{a}\right)^{1/3}$ if $b \leqslant a$.

2) A uniform rod AB of mass m and length $2l$ is smoothly jointed at A to a fixed point. A light elastic string of length l and modulus of elasticity mg connects the end B to a point C distant $2l$ vertically above A. Show that there are two positions in which the rod can rest in equilibrium with the string taut and determine whether the equilibrium at these positions is stable or unstable.

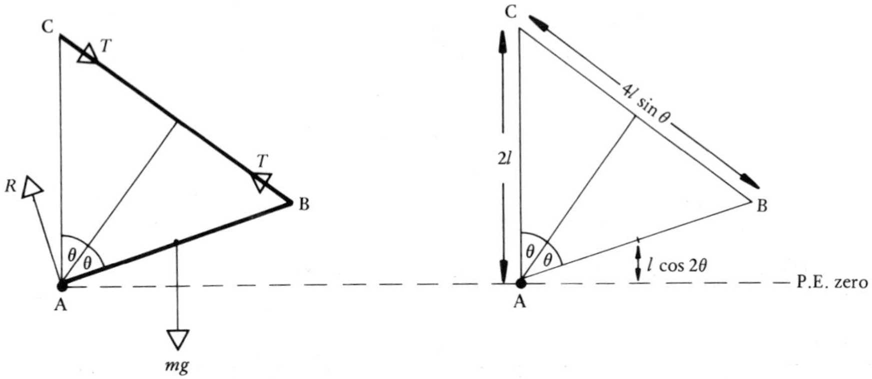

(This is a conservative system since the external reaction at the hinge does no work as the rod rotates about A. So it is valid to use the potential energy method to investigate the equilibrium of the rod.)

Consider the rod in a general position where the angle CAB = 2θ.

In this position $$\text{P.E.} = mgl\cos 2\theta + \frac{mg}{2l}(4l\sin\theta - l)^2$$

Hence $$\frac{1}{mgl}\frac{\mathrm{d(P.E.)}}{\mathrm{d}\theta} = -2\sin 2\theta + (4\sin\theta - 1)4\cos\theta$$

$$= 6\sin 2\theta - 4\cos\theta \qquad [1]$$

Therefore $$\frac{1}{mgl}\frac{\mathrm{d^2(P.E.)}}{\mathrm{d}\theta^2} = 12\cos 2\theta + 4\sin\theta \qquad [2]$$

There are equilibrium positions where $\dfrac{\mathrm{d(P.E.)}}{\mathrm{d}\theta} = 0$

From equation [1] $$6\sin 2\theta - 4\cos\theta = 0$$

$\Rightarrow$ $$4\cos\theta(3\sin\theta - 1) = 0$$

Hence the rod is in equilibrium when

$$\cos\theta = 0 \quad \text{or} \quad \sin\theta = \tfrac{1}{3}$$

i.e. $$\theta = \tfrac{1}{2}\pi \quad \text{or} \quad \arcsin\tfrac{1}{3}$$

Now the stability of equilibrium depends upon whether the P.E. is maximum or minimum, i.e. it depends upon the sign of $\dfrac{d^2(\text{P.E.})}{d\theta^2}$

Referring to equation [2] we have

θ	$\frac{1}{2}\pi$	$\arcsin \frac{1}{3}$
$\dfrac{d^2(\text{P.E.})}{d\theta^2}$	$-ve$	$+ve$
P.E.	Maximum	Minimum
State of equilibrium	Unstable	Stable

So the rod is in equilibrium, with the string taut, in two positions
(a) with B vertically below A (an unstable position),
(b) with the rod inclined at $2 \arcsin \frac{1}{3}$ to the upward vertical (a stable position).

Note. The *positions* of equilibrium could have been determined by the standard consideration of three forces in equilibrium.

3) A uniform rod AB of mass $12m$ is smoothly hinged at A to a fixed point. A light inextensible string attached to B passes over a small smooth peg at a point C and carries at its other end a particle of mass m. AC is horizontal, AC = AB and the angle CAB is θ. Find the value of θ for which the rod is in a position of stable equilibrium.

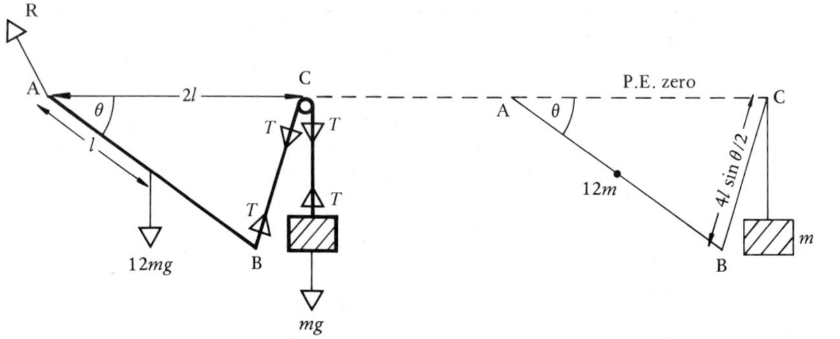

(The system is conservative since the external force R at the hinge does no work when θ varies and the tensions, being internal forces, occur in equal and opposite pairs.)

Let AB = AC = $2l$ and let the length of the string be a.

In a general position, the potential energy of the system is given by

$$\text{P.E.} = -12mgl\sin\theta - mg\left(a - 4l\sin\frac{\theta}{2}\right)$$

$$\frac{1}{mg}\frac{\mathrm{d(P.E.)}}{\mathrm{d}\theta} = -12l\cos\theta + 2l\cos\frac{\theta}{2}$$

$$\frac{1}{mg}\frac{\mathrm{d^2(P.E.)}}{\mathrm{d}\theta^2} = +12l\sin\theta - l\sin\frac{\theta}{2}$$

There is an equilibrium position if, for some value of θ, $\dfrac{\mathrm{d(P.E.)}}{\mathrm{d}\theta} = 0$.

Hence $$12l\cos\theta = 2l\cos\frac{\theta}{2}$$

$$6\left(2\cos^2\frac{\theta}{2} - 1\right) = \cos\frac{\theta}{2}$$

The solutions of this quadratic equation are:

$$\cos\frac{\theta}{2} = \frac{3}{4} \quad \text{or} \quad -\frac{2}{3}$$

Now $$\cos\frac{\theta}{2} = \frac{3}{4} \Rightarrow \cos\theta = \frac{1}{8}$$

and $$\cos\frac{\theta}{2} = -\frac{2}{3} \Rightarrow \cos\theta = -\frac{1}{9}$$

Checking for stability in equation [2] we have:

θ	$\arccos\frac{1}{8}$	$\arccos(-\frac{1}{9})$
$\dfrac{\mathrm{d^2(P.E.)}}{\mathrm{d}\theta^2}$	$+ve$	$-ve$
P.E.	Minimum	Maximum
State of equilibrium	Stable	Unstable

Thus there is a position of stable equilibrium when $\theta = \arccos\frac{1}{8}$.

4) Three equal uniform rods PQ, QR, RS each of length a are smoothly jointed at Q and R. The rods rest symmetrically with PQ and RS supported by two smooth pegs which are distant $(a+b)$ apart in a horizontal line.

(a) If $2a > 3b$ show that there are two equilibrium positions and investigate their stability.

(b) If $2a = 3b$ show that there is only one position of equilibrium.

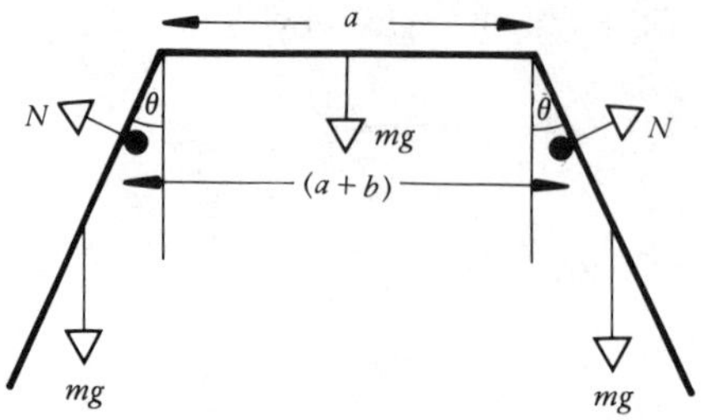

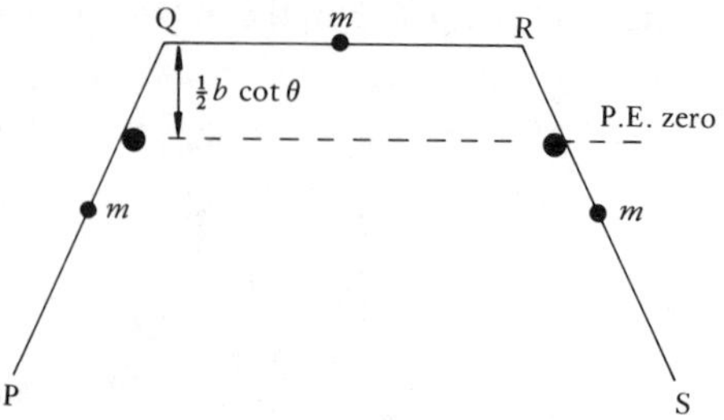

(When the angle between the inclined rods and the vertical varies the forces at the pegs do no work, so the system is conservative.)
Let θ be the angle between PQ (or RS) and the upward vertical in a general position.
The total P.E. of the three rods is then given by

$$\text{P.E.} = mg\,(\tfrac{1}{2}b\cot\theta) - 2mg\,(\tfrac{1}{2}a\cos\theta - \tfrac{1}{2}b\cot\theta)$$

$$\frac{2}{mg}\frac{\mathrm{d}(\text{P.E.})}{\mathrm{d}\theta} = -b\,\mathrm{cosec}^2\theta + 2a\sin\theta - 2b\,\mathrm{cosec}^2\theta$$

$$= 2a\sin\theta - 3b\,\mathrm{cosec}^2\theta \qquad [1]$$

$$\frac{2}{mg}\frac{\mathrm{d}^2(\text{P.E.})}{\mathrm{d}\theta^2} = 2a\cos\theta + 6b\,\mathrm{cosec}^2\theta\cot\theta \qquad [2]$$

For equilibrium $\quad \dfrac{\mathrm{d}(\text{P.E.})}{\mathrm{d}\theta} = 0$

Hence from [1] $\quad 2a\sin\theta = 3b\,\mathrm{cosec}^2\theta \;\Rightarrow\; \sin^3\theta = 3b/2a$

(a) If $2a > 3b$, $\sin\theta < 1$, so there are two positions of equilibrium given by $\theta = \alpha$ and $\theta = \pi - \alpha$ where α is the acute angle arcsin $(3b/2a)^{1/3}$

Now the stability of equilibrium depends upon whether the potential energy is maximum or minimum.

When $\theta = \alpha$ in equation [2], $\cos\alpha$, $\mathrm{cosec}\,\alpha$ and $\cot\alpha$ are all positive so $\dfrac{\mathrm{d}^2(\text{P.E.})}{\mathrm{d}\theta^2}$ is positive.

Hence when $\theta = \alpha$ the rods are in a position of stable equilibrium.

When $\theta = \pi - \alpha$ in equation [2], $\cos(\pi - \alpha)$ and $\cot(\pi - \alpha)$ are negative but $\mathrm{cosec}^2(\pi - \alpha)$ is positive so $\dfrac{\mathrm{d}^2(\text{P.E.})}{\mathrm{d}\theta^2}$ is negative.

Hence when $\theta = \pi - \alpha$ the rods are in a position of unstable equilibrium.

(b) If $2a = 3b$, the condition for equilibrium is $\sin^3\theta = 1$.
Hence the only angle giving a position of equilibrium is $\theta = \frac{1}{2}\pi x$.

Note. In each of the examples above, the level of some *fixed* object was chosen as the potential energy zero level.

EXERCISE 7c

In this exercise use potential energy methods to investigate positions of equilibrium and their stability.

1) The potential energy, V, of a system is given by

$$V = mga(\cos 2\theta + 2\sin\theta)$$

Show that, when θ is acute, the system has exactly two positions of equilibrium.

2) A system has potential energy given by the expression $a^2 \sin\theta\,(1-\sin\theta)$. Show that there is only one position in which the system is in equilibrium and that this position is unstable. (Assume $0<\theta<\frac{1}{2}\pi$.)

3) Two uniform smooth heavy rods AB and BC, each of length $2a$, are smoothly jointed together at B. If they are placed symmetrically in a vertical plane over a fixed cylinder of radius a whose axis is horizontal, show that the rods are in equilibrium if $\cos^3\theta = \sin\theta$ where θ is the inclination of each rod to the horizontal.

4) Four equal uniform heavy rods are smoothly jointed to form a rhombus ABCD. A and C are connected by a light elastic string whose natural length is half the length of each rod and whose modulus of elasticity is twice the weight of a rod. Find the angle BAC when the framework hangs in equilibrium from A.

5)

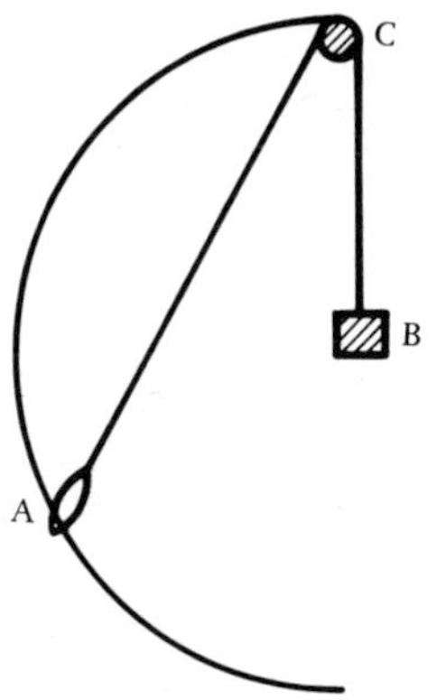

A is a smooth ring of mass m which is threaded on to a wire in the shape of a semicircle in a vertical plane.
The bead is connected by a light inextensible string to a particle B also of mass m. The string passes over a small smooth pulley at the highest point of the wire.

If the length of the string exceeds the diameter of the circle, show that there are two positions of equilibrium for A and B and investigate their stability.

6)

A heavy circular disc of radius a hangs so that it touches a smooth vertical wall and is attached to the wall as shown by a light inextensible string AB of length $2a$.
Show that there is only one position of equilibrium and that it is stable.

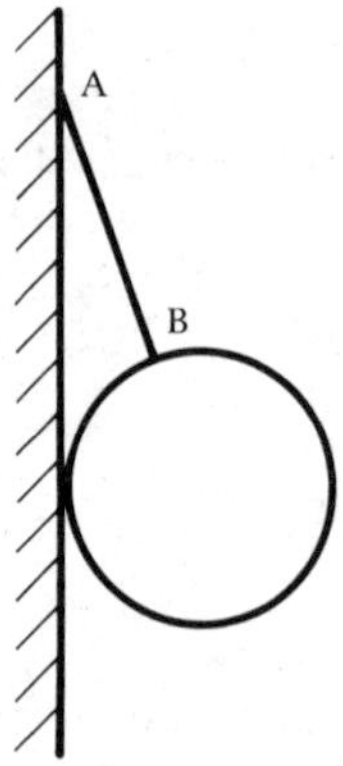

7) Two uniform rods, PQ and PR, each of mass M and length $2l$ are smoothly jointed at P. The rods are placed in a vertical plane over two smooth parallel rails distant $2a$ apart in a horizontal plane. Each rod is inclined at an angle θ to the upward vertical through P (P is the lowest point in the system). Show that the potential energy of the system is given by

$$V = 2Mg(l\cos\theta - a\cot\theta) + K \quad \text{where } K \text{ is a constant.}$$

Show that the system is in equilibrium when $\sin^3\theta = \dfrac{a}{l}$ and determine whether each of the two equilibrium positions is stable or unstable.

8)

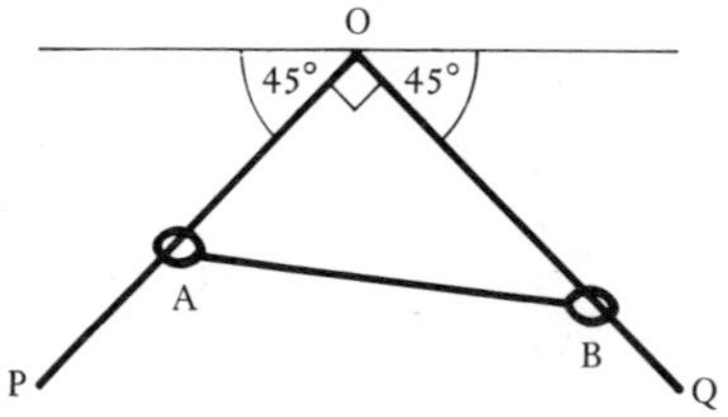

OP and OQ are two fixed smooth wires and angle POQ is a right angle. AB is a uniform heavy rod with a smooth light ring at each end. The rings are threaded on to OP and OQ as shown. Find the position in which AB is in stable equilibrium.
Is there an unstable position?

9) ABC is an equilateral triangular framework of three equal uniform heavy rods each of length $4a$. The framework rests over two smooth fixed pegs P and Q distant $2b$ apart in a horizontal line. Investigate the positions of equilibrium of the framework if

(a) $2b > a$ (b) $2b < a$.

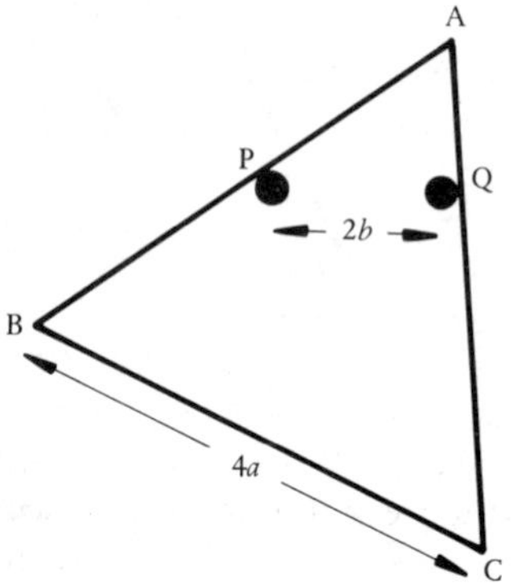

10)

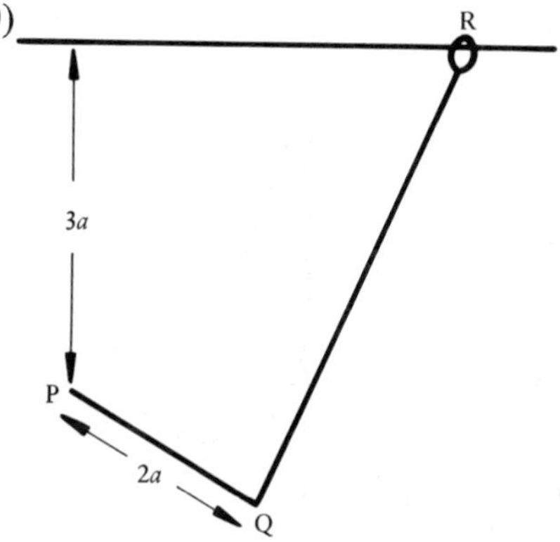

In the diagram PQ is a uniform rod of mass $2m$ and length $2a$. The end P is smoothly pivoted to a fixed point. A light elastic string of natural length a and modulus mg connects Q to a small smooth ring R which slides on a horizontal wire which is distant $3a$ vertically above P. Explain why QR must be vertical in an equilibrium position. Find the positions of equilibrium and investigate their stability.

MISCELLANEOUS EXERCISE 7

1) Define the centre of mass of n particles of masses $m_1, m_2, \ldots, m_n$, placed at points whose position vectors with respect to an origin O are $\mathbf{r}_1, \mathbf{r}_2, \ldots, \mathbf{r}_n$, respectively, and show that it is independent of the choice of origin. Find the position vector of the centre of mass of particles of masses 4, 3, 2, 3 units at rest at the points

$$\mathbf{i}+\mathbf{j}, \quad 2\mathbf{i}-\mathbf{j}, \quad 2\mathbf{i}+\mathbf{j}, \quad 2\mathbf{i}+3\mathbf{j}$$

respectively. If each mass is acted upon by a force directed towards the origin and proportional to its distance from the origin, find the direction of the initial acceleration of the centre of mass. (U of L)

2) Show that, if $\mathbf{F}$ is the resultant of the external forces acting on a system of particles with centroid G, then the motion of the point G is the same as if the force $\mathbf{F}$ acted on a particle of suitable mass placed at G.
Particles of masses 6, 5, 2, 3 units are fixed to the vertices of a light square framework, the position vectors of the vertices being $a\mathbf{i}$, $a\mathbf{j}$, $-a\mathbf{i}$, $-a\mathbf{j}$ respectively. Each particle is acted on by a force represented by the vector drawn from the particle to the point with position vector $3a\mathbf{i}+4a\mathbf{j}$. Find the position vector of the centroid of the particles and the direction in which it will begin to move. (U of L)

3) Three particles are at rest on a smooth horizontal plane, their masses being 2, 2, 3 units and their position vectors respectively being

$$3\mathbf{i}+4\mathbf{j}, \quad \mathbf{i}+\mathbf{j}, \quad \mathbf{i}-\mathbf{j}$$

Find the position vector of their centre of mass G.
Each mass is acted on by a constant force directed away from the origin, the magnitudes of the forces being 5, 1, 7 units respectively. Find the magnitude of the acceleration of G. Find also the resultant linear momentum of the particles at the instant when the total work done by the forces is 150/7 units. (U of L)

4) Two particles P_1 and P_2, of masses m_1 and m_2 respectively, are free to move on a smooth horizontal plane. Let $\mathbf{i}$, $\mathbf{j}$ be perpendicular unit vectors in the plane and let

$$\mathbf{r}_1 = x_1\mathbf{i}+y_1\mathbf{j}, \quad \mathbf{r}_2 = x_2\mathbf{i}+y_2\mathbf{j}$$

be the position vectors of P_1, P_2 relative to the origin. Show that the position vector $\mathbf{r}$ and the velocity $\mathbf{v}$ of the centre of mass G of the particles are given by

$$\mathbf{r} = \frac{m_1\mathbf{r}_1+m_2\mathbf{r}_2}{m_1+m_2}$$

and

$$\mathbf{v} = \frac{m_1\mathbf{v}_1+m_2\mathbf{v}_2}{m_1+m_2}$$

respectively where $\mathbf{v}_1 = \dot{x}_1\mathbf{i}+\dot{y}_1\mathbf{j}$ and $\mathbf{v}_2 = \dot{x}_2\mathbf{i}+\dot{y}_2\mathbf{j}$ are the velocities of P_1 and P_2.
Show that the velocity of particle P_1 relative to the point G is

$$\mathbf{V}_1 = \frac{m_2(\mathbf{v}_1-\mathbf{v}_2)}{m_1+m_2}$$

and write down the velocity $\mathbf{V}_2$ of the particle P_2 relative to G.
If T is the kinetic energy of the pair of particles P_1 and P_2, prove that

$$T = \tfrac{1}{2}(m_1+m_2)\mathbf{v}.\mathbf{v}+\tfrac{1}{2}m_1\mathbf{V}_1.\mathbf{V}_1+\tfrac{1}{2}m_2\mathbf{V}_2.\mathbf{V}_2$$ (O)

5) Two particles of mass m and $2m$ are attached to the ends of an elastic string of natural length $2a$ which would be doubled in length by a tension $4mg$. The particles are held at rest at a distance $3a$ apart on a smooth horizontal plane and are then released. Explain why the centre of mass of the system remains at rest during the motion of the particles. Find how far each particle travels before they collide.
From the moment of release the particle of mass $2m$ travels a distance x in time t. Show that, for $x \leqslant a/3$,

$$\frac{d^2x}{dt^2} = g(a-3x)/a$$

Find the speed of each particle immediately before impact. (U of L)

6) A particle of mass 3 units is acted on by the forces

$$\mathbf{F}_1 = 2\mathbf{i}+3\mathbf{j}, \quad \mathbf{F}_2 = 3\mathbf{j}+4\mathbf{k}, \quad \mathbf{F}_3 = \mathbf{i}+2\mathbf{k}$$

and initially it is at rest at the point $\mathbf{i}-\mathbf{j}-\mathbf{k}$. Find the position and the momentum of the particle after 2 seconds. Find also the work done on the particle in this time. (U of L)

7) A particle of mass m kg is acted upon by a constant force of $21m$ newtons in the direction of the vector $3\mathbf{i}+2\mathbf{j}-6\mathbf{k}$. Initially the particle is at the origin moving with speed 18 m/s in the direction of the vector $7\mathbf{i}-4\mathbf{j}+4\mathbf{k}$. Find the position vector of the particle 4 seconds later and find the work done by the force in this period of 4 seconds. (U of L)

8) (a) Masses $5m$, $3m$ and $2m$ are situated at points with position vectors $2\mathbf{i}+2\mathbf{k}$, $-8\mathbf{i}-12\mathbf{j}+5\mathbf{k}$ and $3\mathbf{i}+3\mathbf{j}+7\mathbf{k}$ respectively. Find the position vector of the centroid of these point masses.

(b) A particle with position vector $2\mathbf{i}$ is kept in equilibrium by the following five forces:

$\mathbf{P}=7\mathbf{i}-15\mathbf{j}+\mathbf{k}$, $\mathbf{Q}=5\mathbf{i}-8\mathbf{j}-7\mathbf{k}$,

$\mathbf{R}_1$ acting along the line $\mathbf{r}=2\mathbf{i}+a(3\mathbf{i}+4\mathbf{k})$,

$\mathbf{R}_2$ acting along the line $\mathbf{r}=2\mathbf{i}+b(2\mathbf{i}-\mathbf{j}+2\mathbf{k})$,

$\mathbf{R}_3$ acting along the line $\mathbf{r}=2\mathbf{i}+c(-3\mathbf{i}+4\mathbf{j})$.

Find the magnitudes of $\mathbf{R}_1$, $\mathbf{R}_2$ and $\mathbf{R}_3$. (U of L)

9) A uniform rod AB is of weight W and length $2a$. The upper end A is attached to a small light ring which can slide on a fixed smooth vertical wire; the end B is attached to a small light ring which can slide on a fixed smooth horizontal wire which intersects the vertical wire in O. One end of a light inextensible string is attached to the rod at a point C where $BC=\frac{1}{2}a$. The string passes over a small smooth pulley at O and, at its other end, carries a weight $\frac{1}{2}W$ which hangs freely. Find the potential energy of the system when $O\widehat{A}B=\theta$.

Show that there is a position of equilibrium given by $\theta=30^\circ$ and determine whether it is stable or unstable. (C)

10) A light rod AB of length $2a$ is rigidly attached at its midpoint C to one end of another light rod CD of length $6a$, the two rods being at right angles. Particles of mass m, $4m$ and $6m$ are attached to A, B and D respectively. The system is free to rotate about the midpoint of CD in the vertical plane ABCD. Find the two positions of equilibrium and discuss their stability. (U of L)

11) A and B are the ends of a smooth wire in the form of a semicircle of radius a. The wire is fixed in a vertical plane and hangs below AB which is horizontal. A bead of mass M slides on the wire and is attached to two light strings, each of length $2a$ which pass over smooth pegs at A and B. To the other ends of the strings are attached particles of mass m which hang vertically under gravity. Show that if the radius to the bead of mass M makes an angle θ with the vertical, the potential energy of the system is

$$V_0-Mga\cos\theta+mga2\sqrt{2}\cos(\theta/2),$$

where V_0 is a constant.

Hence deduce that the equilibrium position in which the bead of mass M is at the lowest point of the wire is stable if $m<M\sqrt{2}$. (WJEC)

12)

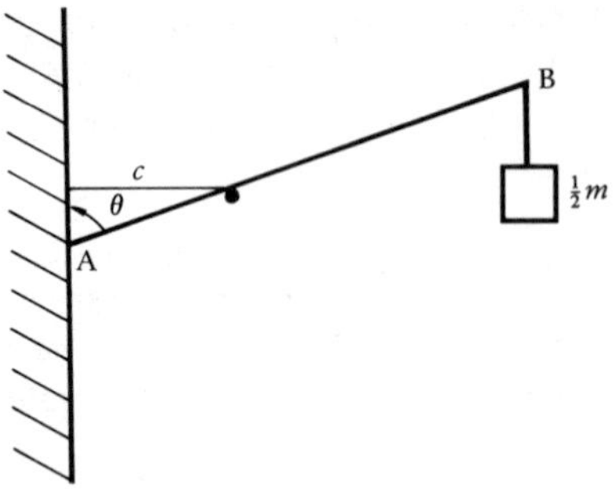

The diagram shows a uniform rod AB, of length $2a$ and mass m, resting with the end A in contact with a smooth vertical wall. The rod is supported by a smooth horizontal rail parallel to the wall at a distance c from the wall. A particle of mass $\frac{1}{2}m$ is attached at B. Show that, when AB makes an angle θ with the vertical, the potential energy is

$$mg(2a\cos\theta - \tfrac{3}{2}c\cot\theta) + \text{constant}.$$

Show also that if $3c > 4a$ there is no equilibrium position, but if $3c < 4a$ there is one equilibrium position. Discuss the stability of this position. (U of L)

13) A uniform rectangular lamina ABCD rests in a vertical plane with A in contact with a smooth vertical wall and B attached by a light inextensible string to a point P of the wall vertically above A. The plane of the lamina is perpendicular to the wall. $AB = BP = 2a$ and $AD = 2b$. If the inclination of the string to the wall is θ, show that the depth of the centre of mass of the lamina below P is $3a\cos\theta + b\sin\theta$.
Hence, or otherwise, show that in equilibrium $\tan\theta = b/(3a)$ and determine whether the equilibrium is stable or unstable.

14) A uniform rod, of mass m and length $2a$, is smoothly hinged at one end to a fixed point A, and to the other end is attached a light inextensible string which passes over a smooth peg B and carries a mass $(m\sqrt{3})/6$. Given that A and B are a distance $2a$ apart on the same horizontal line, express the potential energy of the system in terms of θ, the angle which the rod makes with the line AB.
Show that there is a position of equilibrium when $\theta = \pi/3$, and determine whether this position of equilibrium is stable or unstable. (U of L)

15) A uniform rod OA of weight W and length $2a$ is freely pivoted at O; to the end A is attached a light inextensible string which passes over a small smooth pulley at B, at a distance $4a$ vertically above O. To the other end of the string is attached a weight λW hanging freely. Show that, when $\hat{AOB} = \theta$, the potential energy V of the system is

$$Wa\,[\cos\theta + 2\lambda\sqrt{(5-4\cos\theta)}] + \text{constant}$$

Hence or otherwise show that, in addition to the equilibrium positions given by $\theta = 0°$ and $\theta = 180°$, there is another position of equilibrium for a value of θ between 0° and 180°, provided that $\frac{1}{4} < \lambda < \frac{3}{4}$.
By expanding V in powers of θ, or otherwise, determine whether the equilibrium position given by $\theta = 0$ is stable or unstable when

(a) $\lambda = 1$, (b) $\lambda = \frac{1}{8}$, (c) $\lambda = \frac{1}{4}$.

[Assume throughout that the string is of such length that it never leaves the pulley.] (C)

16) A uniform straight rod OA of mass m and length $2a$ is free to move in a vertical plane about its end O, which is fixed. B and C are fixed points vertically above O and at distances $b(> 2a)$ and $b+c\,(c > 0)$ from O respectively. A light string of natural length c and modulus of elasticity λ passing through a small smooth ring at B has one end attached to C and the other to a point D of the rod at a distance d from O. Assuming that $mgac < \lambda bd$, find all the positions of equilibrium of the rod and investigate their stability. (WJEC)

17) A uniform rod AB, of mass m and length $2a$, is free to rotate in a vertical plane about the end A which is fixed. The end B is attached to one end of a light elastic string of natural length a and modulus $mg/(2\sqrt{3})$, whose other end is attached to a fixed point O, at the same horizontal level as A. If $OA = 2a$, and the angle between AB and the downward vertical at A is $(90° - 2\theta)$, show that, apart from an additive constant, the potential energy of the system is

$$\frac{-mga}{\sqrt{3}}(\sqrt{3}\sin 2\theta + 2\cos 2\theta + 2\sin\theta)$$

provided that the string is not slack.
Hence or otherwise, *verify* that there is a position of equilibrium in which $\theta = 30°$ and determine whether this position is stable or unstable.
Show also that there is another position of equilibrium in which $\theta = \alpha$, where $90° < \alpha < 120°$. (C)

18) A light rod AB can turn freely in a vertical plane about a smooth hinge at A and carries a mass m hanging from B. A light string of length $2a$ fastened to the rod at B passes over a smooth peg at a point C vertically above A and carries a mass km at its free end. If $AC = AB = a$, find the range of values of k for which equilibrium is possible with the rod inclined to the vertical. Given that equilibrium is possible with the rod horizontal find the value of k. If the rod is slightly disturbed when horizontal and in equilibrium, determine whether it will return to the horizontal position or not. (U of L)

19) A uniform rod OA of length $2a$ and weight $8W$ is freely pivoted to a fixed support at O. A light elastic string of natural length a and modulus nW has one end tied to A and the other to a small ring which is free to slide on a smooth horizontal straight wire passing through a point at a height $7a$ above O. Show that when the rod makes an angle θ with the upward vertical at O, and the string is vertical, the potential energy of the system is

$$2Wa[n\cos^2\theta - (6n-4)\cos\theta] + \text{constant}$$

Obtain the positions of equilibrium and discuss their stability in each of the following cases:
(a) $n > 1$, (b) $n = \frac{4}{5}$, (c) $n < \frac{1}{2}$. (C)

20) A uniform circular hoop of weight W and radius a is free to rotate in a vertical plane about a horizontal axis through a fixed point O on its circumference. A light inextensible string AB of length $a\sqrt{7}$ is attached at one end A to the point on the circumference of the hoop diametrically opposite O and at the other end B carries a small smooth ring of weight $\frac{1}{2}W$ which is free to slide on a smooth vertical wire which lies in the plane of the hoop and at a distance $3a$ from O. Write down an expression for the potential energy of the system when OA makes an acute angle θ with the horizontal and prove that there exists a single position of equilibrium which occurs when $\theta = \frac{1}{3}\pi$. Prove that this position is stable. (O)

21) A bead of mass m can slide on a smooth vertical circular loop of wire of radius a. Attached to the bead is one end of a light spring, of natural length a and modulus $3mg$. The other end of the spring is attached to the highest point of the wire. Show that, when the spring makes an angle θ with the downward vertical, the potential energy of the system is

$$\frac{3mga}{2}(2\cos\theta - 1)^2 - 2mga\cos^2\theta + \text{constant}.$$

Show also that possible equilibrium positions occur when $\theta = 0$ and $\cos\theta = \frac{3}{4}$. Investigate the stability of these positions. (U of L)

22) A rod AB, of length $2a$ and weight W, is pivoted to a wall at A, and its end B is connected by an elastic string of natural length c and modulus λ, to a point C at a height $2a$ vertically above A. Show that, provided the string is extended, the potential energy of the system is

$$\frac{\lambda}{2c}(4a\cos\theta - c)^2 - Wa\cos 2\theta + \text{constant},$$

where 2θ is the inclination of the rod to the downward vertical.
Given that,

$$c(W+\lambda) < 4a\lambda,$$

show that there are two positions of equilibrium of which the one with BA vertical is unstable. (U of L)

23) ADB is a rigid wire of length πa bent into the form of a semicircle of radius a and centre O. It is fixed in a vertical plane with AB horizontal and the midpoint D of the wire vertically below O. A smooth ring P of mass m is threaded on the wire and is attached to one end of an elastic string CP whose other end is attached to a fixed point C at a distance a vertically above O. The modulus of the string is $4mg$ and its natural length is $c \leqslant \sqrt{2}a$. Show that when CP makes an angle θ with the vertical the potential energy of the system is

$$V_0 - 2mga\cos^2\theta + \frac{2mg}{c}(2a\cos\theta - c)^2,$$

where V_0 is a constant.

If $\frac{4}{7}(2\sqrt{2}-1)a < c < \frac{4}{3}a$, show

(a) that there are positions of equilibrium in which the string is not vertical,

(b) that the position of equilibrium in which the string is vertical is unstable. (WJEC)

CHAPTER 8

UNITS AND DIMENSIONS

UNITS

Most of the quantities that occur in the solution of mechanics problems require a unit for their measurement.
Some of these units are *fundamental*, that is each can be chosen in an arbitrary manner and is quite independent of the others.
The quantities whose units are fundamental are mass, length and time.
Other quantities are measured in units which are derived from the fundamental ones. For instance the unit of speed is one unit of length described in one unit of time. If we denote unit mass by M, unit length by L and unit time by T, a *dimensional symbol* can be formed for the units of non-fundamental quantities. The dimensional symbol for the unit of speed, for example, is L/T or LT^{-1}.

DIMENSIONS

The *dimensions* of any physical quantity in terms of mass, length and time are indicated by the indices attached to M, L and T.
For example:

(a) The unit of acceleration is a unit increase in speed in one unit of time, i.e. the metre per second per second.
The dimensional symbol for this unit is therefore LT^{-2}.

(b) The unit of force imparts a unit of acceleration to a unit of mass, i.e. the kilogramme metre per second per second. This is too clumsy an expression to use for a unit of force so it is given the simple name newton.
The dimensional symbol, however, is MLT^{-2} and this cannot be simplified in any way.

(c) Work is the product of a force and a distance so its dimensions are given by multiplying the dimensions of force by the dimensions of distance,
i.e. $(MLT^{-2})(L) = ML^2T^{-2}$.

(d) An angle is measured in a unit defined by the ratio of unit arc length to unit radius, so its dimensional symbol is LL^{-1}, i.e. L^0. An angle is therefore said to be *dimensionless*.
Further examples of dimensionless quantities are coefficient of friction and coefficient of restitution.

(e) Moment of inertia has been defined as Σmr^2 so the unit in which it is measured is the kilogramme metre squared and its dimensional symbol is ML^2.

Quantities of different dimensions can neither be collected together nor equated to each other. Hence, whenever an equation is formed in the solution of a mechanics problem, every term must have the same dimensions. Checking this property is a useful aid to accuracy.

EXAMPLES 8a

1) Find the dimensions of impulse and momentum and hence verify the relationship Impulse = change in momentum.

The unit of impulse is a unit force acting for a unit of time, so its dimensional symbol is $(MLT^{-2})(T)$,
i.e. MLT^{-1}.

The unit of momentum is a unit mass moving with unit velocity.
The dimensional symbol is therefore $(M)(LT^{-1})$,

i.e. MLT^{-1}.

The dimensions of impulse and momentum are identical.
Impulse = change in momentum is therefore a dimensionally correct relationship.

2) The period of oscillation of a simple pendulum can be assumed to depend on nothing other than the length l of the string, the mass m of the bob and the acceleration g due to gravity. Use the theory of dimensions to determine a formula for the period.

If the period depends on mass, length and acceleration its dimensional symbol is made up of

$$(M^\alpha)(L^\beta)(LT^{-2})^\gamma$$

where α, β and γ are unknown constants.
But the dimensional symbol of a periodic time must be T, i.e. $M^0L^0T^1$.
Hence $M^0L^0T^1 \equiv M^\alpha L^{(\beta+\gamma)}T^{-2\gamma}$
Therefore $\alpha = 0$, $\gamma = -\frac{1}{2}$ and $\beta = \frac{1}{2}$.
So the structure of the dimensional symbol of the period is

$$M^0L^{1/2}(LT^{-2})^{-1/2}$$

The formula for calculating the period of oscillation of the pendulum can therefore be written as

$$\text{period} = k\sqrt{\frac{l}{g}}$$

(**Note.** The value of the numerical constant k cannot be found by considering dimensions since k is dimensionless.)

3) The formula

$$v^2 = \frac{3mgl(1-\cos\theta)}{M(1+\sin^2\theta)}$$

is obtained as the solution of a problem. Use dimensions to find whether this is a reasonable solution (v is a velocity, m, M are masses, l is a length and g is gravitational acceleration).

Considering the dimensions of the left-hand side we have,

$$(LT^{-1})^2 = L^2T^{-2}$$

On the right-hand side, the following terms are dimensionless: 3, 1, $\cos\theta$, $\sin^2\theta$.
So the dimensions of the right-hand side are:

$$\frac{(M)(LT^{-2})(L)}{(M)} = L^2T^{-2}$$

As the dimensions of both sides of the equation are the same, the formula is a reasonable solution of the problem.

(**Note.** It is not possible to say that the formula is exactly correct because the dimensionless terms may not be numerically accurate.)

EXERCISE 8a

1) Find the dimensional symbol for the following quantities:
Kinetic energy, Momentum, Torque, Angular velocity, Angular acceleration, Area, Volume, Weight, Impulse, Power, Modulus of elasticity.

2) A particle of mass m is attached to the end of a light string of length l. The other end of the string is attached to a fixed point on a smooth table. The particle is travelling in a horizontal circle on the table with angular velocity ω. Assuming that the tension in the string depends only upon m, l and ω find an expression for this tension.

3) Show that the relationship,

$$\text{work done} = \text{change in kinetic energy}$$

is dimensionally correct.

In questions 4 to 12 there are errors in some of the equations. By considering dimensions determine which equations are definitely wrong.

The symbols used in these questions are explained below:

m	mass	θ	angle
F	force	$\dot{\theta}, \omega$	angular velocity
J	impulse	$\ddot{\theta}$	angular acceleration
C	torque	I	moment of inertia
u, v	velocity	λ	modulus of elasticity
l, r	length	μ	coefficient of friction

4) $5\mu g\theta = r\omega^2$

5) $C = I\dot{\theta}$

6) $J/\omega = mgl^2/C$

7) $I = \frac{mr}{v}(ru - l\omega)$

8) $mlg^2 = \lambda v$

9) $l^2 = v^2 r/g + \mu^2 r^2$

10) $F = 2\omega^2 l^2/7\mu g$

11) $Cv = I\dot{\theta}$

12) $J = mv - I\ddot{\theta}$

CHAPTER 9

ROTATION

One of the most fundamental characteristics possessed by an object is its intrinsic reluctance to accept a change in its state of motion, i.e. its inertia.

A body whose linear motion (translation) is changing has a linear acceleration. In order to produce such a change it is necessary to apply a force.

The same force applied to different bodies, however, produces different linear accelerations, indicating that each body has an individual amount of *linear inertia* which controls the degree of change in linear motion.

The measure of a body's linear inertia is better known as *mass*.

The relationship between mass, force and the linear acceleration produced was recognised by Newton whose second law can, for a body of constant mass, be expressed in the familiar form

$$F = ma$$

On the other hand the state of motion of a body can undergo a change in rotation if a torque is applied.

The resulting angular acceleration depends in part on the magnitude of the applied torque.

Experimental evidence shows, however, that the same torque applied to different bodies produces different angular accelerations, indicating that each body has an individual amount of *rotational inertia* which controls the degree of change in rotation.

The measure of a body's rotational inertia is called *moment of inertia* and it is represented by the symbol I.

At this stage we are not in a position to specify the actual constitution of a body's moment of inertia. Experiments can be carried out, however, which show that when a torque is applied to a body, causing it to rotate about an axis, then

for a particular body, free to rotate about a specific axis, the angular acceleration is proportional to the torque applied.

The angular acceleration produced by a particular torque changes if

(a) the mass of the body is altered,
(b) the axis of rotation is changed,
(c) a body of equal mass but different size or shape is used.

These results suggest that the moment of inertia of a body is a function of the mass of the body, the distribution of that mass (e.g. size and shape) and the position of the axis of rotation.

The theoretical analysis of a rotating body will be seen to confirm this hypothesis.

ROTATION OF A RIGID BODY ABOUT A FIXED AXIS

A rigid body is made up of a large number of particles whose relative displacements are fixed.

Consider a rigid body which is freely rotating about a fixed axis and let P be a typical constituent particle, of mass m and distant r from the axis of rotation. The diagram below shows a cross-section of the body, perpendicular to the axis which passes through A.

Because the constituent particles do not move relative to one another, every particle has at any instant the same angular velocity $\dot{\theta}$, the same angular acceleration $\ddot{\theta}$ and is describing a circle about A as centre.

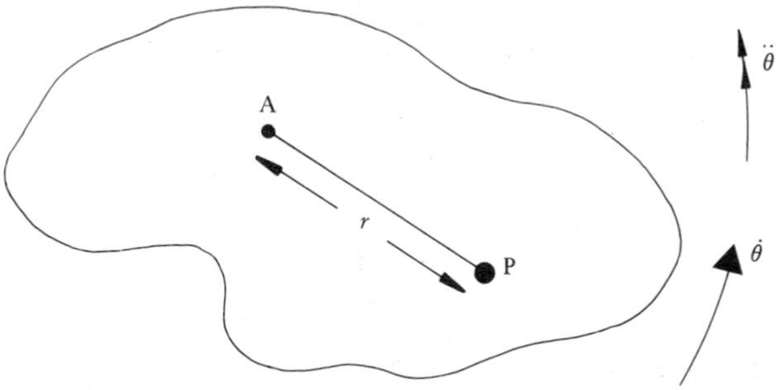

In the direction perpendicular to AP, the linear velocity and acceleration components of P are $r\dot{\theta}$ and $r\ddot{\theta}$ respectively (P also has a linear acceleration component $r\dot{\theta}^2$ in the direction PA).

Now suppose that P is under the action of a force whose components parallel and perpendicular to PA (i.e. normal and tangential) are F_N and F_T.

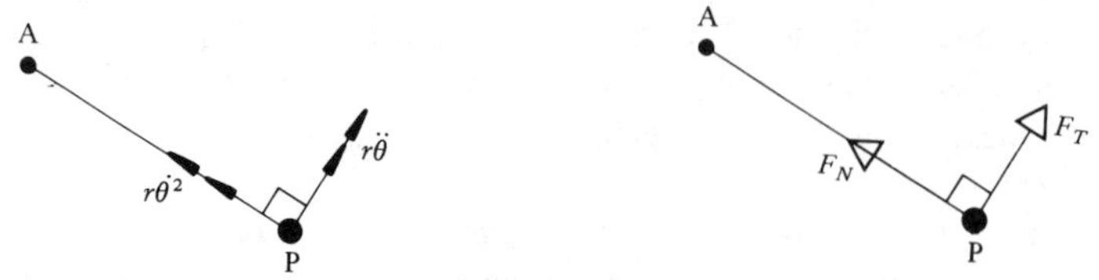

If we now apply Newton's law to the particle in the tangential direction we have:

$$F_T = m(r\ddot{\theta})$$

Therefore $$F_T r = mr^2\ddot{\theta}$$

But $F_T r$ is the moment of F_T about the axis of rotation and F_N has no moment about this axis.

Now for the whole body, $$\sum F_T r = \sum mr^2\ddot{\theta}$$

But $\Sigma F_T r$ is the resultant torque C about the axis of rotation.

Thus $$C = \left(\sum mr^2\right)\ddot{\theta} \qquad [1]$$

In this equation Σmr^2 is a quantity which depends solely on the mass of the body and the distribution of the mass about the axis of rotation, i.e. it has the properties of moment of inertia.
Then, defining the value of moment of inertia as the quantity Σmr^2, we have

$$I = \sum mr^2$$

and equation [1] becomes $$C = I\ddot{\theta}$$

This relationship is, in effect, Newton's second law adapted to rotation, as can be seen by comparing $F = M\ddot{x}$ with $C = I\ddot{\theta}$ since

F is necessary to produce change in translation
C is necessary to produce change in rotation,

M is the intrinsic control over change in translation
I is the intrinsic control over change in rotation,

$\ddot{x}$ is the resulting linear acceleration in a straight line
$\ddot{\theta}$ is the resulting angular acceleration.

PROPERTIES OF A ROTATING BODY

Consider a rigid body which is free to rotate about a smooth fixed axis and whose angular velocity at any instant is represented by $\dot{\theta}$.

A cross-section of the body, perpendicular to the axis of rotation, is shown in the diagram. The axis passes through A and P is a typical constituent particle of mass m and distant r from A.

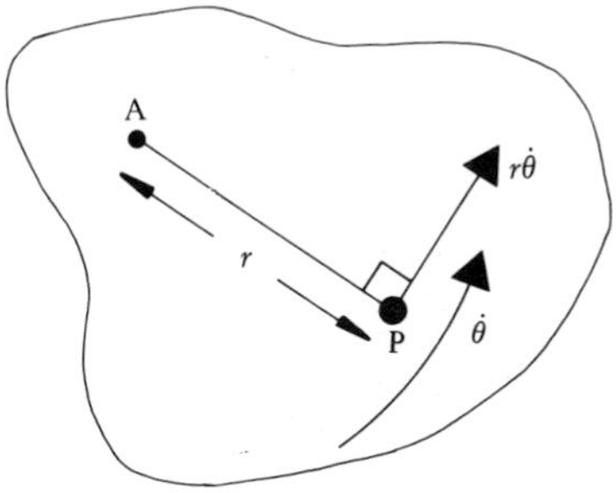

The linear velocity of P is $r\dot{\theta}$ perpendicular to AP as shown.

Kinetic Energy of Rotation

For P, kinetic energy $= \frac{1}{2}m(r\dot{\theta})^2$

So for the whole body

$$\text{kinetic energy} = \sum \tfrac{1}{2}mr^2\dot{\theta}^2$$

$$= \tfrac{1}{2}\left(\sum mr^2\right)\dot{\theta}^2$$

(as every constituent particle has the same angular velocity $\dot{\theta}$).

Hence **the kinetic energy of the rotating body is given by $\frac{1}{2}I\dot{\theta}^2$**

This quantity has dimensions $(ML^2)(T^{-1})^2$ and it is therefore dimensionally equal to linear kinetic energy.

Consequently linear kinetic energy and rotational kinetic energy can be added together in an equation and can also be measured in the same unit, the joule.

Moment of Momentum

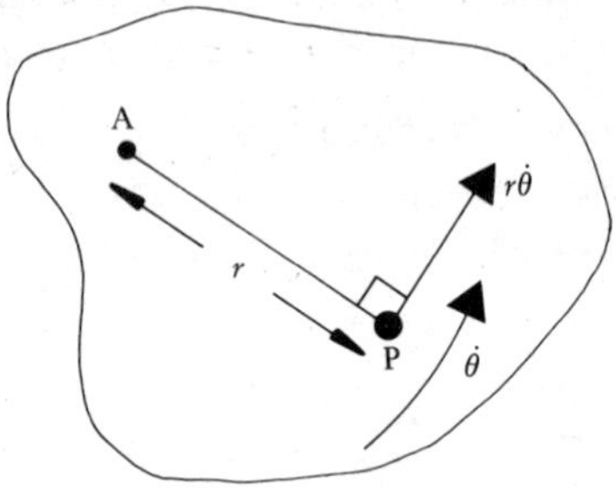

For one particle P,

$$\text{linear momentum} = m(r\dot{\theta})$$

Then the *moment of the momentum* of P about the axis through A is given by

$$\text{moment of momentum} = (mr\dot{\theta})r$$
$$= mr^2\dot{\theta}$$

Again $\dot{\theta}$ is common to all constituent particles, so for the whole body we have

$$\text{total moment of momentum} = \left(\sum mr^2\right)\dot{\theta}$$
$$= I\dot{\theta}$$

i.e. the moment of momentum of a rotating body is $I\dot{\theta}$

An alternative name for this quantity is *angular momentum* and it is measured in a rather clumsy unit, the $\text{kg m}^2\text{s}^{-1}$.

The dimensional symbol for this unit is ML^2T^{-1} or $(MLT^{-1})L$ which can be seen to have the same dimensions as linear momentum times distance. Hence, in forming an equation, the angular momentum of a rotating body can be added to the moment of the linear momentum of another body.

SUMMARY

1. Moment of inertia is the name given to a body's intrinsic control over its change in rotation.
2. The value of a body's moment of inertia, I, is defined by $I = \Sigma mr^2$.
3. When a torque C acts on a body with moment of inertia I, producing an angular acceleration $\ddot{\theta}$, then $C = I\ddot{\theta}$.
4. The kinetic energy of a body whose angular velocity is $\dot{\theta}$ is $\frac{1}{2}I\dot{\theta}^2$.
5. The angular momentum, or moment of momentum, of that body is $I\dot{\theta}$.

PROBLEMS

Detailed methods of evaluating the moment of inertia of a variety of bodies are explained in Chapter 10. At present, whenever a specific moment of inertia is required in the solution of a problem, it will be stated and used without proof.

In some problems the rigid body concerned will be found to have a constant angular acceleration $(\ddot{\theta} = \alpha)$. The motion of such a body can be analysed using the standard formulae for motion of this type.

A summary of these formulae, which are derived in *Mechanics and Probability*, is given below.

If

ω_1 = initial angular velocity (rad s^{-1})

ω_2 = final angular velocity (rad s^{-1})

θ = angular displacement (rad)

α = angular acceleration (rad s^{-2})

t = time (s)

then

$$\omega_2 = \omega_1 + \alpha t$$

$$\theta = \omega_1 t + \tfrac{1}{2}\alpha t^2$$

$$\theta = \omega_2 t - \tfrac{1}{2}\alpha t^2$$

$$\theta = \tfrac{1}{2}(\omega_1 + \omega_2)t$$

$$2\alpha\theta = \omega_2^2 - \omega_1^2$$

EXAMPLES 9a

1) A door swings open with an angular speed of 0.5 rad s^{-1}. If the door has a moment of inertia about its hinges of 160 kg m^2, find its angular momentum.

$$\text{Angular momentum} = I\dot{\theta}$$

Hence the angular momentum of the door is given by

$$(160 \text{ kg m}^2)(0.5 \text{ rad s}^{-1}) = 80 \text{ kg m}^2 \text{ s}^{-1}$$

2) The moment of inertia of a flywheel about its axis is $20\,\text{kg}\,\text{m}^2$. When it is stationary, a constant torque of $40\,\text{Nm}$ is applied to the flywheel. Find its kinetic energy after three seconds assuming the flywheel has smooth bearings. (A flywheel is either a circular disc or a circular rim which can rotate about an axis through its centre perpendicular to the flywheel.)

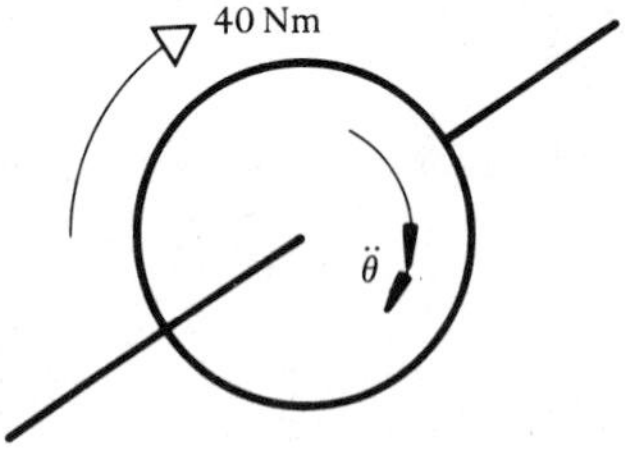

Let $\ddot{\theta}$ represent the angular acceleration of the flywheel at any instant and let ω be the angular velocity after three seconds.

Using $C = I\ddot{\theta}$ we have

$$40 = 20\ddot{\theta}$$

Hence $\ddot{\theta} = 2$

Therefore the flywheel has a constant angular acceleration of $2\,\text{rad}\,\text{s}^{-2}$ and its motion can be investigated using the standard equations for motion of this type. The flywheel starts from rest so, after three seconds,

$$\omega = \ddot{\theta}t = (2)(3)$$

Hence the flywheel has an angular velocity of $6\,\text{rad}\,\text{s}^{-1}$ and therefore possesses kinetic energy given by

$$\text{K.E.} = \tfrac{1}{2}I\omega^2$$

$$= \tfrac{1}{2}(20)(6)^2\,\text{J}$$

$$= 360\,\text{J}$$

3) A crate of mass $100\,\text{kg}$ is dragged across smooth horizontal ground by a light rope whose other end is being wrapped round a pulley of radius $0.4\,\text{m}$. The pulley has a fixed smooth vertical axle and its moment of inertia about its axis is $24\,\text{kg}\,\text{m}^2$.
If the crate moves with an acceleration of $0.2\,\text{m}\,\text{s}^{-2}$, find the magnitude of the torque driving the pulley.
Find also the total kinetic energy of the moving system when the speed of the crate is $2\,\text{m}\,\text{s}^{-1}$.

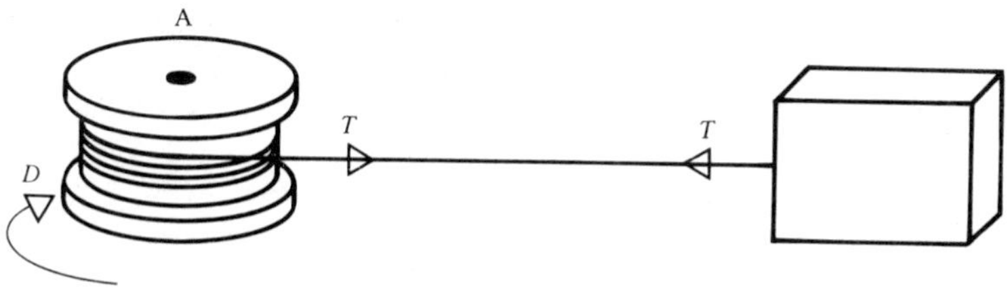

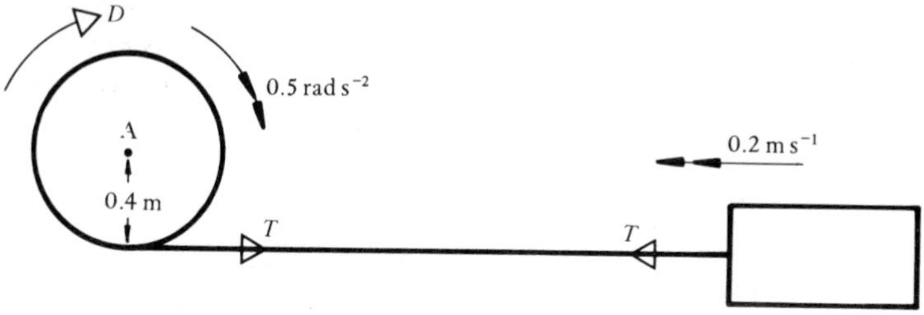

Let D N m be the torque driving the pulley.
(The linear acceleration of the crate is equal to the linear acceleration of the rope on the rim of the pulley. So the angular acceleration of the pulley is given by dividing the linear acceleration of the crate by the radius of the pulley, i.e. $\ddot{\theta} = 0.2/0.4 = 0.5\ \text{rad s}^{-2}$)

Using $F = ma$ for the motion of the crate

$\leftarrow$ $$T = (100)(0.2) \qquad [1]$$

Using $C = I\ddot{\theta}$ for the rotation of the pulley

A↻ $$D - 0.4T = 24(0.5) \qquad [2]$$

Eliminating T from [1] and [2] gives

$$D - 8 = 12$$

Hence the driving torque is 20 N m.

(When the linear speed of the crate is $2\ \text{m s}^{-1}$, this is also the speed of the rope as it touches the pulley so the angular speed of the pulley is $2/0.4\ \text{rad s}^{-1}$, i.e. $5\ \text{rad s}^{-1}$.)

The total kinetic energy is made up of two parts

(a) the linear kinetic energy of the crate $(\frac{1}{2}mv^2)$,
(b) the rotational kinetic energy of the pulley $(\frac{1}{2}I\dot{\theta}^2)$.

Now $\quad$ K.E. of crate $= \frac{1}{2}(100)(2)^2\ \text{J} = 200\ \text{J}$

and $\quad$ K.E. of pulley $= \frac{1}{2}(24)(5)^2\ \text{J} = 300\ \text{J}$

Thus the total kinetic energy of the system is 500 J.

4) A cylinder of radius r is free to rotate about its axis which is horizontal. A light string hangs over the pulley and carries a particle P of mass m at one end and a particle Q of mass km at the other end. The string is rough enough not to slip on the pulley.
When the system is released from rest a constant frictional torque $\frac{1}{5}mgr$ acts on the pulley.
If the acceleration of the particles when $k = 3$ is twice their acceleration when $k = 2$ find the moment of inertia of the pulley.

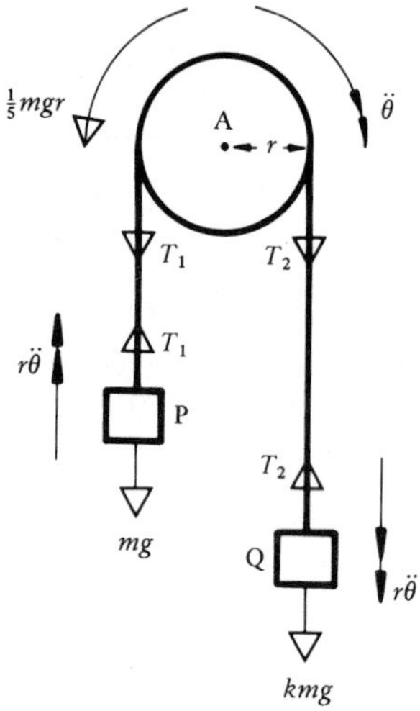

General diagram

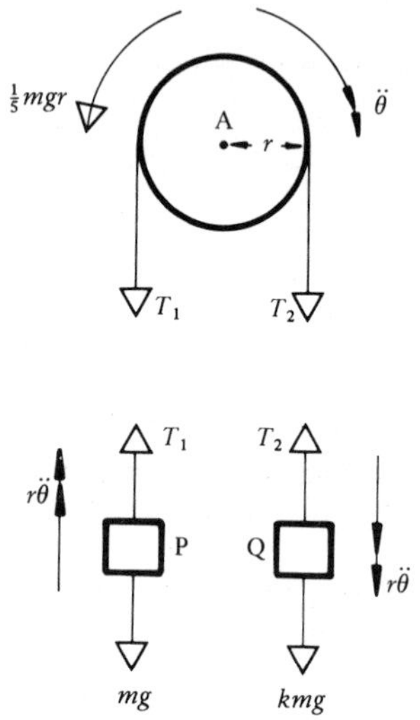

Diagram showing separately the forces acting on the cylinder and on each particle

(When a rough string passes over a pulley, the tensions in the string on either side of the pulley are different.)

Let the moment of inertia of the pulley be I.

Using $F = ma$ for the motion of P and Q we have

for P $\uparrow$ $$T_1 - mg = m(r\ddot{\theta}) \quad [1]$$

for Q $\downarrow$ $$kmg - T_2 = km(r\ddot{\theta}) \quad [2]$$

Using $C = I\ddot{\theta}$ for the motion of the pulley we have

A↻ $$T_2r - T_1r - \tfrac{1}{5}mgr = I\ddot{\theta}$$

$\Rightarrow$ $$T_2 - T_1 - \tfrac{1}{5}mg = I\ddot{\theta}/r \quad [3]$$

$$\text{Adding [1], [2] and [3]} \quad (k-\tfrac{6}{5})mg = (kmr+mr+I/r)\ddot{\theta}$$

$$\text{Now when } k=3, \quad \ddot{\theta}_3 = \tfrac{9}{5}mg/(4mr+I/r)$$

$$\text{and when } k=2, \quad \ddot{\theta}_2 = \tfrac{4}{5}mg/(3mr+I/r).$$

$$\text{But} \quad \ddot{\theta}_3 = 2\ddot{\theta}_2$$

$$\text{therefore} \quad \tfrac{9}{5}mg/(4mr+I/r) = \tfrac{8}{5}mg/(3mr+I/r)$$

$$\text{Hence} \quad 9(3mr+I/r) = 8(4mr+I/r)$$

$$\Rightarrow \quad I = 5mr^2$$

Thus the moment of inertia of the pulley is $5mr^2$.

EXERCISE 9a

1) A flywheel whose moment of inertia is $50\ \text{kg m}^2$ is rotating at $4\ \text{rad s}^{-1}$. Find its kinetic energy and angular momentum.

2) A flywheel loses kinetic energy amounting to 640 J when its angular speed falls from $7\ \text{rad s}^{-1}$ to $3\ \text{rad s}^{-1}$. What is the moment of inertia of the flywheel?

3) Find the kinetic energy of rotation of the earth, taking its moment of inertia as $98 \times 10^{36}\ \text{kg m}^2$.

4) A string is wound round a pulley of radius 0.3 m which is free to rotate about a smooth vertical axis. When the string is pulled horizontally with a constant force of 48 N, the angular velocity of the pulley increases from $2\ \text{rad s}^{-1}$ to $6\ \text{rad s}^{-1}$ in five seconds. Find the moment of inertia of the pulley.

5) A uniform circular flywheel whose moment of inertia is $64\ \text{kg m}^2$ rotates under the action of a constant couple. The flywheel starts from rest and after turning through 200 revolutions its angular speed is 20 revolutions per second. Find the moment of the couple.

6) An engine is running at 160 revolutions per minute and its flywheel has a moment of inertia of $600\ \text{kg m}^2$. When the engine is switched off the flywheel continues to rotate for 40 seconds before coming to rest. Find the magnitude of the constant retarding torque acting on the flywheel.

7) A particle of mass m is attached to one end of a light rough string which is wrapped several times round a cylinder of radius r. The cylinder can rotate about its axis of symmetry which is horizontal, and its moment of inertia about that axis is $4mr^2$.
Assuming that the string is rough enough not to slip on the cylinder find the angular acceleration of the cylinder when the system is released from rest.

8) A string is wrapped round a cylindrical shaft of radius 0.1 m and moment of inertia 200 kg m^2, and is pulled with a constant force of 400 N for five seconds. If there is a frictional couple of 15 N m acting on the cylinder, find the angular acceleration of the cylinder and the angle through which it turns in the five seconds (from rest).

9) Two particles of masses m_1 and m_2 are connected by a light string which passes over a pulley of radius r and moment of inertia I. If the string does not slip on the pulley find the difference between the tensions on the two sides of the pulley when the system is released from rest.
Find also the force which the string exerts on the pulley.

10)

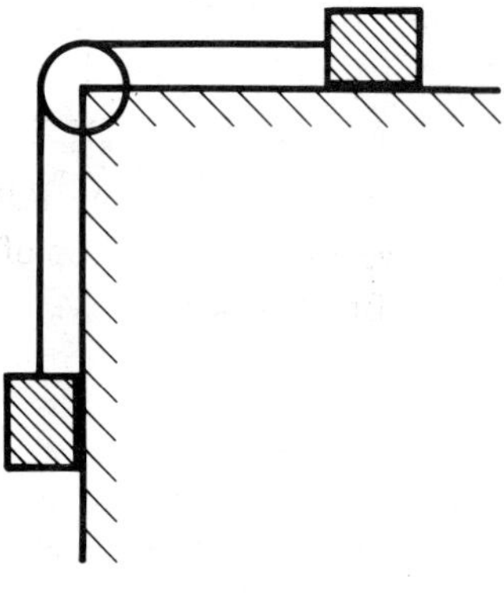

The diagram shows two particles of equal mass m connected by a rough light string which passes over a pulley at the edge of a smooth horizontal table. If the pulley is free to rotate about its axis and has a moment of inertia $4mr^2$ where r is its radius, find its angular acceleration when the system is released from rest.

MISCELLANEOUS EXERCISE 9

1) A uniform solid cylinder of mass $2m$ and radius r is free to rotate about its axis which is fixed and horizontal. A light inextensible string is wrapped round the surface of the cylinder several times and one end of the string is attached to the cylinder. The other end of the string is attached to a particle of mass m which hangs vertically downwards. (For the cylinder, $I = mr^2$.)
If the system is released from rest, prove that when the particle has fallen a distance r the angular velocity of the cylinder is $\sqrt{(g/r)}$. Find the time taken for the cylinder to complete one revolution from rest. (U of L)

2) Two particles of masses M and m are connected by a light inextensible string. The string passes over a pulley of radius a smoothly mounted on a horizontal axis and the particles hang freely. The string does not slip on the pulley and the moment of inertia of the pulley about its axis is na^2. Find the angular acceleration of the pulley while the system is in motion, and prove that the strings exert on the pulley a vertical force of magnitude

$$\frac{(M+m)n+4Mm}{M+m+n}g$$

(O)

3) A thin uniform circular disc, of mass M, radius a and centre C, is smoothly hinged to a fixed pivot at a point A on its circumference and is free to rotate in its own plane. When this plane is horizontal a constant coplanar couple G will bring the disc to rest from an angular velocity ω in exactly two revolutions. Show that $G = 3Ma^2\omega^2/(16\pi)$ and find the time taken in thus coming to rest. If the couple continues to act after bringing the disc momentarily to rest, find the further time taken and the number of revolutions completed whilst the disc reaches an angular velocity -2ω.
[The moment of inertia of the disc about the given axis is $\frac{3}{2}Ma^2$.] (U of L)

4) A uniform square lamina ABCD is free to rotate about a fixed vertical axis which coincides with the diagonal AC. The lamina is given an initial angular velocity ω_0 and, under the action of a constant driving torque G against a constant frictional torque T, completes 10 revolutions in the first second and 20 revolutions in the next second. Show that $\omega_0 = 10\pi \text{ rad s}^{-1}$.
The constant driving torque G is then removed and the lamina is brought to rest by the frictional torque T which has been constant throughout the motion. The lamina is thus brought to rest in a further 15 revolutions. Find T in terms of G.
Find also, in terms of G, the moment of inertia of the square about AC.

5) A cylinder, of radius r, is free to rotate about its axis which is horizontal. A light inextensible string is wrapped round the cylinder and has a load attached to its free end. With a load of W it is found that the load falls from rest a distance l in one second and a load of $4W$ similarly falls a distance $2l$ in one second. Assuming that the string does not slip and the frictional couple opposing the motion of the cylinder remains constant, calculate
(a) the moment of inertia of the cylinder about its axis,
(b) the magnitude of the frictional couple.

6) The diagram shows a uniform solid circular cylinder of mass M and radius a which can rotate freely about its axis of symmetry, which is horizontal. A thin light inextensible string with one end attached to a point on the curved surface of the cylinder is wound around the cylinder, the string lying in a plane perpendicular to the axis. A particle having mass equal to that of the cylinder is attached to the other end of the string and initially hangs, supported in equilibrium, below the cylinder with the string taut (as illustrated).

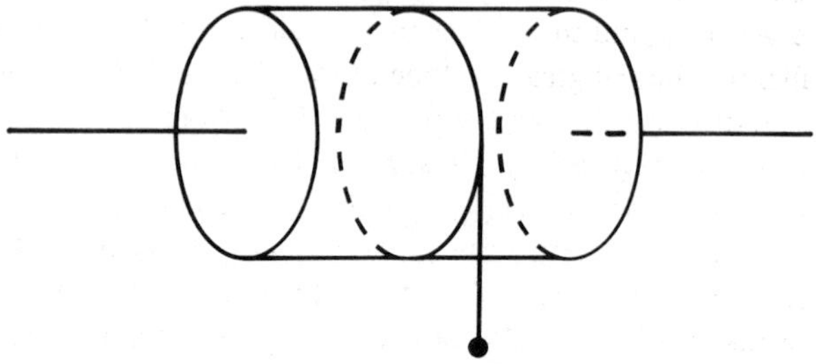

The particle is released and descends under gravity causing the cylinder to rotate. Find the angular velocity of the cylinder when the particle has descended a distance d, the string being still partly wound round the cylinder. (The moment of inertia of the cylinder is $\frac{1}{2}Ma^2$.) (JMB)

7) A heavy pulley, which may be regarded as a uniform circular disc of mass $6m$, centre O and radius a, can turn freely in a vertical plane about a fixed horizontal axis through O. A light inextensible string passes over the pulley and particles of mass $2m$ and $3m$ are attached at its free ends. The system is released from rest and the string does not slip on the pulley. Show that the heavier particle falls with acceleration $g/8$. (For the pulley, $I = 3ma^2$.)
After time t a constant frictional couple is applied to the pulley and in consequence the system comes to rest again in a *further* time t.
Assuming that the lighter particle does not reach the pulley throughout the motion, calculate
(a) the total distance covered by the heavier particle,
(b) the magnitude of the constant frictional couple. (U of L)

8) A rigid body can turn freely about a smooth fixed vertical axis and the moment of inertia of the body about this axis is I. A constant couple of magnitude N acts on the body in a plane perpendicular to the axis of rotation. Find (a) the time taken for the angular velocity of the body to increase from ω_1 to ω_2, (b) the angle through which the body rotates in this time.
A uniform circular disc whose moment of inertia is $20\ \text{kg m}^2$ rotates, under the action of a constant couple, about a fixed axis through its centre perpendicular to its plane.

If the speed of rotation changes from 10 revolutions per minute to 35 revolutions per minute in 15 seconds, find the moment of the couple acting on the disc.
Find also the angular acceleration and the number of revolutions made by the disc in this time.

9) The diagram shows particles A and B, each of mass M, connected by a light inextensible string which passes over a smooth light fixed pulley C. The particle A is on a plane inclined at an angle α to the horizontal, the coefficient of friction between A and the plane being μ. Between A and the pulley the string is parallel to a line of greatest slope of the plane and B is hanging freely. Find the acceleration with which the particle B descends.
The pulley C is now replaced by a heavy one D, of the same radius a which is rough enough to prevent the string slipping over it, all other conditions remaining unchanged. As a result the acceleration of B is halved. Assuming that the bearings of D are frictionless, find the moment of inertia of D about its axis. If the change of pulley also causes a 10% increase in the tension in the vertical portion of the string, show that

$$\mu = \frac{2-3\sin\alpha}{3\cos\alpha}$$

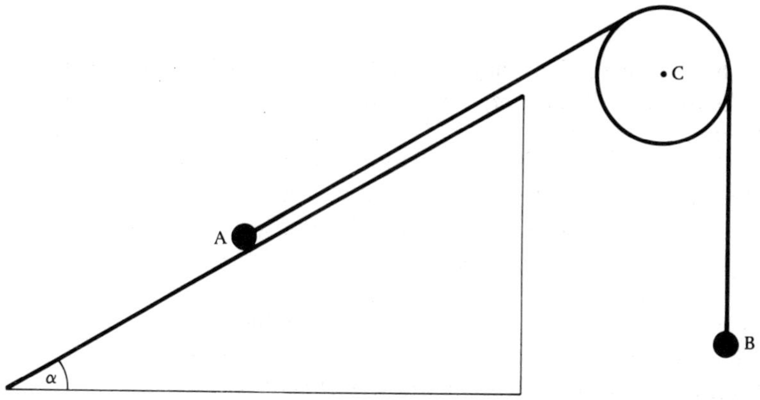

(U of L)

10) When a uniform solid cylinder of radius a and mass M is rotating about its fixed horizontal axis there is a constant resisting torque of magnitude G. A light inextensible string has one end attached to the curved surface of the cylinder and is wound several times round this surface. Hanging from the free end of this string is a particle of mass km. When the system is released from rest the particle has an acceleration $\frac{1}{4}g$ if $k = 1$ and an acceleration $\frac{1}{2}g$ if $k = 2$. Find M and G in terms of a, m and g. Find the acceleration of the particle when $k = 4$.
(The moment of inertia of the cylinder is $\frac{1}{2}Ma^2$.) (U of L)

11) A uniform circular pulley, of radius r and mass m, is in a vertical plane and is free to turn about a horizontal axis passing through its centre, the moment of inertia of the pulley about this axis being $\frac{mr^2}{2}$. A light inextensible string passes over the pulley and hangs vertically on each side, supporting a particle of mass m on one side and a particle of mass $2m$ on the other. The system is released from rest at time $t = 0$, and, in the subsequent motion, the string does not slip on the pulley and the rotation of the pulley is resisted by a frictional couple of constant moment G. If θ is the angular displacement of the pulley at time t, prove that

$$7mr^2 \frac{d^2\theta}{dt^2} = 2(mgr - G)$$

When the pulley has turned through an angle α, the string is completely detached from the pulley without any impulse. If the pulley then turns through a further angle β before being brought to rest by the frictional couple, show that

$$G = \frac{mgr\alpha}{\alpha + 7\beta}$$

(JMB)

12) Two particles of masses m and $3m$ are connected by a light inextensible string which passes over a pulley of radius a. The pulley is free to turn without friction about a fixed horizontal axis through its centre and the moment of inertia about this axis is I. This system is released from rest. If the string does not slip on the pulley, show that the time taken for the mass $3m$ to descend a distance x is

$$\left[\frac{x(4ma^2 + I)}{mga^2}\right]^{1/2}$$

Find the constant couple which must be applied to the pulley in order that the particles reverse their motion and move with an acceleration $g/2$. (U of L)

13) A wheel has a horizontal cylindrical axle of radius a. The system of wheel and axle has a moment of inertia Mk^2 about its axis and rotates without friction. A light thin inextensible string is wound round the axle and has one end attached to a point on the axle. To the other end of the string there is attached a particle P of mass m which hangs vertically. All of the string lies in a vertical plane. If the system is released from rest, prove that the downward acceleration of P is $mga^2/(ma^2 + Mk^2)$.
When P has descended a distance $8a$ it falls off and the wheel is then brought to rest after n further revolutions by a constant braking couple of magnitude G. Calculate G. (U of L)

14) A uniform circular pulley, of radius a, is free to rotate about a smooth fixed horizontal axis about which its moment of inertia is $2ma^2$. A light inextensible string AB passes over the pulley and particles of masses $2m$ and $3m$ are attached to the ends A and B respectively of the string. The rim of the pulley is rough enough to prevent the string slipping. When the system is moving freely, find the angular acceleration of the pulley and the tension in each of the vertical parts of the string.
When the angular speed of the pulley is Ω a constant braking couple of magnitude G is applied to the pulley so that the system is brought to rest in time T without the string going slack or slipping. Find an expression for G in terms of m, a, Ω, g and T. (U of L)

15) A flywheel, whose moment of inertia about an axis perpendicular to its plane and through its centre of mass is I, is initially at rest. The axis is fixed and the flywheel rotates about it under the influence of a constant couple of magnitude C and a resisting couple of magnitude $k\omega^2$, where k is a positive constant and ω is the angular speed. Show that ω can never exceed Ω, where $\Omega = \sqrt{\dfrac{C}{k}}$. Find the time taken for the flywheel to attain the angular speed ω and show that during this time it rotates through an angle

$$\frac{I\Omega^2}{2C}\ln\left\{\frac{\Omega^2}{\Omega^2-\omega^2}\right\}$$

16) Two gear wheels A and B, of radii a and $2a$ and moments of inerta I and $3I$ respectively, are mounted on parallel axes and run permanently in mesh. The system is set in motion with a constant driving torque G applied to A. When in motion, there are constant frictional torques $3L$ and $2L$ acting on A and B respectively. Find
(a) the tangential force between the wheels,
(b) the angular acceleration of A,
(c) the number of revolutions made by B in acquiring from rest an angular velocity Ω.

Prove also that, for motion to be possible, $G > 4L$. (U of L)

CHAPTER 10

CALCULATION OF MOMENT OF INERTIA

When a rigid body rotates about a fixed axis the moment of inertia of that body about that axis is Σmr^2 where m is the mass of a constituent particle of the body and r is the distance of that particle from the axis of rotation.

As moment of inertia is the product of mass and the square of a distance the *unit* in which it is measured is one *kilogram metre squared* (kg m^2). It should also be noted that Σmr^2 is a scalar quantity.

Thus the moment of inertia of a body depends on two properties

(a) the position of the axis of rotation,

(b) the distribution of the mass about that axis.

It must be emphasised that it is inadequate to refer simply to 'the moment of inertia of a body' – this term is meaningless unless the axis of rotation is also specified.

CALCULATION OF MOMENT OF INERTIA

The evaluation of Σmr^2 for a particular body is done in one of two ways. If the body consists of a finite number of particles (e.g. two particles fixed at the ends of a light rod), this summation can be done by simple addition. In the case of a solid body the summation may have to be done by integration.

The Moment of Inertia of a Set of Particles

EXAMPLES 10a

1) Three light rods, each of length $2l$, are joined together to form a triangle. Three particles A, B, C of mass m, $2m$, $3m$ are fixed to the vertices of the triangle. Find the moment of inertia of the resulting body about
(a) an axis through A perpendicular to the plane ABC,
(b) an axis passing through A and the midpoint of BC.

(a)

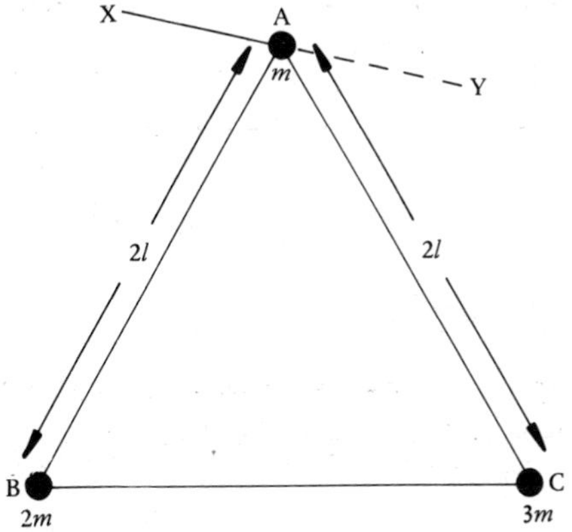

B is distant $2l$ from the axis XY
so the moment of inertia of B (I_B) about XY is $2m(2l)^2$
similarly I_C about XY is $3m(2l)^2$
and I_A about XY is $m(0)^2$

Therefore the moment of inertia of the body about XY is

$$2m(2l)^2+3m(2l)^2+m(0)^2 = 20ml^2$$

(b)

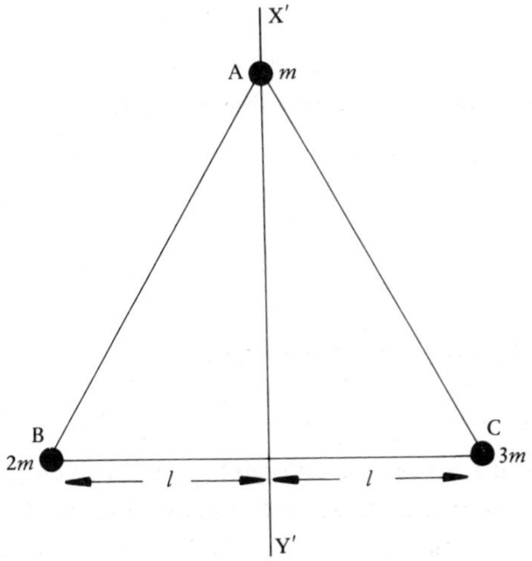

I_A about X'Y' $= m(0)^2$

I_B about X'Y' $= 2m(l)^2$

I_C about X'Y' $= 3m(l)^2$

Therefore the moment of inertia of the body about X'Y' is

$$m(0)^2 + 2m(l)^2 + 3m(l)^2 = 5ml^2$$

(This example emphasises that the moment of inertia of a body is not unique but depends on the axis of rotation.)
The moment of inertia of any object which consists of a *finite* number of particles can be found in a similar way.

We shall now consider bodies which have a continuous distribution of mass.

The Moment of Inertia of a Ring

Consider a uniform ring of mass M and radius a and an axis through its centre and perpendicular to the plane of the ring.

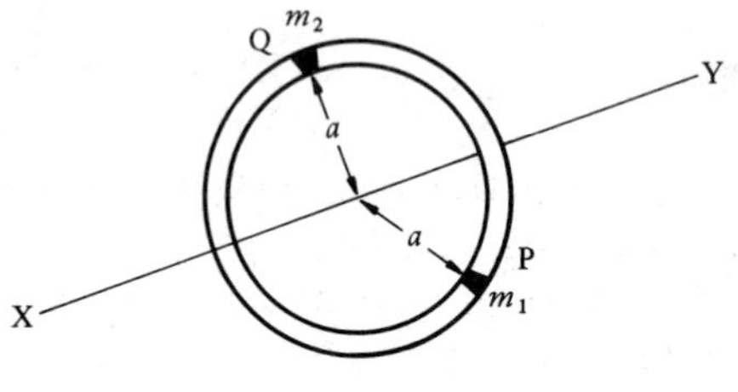

The ring is made up of particles whose masses are $m_1, m_2, m_3, \ldots$

The moments of inertia of these particles about XY are

$$m_1a^2, \quad m_2a^2, \quad m_3a^2, \ldots$$

Therefore the moment of inertia of the ring about XY is

$$m_1a^2 + m_2a^2 + m_3a^2 + \ldots = (m_1 + m_2 + m_3 + \ldots)a^2$$

But $m_1 + m_2 + m_3 + \ldots = M$
Therefore the moment of inertia of the ring about XY is Ma^2

This result also provides a useful 'building block' for finding the moments of inertia of more complex bodies which can be considered to be made up of ring-like elements.

The Moment of Inertia of a Disc

Consider a uniform disc of mass M and radius a and an axis through the centre of the disc perpendicular to the plane of the disc.

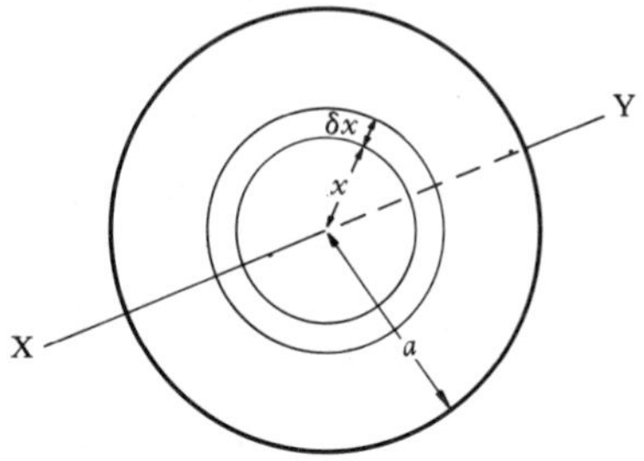

If the disc is divided into annuli each of width δx then, if δx is small, each such annulus is approximately a ring.

Consider a typical such 'ring' of radius x.

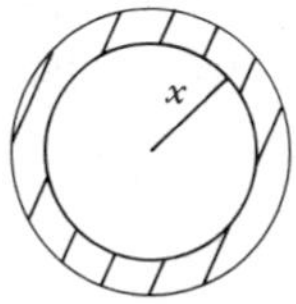

The area of the annulus, which is approximately that of a strip of length $2\pi x$ and width δx, is approximately $2\pi x\,\delta x$.
So the mass of the ring element is approximately $m(2\pi x\,\delta x)$ where m is the mass per unit area of the disc.

Therefore the moment of inertia of the element about $\text{XY} \simeq m(2\pi x\,\delta x)(x^2)$
It follows that the moment of inertia, I of the disc about $\text{XY} \simeq \Sigma\, 2\pi m x^3\,\delta x$

i.e. $$I \simeq 2\pi m \sum x^3\,\delta x$$

$\Rightarrow$ $$I = 2\pi m \int_0^a x^3\,\mathrm{d}x = \tfrac{1}{2}\pi m a^4$$

But $\pi a^2 m = M$

Therefore the moment of inertia of the disc about XY is $\tfrac{1}{2}Ma^2$.

This is another very useful result as it enables us to find the moment of inertia of bodies which can be divided into 'disc-like' elements, particularly solids of revolution which can be divided into such elements perpendicular to the axis of symmetry.

The Moment of Inertia of a Solid of Revolution

Consider a uniform solid which is formed by rotating, about the x axis, the area bounded by the lines $x = a$ and $x = b$, the x axis and part of the curve with equation $y = f(x)$.

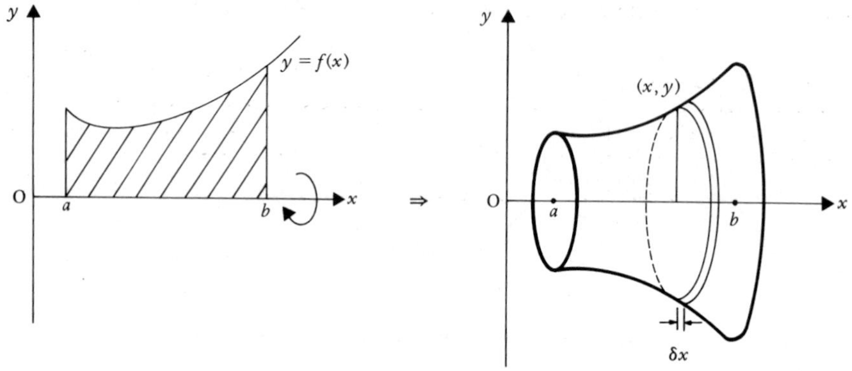

If we divide the solid into elements that are 'slices' parallel to the plane faces, each such element is approximately a disc.

For a typical element, volume $\simeq \pi y^2 \delta x$

$\Rightarrow$ mass $\simeq (\pi y^2 \delta x)m$

(where m is the mass/unit volume)

The element is approximately a disc of radius y so, using the known moment of inertia of this disc about Ox, we have

for a typical element,

$$\text{moment of inertia about } Ox \simeq \tfrac{1}{2}(\pi m y^2 \delta x)(y^2)$$

Hence for the whole solid,

$$\text{moment of inertia about } Ox \simeq \sum_a^b \tfrac{1}{2}\pi m y^4 \delta x$$

$$= \tfrac{1}{2}\pi m \int_a^b y^4 \, dx$$

Up to this stage the analysis applies to *any* solid of revolution formed by rotation about Ox.

For a particular solid, where the values of a and b and the equation of the curve are known, the integration can be carried out and the moment of inertia determined.

EXAMPLES 10a (continued)

2) Find the moment of inertia about a diameter of a uniform solid sphere of mass M and radius a.

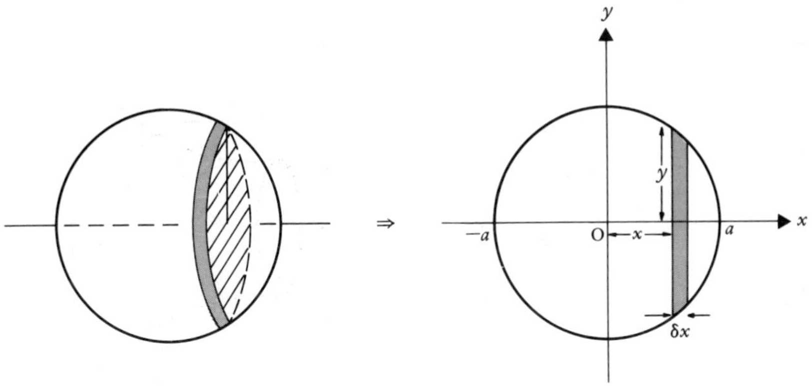

Taking the centre of the sphere as origin, the equation of the generating curve is $x^2+y^2=a^2$.
If the sphere is divided into slices perpendicular to Ox, each slice is approximately a disc of radius y and thickness δx.

Therefore, for a typical element, volume $\simeq \pi y^2 \delta x$

mass $\simeq (\pi y^2 \delta x)m$

moment of inertia about $Ox \simeq \frac{1}{2}(\pi m y^2 \delta x)(y^2)$

Hence the moment of inertia about Ox, I_{Ox}, for the whole sphere is given by

$$I_{Ox} \simeq \sum_{-a}^{a} \tfrac{1}{2}\pi m y^4 \delta x = \tfrac{1}{2}\pi m \int_{-a}^{a} y^4 \, dx$$

But $y^2 = a^2 - x^2$ therefore

$$I_{Ox} = \tfrac{1}{2}\pi m \int_{-a}^{a} (a^4 - 2a^2x^2 + x^4)\, dx$$

$$= \tfrac{1}{2}\pi m \left[a^4 x - \tfrac{2}{3}a^2x^3 + \tfrac{1}{5}x^5\right]_{-a}^{a}$$

$$= \tfrac{8}{15}\pi m a^5$$

Now $M = \frac{4}{3}\pi a^3 m$

Therefore the moment of inertia of the sphere about a diameter is $\frac{2}{5}Ma^2$.

3) Find the moment of inertia about its axis of symmetry, of a uniform solid of mass M, formed by rotating about the y axis the area bounded by the y axis, the line $y=2$ and part of the curve with equation $y=x^2$.

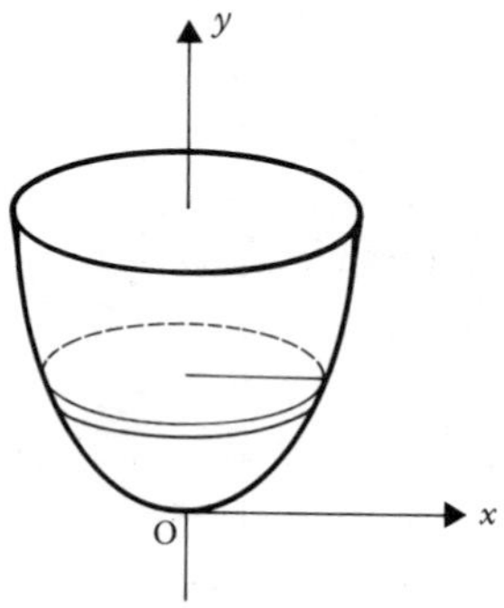

$\Rightarrow$

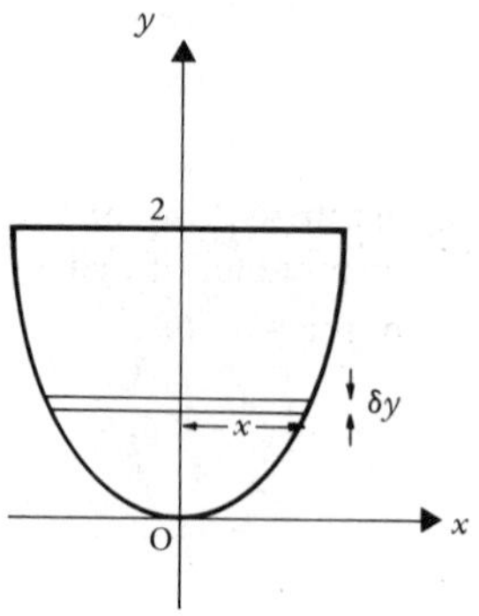

If the solid is divided into horizontal disc-like elements of radius x and thickness δy then, for a typical element,

$$\text{volume} \simeq \pi x^2 \delta y$$

$$\text{mass} \simeq (\pi x^2 \delta y)m$$

$$\text{moment of inertia about Ox} \simeq \tfrac{1}{2}(\pi m x^2 \delta y)x^2$$

Hence I_{Oy}, the moment of inertia of the whole solid about Oy, is given by

$$I_{Oy} \simeq \sum_{y=0}^{y=2} \tfrac{1}{2}\pi m x^4 \delta y$$

$$= \tfrac{1}{2}\pi m \int_0^2 x^4 \,\mathrm{d}y$$

$$= \tfrac{1}{2}\pi m \int_0^2 y^2 \,\mathrm{d}y \qquad (y = x^2)$$

$$= \tfrac{4}{3}\pi m$$

In this problem the relationship between M and m depends on finding the volume of the solid of revolution.

$$\text{Volume} = \int_0^2 \pi x^2 \,\mathrm{d}y = \pi \int_0^2 y \,\mathrm{d}y = 2\pi$$

$$\Rightarrow \qquad M = 2\pi m$$

Therefore the moment of inertia of the solid about Oy is $\tfrac{2}{3}M$.

EXERCISE 10a

1) Four particles, A, B, C and D of mass 2 kg, 5 kg, 6 kg, 3 kg respectively are rigidly joined together by light rods to form a rectangle ABCD, where $AB = 2a$ and $BC = 4a$. Find the moment of inertia of this system of particles about an axis along (a) AB (b) BC (c) AC (d) through the midpoints of AB and CD.

2) A uniform annulus whose inner and outer radii are a and $2a$, is of mass M. Find its moment of inertia about an axis through the centre of, and perpendicular to, the annulus.

3) Find the moment of inertia, about the axis of symmetry, of a uniform solid cylinder of mass M, radius a and length l.

4) A uniform right solid circular cone has a base radius r and height h. Find its moment of inertia, about its axis of symmetry, given that its mass is M.

5) A solid uniform sphere of radius $2a$ is cut into two portions by a plane distant a from the centre of the sphere. Using an axis through the centres of the sphere and the circular cut faces, find, in terms of m and a, the moment of inertia of each portion (m is the mass per unit volume of the sphere).

6) A uniform solid is formed by rotating the portion of the curve $y^2 = 4ax$ between $x = 0$ and $x = 2a$ about the x axis. If the mass of the solid is M and the mass per unit volume is m, find a relationship between M and m and then find the moment of inertia of the solid about the x axis.

7) A uniform solid is formed by rotating, about the x axis, the area bounded by the x axis, the line $x = 1$ and part of the curve $y = x^3$. Find the moment of inertia of the solid, given that its mass is M.

8) A solid right circular cone has base radius a and height $4a$. A cut is made through the midpoint of, and perpendicular to, the axis of symmetry. The upper portion is removed leaving a frustum of mass M. Find the moment of inertia of the frustum about its axis of symmetry.

9) The area bounded by the x and y axes, the line $x = 2$ and part of the curve $y^2 = kx$ is rotated about the x axis to form a uniform solid of revolution of mass M. If the moment of inertia of the solid about the x axis is $\frac{4}{3}M$, find the value of k.

The Moment of Inertia of a Surface of Revolution

When a line is rotated about an axis, it generates a surface which is symmetrical about that axis.
If the surface is then divided into elements by cuts perpendicular to the axis, each element is approximately a ring.

This, then is another case where we can use the known moment of inertia of a ring element, this time to 'build up' the moment of inertia of a uniform surface of revolution about its axis of symmetry.

EXAMPLES 10b

1) Find the moment of inertia of a hollow cylinder, of mass M and radius a, about its axis.

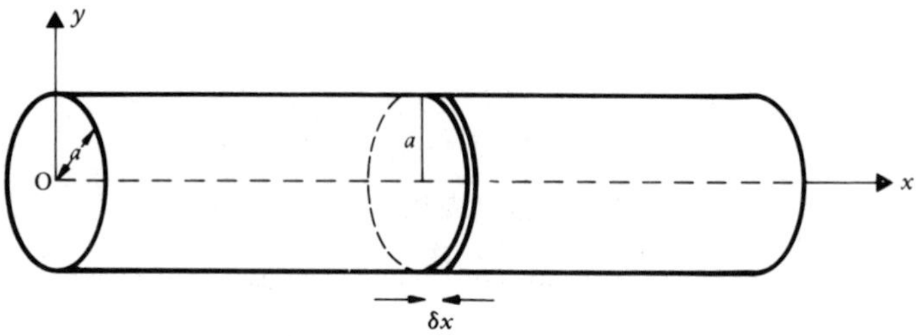

Consider the hollow cylinder to be split into ring-like elements of radius a and width δx and let m be the mass/unit area of the cylinder.

For a typical element,

$$\text{area} = 2\pi a\,\delta x$$

$$\text{mass} = 2\pi ma\,\delta x$$

$$\text{moment of inertia about Ox} = (2\pi ma\,\delta x)a^2$$

Therefore I_{Ox}, the moment of inertia of the hollow cylinder about Ox, is given by

$$I_{Ox} = \sum 2\pi ma^3\,\delta x = 2\pi ma^3 \sum \delta x$$

But $$M = \sum (2\pi a\,\delta x)m = 2\pi ma \sum \delta x$$

Therefore the moment of inertia of the hollow cylinder about its axis is Ma^2.

Note that all expressions in this problem are exact and not approximations.

2) Find the moment of inertia of a spherical shell about a diameter.

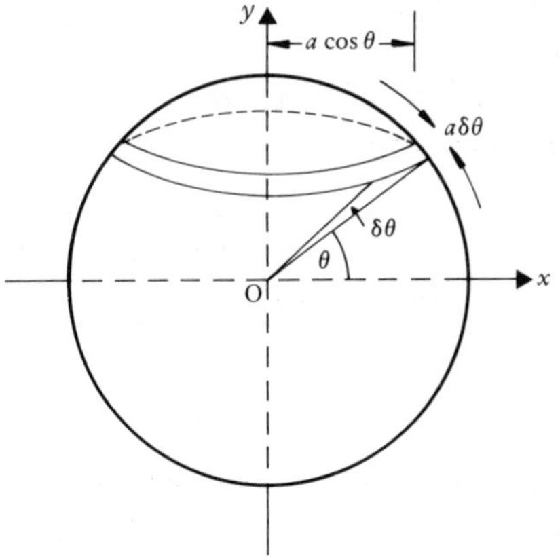

Consider the spherical shell to be divided into ring-like elements, by cuts perpendicular to Oy, and let m be the mass per unit area of the shell.

For a typical element, of radius $a\cos\theta$ and width $a\delta\theta$, we have,

$$\text{area} \simeq 2\pi(a\cos\theta)(a\delta\theta)$$

$$\text{mass} \simeq (2\pi a^2\cos\theta\,\delta\theta)m$$

$$\text{moment of inertia about O}y \simeq (2\pi m a^2\cos\theta\,\delta\theta)(a\cos\theta)^2$$

Therefore the moment of inertia, $I_{\text{O}y}$, of the whole spherical shell about Oy is given by

$$I_{\text{O}y} \simeq 2\pi m a^4 \sum_{-1/2\pi}^{1/2\pi} \cos^3\theta\,\delta\theta$$

$$= 2\pi m a^4 \int_{-1/2\pi}^{1/2\pi} \cos^3\theta\,d\theta$$

$$= 4\pi m a^4 \int_{0}^{1/2\pi} \cos^3\theta\,d\theta$$

$$= 4\pi m a^4(\tfrac{2}{3}) \qquad \text{(Reduction formula)}$$

But $M = (4\pi a^2)m$

Hence the moment of inertia of the spherical shell about a diameter is $\frac{2}{3}Ma^2$.

Note. In this problem θ was chosen as the variable as neither δx nor δy can be used as a reasonable approximation for the width of the element 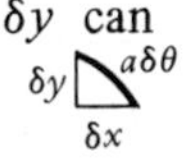

EXERCISE 10b

1) Find the moment of inertia, about its axis of symmetry, of a uniform hollow hemisphere of radius $5a$ and mass M.

2) Find the moment of inertia of a uniform hollow cone of mass M, base radius a and semi-vertical angle α, about the axis of the cone.
(**Note**. Taking the vertex of the cone as the origin O, and the axis of the cone as the x axis, the width of an element is $(\delta x)\sec\alpha$.)

Questions 3 and 4 require knowledge of the method for finding the area of a surface of revolution.

3) A uniform surface of mass m per unit area is formed by rotating the section of the curve $y = e^x$ between $x = 0$ and $x = 1$ about the x axis. Find the moment of inertia of this surface of revolution about its axis in terms of its mass M.

4) The portion of the curve $y = \cosh x$ between $x = 0$ and $x = 1$ is rotated completely about the x axis to form a uniform surface of revolution of mass M. Find its moment of inertia about the x axis.

The Moment of Inertia of a Rod

So far our elements have followed the 'building-up' sequence of

$$\text{particle} \Rightarrow \text{ring} \begin{array}{l} \nearrow \text{disc} \Rightarrow \text{solid of revolution} \\ \searrow \text{surface of revolution.} \end{array}$$

Now we begin another sequence by starting with a rod and we will find its moment of inertia about an axis
(a) through the centre of the rod and perpendicular to the rod,
(b) through the centre of the rod and inclined at an angle α to the rod,
(c) parallel to the rod and distant d from it.

Consider a rod of length $2a$ and mass per unit length m, divided into elements of length δx, each element being approximately a particle.

(a)

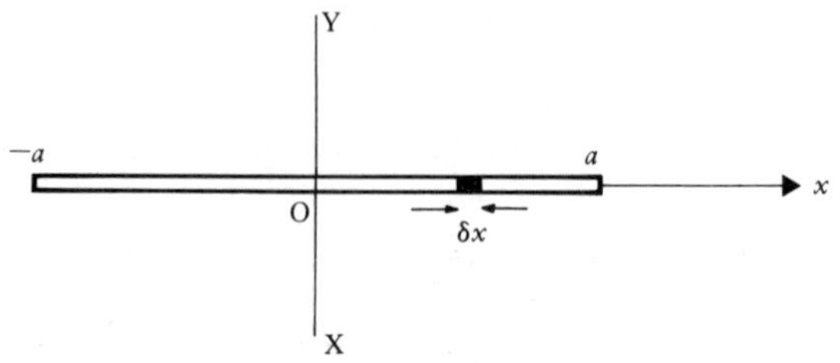

For a typical element,

$$\text{mass} = m\delta x$$

$$\text{moment of inertia about XY} \simeq (m\delta x)(x^2)$$

Therefore I_{XY}, the moment of inertia of the rod about XY is given by

$$I_{XY} \simeq \sum_{-a}^{a} mx^2\,\delta x$$

$$= m\int_{-a}^{a} x^2\,dx$$

$$= \tfrac{2}{3}ma^3$$

$$= \tfrac{1}{3}Ma^2 \qquad (M = 2am)$$

(b)

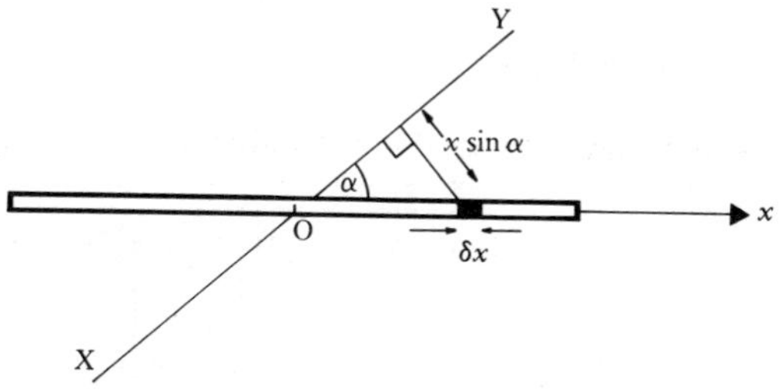

The distance from the axis XY of the particle-like element, of mass $m\delta x$, is $x \sin\alpha$, therefore

$$\text{moment of inertia of element about XY} \simeq (m\delta x)(x \sin\alpha)^2$$

Hence I_{XY}, the moment of inertia of the rod about XY, is given by

$$I_{XY} \simeq \sum_{-a}^{a} mx^2 \sin^2\alpha\,\delta x$$

$$= m\sin^2\alpha\int_{-a}^{a} x^2\,dx$$

$$= \tfrac{2}{3}ma^3\sin^2\alpha$$

$$= \tfrac{1}{3}Ma^2\sin^2\alpha$$

(c)

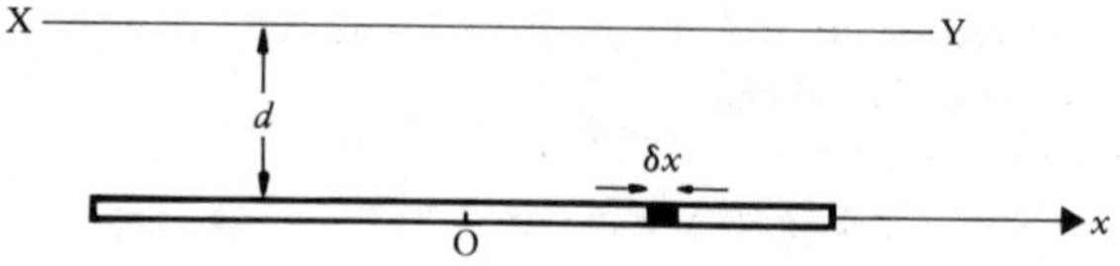

In this case every element of the rod is the same distance, d, from the axis XY.

The moment of inertia of an element about $\text{XY} = (m\delta x)(d^2)$

Therefore the moment of inertia of the rod about $\text{XY} = \sum_{-a}^{a} md^2\delta x$

$$\Rightarrow \qquad d^2 \sum m\,\delta x = Md^2$$

We are now in a position to find the moments of inertia about certain axes of bodies that can be divided into rod-like elements.

EXAMPLES 10c

1) Find the moment of inertia of a uniform lamina, of mass M, in the shape of a rectangle ABCD, where $\text{AB} = 2a$ and $\text{BC} = 2b$, about an axis through the midpoints of AB and CD.

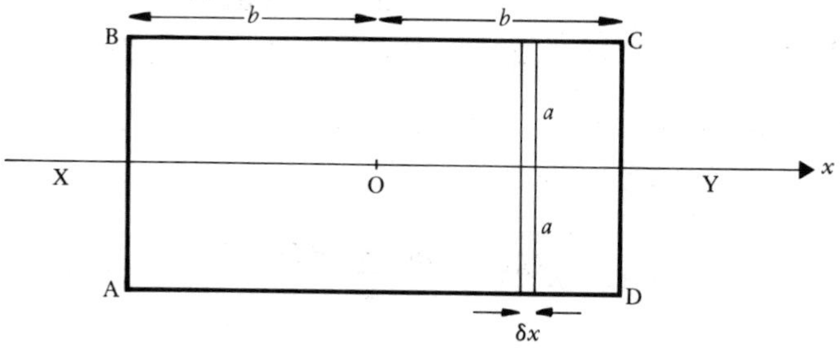

If the lamina is divided into strips parallel to AB then each strip is approximately a rod of length $2a$.

For a typical element, mass $= (2a\delta x)m$

moment of inertia about XY $= \frac{1}{3}(2am\delta x)a^2$

Therefore I_{XY}, the moment of inertia of the lamina about XY is given by

$$I_{\text{XY}} = \sum_{-b}^{b} \tfrac{2}{3}ma^3\delta x = \tfrac{4}{3}ma^3 b = \tfrac{1}{3}Ma^2 \qquad (M = 4ab)$$

2) Find the moment of inertia of a uniform equilateral triangular lamina ABC, of mass M and side $2a$, about
(a) the median through A,
(b) a line through A parallel to BC.

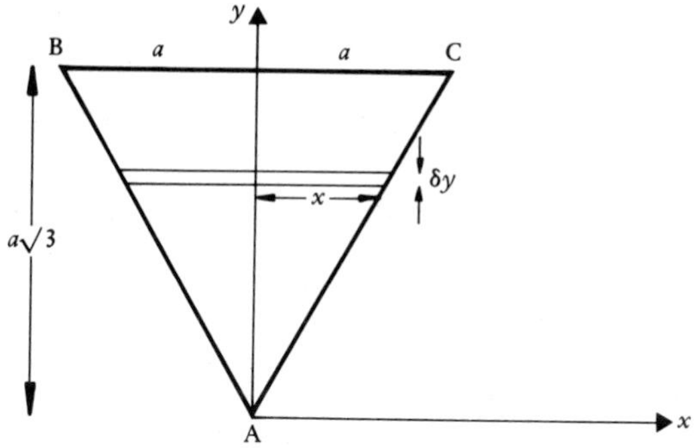

(a) The median through A is the y axis and m is the mass per unit area of the lamina which is divided into strips parallel to BC.

For a typical element which is approximately a rod of length $2x$ perpendicular to Ay, we have,

$$\text{area} \simeq 2x\,\delta y$$

$$\text{mass} \simeq (2x\,\delta y)m$$

$$\text{moment of inertia about A}y \simeq \tfrac{1}{3}(2mx\,\delta y)(x^2)$$

Therefore, for the whole lamina, the moment of inertia, $I_{\text{A}y}$, about Ay is given by

$$I_{\text{A}y} \simeq \tfrac{2}{3}m \sum_{0}^{a\sqrt{3}} x^3\,\delta y$$

$$= \tfrac{2}{3}m \int_0^{a\sqrt{3}} x^3\,\mathrm{d}y$$

The equation of AC is $y = x\tan 60^\circ \Rightarrow y = x\sqrt{3}$

Therefore
$$I_{\text{A}y} = \tfrac{2}{3}m \int_0^{a\sqrt{3}} \frac{y^3}{3\sqrt{3}}\,\mathrm{d}y$$

$$= \frac{2\sqrt{3}}{27}\left(\frac{9a^4}{4}\right)$$

But $M = (a^2\sqrt{3})m$ so

the moment of inertia of the lamina about a median is $\tfrac{1}{6}Ma^2$.

(b) The line through A parallel to BC is the x axis.

We can use the same element, which is now parallel to the axis of rotation, so we have

$$\text{moment of inertia of element about Ax} \simeq (2mx\,\delta y)(y^2)$$

Hence for the whole lamina,

$$I_{Ax} \simeq \sum_{0}^{a\sqrt{3}} 2mxy^2\,\delta y$$

$$= 2m\int_0^{a\sqrt{3}} xy^2\,dy$$

$$= 2m\int_0^{a\sqrt{3}} \frac{y}{\sqrt{3}}\,y^2\,dy$$

$$= \frac{3\sqrt{3}}{2}ma^4$$

$$= \tfrac{3}{2}Ma^2$$

Note. This example shows very clearly that a change of axis produces, for the same body, a completely different moment of inertia.

3) Find the moment of inertia, about the axis of symmetry, of a uniform lamina of mass M in the shape of the area of the parabola $y^2 = 4ax$ cut off by the line $x = a$.

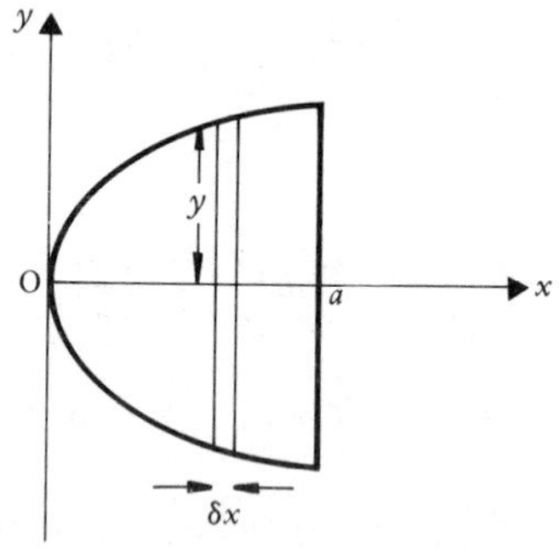

Dividing the lamina into vertical strips perpendicular to Ox we have, for a typical rod-like element,

$$\text{mass} \simeq (2y\,\delta x)m$$

$$\text{moment of inertia about Ox} \simeq \tfrac{1}{3}(2my\,\delta x)(y^2)$$

For the whole lamina, therefore,

$$I_{Ox} \simeq \tfrac{2}{3}m \sum_{0}^{a} y^3 \,\delta x$$

$$= \tfrac{2}{3}m \int_0^a (4ax)^{3/2}\,dx$$

$$= \tfrac{16}{3}ma^{3/2}\left[\tfrac{2}{5}x^{5/2}\right]_0^a$$

$$= \tfrac{32}{15}ma^4$$

To find the relationship between M and m, we need the area of the lamina, which is

$$2\int_0^a y\,dx = 2\int_0^a 2(ax)^{1/2}\,dx = \tfrac{8}{3}a^2$$

$\Rightarrow \quad M = \tfrac{8}{3}ma^2.$

Hence the moment of inertia about Ox of the lamina is $\tfrac{4}{5}Ma^2$.

EXERCISE 10c

1) ABCD is a uniform rectangle of mass M where $AB = 2a$ and $BC = 2b$. Find the moment of inertia of ABCD about the edge AB by dividing the rectangle into rod-like elements parallel to AB.

2) Find the moment of inertia of a uniform lamina in the form of a triangular lamina ABC, of mass M, where $AB = AC = 2a$ and $BC = 2b$, about the median through A.

3) Find the moment of inertia of a uniform equilateral triangular lamina about one side given that the length of each side is $2l$ and the mass of the lamina is M.

4)

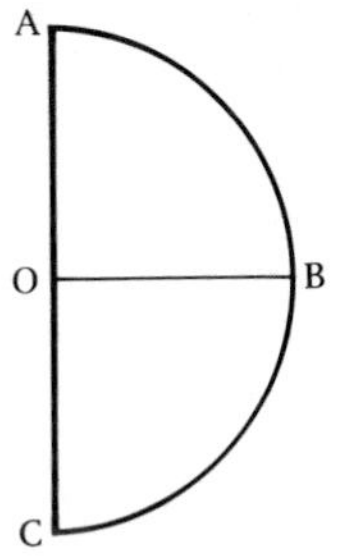

OABC is a uniform semicircular lamina of mass M and radius a. By using strip elements parallel to AC, find the moment of inertia of the lamina about OB.

5) 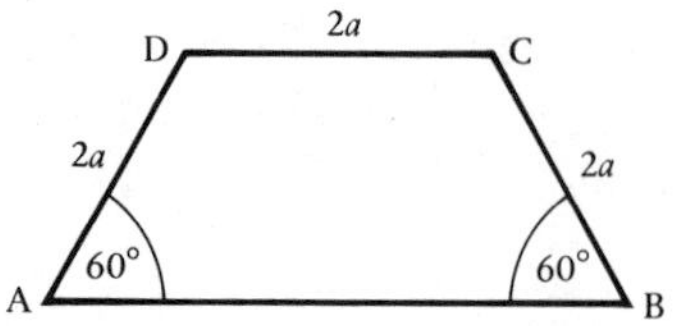

ABCD is a uniform lamina of mass M in the shape of a trapezium. By using strip elements parallel to AB, find the moment of inertia of the lamina about AB.

6) Find, by integration, the moment of inertia of a uniform rod of mass M and length $2l$ about an axis through one end of the rod and which is (a) perpendicular to the rod, (b) inclined at an angle θ to the rod.

7) Use the result obtained in question 6 (b) to find the moment of inertia of a uniform lamina in the form of an equilateral triangle of side $2a$ and mass M, about one side. Compare this method with the one used in question 3.

Non-Uniform Bodies

All the bodies we have considered so far have been uniform. Some of the results obtained apply equally well when the body concerned is not uniform. This is so in the case of bodies whose constituent particles are all at the same distance from the axis of rotation. In the case of a ring, for example, the moment of inertia about an axis through the centre and perpendicular to the ring is Ma^2 whether the ring is uniform or not.

This property is *not* true for any other axis.

In general, the moment of inertia of a non-uniform body can be determined if the mass of a particle is a function of its position in the body.

EXAMPLE 10d

A semicircular lamina is of radius a and O is the midpoint of its diameter. The mass per unit area at a point distant x from O is mx. If M is the total mass, find the moment of inertia of the lamina about an axis through O and perpendicular to the lamina.

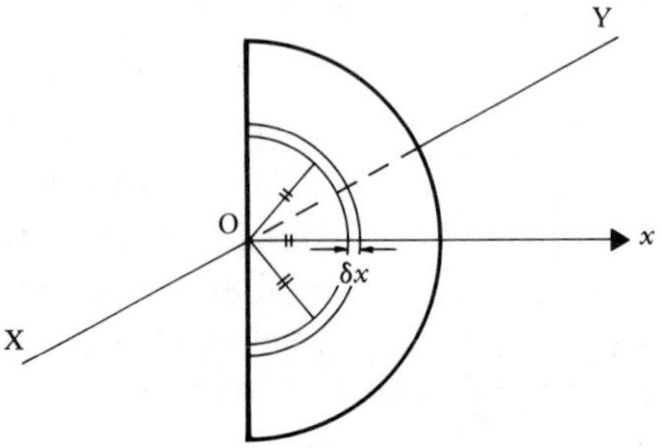

Let the lamina be divided into semicircular elements of radius x and width δx. Each point on the element is distant x from O so, for a typical element, we have,

$$\text{area} \simeq \pi x\,\delta x$$

$$\text{mass} \simeq (\pi x\,\delta x)(mx)$$

$$\text{moment of inertia about XY} \simeq (\pi m x^2\,\delta x)(x^2)$$

Hence the moment of inertia, I_{XY}, of the whole lamina is given by

$$I_{XY} \simeq \sum_0^a \pi m x^4\,\delta x$$

$$= \pi m \int_0^a x^4\,dx$$

$$= \tfrac{1}{5}\pi m a^5$$

We must now find the mass of the non-uniform lamina.

The mass of an element $\simeq \pi m x^2\,\delta x$

The mass M of the lamina $= \displaystyle\int_0^a \pi m x^2\,dx$

Hence $M = \frac{1}{3}\pi m a^3$

Therefore the moment of inertia of the lamina about XY is $\frac{3}{5}Ma^2$.

EXERCISE 10d

1) A non-uniform rod AB has a mass M and length $2l$. The mass per unit length of the rod is mx at a point of the rod distant x from A. Find the moment of inertia of this rod about an axis perpendicular to the rod passing (a) through A, (b) through the midpoint of AB.

2) A circular lamina of radius a and centre O has a mass per unit area of kx^2 where x is the distance from O and k is a constant. If the mass of the lamina is M find, in terms of M and a, the moment of inertia of the lamina about an axis through O and perpendicular to the lamina.

3) The surface of a cone is generated by rotating, about Ox, that part of the line $4y = 3x$ between the origin and the point $(4, 3)$. The mass per unit area of the surface is given by λy and the total mass is M. Find, in terms of M, the moment of inertia of the hollow cone about Ox.

4) A non-uniform solid sphere of radius a and mass M has a mass mx per unit volume at all points distant x from its centre. By dividing the sphere into elements which are approximately hollow spheres, find the moment of inertia of the sphere about a diameter. (Use the result of Example 10b, No. 2 for the moment of inertia of the element.)

5) The sides AB and BC of a rectangular lamina ABCD are of length $2a$ and $2b$ respectively. The density of a point on the lamina distant x from AB is kx^3. Show that the moment of inertia of the lamina about the side AB is $\frac{8}{3}Mb^2$ where M is the mass of the lamina.

RADIUS OF GYRATION

The moment of inertia of any rigid body about a specified axis can be expressed in the form Mk^2 where M is the mass of the body and k is a length. This is the same as the moment of inertia of a particle of mass M distant k from the axis, and k is called the *radius of gyration* of the body about that axis.

Consider, for example, a uniform rod of mass M and length $2l$ rotating about an axis through its centre and perpendicular to the rod. If I is the moment of inertia of the rod about this axis then

$$I = \frac{Ml^2}{3} = M\left(\frac{l}{\sqrt{3}}\right)^2$$

Therefore $$k = \tfrac{1}{3}l\sqrt{3}$$

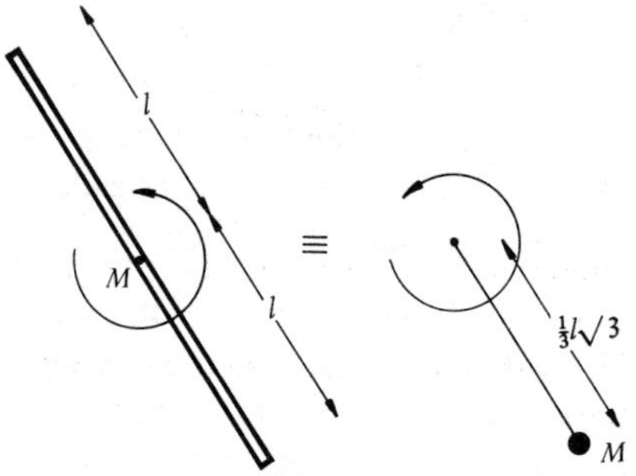

Many students are tempted to replace a rotating rigid body by a particle of equal mass at the centre of gravity, but the above example shows that this does *not* give the correct result for the moment of inertia.

Second Moment of Area

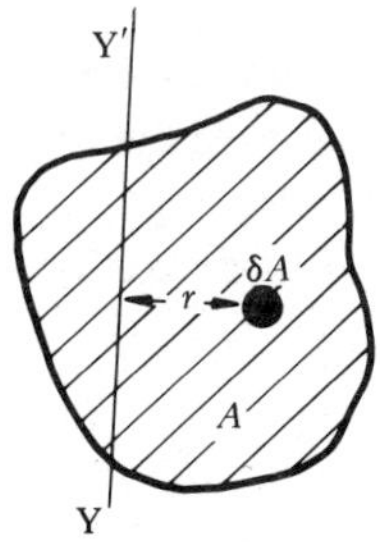

If δA is a small element of an area A, the second moment of the area A about an axis YY′ is defined as $\Sigma r^2 \delta A$ where r is the distance of δA from YY′.
If the area A represents a uniform lamina of mass m per unit area, the moment of inertia of the lamina about YY′ is

$$\sum r^2(m\,\delta A) = m \sum r^2 \delta A = m(\text{the second moment of its area about YY}').$$

Thus the second moment of area is calculated in the same way as the moment of inertia of a uniform lamina except that the mass per unit area is not included.

Second Moment of Volume

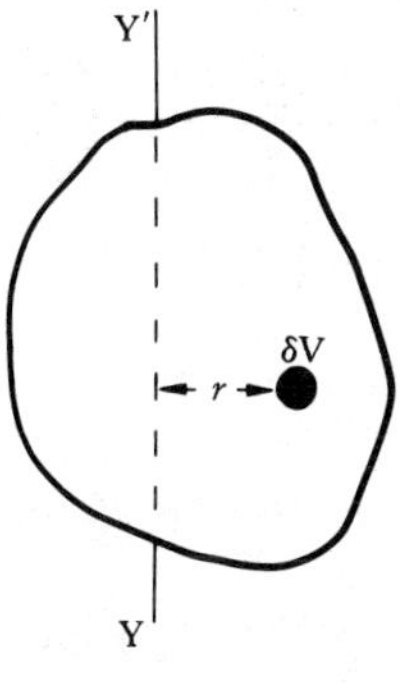

If δV is a small element of a volume V, the second moment of the volume about YY′ is defined as $\Sigma r^2 \delta V$ where r is the distance of δV from YY′.
If V represents a uniform solid of mass m per unit volume, the moment of inertia of that solid about YY′ is

$$\sum r^2(m\,\delta v) = m \sum r^2 \delta v = m(\text{the second moment of volume about YY}').$$

Thus the second moment of volume is calculated in the same way as the moment of inertia of a uniform solid except that the mass per unit volume is not included.

MOMENTS OF INERTIA OF STANDARD BODIES

Not all problems on rotating bodies require the calculation of the appropriate moment of inertia. The table below gives the moments of inertia of some standard bodies about a specified axis.
These results are quotable unless their derivation is asked for.

Standard Results

Uniform Body of Mass M	Axis	Moment of Inertia
Rod of length $2l$	Perpendicular to the rod through the midpoint	$\frac{1}{3}Ml^2$
Rod	Parallel to the rod and distant d from it	Md^2
Rectangle of length $2l$ and width $2d$	Perpendicular to the sides of length $2l$ and passing through their midpoints	$\frac{1}{3}Ml^2$
Ring of radius a	Perpendicular to the ring and through its centre	Ma^2
Disc of radius a	Perpendicular to the disc and through its centre	$\frac{1}{2}Ma^2$
Solid sphere of radius a	A diameter	$\frac{2}{5}Ma^2$
Hollow sphere of radius a	A diameter	$\frac{2}{3}Ma^2$
Solid cylinder of radius a	The axis of symmetry	$\frac{1}{2}Ma^2$
Hollow cylinder radius a	The axis of symmetry	Ma^2

CHANGE OF AXIS

Up to this point we have usually calculated the moment of inertia of a body about an axis which passes through its centre of mass. If the moment of inertia about a different axis is required, we do not always have to go back to first principles. If the specified axis is parallel to the standard axis through the centre of mass, the following theorem provides an easy way to find the required moment of inertia.

The Parallel Axis Theorem

If the moment of inertia of a uniform body of mass M about an axis through G, its centre of mass, is I_G, and I_A is the moment of inertia about a parallel axis through a point A, then

$$I_A = I_G + Md^2$$

where d is the distance between the parallel axes.

This important theorem can be proved as follows.

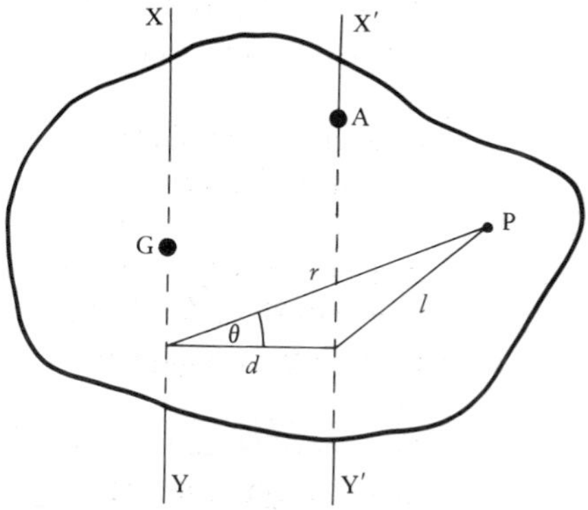

Let XY be an axis passing through the centre of mass G of the body and X′Y′ be a parallel axis distant d from XY.
Consider a typical constituent particle P of mass m where P is distant l from X′Y′ and distant r from XY.
The moment of inertia of P about X′Y′ is

$$ml^2 = m(r^2 + d^2 - 2rd\cos\theta) \qquad \text{(Cosine rule)}$$

Therefore the moment of inertia of the whole body about X′Y′ is I_A where

$$\begin{aligned} I_A &= \sum m(r^2 + d^2 - 2rd\cos\theta) \\ &= \sum mr^2 + \sum md^2 - \sum 2mrd\cos\theta \\ &= I_G + d^2\sum m - 2d\sum mr\cos\theta \end{aligned}$$

Now $mr\cos\theta$ is the moment of the mass of the particle P about the axis XY. But XY passes through the centre of mass, hence $\Sigma mr\cos\theta = 0$.

Therefore $$I_A = I_G + Md^2$$

Note that this can be expressed in the form,

$$I_A = M(k^2 + d^2)$$

where k is the radius of gyration about the axis through G.

EXAMPLES 10e

1) Use the parallel axis theorem to find the moment of inertia of a uniform rod of mass M and length $2a$, about a perpendicular axis through one end.

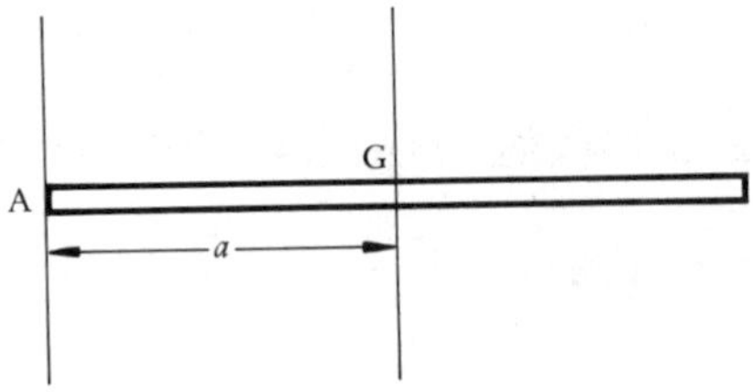

The moment of inertia, I_G, about an axis through G and perpendicular to the rod is $\frac{1}{3}Ma^2$.
The axis through the end A is a parallel axis, therefore

$$\begin{aligned} I_A &= I_G + Ma^2 \\ &= \tfrac{1}{3}Ma^2 + Ma^2 \\ &= \tfrac{4}{3}Ma^2 \end{aligned}$$

It must be emphasised that this theorem gives the relationship between the moments of inertia of a uniform body about two parallel axes, *one of which passes through the centre of mass of the body*. It is possible, given the moment of inertia of a body about an axis not through the centre of mass, to find the moment of inertia about a parallel axis, again not through the centre of mass, but the calculation has to be done in two stages.

Consider the body shown. If M is the mass of the body and its moment of inertia about XY is I then

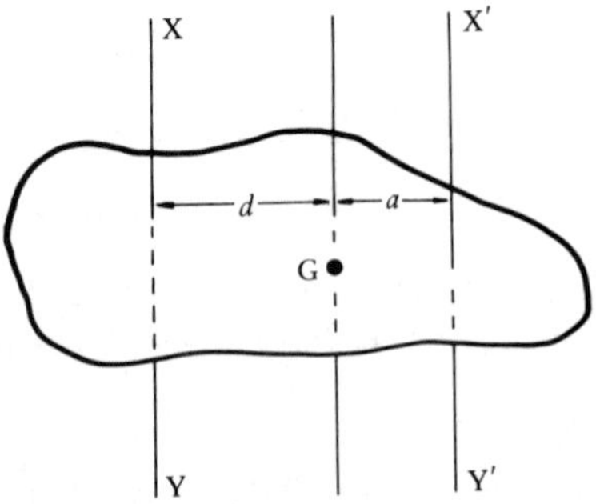

$$I = I_G + Md^2$$

Therefore $I_G = I - Md^2$

Then using the parallel axis theorem again

$$\begin{aligned} I_{X'Y'} &= I_G + Ma^2 \\ &= (I - Md^2) + Ma^2 \\ &= I + M(a^2 - d^2) \end{aligned}$$

Note. It follows from the parallel axis theorem that the moment of inertia about an axis through the centre of mass of a uniform body is less than that about any other parallel axis.

The Perpendicular Axes Theorem

If a plane body has moments of inertia I_{Ox} and I_{Oy} about two perpendicular axes, Ox and Oy, in the plane of the body then its moment of inertia about an axis Oz, perpendicular to the plane, is $I_{Ox}+I_{Oy}$.

The validity of this theorem can be proved as follows.

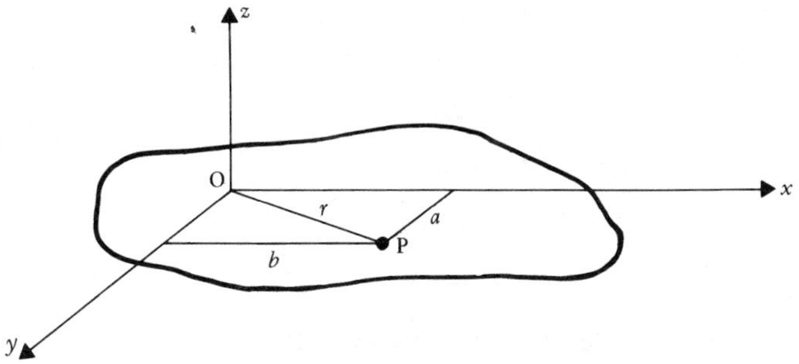

The diagram shows a lamina in the plane containing Ox and Oy and P is a constituent particle of the lamina. If P is of mass m, the moment of inertia of P about Oz is I_{Oz} where

$$I_{Oz} = \sum mr^2$$

$$= \sum m(a^2+b^2)$$

$$= \sum ma^2 + \sum mb^2$$

$$\Rightarrow \quad I_{Oz} = I_{Ox} + I_{Oy}$$

Note. The three axes under consideration must be mutually perpendicular *and* concurrent, although none of them needs to pass through the centre of mass of the body.
We can now use this theorem to extend some of the results already obtained for plane bodies.

Note. This theorem *cannot* be applied to three dimensional bodies.

EXAMPLES 10e (continued)

2) Find the moment of inertia, about a diameter, of a uniform ring of mass M and radius a.

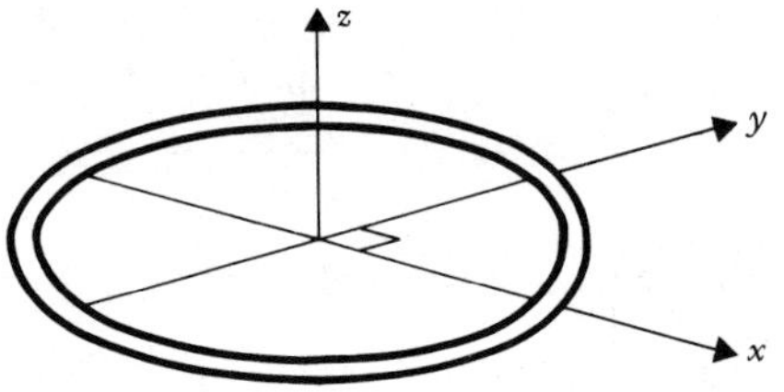

We know that the moment of inertia, I_{Oz}, of the ring about Oz is Ma^2. We also know that, from symmetry, the moment of inertia about any one diameter is the same as that about any other diameter,

i.e. $$I_{Ox} = I_{Oy}$$

Using the perpendicular axes theorem gives

$$I_{Oz} = I_{Ox} + I_{Oy}$$

$$\Rightarrow \quad Ma^2 = 2I_{Ox} = 2I_{Oy}$$

The moment of inertia of the ring about any diameter is $\frac{1}{2}Ma^2$.

3) Find the moment of inertia of a uniform disc, of mass M and radius a, about a tangent.

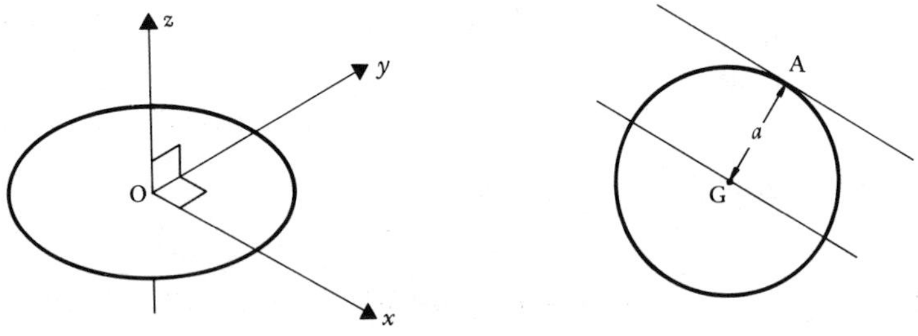

For the disc we have the quotable result, $I_{Oz} = \frac{1}{2}Ma^2$

Then $$I_{Oz} = I_{Ox} + I_{Oy} \qquad \text{(Perpendicular axes)}$$

and $$I_{Ox} = I_{Oy} \qquad \text{(Symmetry)}$$

$$\Rightarrow \quad I_{Ox} = \tfrac{1}{4}Ma^2$$

Now we use the parallel axis theorem,

$$I_A = I_G + Ma^2 \qquad \text{(G is at O)}$$

$$= \tfrac{1}{4}Ma^2 + Ma^2$$

Therefore the moment of inertia of the disc about a tangent is $\frac{5}{4}Ma^2$.

4) Find the moment of inertia of a uniform cube of mass M and edge $2a$, about an axis along one edge.

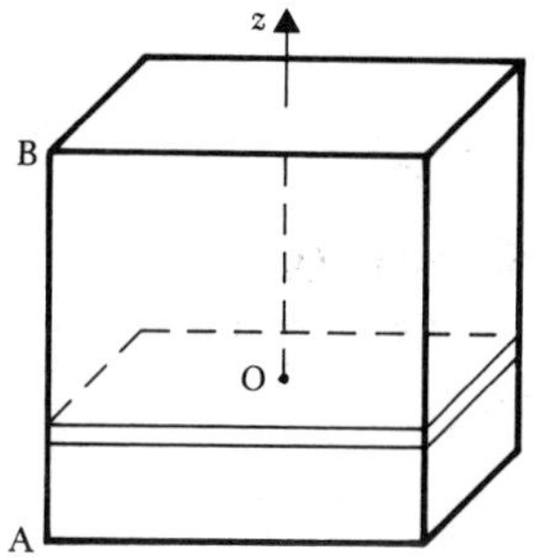

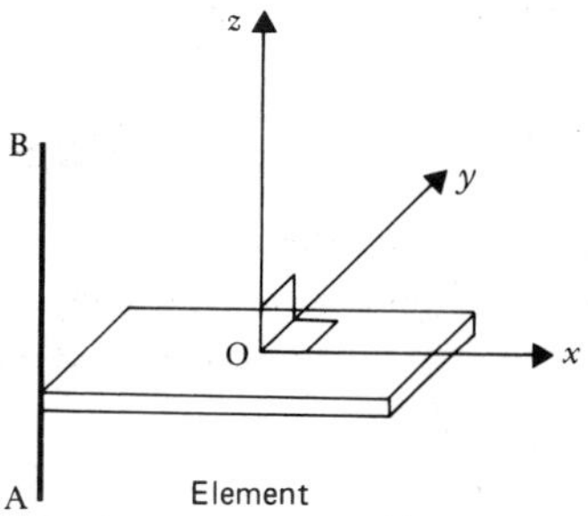

First the cube is divided into elements that are approximately square laminas of side $2a$.
If the mass of one such element is m, we can quote its moment of inertia about Ox,

i.e. $$I_{Ox} = \tfrac{1}{3}ma^2 = I_{Oy} \qquad \text{(From symmetry)}$$

But $$I_{Oz} = I_{Ox} + I_{Oy} \qquad \text{(Perpendicular axes theorem)}$$
$$= \tfrac{2}{3}ma^2$$

Then $$I_{AB} = I_{Oz} + m(AO)^2 \qquad \text{(Parallel axis theorem)}$$
$$= \tfrac{2}{3}ma^2 + m(a^2 + a^2)$$
$$= \tfrac{8}{3}ma^2$$

For the whole cube,

$$I_{AB} = \sum \tfrac{8}{3}ma^2$$
$$= \tfrac{8}{3}a^2 \sum m$$
$$= \tfrac{8}{3}Ma^2$$

5) Find the moment of inertia of a solid cone of mass M, base radius r and height h, about an axis through the vertex, perpendicular to the axis of symmetry.

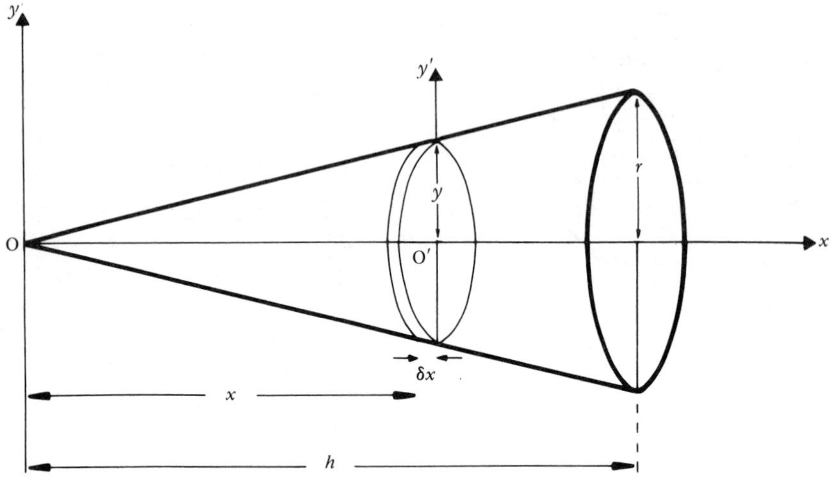

The cone is divided into disc-like elements by cuts parallel to the base. Each element has radius y, thickness δx and approximate mass $\pi m y^2 \delta x$ where m is the mass/unit volume.
Therefore, for a typical element,

$$I_{O'x} \simeq \tfrac{1}{2}(\pi m y^2 \delta x)(y^2)$$

$$I_{O'y'} \simeq \tfrac{1}{4}(\pi m y^2 \delta x)(y^2) \qquad \text{(Perpendicular axes theorem)}$$

$$I_{Oy} \simeq \pi m y^2 \delta x\,(\tfrac{1}{4}y^2 + x^2) \qquad \text{(Parallel axis theorem)}$$

Therefore, for the whole cone,

$$I_{Oy} \simeq \sum_{0}^{h} \pi m\,(\tfrac{1}{4}y^4 + x^2y^2)\,\delta x$$

$$= \int_0^h \pi m\,(\tfrac{1}{4}y^4 + x^2y^2)\,\mathrm{d}x$$

But $\dfrac{x}{h} = \dfrac{y}{r} \Rightarrow y = \dfrac{rx}{h}$

Hence
$$I_{Oy} = \pi m \int_0^h \left(\frac{r^4}{4h^4} + \frac{r^2}{h^2}\right) x^4 \,dx$$
$$= \frac{\pi m r^2}{4h^2}(r^2 + 4h^2)\left(\frac{h^5}{5}\right)$$
$$= \tfrac{1}{20}\pi m r^2 h\,(r^2 + 4h^2)$$

But $M = \frac{1}{3}\pi r^2 hm$

Therefore, for the cone,
$$I_{Oy} = \tfrac{3}{20}M(r^2 + 4h^2)$$

Note that the moment of inertia of each element must be found *about the required axis* before summation over the whole body is attempted.

EXERCISE 10e

In this exercise, standard results may be quoted.

1) Find the moment of inertia of a uniform rectangular lamina of sides $2a$ and $2b$ and mass M,
(a) about an axis along one edge of length $2b$,
(b) about an axis through one corner and perpendicular to the rectangle.

2) Find the moment of inertia of a uniform ring of radius r and mass M about a tangent.

3) Find the moment of inertia of a uniform disc of radius a about an axis perpendicular to its plane passing through a point on its circumference.

4) Find the moment of inertia of a uniform cube of edge l and mass M, about an axis through the centre of the cube and perpendicular to two faces.

5) Find the moment of inertia of a uniform solid cylinder of radius r and mass M about a generator of the cylinder.

6) Find the moment of inertia of a solid uniform sphere of mass $5M$ and radius a, about a tangent.

7) A uniform lamina of mass M has the shape of an annulus of inner and outer radii $2a$ and $4a$. Find its moment of inertia about a tangent to the inner circle.

8) A uniform circular disc has centre O and B is the midpoint of a radius OA. About an axis through B, perpendicular to the disc, the moment of inertia of the disc is I. Find, in terms of I alone, the moment of inertia of the disc about a parallel axis through A.

9) Find the moment of inertia of a uniform hollow cylinder of radius r, height h and mass M about a diameter of one end.

10) A uniform lamina is in the form of an isosceles triangle ABC and has a mass M. $AB = AC = 5l$, and $BC = 6l$. Find the moment of inertia of this lamina about an axis (a) passing through A and the midpoint of BC, (b) perpendicular to the triangle through A.

COMPOUND BODIES

Consider two bodies A and B, rigidly joined together. The moment of inertia of this compound body, about an axis XY, is required.

If I^{A} is the moment of inertia of body A about XY,
I^{B} is the moment of inertia of body B about XY,
and $I^{\mathrm{A+B}}$ is the moment of inertia of the compound body about XY,

then
$$I^{\mathrm{A+B}} = \sum mr^2 \quad \text{for all the particles in A and B}$$
$$= \sum mr^2 \quad \text{for all the particles in A}$$
$$+ \sum mr^2 \quad \text{for all the particles in B}$$
$$= I^{\mathrm{A}} + I^{\mathrm{B}}$$

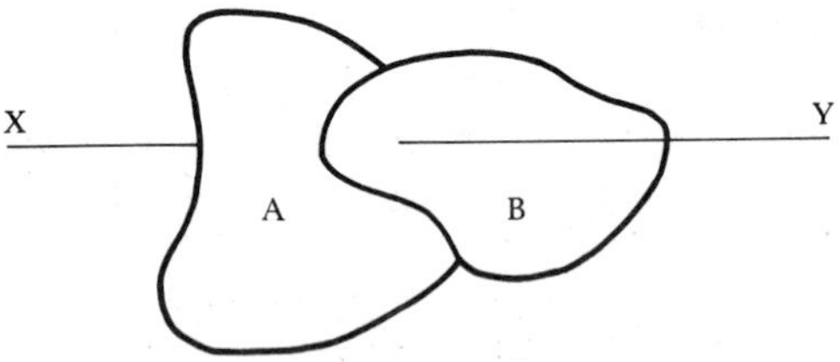

Extending this argument to cover any number of bodies rigidly joined together, we see that the moment of inertia of a compound body, about a specified axis, is the sum of the moments of inertia of the separate parts of the body about the same axis.

EXAMPLES 10f

1) Three uniform rods, each of length $2l$ and mass M are rigidly joined at their ends to form a triangular framework. Find the moment of inertia of the framework about an axis passing through the midpoints of two of its sides.

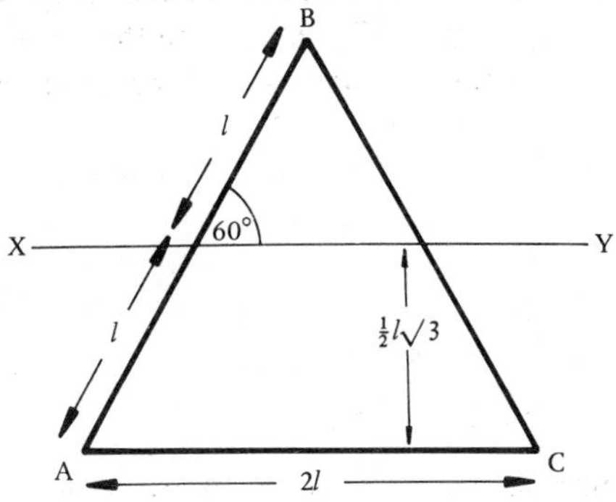

The rod AB is rotating about an axis through its midpoint and inclined to AB at 60°, therefore

for rod AB, $$I_{XY} = \tfrac{1}{3}Ml^2\sin^2 60° = \tfrac{1}{4}Ml^2$$

Similarly for rod BC $$I_{XY} = \tfrac{1}{4}Ml^2$$

The rod AC is rotating about an axis parallel to AC and distant $\frac{1}{2}l\sqrt{3}$ from AB, therefore

for rod AC, $$I_{XY} = M(\tfrac{1}{2}l\sqrt{3})^2 = \tfrac{3}{4}Ml^2$$

Hence for the whole framework

$$\begin{aligned} I_{XY} &= \tfrac{1}{4}Ml^2 + \tfrac{1}{4}Ml^2 + \tfrac{3}{4}Ml^2 \\ &= \tfrac{5}{4}Ml^2 \end{aligned}$$

Warning. Hand-drawn diagrams of a triangular lamina, a triangular framework of rods and the cross-section of a cone, all look the same. When dealing with any of these, great care must be taken not to mistake it for one of the others during the working.

EXERCISE 10f

1) A uniform rod of mass M and length $2l$ has a particle of mass $\frac{1}{2}M$ fixed to one end. Find the moment of inertia of this system about an axis through the centre of the rod and perpendicular to the rod.

2) A uniform ring of radius a and mass M has two particles each of mass M attached to it. Find the moment of inertia of the resulting body about an axis through the centre of the ring and perpendicular to the ring.

3) A uniform solid consists of a solid cylinder of radius a and length l surmounted by a solid hemisphere of radius a. If the total mass of the solid is M find the moment of inertia of the solid about its axis.
(*Hint*. Work in terms of mass per unit volume until both the total moment of inertia and the total mass have been found.)

4) A triangular framework ABC consists of three uniform rods each of mass M where $AB = BC = 2l$, $AC = 2\sqrt{2}l$. Find the moment of inertia of the framework about an axis passing through the midpoints of AB and BC.

5)

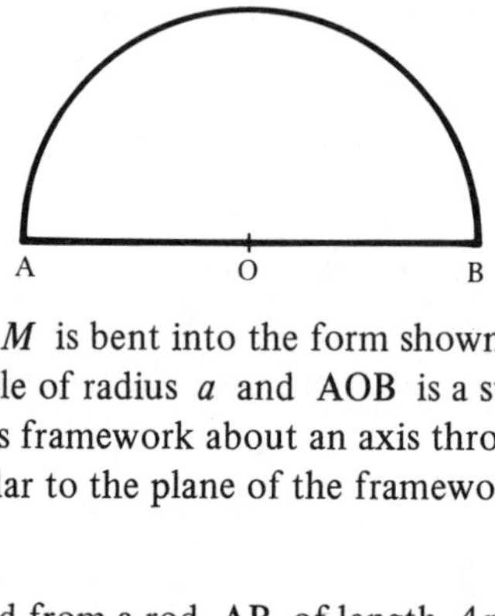

A uniform wire of mass M is bent into the form shown in the diagram where the arc AB is a semicircle of radius a and AOB is a straight line. Find the moment of inertia of this framework about an axis through O (the midpoint of AB) and perpendicular to the plane of the framework.

6) A pendulum is formed from a rod AB of length $4a$ and mass M, and a disc of radius a and mass $4M$. The end B of the rod is attached at the centre of the disc. Find the moment of inertia of the pendulum about an axis through A, perpendicular to the plane of the disc and rod.

7) The governor of an engine comprises two identical spheres each of radius a and mass M. They are joined by a light rod so that their centres are distant $12a$ apart. Find the moment of inertia of the governor about the axis of symmetry perpendicular to the rod.

8) 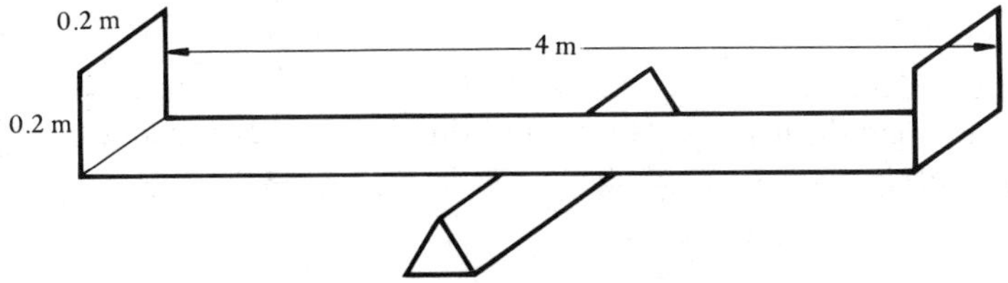

A see-saw is made from a sheet of metal 0.2 m wide and 4.4 m long. At a distance of 0.2 m from each end, the metal is bent up at right angles to form a back rest. If the mass per unit area of the metal is 24.8 kg, find the moment of inertia of the see-saw about its pivot (give the result in kg m^2).

MULTIPLE CHOICE EXERCISE 10

(The instructions for answering these questions are given on page ix.)

TYPE I

1) 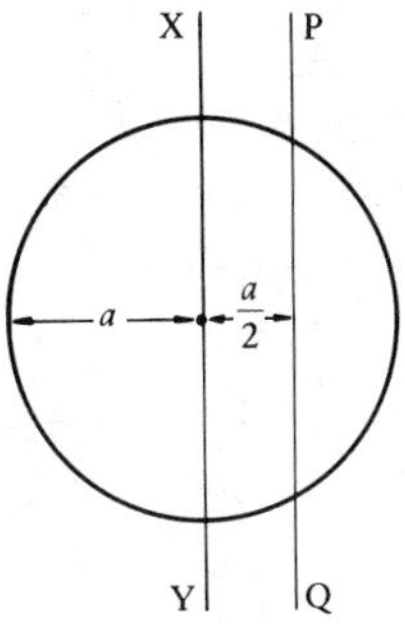

The diagram shows a uniform ring of mass M and radius a. If the moment of inertia of the ring about the axis PQ is I, the moment of inertia of the ring about the axis XY is

(a) $I+\frac{1}{4}Ma^2$ (b) $I+\frac{1}{2}Ma^2$ (c) $I-\frac{1}{4}Ma^2$ (d) $I-\frac{1}{2}Ma^2$.

2) 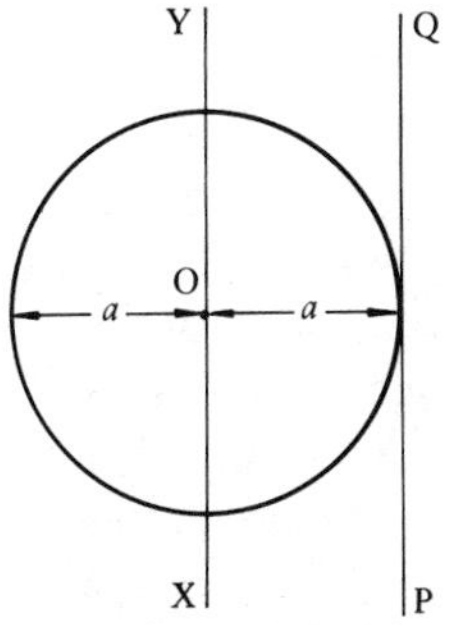

The diagram shows a non-uniform disc of radius a and mass M whose centre of mass is not at O. If the moment of inertia of the disc about XY is I, its moment of inertia about PQ is

(a) $I-Ma^2$ (b) $I+Ma^2$ (c) Ma^2 (d) none of these.

3)

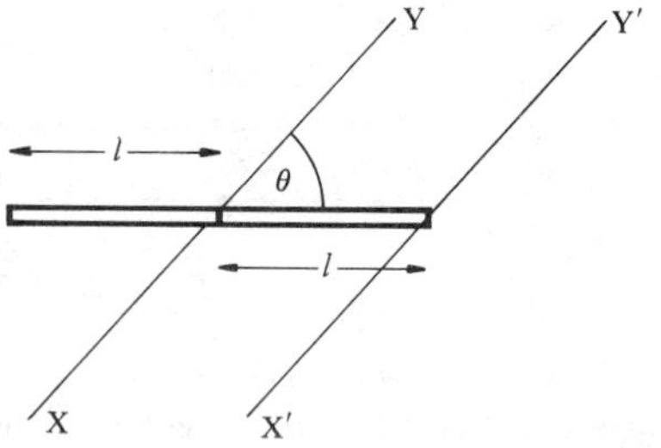

The diagram shows a uniform rod of mass M and length $2l$. The moment of inertia of the rod about XY is I. The moment of inertia of the rod about the axis X′Y′ (which is parallel to XY) is

(a) $I+Ml^2\sin^2\theta$ (b) $I+Ml^2$ (c) $\frac{1}{3}Ml^2\sin^2\theta$ (d) $I-Ml^2\sin^2\theta$.

4)

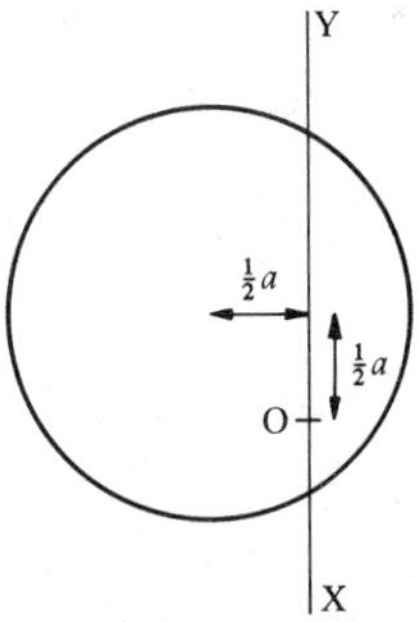

The diagram shows a uniform disc of mass M and radius a. If the moment of inertia of the disc about the axis XY is I, its moment of inertia about an axis through O perpendicular to the plane of the disc is

(a) $I+\frac{1}{2}Ma^2$ (b) $I+\frac{1}{4}Ma^2$ (c) $2I$ (d) $\frac{1}{2}I$ (e) $I-\frac{1}{2}Ma^2$.

5)

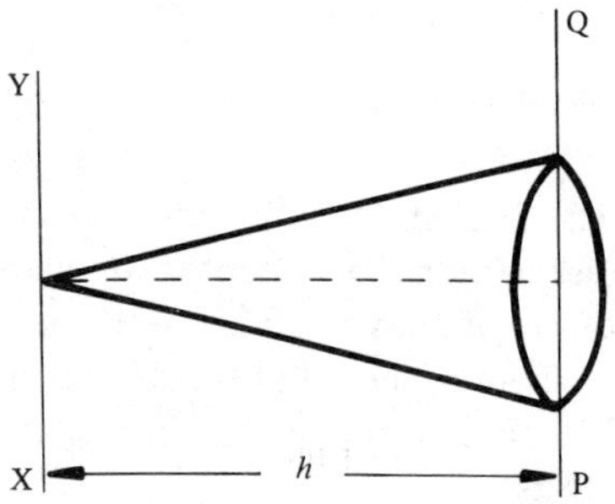

The diagram shows a uniform solid cone of height h and mass M. If the moment of inertia of the cone about XY is I, its moment of inertia about PQ is

(a) $I+Mh^2$ (b) Mh^2-I (c) $I+\frac{1}{2}Mh^2$ (d) $I+\frac{1}{4}Mh^2$ (e) $I-\frac{1}{2}Mh^2$.

TYPE II

6)

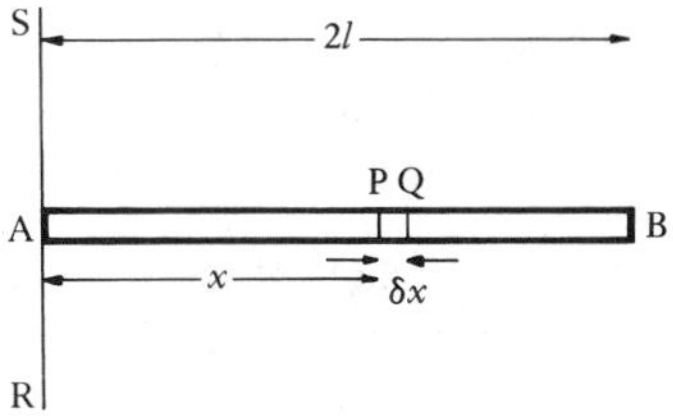

AB is a rod of length $2l$ and mass mx per unit length at a point distant x from AB.

(a) The element PQ has a mass of approximately $mx\,\delta x$.
(b) The moment of inertia of the element PQ about the axis RS is approximately $mx^3\,\delta x$.
(c) The moment of inertia of the rod about RS is given by $\int_0^l mx^3\,dx$.

7)

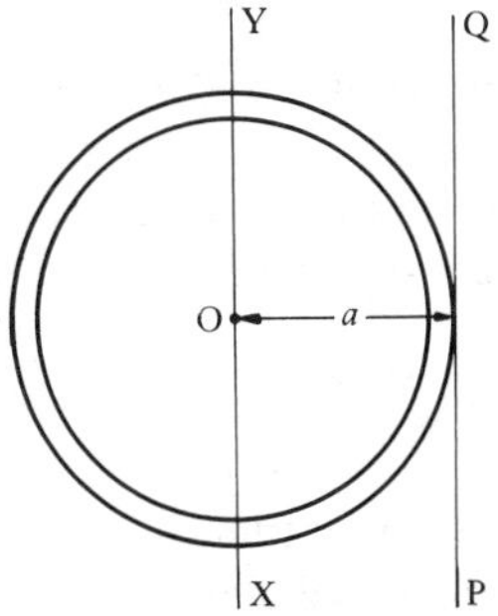

The diagram shows a non-uniform ring of radius a and mass M, whose moment of inertia about the diameter XY is I.

(a) The moment of inertia of the ring about an axis through O perpendicular to the plane of the ring is Ma^2.
(b) The moment of inertia about any diameter other than XY is also I.
(c) The moment of inertia about the tangent PQ is $I + Ma^2$.

8)

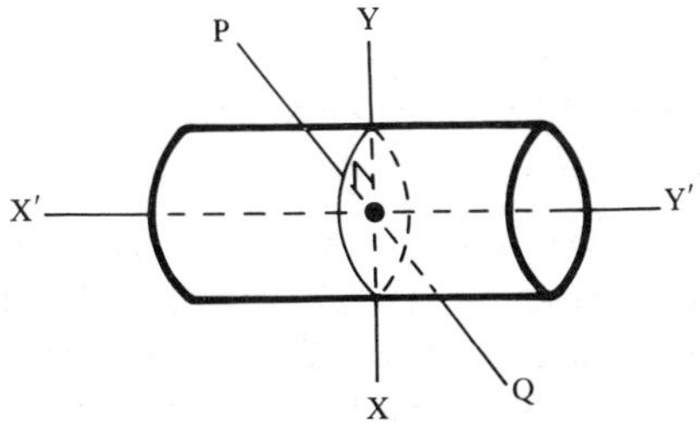

The diagram shows a uniform solid cylinder of mass M and length $2l$. XY and PQ are axes passing through the centre of the cylinder and parallel to its plane face.

(a) If the moment of inertia of the cylinder about the axis XY is I, the moment of inertia about PQ is also I.
(b) The moment of inertia of the cylinder about its axis X′Y′ is $2I$.
(c) The moment of inertia about a diameter of one end is $I + Ml^2$.

TYPE IV

9) Find the moment of inertia of a hollow cylinder about its axis.
(a) The mass of the cylinder is M.
(b) The radius of the cylinder is a.
(c) The length of the cylinder is l.

10) Find the moment of inertia of a rectangular lamina about an axis passing through its centre and the midpoints of the longer sides.
(a) The length of the rectangle is $2l$.
(b) The width of the rectangle is $2a$.
(c) The rectangle is uniform and of mass M.

11) Find the moment of inertia of a solid sphere about a diameter.
(a) The mass of the sphere is M.
(b) The radius of the sphere is r.
(c) The sphere is uniform.

12) Find the moment of inertia of a uniform triangular lamina ABC about the median passing through A.
(a) $AB = AC = 2l$.
(b) The triangle has a mass M.
(c) $BC = l$.

TYPE V

13) The moment of inertia of a hollow cylinder about its axis does not depend on whether the cylinder is uniform or not.

14) The perpendicular axis theorem can be applied to any solid body.

15) The parallel axis theorem can be applied to any solid body provided that the centre of mass is known.

MISCELLANEOUS EXERCISE 10

1) A solid consists of a uniform solid cylinder of radius a and two uniform discs of radius $2a$. The discs are placed one on each end of the cylinder such that the centres of the discs are on the axis of the cylinder. If the mass of the cylinder is M and the mass of each disc is $\frac{1}{2}M$ find the moment of inertia of the complete solid about the axis of the cylinder.

2) A uniform rod is of mass M and length l. Find, by integration, the moment of inertia of the rod about an axis perpendicular to the rod and passing through one end.

3) A uniform lamina of mass M consists of two uniform discs of equal density and radii a and b, $(a<b)$. They are fixed together concentrically. Find the moment of inertia of the lamina about an axis perpendicular to its plane and passing through the centre.

4) Prove, by integration, that the moment of inertia of a uniform rectangular lamina of mass M about an axis along one side is $\frac{4}{3}Mb^2$ where $2b$ is the length of the sides perpendicular to the axis.

5) A uniform lamina is in the form of a semicircle of radius a and mass M. Find, by integration, its moment of inertia about an axis along its straight edge.

6) A rod AB of length l has a mass per unit length mx^2 at a point distant x from A. Find the moment of inertia of the rod about a perpendicular axis through A.

7) A uniform lamina of mass M is in the form of an isosceles triangle ABC in which AB = AC and A is distant h from BC. Show that the moment of inertia of the lamina about BC is $\frac{1}{6}Mh^2$.
A cut is made, parallel to BC, at a distance $\frac{1}{2}h$ from BC and the upper triangle removed. Find the moment of inertia about BC of the remaining trapezium.

8) Show that the moment of inertia of a uniform solid sphere of mass m and radius r about a diameter AB is $\frac{2}{5}mr^2$.
A uniform rod CB of length $4r$ and mass $\frac{1}{2}m$ is rigidly attached to the sphere at B so that ABC is a straight line. The system is free to rotate about a fixed horizontal axis at C so that ABC moves in a vertical plane. Show that the moment of inertia of the system about this axis is $(421/15)mr^2$. (AEB)

9) Show, by integration, that the moment of inertia of a uniform triangular lamina about an edge is $mh^2/6$, where m is the mass of the lamina and h is the length of the altitude from that edge.
Deduce the radius of gyration of a uniform square lamina of edge $2a$ about a diagonal.
The points E and F are the midpoints of the sides CB and CD respectively of a square ABCD of side $2a$. A uniform lamina of mass M is in the shape ABEFD. Find the moment of inertia of this lamina about EF. (U of L)

10) Prove by integration that the moment of inertia of a uniform solid sphere of radius r and mass M about a diameter is $2Mr^2/5$. (The formula for the moment of inertia of a uniform circular disc may be assumed.)
A solid consists of two such spheres joined by a thin uniform rod AB of mass $\frac{1}{2}M$ and length $2r$ so that the centres of the spheres are a distance $4r$ apart. The solid is pivoted at a point C of AB where $CB = \frac{1}{2}r$ and performs small oscillations with AB in a vertical plane. Show that the moment of inertia of the solid about an axis through C perpendicular to AB is $1151Mr^2/120$.
(U of L)

11) A uniform rod AB, of length $2a$ and mass $6m$, has a particle of mass $2m$ attached at B. The rod is free to rotate in a vertical plane about a smooth fixed horizontal axis perpendicular to the rod and passing through a point X of the rod so that $AX = x$, where $x < a$. Show that the moment of inertia of the system about this axis is

$$4m(4a^2 - 5ax + 2x^2).$$

(U of L)

12) Show, by integration, that
(a) the moment of inertia of a uniform rod, of mass m and length $2l$, about an axis through its centre and perpendicular to its length is $\frac{1}{3}ml^2$,
(b) the moment of inertia of a uniform circular disc, of mass M and radius a, about an axis through its centre and perpendicular to its plane is $\frac{1}{2}Ma^2$.
A pendulum is made up of a thin uniform rod AB, of mass m and length $4a$, and a circular disc of mass $2m$, centre C and radius a. A point on the circumference of the disc is rigidly attached to the end B of the rod so that the rod and the disc are in the same plane and such that ABC is a straight line. The pendulum is suspended in a vertical plane about a smooth fixed pivot at A and swings in the plane of the disc. Show that the moment of inertia about the axis of the pivot at A is $\frac{169}{3}ma^2$. (AEB)

13) A thin circular wire of radius a, mass m has fastened to it three particles each of mass m at the corners of an equilateral triangle ABC. Find the moment of inertia of the system about the tangent to the circle at A.
If P is a point on the circle such that the arc AP subtends an angle θ at the centre of the circle, show that the moment of inertia of the system about the tangent at P is independent of θ. (U of L)

14) A uniform lamina of mass m is bounded by two concentric circles of radii a_1, a_2 ($a_2 < a_1$) respectively. Show that the moment of inertia of the lamina about an axis perpendicular to its plane through its centre of mass is

$$\tfrac{1}{2}m(a_1{}^2 + a_2{}^2).$$

A solid of mass M consists of what remains of a uniform solid sphere of radius $2a$ after a circular hole of radius a, with axis along a diameter, has been drilled through it. Show that the moment of inertia of the solid about the axis of the hole is $\frac{11}{5}Ma^2$. (WJEC)

15) Show that the radius of gyration of a uniform triangular lamina PQR about the edge QR is $PS/\sqrt{6}$, where PS is an altitude of the triangle PQR. In a kite-shaped uniform lamina ABCD of mass M the angles B and D are right angles, $AB = AD = 2a$ and $BC = DC = a$. Show that the moment of inertia of the lamina about BD is $13Ma^2/30$.
Find also the moment of inertia about the axis through the point of intersection of AC and BD perpendicular to the plane of the lamina. (U of L)

16) Find the radius of gyration of a uniform right circular solid cone, of height h and base radius a, about (a) the axis of the cone, (b) a diameter of the base. Show that these two radii of gyration are equal if $2h^2 = 3a^2$. (U of L)

17) The radius of gyration of a uniform triangular lamina PQR about an axis through P parallel to QR is k. Show that $k^2 = \frac{1}{2}h^2$, where h is the perpendicular distance from P to QR. Find the radius of gyration about QR. Hence, or otherwise, show that the radius of gyration of a uniform regular hexagonal lamina ABCDEF of side $2a$ about AD is $a\sqrt{(\frac{5}{6})}$. If the mass of the hexagonal lamina is M, find the moment of inertia of this lamina about an edge. (U of L)

18) The radius of gyration of a uniform thin circular disc, of radius a, about a line through its centre and perpendicular to its plane is $a/\sqrt{2}$. Deduce that the radius of gyration of a uniform solid sphere of radius r about a diameter is $r\sqrt{(\frac{2}{5})}$. A spherical cavity of radius r is made inside a uniform solid sphere of radius R, in such a way that the two spherical surfaces touch at the point P. If G is the centre of mass of the remaining material, find (a) the distance PG, and (b) the radius of gyration of the remaining material about a common tangent at P to the spherical surfaces. (JMB)

CHAPTER 11

ROTATION ABOUT A FIXED AXIS

CONSERVATION OF MECHANICAL ENERGY

The Principle of Conservation of Mechanical Energy can be expressed as follows:

> The total mechanical energy of a system remains constant so long as no external work is done (other than by the weight of the body) and no mechanical energy is converted into another form of energy.

This principle, which has already been used to advantage in solving many problems involving linear motion, applies equally well to rotating bodies.
If the axis about which a body is turning offers no resistance to the rotation, the axis is said to be smooth and is not responsible for any external work being done.

Consequently, when a body is rotating about a smooth axis and no work is done by any external force other than its own weight, the total mechanical energy of that body remains constant. In these circumstances the body is said to *rotate freely* about the axis.

EXAMPLES 11a

1) A uniform circular disc of mass m, radius r and centre O is free to turn in its own plane about a smooth horizontal axis passing through a point A on the rim of the disc. The disc is released from rest in the position in which OA is horizontal and the disc is vertical. Find the angular velocity of the disc when OA first becomes vertical.

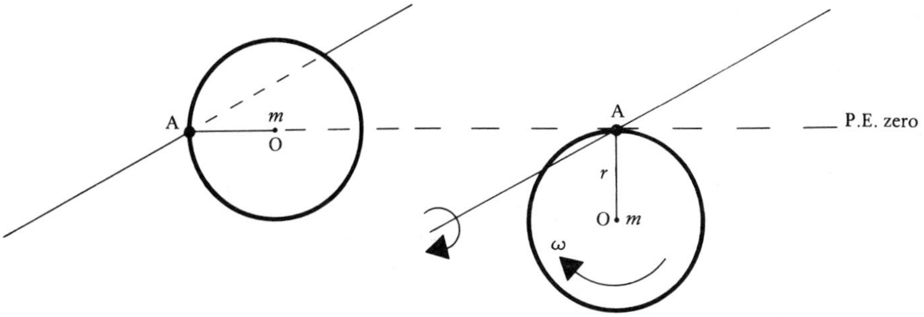

Initial position

(i)

Position with AO vertical

(ii)

The moment of inertia of the disc about the axis through A perpendicular to the disc is given by

$$I = \tfrac{1}{2}mr^2 + mr^2 \qquad \text{(Parallel axis theorem)}$$

i.e. $$I = \tfrac{3}{2}mr^2$$

Initially (diagram (i))

$$\text{K.E.} = 0$$

$$\text{P.E.} = 0$$

When AO is vertical (diagram (ii)) and the angular velocity is ω

$$\text{K.E.} = \tfrac{1}{2}I\omega^2 = \tfrac{1}{2}(\tfrac{3}{2}mr^2)\omega^2$$

$$\text{P.E.} = -mgr$$

Using the principle of conservation of mechanical energy we have:

$$0+0 = \tfrac{3}{4}mr^2\omega^2 - mgr$$

Hence $$\omega^2 = 4g/3r$$

i.e. $$\omega = 2\sqrt{g/3r}$$

2) A uniform rod AB of mass $3m$ and length $2l$ is free to turn in a vertical plane about a smooth horizontal axis through A. A particle of mass m is attached to the rod at B. When the rod is hanging in equilibrium, it is set moving with an angular velocity $\sqrt{kg/l}$.

(a) If $k = 2$, find the height of B above the level of A when the rod first comes to instantaneous rest.

(b) Find the range of values of k for which the particle describes complete circles about A.

About the given axis, the moment of inertia of the rod is $4ml^2$ and that of the particle is $m(2l)^2$.
Therefore the moment of inertia of the rod with attached particle is I where

$$I = 4ml^2 + 4ml^2 = 8ml^2$$

Consider the rod in a general position, making an angle θ with the downward vertical.
The angular velocity of the rod in this position is $\dot{\theta}$.

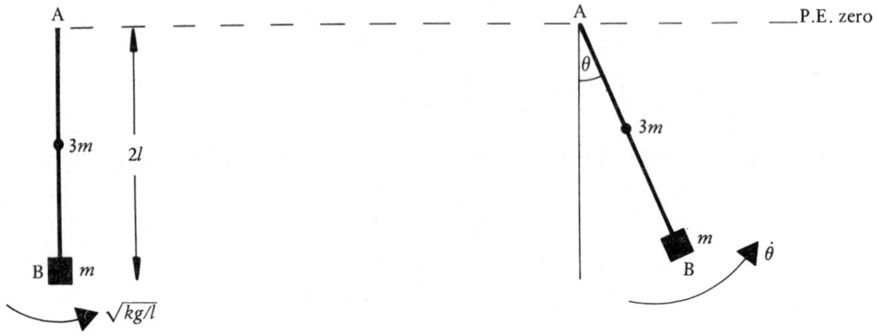

Initial position
(i)

General position
(ii)

In the initial position

$$\text{P.E.} = -3mgl - mg(2l) = -5mgl$$

$$\text{K.E.} = \tfrac{1}{2}I(\sqrt{kg/l})^2 \qquad = 4kmgl$$

In the general position

$$\text{P.E.} = -3mg(l\cos\theta) - mg(2l\cos\theta) = -5mgl\cos\theta$$

$$\text{K.E.} = \tfrac{1}{2}I\dot{\theta}^2 = 4ml^2\dot{\theta}^2$$

Using conservation of mechanical energy we have

$$4kmgl - 5mgl = 4ml^2\dot{\theta}^2 - 5mgl\cos\theta$$

$$\Rightarrow \qquad 4l\dot{\theta}^2 = 4kg - 5g(1-\cos\theta) \qquad [1]$$

(a) When the rod comes to instantaneous rest, $\dot{\theta} = 0$, therefore

$$4kg - 5g(1-\cos\theta) = 0$$

But $\quad k = 2$

Hence $\quad 1-\cos\theta = \frac{8}{5}$

i.e. $\quad \cos\theta = -\frac{3}{5}$

In this position θ is an obtuse angle so the height h of B above A is given by

$$h = 2l\cos(\pi - \theta) = 2l(\tfrac{3}{5})$$

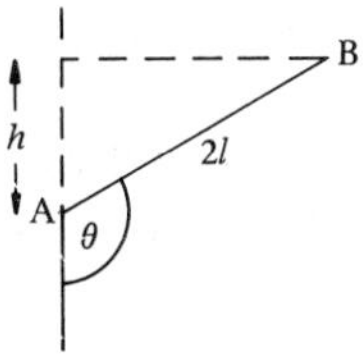

Hence B is distant $6l/5$ above the level of A.

(b) Since the particle is attached to a rigid rod, its motion is restricted to a circular path. It will therefore describe complete circles if its speed at the highest point is greater than zero.
Thus, in equation [1] $\dot{\theta}$ has a real value when $\theta = \pi$.

Hence $$4kg - 5g(1 - \cos\pi) > 0$$

The particle will therefore describe complete circles provided that

$$k > \tfrac{5}{2}$$

3) ABC is a triangular framework of three uniform rods each of mass m and length $2l$. It is free to rotate in its own plane about a smooth horizontal axis through A which is perpendicular to ABC. If it is released from rest when AB is horizontal and C is above AB, find the maximum velocity of C in the subsequent motion.

(The velocity of any point in the framework is greatest when the angular velocity is maximum, i.e. when the kinetic energy is greatest. Because the total mechanical energy is constant, the body has maximum kinetic energy when its potential energy is least, i.e. when the centre of mass is in its lowest position, which is when BC is horizontal and below A.)

The moment of inertia about an axis through A perpendicular to the framework is

for the rod AB $\quad \frac{4}{3}ml^2$

for the rod AC $\quad \frac{4}{3}ml^2$

for the rod BC $\quad \frac{1}{3}ml^2 + m(l\sqrt{3})^2 = \frac{10}{3}ml^2$

Hence for the framework $\quad I = 6ml^2$

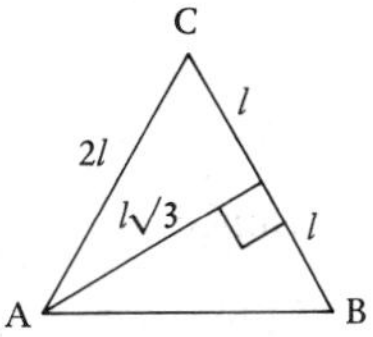

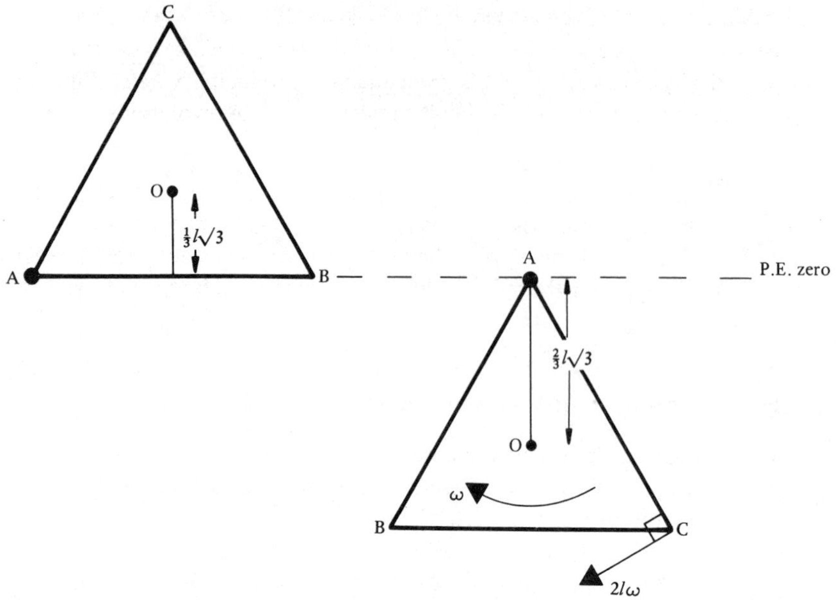

If ω is the angular velocity in the second position then, using conservation of mechanical energy, we have

$$3mg\left(\frac{l\sqrt{3}}{3}\right) = \tfrac{1}{2}(6ml^2)\omega^2 + 3mg\left(-\frac{2l\sqrt{3}}{3}\right)$$

Hence $$\omega^2 = g\sqrt{3}/l$$

Now the velocity of C at this instant is $2l\omega$, i.e. $2l\sqrt{\dfrac{g\sqrt{3}}{l}}$

Note. When drawing diagrams showing various positions of a rotating body, the level of the axis of rotation should be kept constant and potential energy measured from this level.

EXERCISE 11a

1) A uniform rod of length $2l$ and mass m is free to rotate in a vertical plane about a smooth fixed horizontal axis through one end of the rod. The rod is held in a horizontal position and is then released. Find the maximum angular velocity of the rod in the subsequent motion.

2) A uniform ring of radius a and mass m is free to rotate in a vertical plane about a fixed smooth axis which is perpendicular to the plane of the ring and passes through a point A on the ring. A particle of mass m is attached to the ring at B, where AB is a diameter. When the ring is hanging in a position of stable equilibrium the particle is struck a blow which gives it a velocity $3\sqrt{ga}$. Find the height above A to which the particle rises.

3) A uniform square lamina of side $2a$ and mass m is free to rotate about a fixed smooth horizontal axis along one edge of the lamina. When the lamina is hanging vertically below the axis it is given an angular velocity $\sqrt{kg/a}$. Find the range of values of k for which the lamina makes complete revolutions.

4) A uniform circular disc of mass m, radius r and centre O is free to rotate about a smooth, horizontal axis which is tangential to the disc at a point A. The disc is held in a vertical plane with A below O and is then slightly displaced from this position. Find the angular velocity of the disc when its plane is next vertical.

5) Four uniform rods of equal length l and mass m are rigidly joined together at their ends to form a square framework ABCD. The framework is free to rotate in a vertical plane about a fixed smooth horizontal axis passing through A. The framework is slightly displaced from its position of unstable equilibrium. Find the maximum angular velocity reached in the subsequent motion.

6) A uniform rod of length $6a$ and mass m is free to rotate in a vertical plane about a fixed smooth axis through one end, A. A light elastic string of natural length a and modulus of elasticity mg connects the midpoint B of the rod to a fixed point C which is level with A, distant $3a$ from A and in the vertical plane containing the rod. The rod is released from rest with AB horizontal and when it first comes to instantaneous rest, the angle BAC $= 2\theta$. Show that

$$6 \sin 2\theta = (6 \sin \theta - 1)^2$$

Find also the angular velocity of the rod when $\theta = \pi/6$.

7) Three uniform rods, each of length $2l$ and mass m, are rigidly joined together to form a triangular framework ABC. The framework is free to rotate in a vertical plane about a fixed smooth horizontal axis through A. B is attached to one end of a light elastic string of natural length l and modulus mg, the other end of the string is attached to a point D, level with A and distant $2l$ from A such that A, B, C and D are all in the same vertical plane. The framework is held with the vertex B at D and is then released. Find the angular velocity of the framework when BC is first horizontal.

THE EFFECT OF AN EXTERNAL COUPLE

When the state of rotation of a body is changed by the action of an external couple, the total mechanical energy is no longer constant but undergoes a change equal to the amount of work done by the couple.

CALCULATION OF THE WORK DONE BY A COUPLE

Consider a body which is rotating about an axis XY, under the action of a couple comprising two equal and opposite forces of magnitude F, each distant a from XY.

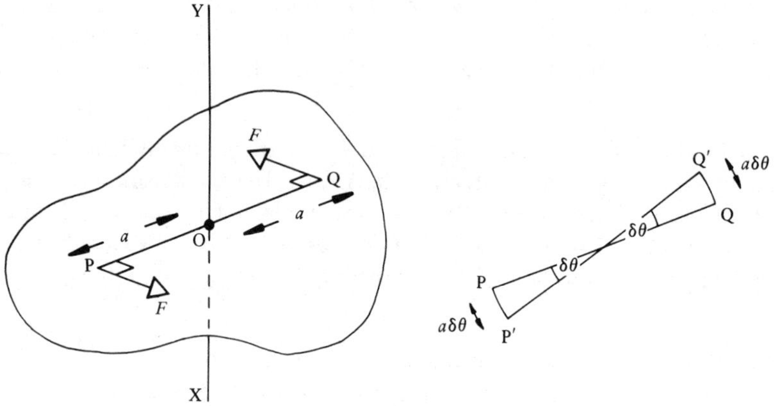

When the body rotates through a small angle $\delta\theta$, the point of application of each force moves through a distance $a\delta\theta$.
The work done by each force, therefore, is given approximately by $Fa\,\delta\theta$.
Then, if the work done by the couple is δW, we have

$$\delta W \simeq 2Fa\,\delta\theta$$

But $2Fa$ is the moment of the couple and can be represented by C where

$$C = 2Fa$$

Therefore $$\delta W \simeq C\,\delta\theta$$

or $$\frac{\delta W}{\delta\theta} \simeq C$$

Hence, as $\delta\theta \to 0$ we have $$\frac{dW}{d\theta} = C$$

Thus W, the work done by the couple as the body rotates through an angle α, is given by

$$W = \int_0^{\alpha} C\,d\theta$$

When C is constant, this result becomes

$$W = C\alpha$$

EXAMPLES 11b

1) A uniform rod AB of mass m and length $2l$ can rotate in a vertical plane about a horizontal axis through A. It is released from rest with AB horizontal and its angular velocity when AB first becomes vertical is $\sqrt{g/l}$.
Find the magnitude of the constant frictional couple exerted by the axis.

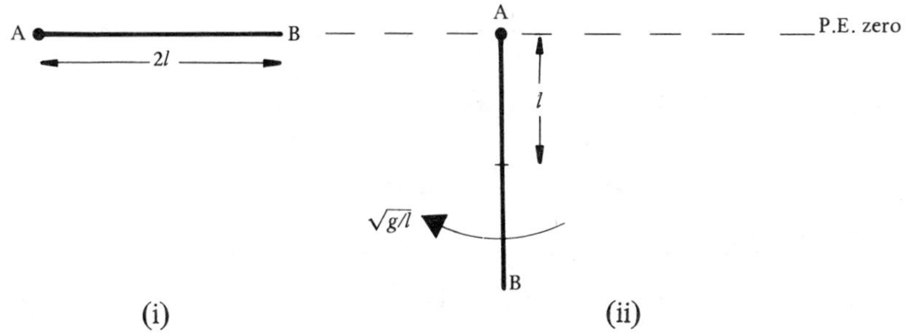

The moment of inertia of the rod about the axis through A is $\frac{4}{3}ml^2$
The initial mechanical energy (diagram (i)) is zero
The final mechanical energy (diagram (ii)) is

$$\tfrac{1}{2}(\tfrac{4}{3}ml^2)\,(\sqrt{g/l})^2 - mgl$$

The loss in mechanical energy during the displacement from (i) to (ii) is therefore

$$0-(\tfrac{2}{3}mgl-mgl) = \tfrac{1}{3}mgl$$

But the loss in mechanical energy is equal to the work done by the constant frictional couple C i.e. $C\times\frac{1}{2}\pi$.

Hence $$\tfrac{1}{2}\pi C = \tfrac{1}{3}mgl$$

$\Rightarrow$ $$C = 2mgl/3\pi$$

2) A disc of mass m and radius a can rotate in a horizontal plane about a smooth vertical axis perpendicular to the disc and passing through the centre of the disc. The disc is made to rotate from rest by a torque whose magnitude, when the disc has rotated through an angle θ, is $4m\theta$. If the disc is uniform find the angular velocity when it has turned through one revolution.

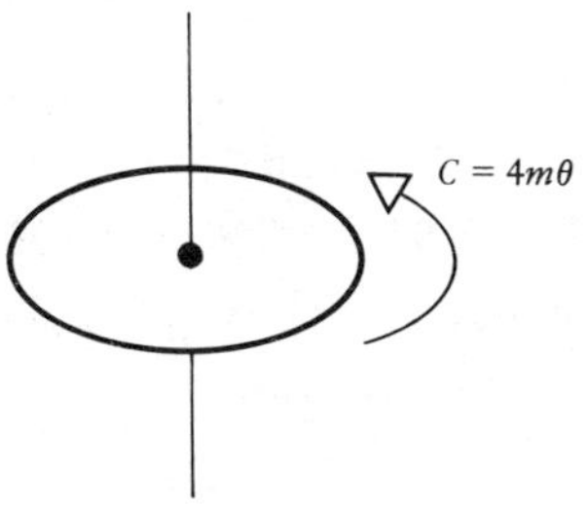

The work done by the couple is given by

$$\int_0^{2\pi} 4m\theta \, d\theta = \left[2m\theta^2\right]_0^{2\pi}$$

$$= 8m\pi^2$$

The increase in mechanical energy of the disc is equal to the work done by the couple.
Thus, if ω is the angular velocity after rotation through one revolution, we have

$$\tfrac{1}{2}I\omega^2 = 8m\pi^2$$

where I is the moment of inertia of the disc about the given axis.

Now $$I = \tfrac{1}{2}ma^2$$

Hence $$\omega^2 = \frac{32\pi^2}{a^2}$$

$\Rightarrow$ $$\omega = 4\pi\sqrt{2}/a$$

EXERCISE 11b

1) A uniform solid cylinder of mass m and radius a is free to rotate about its axis which is smooth and vertical. A light inextensible string is wound round the cylinder and its free end is pulled horizontally with a constant force $2mg$. Find the angular velocity of the cylinder when the free end of the string has moved through a distance $4a$. At this instant the string slips off and a retarding torque C is applied to the cylinder bringing it to rest after it has turned through two revolutions. Find the magnitude of C.

2) A uniform circular disc of radius r and mass m is free to rotate in a vertical plane about a horizontal axis through a point A on the circumference of the disc. The disc is released from rest with the diameter AB horizontal and it next comes to rest when the diameter AB has turned through an angle $\pi/3$. Find the moment of the constant frictional couple acting at the axis.

3) A uniform rod of length $2l$ and mass m is free to rotate in a vertical plane about a horizontal axis through one end A of the rod. When the other end B of the rod is hanging vertically below A it is given an angular velocity $\sqrt{\dfrac{5g}{l}}$. The rod next comes to rest when AB is horizontal. Find the moment of the constant frictional couple acting at the axis.

4) A uniform rod of length $4l$ and mass m is free to rotate in a vertical plane about a horizontal axis through one end of the rod. A constant frictional couple of moment mgl acts at the axis. The rod is released from rest in a horizontal position. If θ is the angle the rod turns through before it next comes to instantaneous rest show that $2\sin\theta = \theta$.

5) A uniform solid cylinder of radius a and mass m is free to rotate about its axis which is smooth and fixed horizontally. The cylinder rotates from rest under the action of a couple whose magnitude is given by $3m\theta^2$ where θ is the angle through which the cylinder has turned. Find the angular velocity of the cylinder when it has completed one revolution.

ANGULAR ACCELERATION

Consider a body of mass m which is rotating about a smooth horizontal axis passing through a point A. The centre of mass is at a point C distant h from the axis of rotation and I is the moment of inertia of the body about that axis.

The system is such that no work is done by any external torque or force (other than the weight of the body).

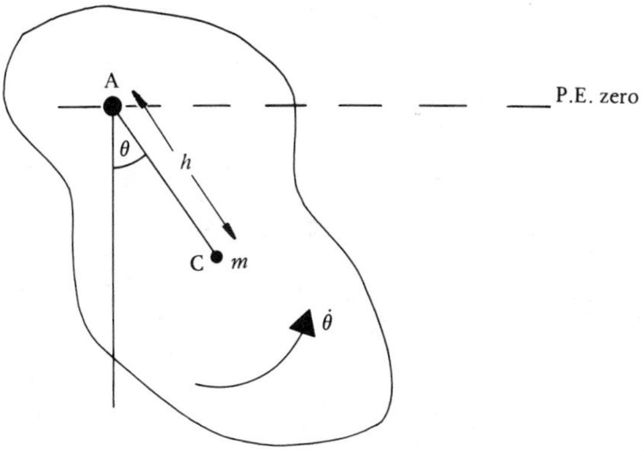

If, at any instant, the line AC is inclined at an angle θ to the downward vertical, the angular velocity of the body at this instant is $\dot{\theta}$.
Taking the initial mechanical energy of the body as K, then applying the principle of conservation of energy gives

$$\tfrac{1}{2}I\dot{\theta}^2 - mgh\cos\theta = K$$

Differentiating this expression with respect to time we have

$$I\dot{\theta}\ddot{\theta} + mgh\sin\theta\,\dot{\theta} = 0$$

Hence $$I\ddot{\theta} = -mgh\sin\theta \qquad [1]$$

In this way the general energy equation for the body can be used to derive an expression for its angular acceleration at any instant. The equation [1] on the previous page is called the equation of motion of the body.

It is not always convenient, or even possible, to use this method to find the angular acceleration of a rotating body. It would not be used, for instance, if work is done by forces other than the weight of the body. The equation of motion of a rotating body can, in such cases, be determined by applying Newton's law in the form $C = I\ddot{\theta}$ (see Example 9a).

The application of Newton's law also provides an alternative method for analysing the rotation of a system whose mechanical energy *is* constant. For example, consider again the body rotating about a smooth horizontal axis through a point A.

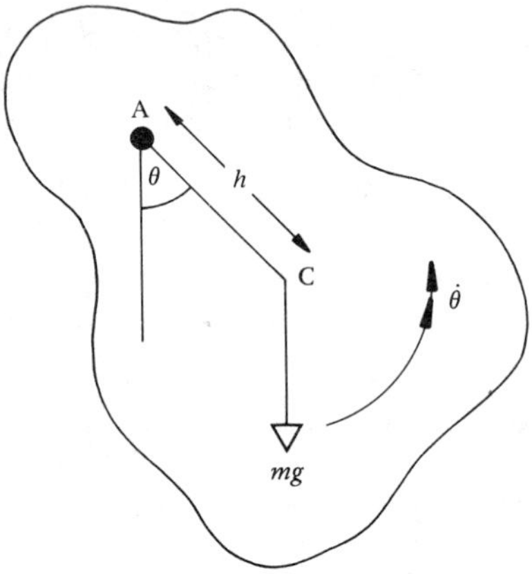

When CA is inclined at an angle θ to the downward vertical, the torque exerted on the body by its weight is $mgh \sin\theta$ clockwise.

Hence $$mgh \sin\theta = -I\ddot{\theta}$$ (see equation [1])

SMALL OSCILLATIONS

Compound Pendulum

When a body supported by a smooth horizontal axis swings through a small angle on either side of its stable equilibrium position, it is said to perform *small oscillations*, and is called a *compound pendulum*. In this case the angle θ in equation [1] on the previous page is such that $\sin\theta \simeq \theta$.
Therefore equation [1] becomes

$$I\ddot{\theta} \simeq -mgh\theta$$

Now this approximates to the equation of angular simple harmonic motion about the vertical through A and the constant of proportion is $\dfrac{mgh}{I}$

The period T of such oscillations is therefore given by

$$T = 2\pi\sqrt{\frac{I}{mgh}}$$

Equivalent Simple Pendulum

The period of small oscillations of a simple pendulum of length l is $2\pi\sqrt{\dfrac{l}{g}}$

Comparing this with the period of oscillation of a compound pendulum,

we see that if $\dfrac{I}{mh} = l$ the periods are equal.

The quantity $\dfrac{I}{mh}$ is therefore known as *the length of the equivalent simple pendulum.*

Note: When solving problems which require the equation of motion, the period of small oscillations or the length of the equivalent simple pendulum, these should be derived from first principles in each case. The *methods* demonstrated in the preceding paragraphs can be used but the expression derived there should not be quoted.

EXAMPLES 11c

1) A ring of mass m and radius r has a particle of mass m attached to it at a point A. The ring can rotate about a smooth horizontal axis which is tangential to the ring at a point B diametrically opposite to A. The ring is released from rest when AB is horizontal. Find the angular velocity and the angular acceleration of the body when AB has turned through at angle $\pi/3$.

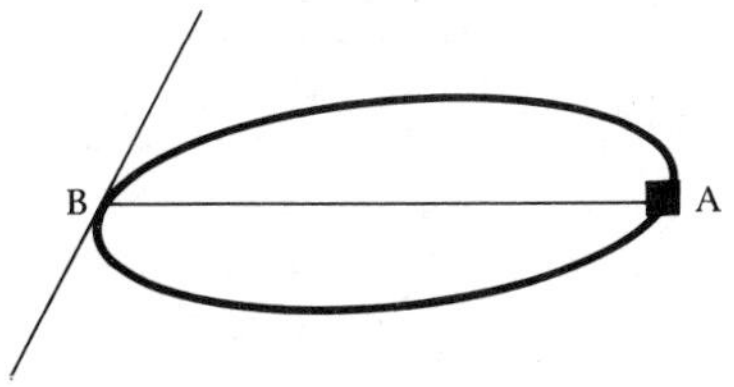

The moment of inertia about the tangent at B is

for the ring alone	$\frac{3}{2}mr^2$
for the particle	$m(2r)^2$

Hence for the whole body $\qquad I = \frac{11}{2}mr^2$

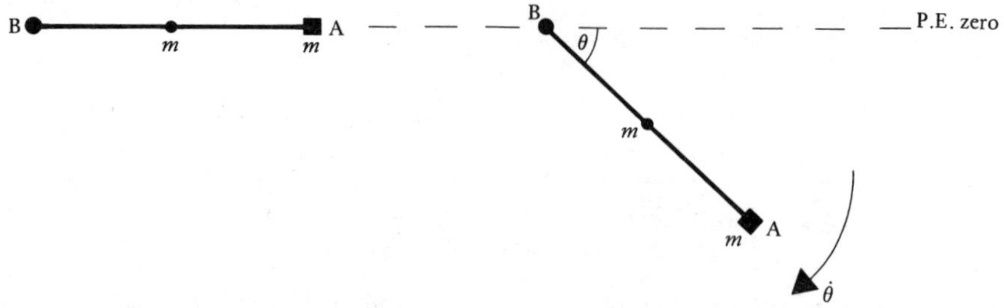

(i) (ii)

Initially (diagram (i)) the total mechanical energy is zero.

In position (ii)

$$\text{P.E. of ring alone} = -mgr\sin\theta$$

$$\text{P.E. of particle} = -mg(2r)\sin\theta$$

$$\text{P.E. of whole body} = -3mgr\sin\theta$$

$$\text{K.E. of whole body} = \tfrac{1}{2}(\tfrac{11}{2}mr^2)\dot{\theta}^2$$

Using conservation of mechanical energy gives

$$0 = \tfrac{11}{4}mr^2\dot{\theta}^2 - 3mgr\sin\theta$$

Hence $$\dot{\theta}^2 = \frac{12g}{11r}\sin\theta \qquad [1]$$

When $\theta = \dfrac{\pi}{3}$, $$\dot{\theta}^2 = \frac{12g}{11r}\left(\frac{\sqrt{3}}{2}\right)$$

Hence $$\dot{\theta} = \sqrt{\frac{6g\sqrt{3}}{11r}}$$

Differentiating equation [1] with respect to time gives

$$2\dot{\theta}\ddot{\theta} = \frac{12g}{11r}(\cos\theta)\dot{\theta}$$

When $\theta = \dfrac{\pi}{3}$, $$\ddot{\theta} = \frac{3g}{11r}$$

Thus after turning through the angle $\frac{1}{3}\pi$, the body has an angular velocity of $\left(\dfrac{6g\sqrt{3}}{11r}\right)^{1/2}$ and an angular acceleration of $\dfrac{3g}{11r}$

2) A disc of mass m and radius a performs small oscillations about a smooth horizontal axis which is tangential to the disc. Find the length of the equivalent simple pendulum.

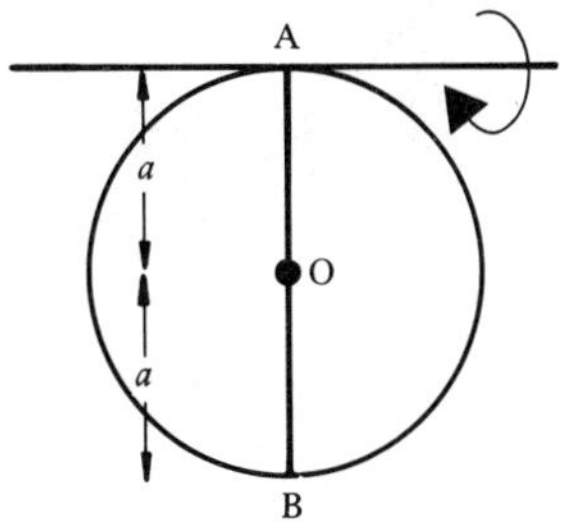

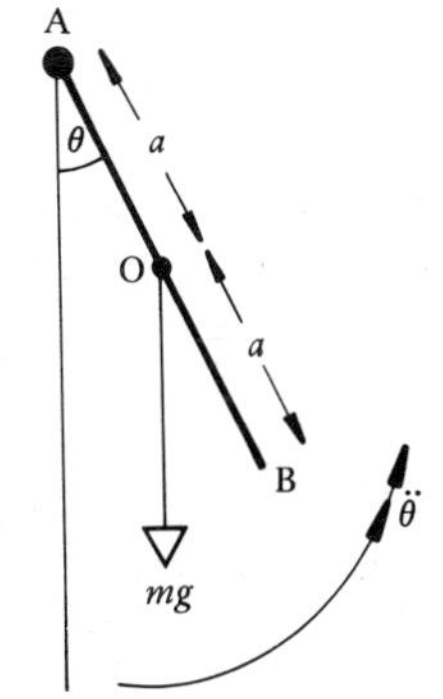

(In order to achieve a clear diagram, θ is not shown as a very small angle.)

The moment of inertia of the disc about the tangent at A is given by

$$\tfrac{1}{4}ma^2 + ma^2 = \tfrac{5}{4}ma^2$$

Using Newton's law for rotation we have

A) $$mga\sin\theta = -\tfrac{5}{4}ma^2\ddot{\theta}$$

But for small oscillations, $\sin\theta \simeq \theta$

Hence $$\ddot{\theta} \simeq -\frac{4g}{5a}\theta$$

This is the equation for angular simple harmonic motion of period T where

$$T = 2\pi\sqrt{\frac{5a}{4g}}$$

Comparing this with the time of oscillation of a simple pendulum of length l we have

$$2\pi\sqrt{\frac{5a}{4g}} = 2\pi\sqrt{\frac{l}{g}}$$

Hence $$l = 5a/4$$

The length of the equivalent simple pendulum is therefore $5a/4$.

3) A uniform rod AB of mass m and length $4l$ is free to rotate in a vertical plane about a smooth horizontal axis through a point P, distant x from the centre of the rod.

If the rod performs small oscillations about its equilibrium position find the period of oscillation and show that this is least when $x = 2l/\sqrt{3}$.

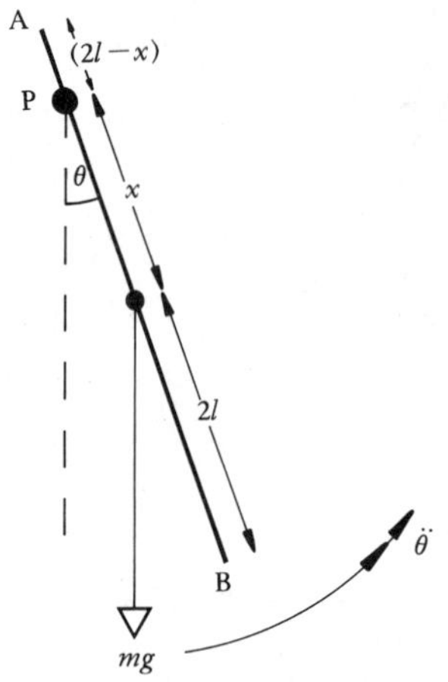

The moment of inertia of the rod about the axis through P is given by

$$I = \tfrac{1}{3}m(2l)^2 + mx^2 \qquad \text{(Parallel axis theorem)}$$

Using Newton's law for rotation we have:

P↻ $$mgx\sin\theta = -I\ddot{\theta}$$

i.e. $$gx\sin\theta = -\tfrac{1}{3}(4l^2+3x^2)\ddot{\theta}$$

Hence $$\ddot{\theta} \simeq -\frac{3gx}{(4l^2+3x^2)}\theta \qquad (\sin\theta \simeq \theta)$$

This is the equation of angular simple harmonic motion about the vertical through P in which $\dfrac{3gx}{4l^2+3x^2}$ is the constant of proportion.

Hence T, the period of oscillations, is given by

$$T = 2\pi\sqrt{\frac{4l^2+3x^2}{3gx}}$$

The period will be least when $\dfrac{4l^2+3x^2}{x}$ is least

i.e. when $$\frac{d}{dx}\left(\frac{4l^2}{x}+3x\right) = 0 \quad \Rightarrow \quad -\frac{4l^2}{x^2}+3 = 0$$

Taking x to be positive $\qquad x = 2l/\sqrt{3}$

The minimum period therefore occurs when $x = 2l/\sqrt{3}$.

(To check that this value of x gives a minimum rather than a maximum period, note that the second derivative, $+\dfrac{8l^2}{x^3}$, is positive when $x = 2l/\sqrt{3}$.)

EXERCISE 11c

1) A uniform rod of length $2l$ and mass m is free to rotate in a vertical plane about a fixed smooth horizontal axis passing through the rod at a point distant $l/2$ from one end. Find general expressions for the angular velocity and angular acceleration of the rod, if it is released from rest when in a horizontal position.

2) A uniform square lamina of side $2a$ and mass m is free to rotate about a fixed, smooth, horizontal axis along one edge of the lamina. Find an expression for the angular acceleration of the lamina when its angular displacement from the vertical plane through the axis is θ. If the lamina swings through a small angle on either side of the vertical through the axis, find the period of its oscillations.

3) A uniform rod of length $2l$ and mass m has a particle of mass $2m$ attached to one end B of the rod. The rod is free to rotate in a vertical plane about a smooth, fixed, horizontal axis through its other end A. The rod performs small oscillations about its position of stable equilibrium. Find the period of these oscillations and hence the length of the equivalent simple pendulum.

4) A uniform circular disc of radius a and mass m is free to rotate in a vertical plane about a smooth, fixed, horizontal axis passing through a point B on the disc, distant h from O, the centre of mass of the disc. A particle of mass m is attached to the circumference of the disc at the end of the diameter through B. The combined body performs small oscillations about its position of stable equilibrium. Find the period of these oscillations and the value of h for which this period is least.

5) A uniform rod AB of length $2l$ and mass m, has a particle of mass m attached at A and another particle, also of mass m, attached at its midpoint C. The rod is free to rotate in a vertical plane about a horizontal axis. If the axis passes through A, the period of small oscillations about the stable equilibrium position is T_1 and if the axis passes through C, the period of small oscillations about the stable equilibrium position is T_2. Prove that $8T_1^2 = 7T_2^2$.

6) A compound pendulum consists of a uniform rod AB of length $2l$ and mass m with a particle of mass m fixed at the end B. It is free to rotate about a smooth horizontal axis through a point C of the rod where $AC = x$. Find x if the length of the equivalent simple pendulum is $3l$.

7) Three uniform rods, each of length $2l$, are rigidly joined together at their ends to form a triangular framework ABC. The rod CA is light, and the rods AB and BC are each of mass m. The framework is free to rotate in a vertical plane about a fixed horizontal axis and performs small oscillations about its stable equilibrium position. Prove that the period of oscillation about an axis through B is equal to the period of oscillation about an axis through the midpoint of AC.

FORCE EXERTED BY THE AXIS

We saw in Chapter 7 that the acceleration of the centre of mass of a system of particles is the same as that of a single particle whose mass is the total mass of the system, acted upon by the resultant of the forces acting on the system. This principle applies to a rigid body, which is made up of a set of particles. Consider a body of mass M, rotating freely about a smooth horizontal axis. Diagram (i) shows a cross-section of the body containing the centre of mass C, which is distant h from the axis through A. The general angular displacement of AC from the downward vertical is θ.

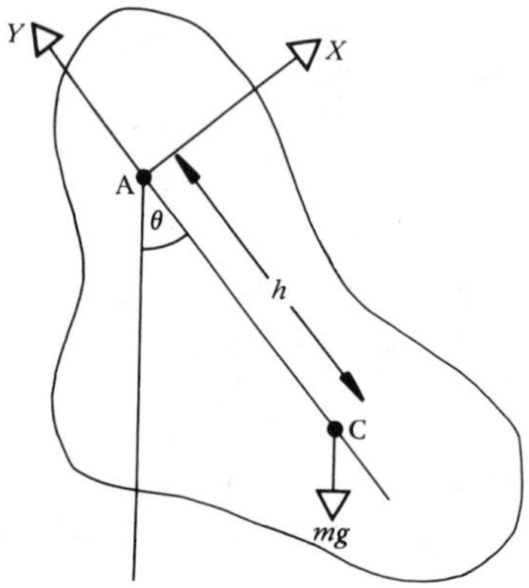

(i)

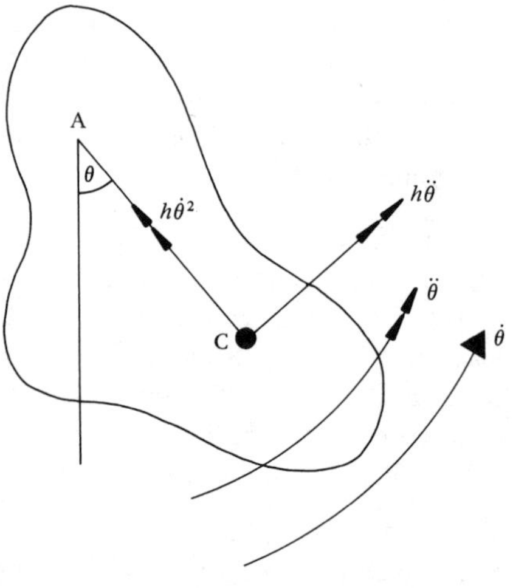

(ii)

The forces acting on the body are its weight and an unknown force exerted by the axis. The force at the axis is shown in the diagram as a pair of perpendicular components acting parallel and perpendicular to CA.

The acceleration components of C are shown in diagram (ii). These are the acceleration components which would be given to a particle of mass M placed at C and acted upon by the resultant of all the forces acting on the system, see diagram (iii).

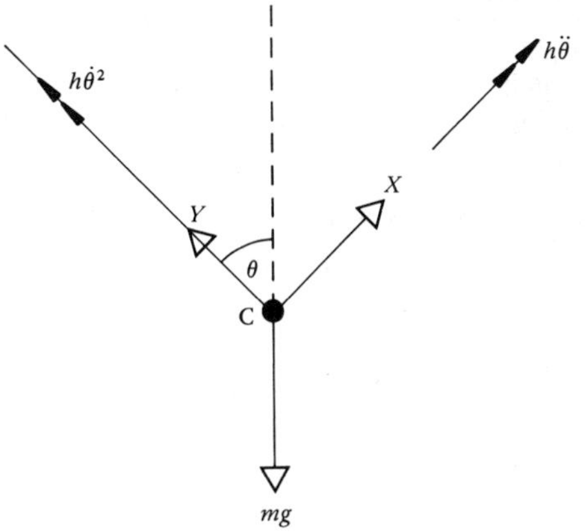

(iii)

Newton's law gives

$$\nwarrow \qquad Y - mg\cos\theta = mh\dot{\theta}^2 \qquad [1]$$

$$\nearrow \qquad X - mg\sin\theta = mh\ddot{\theta} \qquad [2]$$

In order to determine the force exerted on the body by the axis, both X and Y must be found. This can be done only if the angular velocity $(\dot{\theta})$ and the angular acceleration $(\ddot{\theta})$ of the body are known.

Now the value of $\dot{\theta}$ at any instant can be found using conservation of mechanical energy. Then, by differentiating the energy equation (or by applying Newton's law for rotation) $\ddot{\theta}$ can also be determined.

Note: The angle θ need not, in every problem, be measured from the downward vertical, but $\ddot{\theta}$ and $\dot{\theta}$ must always be marked in the sense in which θ increases.

EXAMPLE 11d

A uniform rod AB of mass m and length $2a$ has a particle of mass m attached to the end B. The rod can rotate in a vertical plane about a smooth axis through A. If the body is slightly displaced from the position in which B is vertically above A, find the magnitude of the reaction at the axis, which is horizontal, when the rod has rotated through

(a) $\pi/3$ radians, (b) π radians.

The centre of mass of the body is at a point C midway between B and the midpoint of the rod.

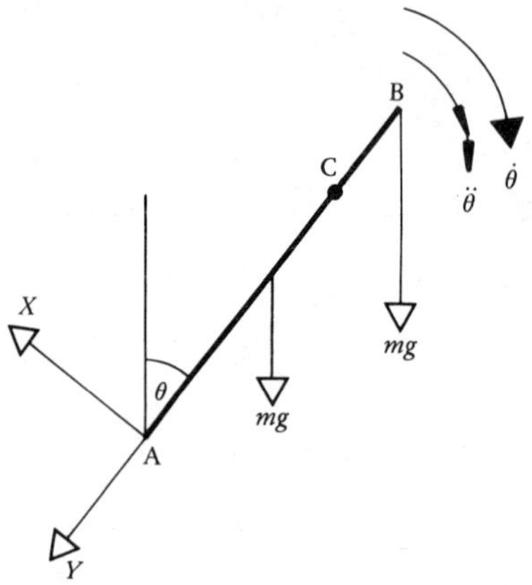

(i)

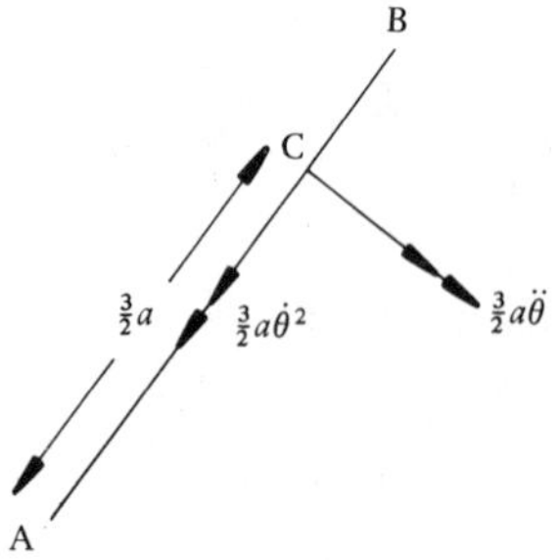

(ii)

Diagram (i) shows the forces acting on the body when it has rotated through an angle θ. Diagram (ii) shows the acceleration components of the point C in this position. Diagram (iii) shows the same forces acting on a particle of mass $2m$ (the total mass of the system) placed at C.

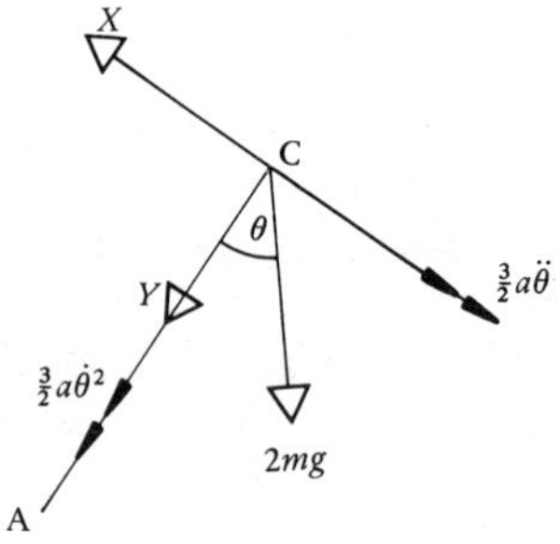

(iii)

Applying Newton's law we have:

$$\swarrow \qquad Y + 2mg\cos\theta = 2m(\tfrac{3}{2}a\,\dot{\theta}^2)$$

$$\nwarrow \qquad 2mg\sin\theta - X = 2m(\tfrac{3}{2}a\,\ddot{\theta})$$

Thus $$Y = 3ma\dot{\theta}^2 - 2mg\cos\theta \qquad [1]$$

and $$X = 2mg\sin\theta - 3ma\ddot{\theta} \qquad [2]$$

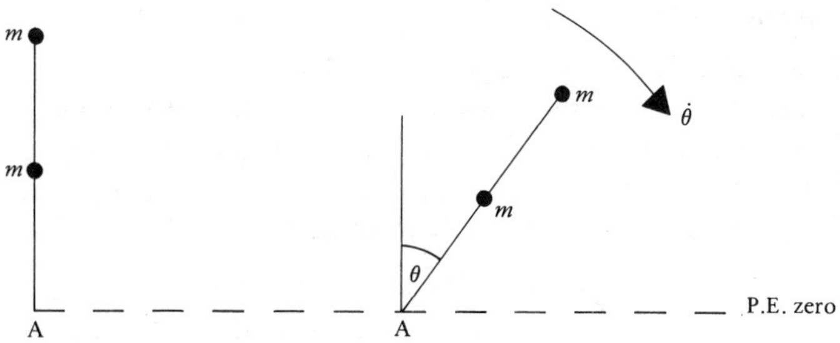

Now comparing the initial and general positions of the body, conservation of mechanical energy gives

$$mga + mg(2a) = mga\cos\theta + mg(2a\cos\theta) + \tfrac{1}{2}I\dot{\theta}^2$$

where I is the moment of inertia of the body about the axis through A.

Now $$I = \tfrac{4}{3}ma^2 + m(2a)^2 = \tfrac{16}{3}ma^2$$

Hence $$3mga(1-\cos\theta) = \tfrac{1}{2}(\tfrac{16}{3}ma^2)\dot{\theta}^2$$

i.e. $$\dot{\theta}^2 = \frac{9g}{8a}(1-\cos\theta)$$

Differentiating with respect to time gives

$$2\dot{\theta}\ddot{\theta} = \frac{9g}{8a}(\sin\theta)\dot{\theta}$$

i.e. $$\ddot{\theta} = \frac{9g}{16a}\sin\theta$$

Substituting these values into equations [1] and [2] we have

$$Y = \tfrac{1}{8}mg(27-43\cos\theta)$$

$$X = \frac{5mg\sin\theta}{16}$$

(a) When $\theta = \pi/3$, $Y = 11mg/16$ and $X = 5\sqrt{3}mg/32$

The magnitude of the resultant force R at the axis is given by

$$R = \sqrt{X^2+Y^2}$$

$$\Rightarrow \quad R = \tfrac{1}{32}mg\sqrt{75+484} = \tfrac{1}{32}mg\sqrt{559}$$

(b) When $\theta = \pi$, $Y = \frac{70}{8}mg$ and $X = 0$.

Hence $R = \frac{70}{8}mg$

Note: The value of $\ddot{\theta}$, and hence the value of X, is zero whenever the rod is vertical.

EXERCISE 11d

1) A uniform circular disc with centre O, radius r and mass m is free to rotate in a vertical plane about a horizontal axis passing through a point A on its circumference. The disc is held at rest with AO horizontal and is then released. Find the component perpendicular to AO of the reaction at the axis when AO makes an angle θ with the downward vertical through A.

2) A uniform rod AB of length $2l$ and mass m is free to rotate in a vertical plane about a horizontal axis through the end A of the rod. The rod is released from rest when AB is horizontal. Find the component along AB of the reaction at the axis when AB makes an angle θ with the downward vertical through A and hence find the total reaction at the axis when AB is vertical.

3) A uniform square lamina ABCD has sides of length $2a$ and is of mass m. The lamina is free to rotate in a vertical plane about a horizontal axis through A. The lamina is slightly displaced from its position of unstable equilibrium. Find the components, parallel and perpendicular to AC, of the reaction at the axis when AC makes an angle θ with the upward vertical through A. Find the total reaction at the axis when the lamina has rotated through an angle $2\pi/3$ from rest.

4) A uniform rod AB of length $2l$ and mass $2m$ has a particle of mass m attached at B. The rod is free to rotate in a vertical plane about a horizontal axis through A. When the rod is hanging at rest, with B below A, it is given an angular velocity $\frac{7}{2}\sqrt{g/l}$. Find the reaction at the axis when the rod first becomes horizontal.

MULTIPLE CHOICE EXERCISE 11

(Instructions for answering these questions are given on page ix)

TYPE II

1) A body rotates freely about a smooth fixed axis. The body is slightly displaced from the position in which the centre of mass is vertically above a point A on the axis.
(a) The potential energy is least when the centre of mass is level with A.
(b) The body turns through a complete revolution.
(c) The angular acceleration is constant.
(d) The angular velocity is maximum when the centre of mass is vertically below A.

2) A body rotates freely about a smooth horizontal axis for which its moment of inertia is I. When the body has turned through an angle θ from rest in the position of unstable equilibrium,
(a) the kinetic energy is $\frac{1}{2}I\dot{\theta}^2$,
(b) the angular acceleration is $\ddot{\theta}$,

(c) the loss in potential energy is $\frac{1}{2}I\dot{\theta}^2$,
(d) the change in mechanical energy is $I\ddot{\theta}$.

3) A uniform rod AB of length $2l$ and mass m performs small oscillations in a vertical plane about a smooth horizontal axis through A.
(a) The period of oscillation is independent of m.
(b) The length of the equivalent simple pendulum is l.
(c) The moment of inertia of the rod about the axis of rotation is $\frac{1}{3}ml^2$.
(d) The total mechanical energy of the rod is constant.

4) A disc is rotating about an axis under the action only of a torque C. The angular velocity changes from ω_1 to ω_2 while the disc turns through an angle θ in time t. If I is the moment of inertia of the disc about the axis of rotation,
(a) $C\theta = \frac{1}{2}I(\omega_2^2 - \omega_1^2)$,
(b) the initial angular momentum of the disc is $I\omega_1$,
(c) the mechanical energy of the disc is constant,
(d) $2\theta = (\omega_1 + \omega_2)t$.

TYPE III

5) (a) Every point in a solid body is rotating in a horizontal plane.
(b) A solid body is rotating about a vertical axis.

6) (a) A rod rotates freely, about a fixed horizontal axis, through an angle θ on either side of its stable equilibrium position.
(b) A rod, which can rotate freely about a fixed horizontal axis, performs angular simple harmonic motion.

7) (a) A disc rotates freely in its own plane about a horizontal axis.
(b) The total mechanical energy of a rotating disc is constant.

8) (a) A compound pendulum has a periodic time $\pi\sqrt{\frac{a}{g}}$
(b) A compound pendulum has a periodic time equal to that of a simple pendulum of length $4a$.

9) A rigid body rotates about a fixed axis through its centre of mass. Apart from the friction at the axis of rotation, no external force exerts any torque on the body. If I is the moment of inertia about the axis of rotation,
(a) the angular velocity decreases from $20\ \text{rad s}^{-1}$ to $10\ \text{rad s}^{-1}$ in one revolution,
(b) the work done by the frictional couple at the axis in one revolution is $150I$ joules.

TYPE IV

10) A uniform ring rotates about a horizontal tangential axis. Find the work done in one complete revolution by the frictional couple acting at the axis if
(a) the mass of the ring is m,
(b) the radius of the ring is a,
(c) apart from forces at the axis the only force acting on the ring is its own weight,
(d) the loss in kinetic energy in one revolution is $8m$.

11) A body free to rotate about a horizontal axis, performs small oscillations about its position of stable equilibrium. Find the period of the oscillations.
(a) The moment of inertia about the axis is I.
(b) The distance of the centre of mass from the axis is l.
(c) The mass of the body is m.
(d) The body swings through an angle of $4°$ on either side of the downward vertical through the axis of rotation.

12) A body is free to rotate about a smooth horizontal axis. When it is at rest it is given a blow which sets it rotating. Find the angle through which it turns before first coming to instantaneous rest.
(a) The body has a mass M.
(b) The body has a radius of gyration k.
(c) The distance of the centre of mass from the axis of rotation is h.
(d) The body is initially in a position of stable equilibrium.

TYPE V

13) When a flywheel, rotating about a fixed horizontal axis through its centre of mass, is brought to rest by a frictional couple acting at the axis, the mechanical energy of the flywheel remains constant.

14) A uniform rod of length $2l$ and mass m is free to rotate in a vertical plane about an axis through its centre of mass. If the rod is held horizontally and then released it will perform complete revolutions.

15) A uniform rod is free to rotate in a vertical plane about a horizontal axis through one end A. If the rod is held at an angle of $4°$ to the downward vertical through A and is then released, it will perform angular simple harmonic motion.

16) Whenever a torque C acts on a body causing it to rotate through an angle θ about a fixed axis, the work done by the torque is $C\theta$.

17) If the total mechanical energy of a rotating body is constant, the axis of rotation must be smooth.

MISCELLANEOUS EXERCISE 11

1) A uniform rod AB, of mass m and length $2a$, is free to rotate in a vertical plane about the end A. The rod is released from rest when AB is horizontal. Find the angular speed of the rod, in terms of g and a, when B passes through its lowest position.
(U of L)

2) A uniform square lamina ABCD, of mass m and side $2a$, is free to rotate in a vertical plane about an axis through its centre O. Particles each of mass m, are attached at the points A and B. The system is released from rest with AB vertical. Show that the angular speed of the square when AB is horizontal is $\sqrt{(6g/7a)}$.
(U of L)

3) A uniform rod AB, of length $2a$ and mass m, is free to rotate in a vertical plane about a horizontal axis through A. Given that a constant frictional couple of magnitude $\frac{1}{2}mga$ acts at the axis and that the rod is released from rest with AB horizontal, show that the angular speed of the rod when B first passes through its lowest position is given by ω where

$$\omega^2 = 3(4-\pi)g/(8a)$$

(U of L)

4) A flywheel has a horizontal axle of radius r. The system has a mass M and radius of gyration k about its axis and rotates without friction. A string is wound around the axle and carries a mass m hanging freely. If the system is released from rest, prove that the acceleration of the mass m is $g\dfrac{mr^2}{Mk^2+mr^2}$.
If the string slips off the axle after the mass m has descended a distance $8r$, find the magnitude of the constant retarding couple which is necessary to bring the flywheel to rest in n more revolutions.
(AEB)

5) A uniform square lamina ABCD is of mass $6m$ and side $2a$. Particles of mass m, $2m$, $3m$, $4m$ are fixed to the lamina at the vertices A, B, C, D respectively. The lamina is rotating freely, with its plane vertical, about a fixed horizontal axis through its centre. Prove that the angular velocity of the lamina is a maximum when CD is horizontal and below AB.
If the maximum kinetic energy of the system is three times its minimum kinetic energy, find the angular velocity of the lamina when the side BC is horizontal. Find also at this instant the rate of change of the moment of momentum of the system about the axis of rotation.
(U of L)

6) A uniform solid circular cylinder of mass M can rotate freely about a horizontal axis which coincides with a diameter of an end face of the cylinder. The length and the radius of the cylinder are both equal to a. Show that its moment of inertia about the axis is $7Ma^2/12$.
The cylinder is slightly disturbed from rest when its centre of mass is vertically above the axis. Show that, when the axis of the cylinder makes an angle θ with the upward vertical,

$$7a\dot{\theta}^2 = 12g(1-\cos\theta)$$

and obtain an expression for the angular acceleration $\ddot{\theta}$. (U of L)

7) A uniform rod AB, of length $2a$ and mass M, is fixed to a smooth pivot at A. One end of a light elastic string of natural length $2a$ and modulus kMg is fastened at the end B and the other end of the string is fastened at a point P vertically above A, where $AP = 2a$. The rod is held at rest with B at P and is slightly displaced. If the rod next comes momentarily to rest when angle PAB lies between $\frac{1}{2}\pi$ and π, find the range of possible values of k. (U of L)

8) Show that the centre of gravity of a uniform lamina in the form of a semi-circle of radius a is at a distance $4a/(3\pi)$ from the straight edge. Show also that the radius of gyration about this edge is $\frac{1}{2}a$.
This lamina, whose mass is m, is hinged so that it can rotate about its straight edge which is fixed in a horizontal position. The lamina is held at rest in a horizontal position and released. The motion of the lamina is opposed by a constant resisting torque. The lamina comes momentarily to rest for the first time when it has turned through an angle of $150°$. Find the magnitude of the resisting torque and the angular speed of the lamina when it is vertical for the first time. (AEB)

9) A uniform solid circular cylinder of radius a and mass m can turn freely about its axis, which is fixed in a vertical position. A horizontal force of magnitude mg acts on the cylinder, its line of action being at a distance a from the axis. Find the time taken to make n complete revolutions from rest.
When the angular velocity is ω radians per second the horizontal force ceases to act and a braking couple is applied. This couple is proportional in magnitude to the angular velocity of the cylinder, and in T seconds the angular velocity is reduced to $\frac{1}{2}\omega$ radians per second. Find the angular velocity after a further T seconds. (U of L)

10. A uniform lamina of mass m is in the form of a quadrant of a circle of radius a. Prove, by integration, that the moment of inertia of the lamina about an axis L, which is perpendicular to its plane and passes through the centre of the circle of which the lamina is a part, is $\frac{1}{2}ma^2$.
Prove, by integration, that the centre of gravity of the lamina is at a distance $\frac{4\sqrt{2}}{3\pi}a$ from L.
The lamina is free to rotate in its own vertical plane about L. Find the period of small oscillations about the stable equilibrium position. (AEB)

11) Prove, by integration, that the moment of inertia of a uniform rod, of mass M and length $2a$, about an axis perpendicular to its length and passing through one end is $\frac{4}{3}Ma^2$.
A uniform rod AB of mass M and length $2a$ can rotate freely about a fixed horizontal axis through A. A particle of mass $\frac{7}{12}M$ is fixed to the rod at a distance x from A. Find the period of small oscillations of the system about its position of stable equilibrium.
Find the value of x for which the period is a minimum. (U of L)

12) A uniform thin rod has length $2a$ and mass M. Prove by integration that the moment of inertia of the rod about an axis through its centre O and perpendicular to its length, is $\frac{1}{3}Ma^2$.
A small nut of mass $\frac{1}{2}M$ can be screwed to any point of the rod. The system is free to rotate in a vertical plane about a fixed frictionless horizontal axis through O. Find the period T of a small oscillation when the nut is at a distance x from O and the value of x which makes T a minimum. (AEB)

13) Prove that the moment of inertia of a uniform rod, of length $2a$ and mass m, about an axis perpendicular to the rod at a distance x from its midpoint is

$$\frac{1}{3}m(a^2+3x^2)$$

A uniform rod, of weight W and length $6b$, is smoothly pivoted to a fixed point at a distance $2b$ from one end and is free to swing in a vertical plane. The rod is held horizontal and released. Prove that after it has rotated through an angle θ its angular velocity is

$$\sqrt{\left(\frac{g\sin\theta}{2b}\right)}$$

and that when the rod is vertical the reaction on the pivot is $3W/2$. (JMB)

14) Prove, by integration, that the moment of inertia of a uniform rod of mass m and length $2l$ about an axis through its midpoint and perpendicular to its length is $\frac{1}{3}ml^2$.
Three uniform rods BC, CA, AB, each of length $2l$, are fastened together at their ends to form an equilateral triangle ABC. The rod BC is of mass M and the rods CA, AB are each of mass m. Show that the moment of inertia of the triangle about an axis through A and perpendicular to its plane is

$$\frac{2}{3}(5M+4m)l^2.$$

The triangle is free to rotate in its own vertical plane about a smooth fixed pivot at A. Find the period of small oscillations about the position of stable equilibrium. Given that L is the length of a simple pendulum of the same period, show that, whatever the values of M and m,

$$8l \leqslant 3\sqrt{3}L \leqslant 10l.$$ (JMB)

15) Show, by integration, that the moment of inertia of a circular uniform lamina of mass m and radius a about its axis is $\frac{1}{2}ma^2$.
The lamina can turn in a vertical plane about a fixed frictionless axis through a point P on its circumference. The lamina is released from rest when the diameter through P is horizontal. Prove that, when the diameter makes an angle θ with the horizontal, then

$$a(\dot{\theta})^2 = \frac{4}{3}g\sin\theta.$$

Find, in terms of m, g and θ, the vertical and the horizontal components of the reaction on the axis. (AEB)

16) A uniform square lamina ABCD of mass m and side $2a$ can rotate freely in a vertical plane about a smooth pivot at the midpoint of the side AB.
A particle of mass m is attached to the lamina at A. The lamina is held with its centre vertically below the pivot and is then released from rest. Show that, when AB makes an acute angle θ with the horizontal, the angular acceleration of the lamina is given by

$$\frac{d^2\theta}{dt^2} = \frac{3g}{8a}(\cos\theta - \sin\theta).$$

Find the maximum angular velocity of the lamina.
Find also the horizontal and vertical components of the force exerted by the pivot on the lamina immediately after it is released. (U of L)

17) A uniform rod AB of length $2a$ and mass $3m$ has a particle of mass m attached to it at B. The rod is free to rotate in a vertical plane about a horizontal axis perpendicular to the rod through a point X of the rod at a distance x ($<a$) from A. Find the length of the simple equivalent pendulum when the rod is slightly displaced from its equilibrium position with B below A.
Show that the length is least when $x = \frac{1}{4}a(5-\sqrt{7})$. (U of L)

18) A uniform circular wire of mass m and radius a is freely pivoted at a fixed point A on its circumference; a particle of mass m is fixed to the point B on the wire diametrically opposite to A. When the system is hanging freely with B below A, an initial angular velocity ω is given to the system which is just sufficient to make it perform complete revolutions in its own vertical plane. Prove that $a\omega^2 = 2g$.
If during the motion, AB makes an angle θ with the downward vertical, R is the component of the reaction of the pivot in the direction BA and S is the component in the direction perpendicular to BA in the sense of θ increasing, prove that

$$R = mg(3+5\cos\theta),$$

and find S. (JMB)

19) A uniform straight rod OA of length $2a$ and mass m can rotate freely in a vertical plane about its end O, which is fixed. The rod is released from rest when horizontal. Show that when OA makes an angle θ with the downward vertical, the component perpendicular to OA of the reaction of OA on the pivot at O is of magnitude $\frac{1}{4}mg\sin\theta$, and find the magnitude of the component along OA.
Find the greatest and least magnitudes of the reaction of OA on the pivot, and find also the angle θ at which the direction of the reaction makes an angle $\frac{1}{4}\pi$ with OA. (WJEC)

20) A uniform circular hoop has radius a, mass M and centre G. It makes complete revolutions in its own plane, which is vertical, being smoothly hinged at a fixed point O of its circumference. If the least angular velocity during its motion is Ω, prove that, when OG makes an angle θ with the upward vertical at O,

$$a\dot{\theta}^2 = g(k-\cos\theta)$$

where $$k = 1+\frac{a\Omega^2}{g}$$

Prove also that, if R is the magnitude of the reaction at O, then

$$R^2 = \tfrac{1}{4}M^2g^2(15\cos^2\theta-16k\cos\theta+4k^2+1)$$

The hinge will break if $R > 4Mg$. Find the maximum value of Ω for which complete revolutions are possible. (C)

21) A uniform rod AB, of mass $6m$ and length $2a$, has a particle of mass m attached at B. The system is free to rotate in a vertical plane about a smooth horizontal axis through A. The rod is released from rest when AB is horizontal. Show that in the subsequent motion

$$3a\left(\frac{d\theta}{dt}\right)^2 = 4g\sin\theta,$$

where θ is the angle made by AB with the horizontal.
Find the horizontal and vertical components of the force exerted by the rod on the axis of rotation when $\theta = \pi/4$. (U of L)

22) A uniform rod can rotate freely in a vertical plane about a fixed horizontal axis passing through a point C in the rod, where C does not coincide with the mid-point of the rod. The rod is given a small displacement from its vertical position of unstable equilibrium and performs a complete revolution in a vertical plane. When the rod has turned through an angle θ the horizontal component of the reaction at C is zero. Show that θ is independent of the position of C and find $\cos\theta$.
If C is at one end of the rod, find the position of the rod when the vertical component of the reaction at C vanishes. (U of L)

23) A thin uniform rod, of mass m and length $2a$, is attached to a pivot at one end. The rod is held in a horizontal position and then released from rest. Show that when the rod makes an angle θ with the downward vertical, its angular velocity is $\left(\dfrac{3g\cos\theta}{2a}\right)^{1/2}$.

The reaction of the hinge on the rod has components along and perpendicular to the rod. Find the magnitudes of these components in terms of m, g and θ. The reaction of the hinge on the rod may also be expressed in terms of horizontal and vertical components. Show that the horizontal component is greatest when $\theta = \pi/4$ and determine the magnitudes of both the horizontal and the vertical components when the rod is in this position. (AEB)

24) A lamina of mass m is free to rotate in a vertical plane about a fixed horizontal axis which passes through it at a point C. The moment of inertia of the lamina about this axis is λma^2, where a is the distance of C from G, its centre of mass, and λ is a constant. Initially the lamina is at rest with G vertically above C. It is then slightly displaced and at time t it has rotated through an angle θ, where $0 < \theta < \pi$.

(a) Express $\dot{\theta}^2$ and $\ddot{\theta}$ in terms of θ, g, a and λ.
(b) Show that the downward vertical component of the acceleration of the centre of mass of the lamina at time t is $g(1+2\cos\theta-3\cos^2\theta)/\lambda$.
(c) Given that the vertical component of the reaction at C is first zero when $\theta = \frac{\pi}{3}$, find λ and the value of $\cos\theta$ for which the vertical component of the reaction at C is next zero. (JMB)

25) A uniform circular disc of mass m and radius a is free to rotate about a fixed horizontal axis which is perpendicular to its plane and passes through a point A on its circumference. A particle of mass $\frac{1}{2}m$ is attached to the disc at the other end B of the diameter through A.
The system is released from rest with AB horizontal and rotation of the disc is opposed by a frictional couple of magnitude $kmga$, where k is a positive constant. Given that ϕ is the angle turned through by AB at time t after the start of the motion, show that, until the system first comes to rest,

$$\frac{d^2\phi}{dt^2} = \frac{2g}{7a}(2\cos\phi - k).$$

Given that the system first comes to rest when $\phi = \frac{5\pi}{6}$, show that $k = \frac{6}{5\pi}$.
For this value of k, find the magnitude of the vertical component of the reaction at A when AB becomes vertical for the first time.
Explain briefly why the differential equation for ϕ is not valid for all $t > 0$.
(JMB)

26) Prove, by integration, that the moment of inertia of a uniform rod, of length $2a$ and mass M, about an axis perpendicular to the rod and passing through one end is $\frac{4}{3}Ma^2$.
A body is made up of two equal uniform rods OA and CD, of mass M and length $2a$, which are joined together to form a letter T. The end A of the rod OA is rigidly attached to the midpoint of the rod CD so that OA and CD are perpendicular. The body is suspended from O, with A vertically below O, and is free to rotate in a vertical plane about a horizontal axis through O perpendicular to the plane of the body. The point A is now given a velocity $\sqrt{(nga)}$, where n is a constant, in the direction CD. Show that, when OA makes an angle θ with the downward vertical, the angular speed of the body, $\dot{\theta}$, is given by

$$a\dot{\theta}^2 = \frac{18g}{17}\left(\cos\theta - 1 + \frac{17n}{72}\right).$$

Deduce the value of n for which the point A just reaches the position vertically above O. For this value of n, find the reaction at the axis when A is vertically above O.

27) A rough uniform rod, of mass m and length $4a$, is held on a horizontal table perpendicularly to an edge of the table, with a length $3a$ projecting horizontally over the edge. If the rod is released from rest and allowed to turn about the edge, show, by using the principle of energy, that its angular speed after turning through an angle θ is

$$\sqrt{\left(\frac{6g\sin\theta}{7a}\right)}$$

assuming that the rod has not started to slip.
Deduce an expression, in terms of θ, for the angular acceleration, and hence determine the reaction normal to the rod. Show that the rod begins to slip when $\tan\theta = 4\mu/13$, where μ is the coefficient of friction.
(The moment of inertia of the rod about the edge of the table is $7ma^2/3$.)

(JMB)

28) State the perpendicular axis theorem relating to moments of inertia.
A uniform square lamina is of mass m and its edges are of length $2a$. Write down the moment of inertia of the lamina about one edge. Hence, or otherwise, show that the moment of inertia about a line through a corner and perpendicular to its plane is $\frac{8}{3}ma^2$.
Such a lamina is free to rotate in a vertical plane about a horizontal axis through one of its corners. It is released from rest with the diagonal through the point of suspension horizontal. Denoting the angle between this diagonal and the downward vertical by θ,

(a) find the horizontal and vertical components of the force exerted by the lamina on the axis when $\theta = \pi/4$
(b) show that the time taken for the diagonal to move from the position $\theta = \pi/4$ to the position when it first becomes vertical is

$$\left(\frac{2a\sqrt{2}}{3g}\right)^{\frac{1}{2}}\int_0^{\pi/4}\sec^{\frac{1}{2}}\theta\,d\theta.$$

(JMB)

CHAPTER 12

FURTHER ROTATION ABOUT A FIXED AXIS

THE IMPULSE OF A TORQUE

Consider a rigid body which is free to rotate about a fixed axis. The moment of inertia of the body about the axis of rotation is I and $\ddot{\theta}$ is the angular acceleration of the body at any instant.
Now suppose that a torque C acts on the body for a time t causing the angular velocity to change from ω_1 to ω_2 in that time.
It was established in Chapter 11 that

$$C = I\ddot{\theta}$$

Integrating this relationship with respect to time we have

$$\int C\,\mathrm{d}t = \int I\ddot{\theta}\,\mathrm{d}t$$

Hence, using corresponding limits,

$$\int_0^t C\,\mathrm{d}t = \left[I\dot{\theta}\right]_{\omega_1}^{\omega_2}$$

$$= I\omega_2 - I\omega_1$$

When C is constant, this result becomes

$$Ct = I\omega_2 - I\omega_1$$

But $I\omega$ is the angular momentum or *moment of momentum* of the body and $\int C\,\mathrm{d}t$ is called the *impulse of the torque* C.

Hence impulse of torque = increase in angular momentum

In some cases the torque C results from the action of a single force F applied at a distance r from the axis of rotation, i.e. $C = Fr$

Then $$\int C\,dt = \int Fr\,dt = r\int F\,dt$$

But $\int F\,dt$ is the impulse of the force F

So $r\int F\,dt$ is the moment of that impulse.

Hence **impulse of torque ≡ moment of impulse of force.**

INSTANTANEOUS IMPULSE OF TORQUE

If a torque is suddenly applied to a rotating body the angular momentum of the body undergoes a sudden change.

Such a torque exerts an instantaneous impulse whose magnitude can be determined only from the change in angular momentum produced,

(since $\int C\,dt$ cannot be evaluated when the time of application is infinitesimal).

EXAMPLES 12a

1) A uniform disc of mass 20 kg and radius 0.5 m can turn about a smooth axis through its centre and perpendicular to the disc. A constant torque is applied to the disc for 3 seconds from rest and the angular velocity at the end of that time is $240/\pi$ revolutions per minute. Find the magnitude of the torque. If the torque is then removed and the disc is brought to rest in t seconds by a constant force of 10 N applied tangentially at a point on the rim of the disc, find t.

The moment of inertia of the disc about the axis of rotation is given by

$$I = \tfrac{1}{2}ma^2$$

Therefore $$I = \tfrac{20}{2}(0.5)^2\,\text{kg m}^2 = \tfrac{5}{2}\,\text{kg m}^2$$

The angular velocity after 3 seconds is ω where

$$\omega = \frac{240 \times 2\pi}{\pi \times 60}\,\text{rad s}^{-1} = 8\,\text{rad s}^{-1}$$

(a)

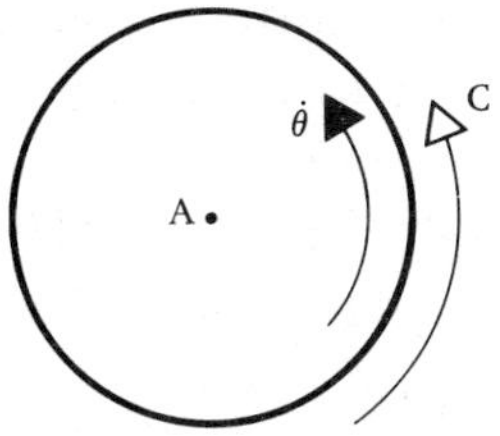

While the torque C is applied

using impulse of torque = increase in angular momentum we have

$\curvearrowleft A$ $$\int_0^3 C\,dt = I\omega - 0$$

Hence $$C = \tfrac{20}{3}\ \text{N m}$$

(b)

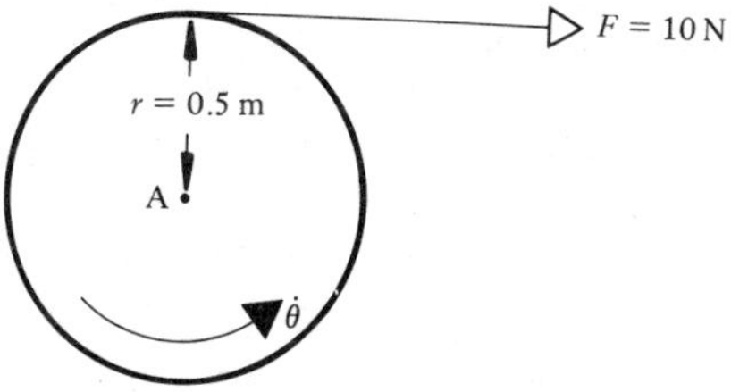

While the brake is applied

using moment of impulse = increase in angular momentum we have

$\curvearrowleft A$ $$-r\int_0^t F\,dt = 0 - I\omega$$

Hence $$rFt = I\omega$$

or $$\tfrac{1}{2}(10)t = (\tfrac{5}{2})(8)$$

$\Rightarrow$ $$t = 4$$

2) A flywheel whose moment of inertia about its axis of rotation is $20\ \text{kg m}^2$ is turning freely with an angular velocity of $3\ \text{rad s}^{-1}$ when a braking torque is applied. The magnitude of the torque, C N m, whose initial value is 32 N m, decreases uniformly with time in such a way that after 8 seconds the magnitude would be zero. Find the time which elapses before the flywheel is brought to rest.

The magnitude of the braking torque after t seconds is given by

$$C = 32-kt$$

But $C = 0$ when $t = 8$ so $k = 4$

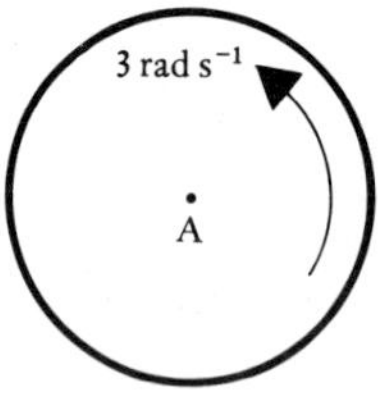

When $t = 0$

(i)

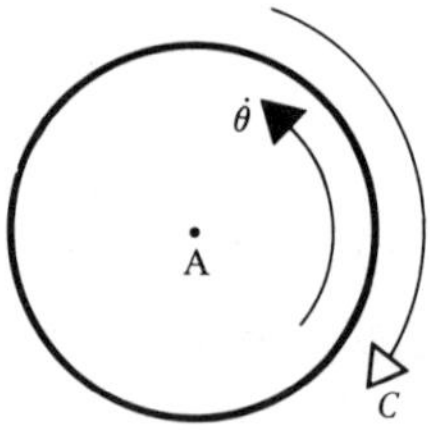

At time t

(ii)

Using impulse of torque = increase in angular momentum we have

A↺ $$-\int_0^t C\,\mathrm{d}t = I\dot{\theta}-I(3)$$

Hence $$\int_0^t (32-4t)\,\mathrm{d}t = 20(3-\dot{\theta})$$

$\Rightarrow$ $$32t-2t^2 = 20(3-\dot{\theta})$$

When the disc comes to rest, $\dot{\theta}=0$ and $32t-2t^2=60$

The values of t that satisfy this equation are $8\pm\sqrt{34}$ (both positive). Rotation ceases when the lower of these two values is reached, and the braking force no longer acts.
Hence the flywheel is brought to rest in $8-\sqrt{34}$ seconds.

3) A uniform rod AB of mass $3m$ and length $4l$, which is free to turn in a vertical plane about a smooth horizontal axis through A, is released from rest when horizontal. When the rod first becomes vertical a point C of the rod, where $AC = 3l$, strikes a fixed peg. Find the impulse exerted by the peg on the rod if
(a) the rod is brought to rest by the peg,
(b) the rod rebounds and next comes to instantaneous rest inclined to the downward vertical at an angle $\pi/3$ radians.

The moment of inertia of the rod about the axis through A is given by

$$I = \tfrac{4}{3}(3m)(2l)^2 = 16ml^2$$

Let ω_1 be the angular velocity of the rod when it first becomes vertical

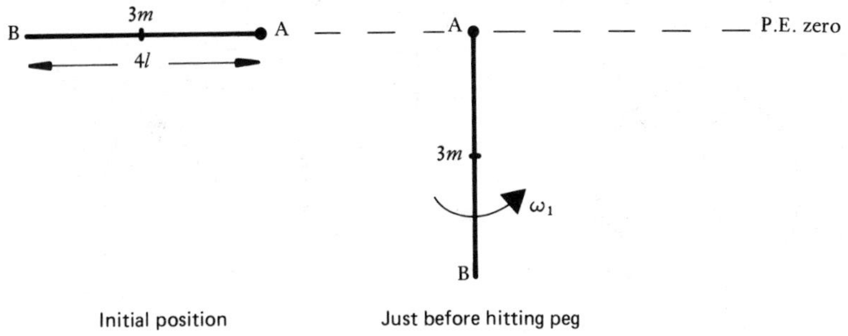

Using conservation of mechanical energy gives

$$0 = \tfrac{1}{2}I\omega_1^2 - 3mg(2l)$$

$$\Rightarrow \quad \omega_1^2 = 3g/4l$$

(a) Let J_1 be the impulse exerted by the peg in bringing the rod to rest.

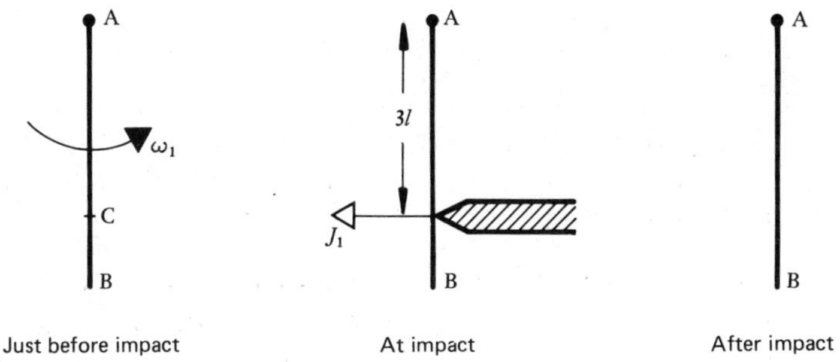

Using moment of impulse = increase in angular momentum we have

$$\text{A}\circlearrowright \quad -J_1(3l) = 0 - I\omega_1$$

Hence

$$3J_1 l = 16ml^2\sqrt{\frac{3g}{4l}}$$

$$J_1 = \tfrac{8}{3}m\sqrt{3gl}$$

(b) Let J_2 be the impulse exerted by the peg in causing the rod to rebound, and let ω_2 be the initial angular velocity of rebound.

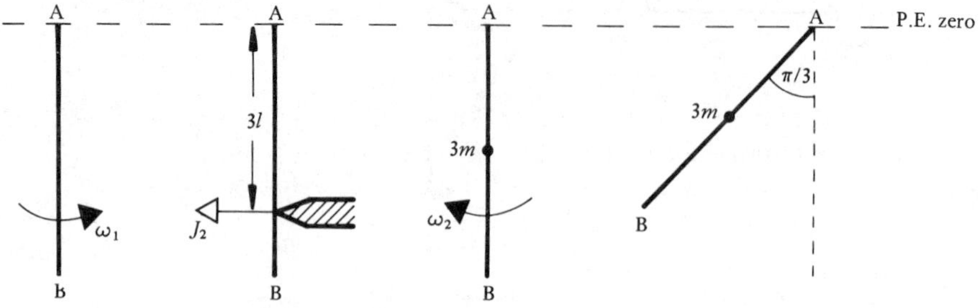

Using moment of impulse = increase in angular momentum we have,

$$\text{A}\circlearrowright \quad J_2(3l) = I\omega_2 - (-I\omega_1)$$

Hence $$3lJ_2 = 16ml^2(\omega_1 + \omega_2)$$

But ω_2 can be found by using conservation of mechanical energy from positions (iii) to (iv), i.e.

$$\tfrac{1}{2}I\omega_2^2 - 3mg(2l) = -3mg(2l\cos\pi/3)$$

Hence $$\omega_2^2 = 3g/8l$$

Then $$3lJ_2 = 16ml^2\left(\sqrt{\frac{3g}{4l}} + \sqrt{\frac{3g}{8l}}\right)$$

$$= 16ml^2\sqrt{\frac{3g}{8l}}\,(\sqrt{2}+1)$$

$$J_2 = \tfrac{4}{3}m\sqrt{6gl}\,(\sqrt{2}+1)$$

Note. Because the form of the expressions derived for J_1 and J_2 is unfamiliar, a check on the dimensions of these results is advisable.

e.g. $\frac{8}{3}m\sqrt{3gl}$ has dimensions $M\sqrt{\frac{L}{T^2}\times L} = MLT^{-1}$.

Now impulse (force × time) has dimensions $\frac{ML}{T^2}\times T = MLT^{-1}$.

Hence the results obtained for J_1 and J_2 are dimensionally correct.

EXERCISE 12a

1) A uniform solid cylinder of mass m and radius a rotates freely about its axis of symmetry under the action of a constant torque $4mga$. Find the angular velocity of the cylinder after 4 seconds if it started from rest.

2) Find the magnitude of the constant torque required to increase the kinetic energy of a uniform solid sphere, rotating freely about a diameter, from 400 J to 9025 J in 6 seconds if the mass of the sphere is 20 kg and its diameter is 0.5 m.

3) A flywheel whose moment of inertia about its axis of rotation is 16 kg m^2 is rotating freely in its own plane about a smooth axis through its centre. Its angular velocity is 9 rad s^{-1} when a torque is applied to bring it to rest in T seconds. Find T if

(a) the torque is constant and of magnitude 4 N m,

(b) the magnitude of the torque after t seconds is given by kt.

4) A uniform rod AB of length $2l$ and mass m is rotating in a horizontal plane about a vertical axis through A, with angular velocity Ω, when the midpoint of the rod strikes a fixed rail and is brought immediately to rest. Find the impulse exerted by the rail.

5) A uniform square trap door ABCD is free to rotate about an axis along the side AB which is horizontal. When it is hanging at rest, the midpoint of CD is struck a blow which causes the trap door to rotate through an angle of $60°$ before next coming to instantaneous rest. Find the impulse of the blow if the trap door is 1 m square and its mass is 40 kg.

6) ABC is a triangular framework of three uniform rods each of mass $3m$ and length $2l$. The framework is free to rotate about a smooth horizontal axis along AB. Initially the framework is in a position of unstable equilibrium and is slightly disturbed from rest. When it is first in a horizontal plane the point C collides with a fixed peg. Find the impulse exerted by the peg if

(a) the framework is brought to rest on impact,

(b) the framework rebounds so that C rises to a height $l\sqrt{3}/2$ above the level of AB before coming to instantaneous rest.

CONSERVATION OF ANGULAR MOMENTUM (MOMENT OF MOMENTUM)

If a torque acts on a rotating object it causes a change in angular momentum equal in magnitude to the impulse of the torque.
When two objects collide they exert equal and opposite instantaneous impulses upon each other.
About any specified axis, therefore, the moments of these two instantaneous impulses are equal and opposite.
If neither object is fixed the change in the angular momentum of each is equal to the moment of the impulse exerted on it, so the two objects undergo equal and opposite changes in angular momentum.
Thus the total angular momentum of the system is unchanged.
This is the Principle of Conservation of Angular Momentum and it can be stated in the following general form.

> Unless some external force exerts a torque on a system the total angular momentum (moment of momentum) of that system is constant.

EXAMPLES 12b

1) A uniform circular disc of mass m and radius a is rotating with constant angular velocity ω in a horizontal plane about a vertical axis through its centre A. A particle P of mass $2m$ is placed gently on the disc at a point distant $\frac{1}{2}a$ from A. If the particle does not slip on the disc find the new angular velocity of the rotating system.

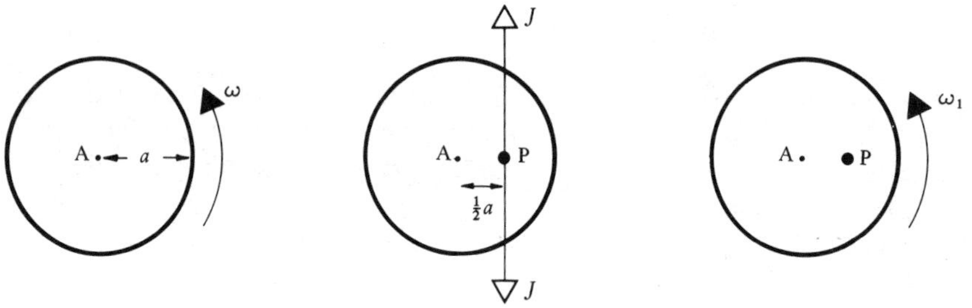

Rotation of disc alone (i) | Instantaneous addition of particle (ii) | Rotation of disc and particle (iii)

The moment of inertia of the disc about the axis through A is $\frac{1}{2}ma^2$

Hence the angular momentum of the rotating disc is $(\frac{1}{2}ma^2)(\omega)$

(At the instant when the particle is placed on the disc a pair of equal and opposite frictional impulses act, one on the disc and one on the particle, causing equal and opposite changes in moment of momentum about the axis through A.)

The particle stays in contact with a fixed point on the disc so it has the same angular velocity as the disc.

The moment of inertia of the disc and particle about the same axis is

$$\tfrac{1}{2}ma^2 + 2m(\tfrac{1}{2}a)^2$$

Hence the angular momentum of the rotating system is $(ma^2)(\omega_1)$

Using the principle of conservation of angular momentum gives

$$\tfrac{1}{2}ma^2\,\omega = ma^2\omega_1$$

$$\Rightarrow \qquad \omega_1 = \tfrac{1}{2}\omega$$

Thus the system rotates with an angular velocity $\frac{1}{2}\omega$.

2) A uniform rod AB of length $2l$ and mass m is turning freely in a horizontal plane about a vertical axis through A, with angular velocity $3v/l$. A particle P, also of mass m, is moving with constant velocity v in the same horizontal plane. The particle and the rod are moving towards each other and when AB is perpendicular to the path of the particle, P collides with point B of the rod. If the coefficient of restitution between the rod and the particle is $\frac{1}{2}$ find the speed of the particle after impact.

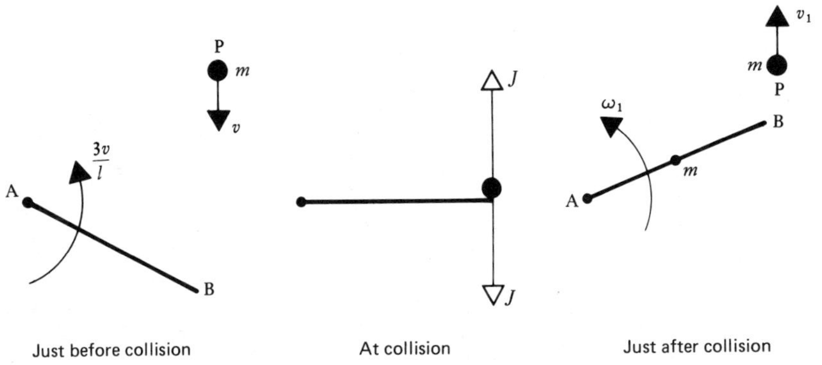

(This problem involves a particle moving with linear momentum as well as a rotating rod. In this context the use of the term moment of momentum is more logical than reference to angular momentum.)

At the moment of collision, equal and opposite impulses of magnitude J act on the rod and the particle, so moment of momentum is conserved.

Just before collision the moment of momentum about the axis through A, taking the clockwise sense as positive, is

for the rod	$\frac{4}{3}ml^2\left(\frac{3v}{l}\right)$
for the particle	$-(mv)(2l)$
hence for the system	$4mlv-2mlv$

Just after impact the moment of momentum about the axis through A is

for the rod	$\frac{4}{3}ml^2\omega_1$
for the particle	$(mv_1)(2l)$
hence for the system	$\frac{4}{3}ml^2\omega_1+2mlv_1$

Using conservation of moment of momentum gives

$$2mlv = \tfrac{4}{3}ml^2\omega_1+2mlv_1$$

i.e. $$3v = 2l\omega_1+3v_1 \qquad [1]$$

Now the law of restitution can be applied to the relative speeds of the particle P and the *point* B with which it collides. The speed of the point B on the rod is given by multiplying the angular velocity of the rod by the distance AB.

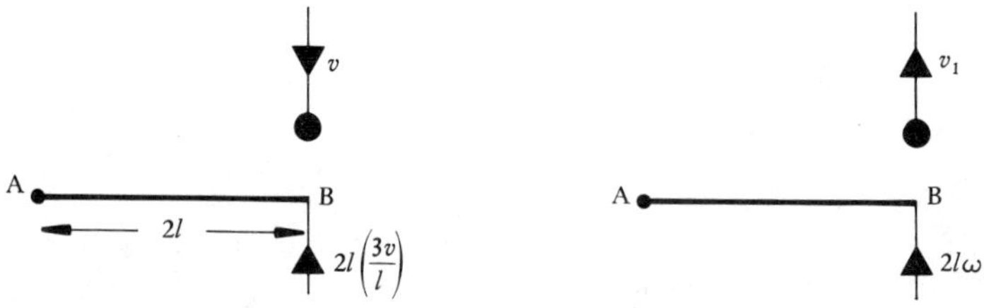

The speed of approach is $6v+v$

The speed of separation is $v_1-2l\omega_1$

Hence the law of restitution gives $\frac{1}{2}(7v) = v_1-2l\omega_1$ [2]

Then solving equations [1] and [2] we have

$$\omega_1 = -\frac{15v}{16l} \quad \text{and} \quad v_1 = \frac{13}{8}v$$

Thus, after impact the rod begins to rotate with angular velocity $15v/16l$ in the opposite sense and the particle begins to move in the reverse direction with speed $13v/8$.

3) A pulley in the form of a uniform disc of mass $2m$ and radius r, is free to rotate in a vertical plane about a fixed horizontal axis through its centre. A light inextensible string has one end fastened to a point on the rim of the pulley and is wrapped several times round the rim. The portion of string not wrapped round the pulley is of length $8r$ and carries a particle of mass m at its free end. The particle is held close to the rim of the pulley and level with its centre. If the particle is released from this position find the initial angular velocity of the pulley and the impulse of the sudden tension in the string when it becomes taut.

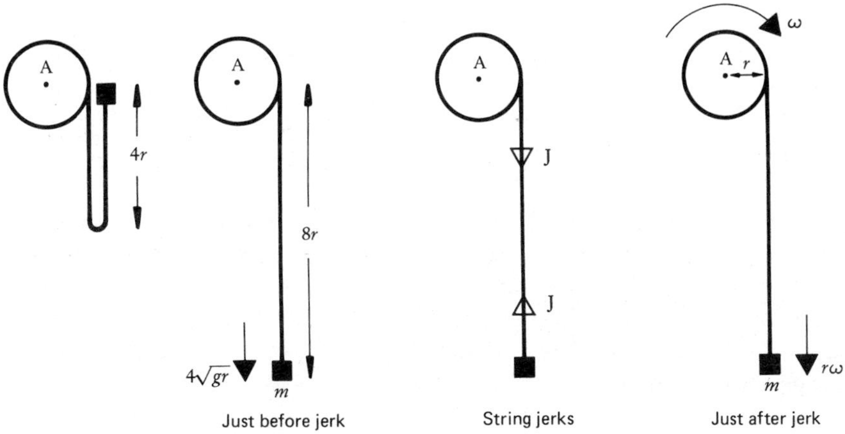

The particle falls freely through a distance $8r$ before the string becomes taut.

By then its velocity is $\sqrt{2g(8r)} = 4\sqrt{gr}$

At this instant the sudden tension in the string exerts equal and opposite impulses J on the particle and the pulley.
If the pulley then begins to move with an angular velocity ω the linear velocity of the particle suddenly changes to $r\omega$.

Therefore, applying conservation of moment of momentum, we have

$$(mv)r = (mr\omega)r + I\omega$$

where I is the moment of inertia of the pulley about the axis of rotation,

i.e. $$I = \tfrac{1}{2}(2m)r^2$$

Hence $$(4m\sqrt{gr})r = (mr\omega)r + \tfrac{1}{2}(2m)r^2\,\omega = 2mr^2\omega$$

The pulley therefore begins to move with angular velocity $2\sqrt{g/r}$

Considering the sudden change in the linear momentum of the particle caused by the impulsive tension J we have

$$J = -mr\omega - (-4m\sqrt{gr}) = 2m\sqrt{gr}$$

EXERCISE 12b

1) A uniform rod AB of mass $3m$ and length $2l$ is lying at rest on a smooth horizontal table with a smooth vertical axis through the end A. A particle of mass $2m$ moves with speed $2u$ across the table and strikes the rod at its midpoint C. If the impact is perfectly elastic find the speed of the particle after impact if
(a) it strikes the rod normally,
(b) its path before impact was inclined at $60°$ to AC.

2) A light string hangs over a pulley which is a uniform disc of mass $4m$ and radius a, and carries a scale pan of mass m at each end. If a particle of mass m is dropped into one of the scale pans from a height $10a$ above it, find the initial angular velocity of the pulley assuming that the string does not slip on the pulley and that the particle does not rebound from the scale pan.

3) A uniform rod AB, of mass m and length $4a$, is smoothly pivoted at a point O of its length, where $AO = a$, and hangs at rest with A uppermost. The rod receives a horizontal impulse of magnitude J at its centre of mass. Find the initial angular velocity of the rod.
If the rod describes complete revolutions in the subsequent motion, find an inequality for J in terms of a, m and g.

4) A uniform circular disc is rotating in its own plane about an axis through its centre. Its mass is M, its radius is a and it is rotating with angular velocity ω. A lamina of mass $6M$ in the shape of an annulus (the area between two concentric circles) with inner radius a and outer radius $2a$ is held with its centre vertically over the centre of the rotating disc and is then dropped. When the annulus circumscribes the disc there is no slipping at the circle of contact. Find the angular velocity of the compound disc.

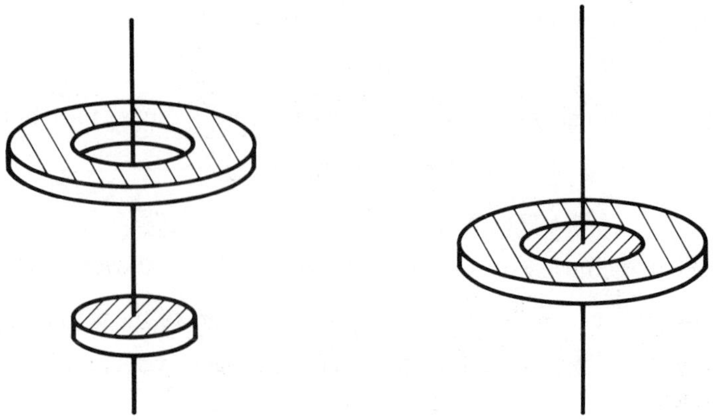

5) A uniform circular disc of mass M and radius a is pivoted at a point O on its circumference so that it can rotate about the tangent at O, which is horizontal, the centre of the disc describing a vertical circle of centre O in a plane perpendicular to the tangent. The point diametrically opposite to O is A and the disc is just displaced from rest when A is vertically above O. Find the angular velocity of the disc when A is vertically below O.
At this instant a particle of mass M travelling with velocity u in the direction of motion of the centre of the disc hits the disc at its centre, and adheres to it. Find the angular velocity of the system immediately after the impact.
If the disc just reaches its initial position show that

$$u = (3\sqrt{2}-\sqrt{5})\sqrt{(ag)}$$

MULTIPLE CHOICE EXERCISE 12

Multiple choice questions for this chapter are included in the exercise at the end of Chapter 13.

MISCELLANEOUS EXERCISE 12

1) Show that the moment of inertia of a uniform circular disc of radius a and mass m about an axis through its centre perpendicular to the disc is $\frac{1}{2}ma^2$.
The disc can rotate freely in a horizontal plane about a fixed vertical axis through a point in its circumference. Find the moment of the constant couple required to turn the disc through three complete revolutions in t seconds from rest, and find the moment of momentum of the disc about the axis at the end of this time. (U of L)

2) A uniform circular disc of mass $4m$ and radius $2a$ is rotating with angular velocity ω about a vertical axis through its centre perpendicular to its plane, which is horizontal. Write down expressions for its kinetic energy and its angular momentum about the axis.
The disc is brought to rest by a tangential force of magnitude $4mg$ applied at its rim. Find the time taken and the angle the disc turns through in that time.
When this disc is rotating as before with constant angular velocity ω, a small uniform disc of mass m and radius a mounted on the same axis is allowed to fall gently upon it from rest. Find their common angular velocity when slipping has finished. (U of L)

3) A flywheel, of radius a and moment of inertia $4Ma^2$ about its axis, is free to rotate about its axis which is horizontal and fixed. A light inextensible string has one end attached to the curved surface of the wheel and is wrapped round this surface and carries a mass M at its free end. This mass is released from rest and falls freely a distance h until the string becomes taut. The mass then continues in the same line and the wheel starts to rotate. The rotation of the flywheel is opposed by a constant resisting torque of magnitude G and the system comes to rest after the mass M has fallen a further distance h. Find the value of G and the impulsive tension in the string when it becomes taut. (AEB)

4) A uniform rod AB of mass m and length $2a$ is suspended from a smooth fixed pivot at its end A. The rod is hanging at rest when a particle of mass m, moving horizontally with speed u, collides with it at a point at a distance x below A. The particle adheres to the rod.
Prove that immediately after the impact the rod starts to move with angular velocity $3ux/(3x^2+4a^2)$. Find the value of x for which the angular velocity is a maximum and find, with this value for x, the value of u for which the rod just becomes horizontal in the subsequent motion. (JMB)

5) A uniform square lamina has mass M and side $2a$. Prove that the moment of inertia about a side is $4Ma^2/3$.
The lamina is smoothly hinged along a horizontal side and hangs vertically; a bullet of mass m, moving horizontally with speed v, strikes the lamina perpendicularly at the midpoint of the lower horizontal side and adheres to the lamina. Show that the whole starts to rotate about the hinge with an angular speed ω given by

$$a\omega = \frac{3mv}{2(M+3m)}$$

Find, in terms of M, m, g and a, the least value of v^2 which would cause the whole to make complete revolutions. (JMB)

6) A uniform rod AB, of mass m and length $2a$, is freely pivoted to a fixed point at A and is initially hanging in equilibrium. A particle of mass m, moving horizontally with speed u, strikes the rod at its middle point and rebounds from it. If there is no loss of energy at the impact, show that immediately afterwards the speed of the particle is $u/7$, and find the angular velocity with which the rod begins to rotate. Deduce that, if $u^2 < 49ag/12$, the rod will come to rest at an inclination θ to the downward vertical at A, where

$$\cos\theta = 1 - \frac{24}{49}\frac{u^2}{ag}$$

(JMB)

7) Prove, by integration, that the moment of inertia of a uniform rod, of length $2a$ and mass M, about an axis perpendicular to the rod and passing through one end is $\frac{4}{3}Ma^2$.
The rod is freely pivoted at one end and a particle of mass M is attached to the other end. The rod is held at rest in a horizontal position and is then released. Show that when the rod makes an angle θ with the downward vertical its angular velocity is given by

$$(\dot{\theta})^2 = \frac{9g\cos\theta}{8a}$$

When the rod reaches the vertical position, its lower end encounters a fixed inelastic stop which brings it to rest. Find the impulse of the blow on this fixed stop and the impulsive reaction at the pivot. (AEB)

8) A uniform circular lamina of mass m and radius a has a light hook attached at a point A on the rim. The lamina is rotating freely in a vertical plane with angular velocity Ω about a fixed horizontal axis through its centre O. When the radius OA is horizontal, with A moving upwards, the hook picks up a small ring, also of mass m, previously at rest. If the lamina just comes to rest with A vertically above O, find Ω.
Find also the horizontal and vertical components of the force exerted by the lamina on the axis immediately after the ring is picked up. (U of L)

9) A uniform triangular lamina ABC of mass M has $BC = 2a$ and $AB = AC = a\sqrt{10}$. It is free to rotate about a horizontal axis through A at right angles to its plane.
Prove that its moment of inertia about this axis is $\frac{14}{3}Ma^2$.
A uniform rod AD of mass M and length $a\sqrt{10}$ can rotate freely about the same axis through A, in the same vertical plane as the lamina. The lamina hangs in equilibrium with BC below A, and the rod is held horizontally and released from rest. When the rod strikes the lamina it adheres to it. Show that the kinetic energy lost in this impact is $\frac{7}{8}Mga$. (C)

10) A uniform circular disc, of mass m, radius a and centre O, is free to rotate in a vertical plane about a fixed smooth horizontal axis through O. A particle of mass m is attached at a point A of the rim of the disc. The system is released from rest when OA is horizontal. Find the angular speed and the angular acceleration of the disc when it has turned through an angle θ.
When A is vertically below O a second particle of mass $2m$, falling vertically, strikes and adheres to the rim of the disc at the point B, where AB is a diameter of the disc. Find the angular speed of the disc when A is vertically above O. (U of L)

11) Prove, by integration, that the moment of inertia of a uniform rod, of mass M and length $2a$, about an axis through its centre and perpendicular to its length is $\frac{1}{3}Ma^2$.

A square ABCD, of side $2a$, comprises four identical uniform rods rigidly joined together at their end points. The square is free to rotate with its plane vertical, about a fixed horizontal axis through A. Show that the moment of inertia of the square about this axis is $\frac{40}{3}Ma^2$, where M is the mass of each rod. The square hangs in equilibrium with C vertically below A when a horizontal impulse in the plane of the square is applied at C. Find the magnitude of this impulse given that, in the subsequent motion, C comes to rest vertically above A. (AEB)

12 A uniform disc has mass $16M$, radius $4a$ and centre O, and the radii OP, OQ are such that $P\hat{O}Q = 120°$. Two discs, each of radius a, with centres at the midpoints of OP, OQ respectively, are removed. Find the mass of the remaining body and the distance of its centre of mass from O.

This body can rotate freely in its plane about a horizontal axis through O. Show that its moment of inertia about this axis is $119Ma^2$.

The disc rests in equilibrium with PQ above O. A particle of mass M moving with speed u in the direction PQ strikes and adheres to the disc at P. Show that the centre of mass of the combined body, when in its initial position, is at the same horizontal level as O and at a distance $\frac{2}{15}a\sqrt{3}$ from O. Deduce that the disc will rotate through complete circles provided that

$$u^2 > 135ga\sqrt{3}$$

(C)

13) A uniform rod AB, of length $2a$ and mass m, is lying at rest on a smooth horizontal table with the end A smoothly hinged at a fixed point. A particle of mass m, moving on the table at right angles to AB with speed u, strikes the rod at the point C, where $AC = 2a/3$. Given that the system loses two thirds of its kinetic energy due to the collision, find the angular velocity of the rod and the velocity of the particle immediately after the collision.

Shortly after the rod has been set rotating, the end B strikes a fixed elastic stop which causes the rod to rebound with one half the angular speed it had just before striking the stop. Find the impulse on the stop in terms of m and u. (JMB)

14) A uniform thin rod AB is of mass m and length $2a$. Prove by integration that the moment of inertia of the rod about a line through A perpendicular to AB is $4ma^2/3$.

The rod is free to rotate about a fixed smooth horizontal axis which passes through A and is released from rest when AB is horizontal. When AB is vertical the rod picks up a stationary particle, also of mass m, which adheres to the rod at B. Show that the rod first comes to instantaneous rest in the subsequent motion when B is at a distance $11a/6$ below the level of A.

(U of L)

15) Show that the moment of inertia of a uniform square lamina of side $2a$ about an axis through its centre parallel to one of the sides is $\frac{1}{3}ma^2$, where m is the mass of the lamina.

This square lamina ABCD is free to rotate with its plane vertical about a horizontal axis through one corner A. Initially the lamina rests in stable equilibrium with AC vertical. A horizontal impulse I, acting in the plane ABCD, is applied to the lamina at C. Show that the angular velocity of the lamina immediately after the impulse is $\dfrac{3I\sqrt{2}}{4ma}$.

Find the minimum value of I such that the lamina will make a complete revolution.

(AEB)

16) A compound pendulum consists of a uniform rod OA of length $2a$ and mass $3M$ fastened at A to the circumference of a solid uniform sphere of radius a, mass $5M$ and centre C. The join is such that OAC is a straight line. The pendulum is free to rotate about a fixed smooth horizontal axis through O perpendicular to OAC. Show that the moment of inertia about this axis is $51Ma^2$.

The pendulum hangs at rest with C vertically below O. It is given a small displacement and then released. Prove that the ensuing motion is approximately simple harmonic and state its period.

While the pendulum is hanging at rest it is struck by a particle of mass $3M$ moving in the same vertical plane as that in which OA is free to move. The particle strikes the sphere at the end of a horizontal diameter. At that instant the particle is moving horizontally with speed v and after impact it adheres to the sphere. Show that the initial angular velocity of the pendulum is $\frac{1}{9}v/a$.

If $v^2 = 18ag$, show that when the rod comes to instantaneous rest it makes an angle α with the vertical where

$$9\cos\alpha + \sin\alpha = 6$$

(C)

17) A uniform rod of mass M and length $2l$ is free to rotate in a vertical plane about a fixed horizontal axis through one end of the rod. When hanging in equilibrium, the rod receives an impulse of magnitude $2M\sqrt{(gl/3)}$ applied to its lower end in the plane of rotation and in a direction making an angle $\pi/3$ with the horizontal. Prove that when next at rest the rod makes an angle $\pi/3$ with the vertical. Prove also that, at this instant, the component perpendicular to the rod of the reaction at the axis is $\sqrt{3}Mg/8$. (U of L)

18) A particle P of mass m rests on a rough horizontal table, the coefficient of friction being μ, and is connected to a particle Q of mass M by a light inextensible string which passes over a circular pulley of radius a and moment of inertia I about its axis. The string from P to the pulley is horizontal, and from the pulley to Q is vertical. The pulley can rotate freely about its axis which is fixed and horizontal. The system is released from rest. Assuming that the system moves and that the string does not slip on the pulley, prove that the magnitude of the acceleration of each particle is

$$\frac{(M-\mu m)g}{M+m+(I/a^2)}$$

and find the tension in the vertical portion of the string.

When the speed of the particles is V the particle P is seized and held fixed, causing Q and the pulley to stop also. Find the impulse which must be applied to P to achieve this. (JMB)

19) Prove, by integration, that the moment of inertia of a uniform square lamina, of mass M and side $2l$, about one of its sides is $\frac{4}{3}Ml^2$.

A uniform square lamina ABCD, of mass M and side $2\sqrt{2}a$ is suspended from A and hangs in equilibrium with the diagonal AC vertical and C below A. Show that the moment of inertia of the lamina about a horizontal axis through A perpendicular to its plane is $\frac{16}{3}Ma^2$.

The lamina, which is free to rotate in its own plane about this horizontal axis through A, receives a horizontal impulse J applied in the plane of the lamina at the point C. Given that the lamina rotates so that C just reaches a position vertically above A, find J.

Also show that the horizontal and vertical components of the reaction at A, when AC is horizontal, are $\frac{3}{2}Mg$ and $\frac{1}{4}Mg$ respectively. (U of L)

20) A thin rod AB of length a has variable mass per unit length $\rho_\circ(a+x)/a$, where x is the distance measured from A and $\rho_\circ$ is a constant. Show that
(a) the mass M of the rod is $3a\rho_\circ/2$,
(b) the centre of mass of the rod is at the point $x = 5a/9$,
(c) the moment of inertia of the rod about an axis through A perpendicular to AB is $7Ma^2/18$.
The rod is freely pivoted at A and is hanging in equilibrium when it is struck by a horizontal impulse of magnitude P at the point B. Find the angular velocity with which the rod begins to rotate and show that if B passes through a point vertically above A then

$$P > \tfrac{1}{9}M\sqrt{(70ag)}$$ (JMB)

21) Show that when a rigid body is rotating about a fixed axis with angular speed ω, its kinetic energy is $\frac{1}{2}I\omega^2$, where I is the moment of inertia about the axis.
A uniform rod AB of mass m and length l can rotate freely in a vertical plane π about a fixed pivot at A. A light elastic string of natural length l and modulus $2mg/3$ has one end fastened to the rod at B and the other end fastened to a fixed point C in the plane π and at the same horizontal level as A. The rod is released from rest in the horizontal position with B on the same side of the pivot as C. Given that $AC = l\sqrt{3}$, show that the angular speed attained by the rod just as it reaches the vertical position is $\sqrt{(g/l)}$.
When it reaches the vertical position, the rod AB strikes another uniform rod AD which is also freely pivoted at A and is hanging vertically at rest. The mass of AD is m and its length is $2l$. Given that immediately after impact the two rods begin to rotate in the same direction with the same angular speed ω, show that

$$\omega = \frac{1}{5}\sqrt{\frac{g}{l}}$$

Suppose now that at the instant of impact an impulse is applied to the rod AD at D and is directed horizontally in the plane π. Given that this impulse reduces the system to rest, find its magnitude. (JMB)

CHAPTER 13

ROTATION AND TRANSLATION

ROTATION ABOUT AN AXIS MOVING IN A STRAIGHT LINE

Motion of a Lamina in its own Plane

Consider a lamina which is moving under the action of a set of coplanar forces so that a point A of the lamina moves in a straight line while the lamina rotates about an axis through A.

If, initially, A is at a point O on the plane and Ox is the line along which A moves, then the linear velocity and acceleration of A at any instant are $\dot{x}$ and $\ddot{x}$.

Consider any other point P on the lamina where AP is of constant length l, and θ is the inclination of AP to Ox at any instant.

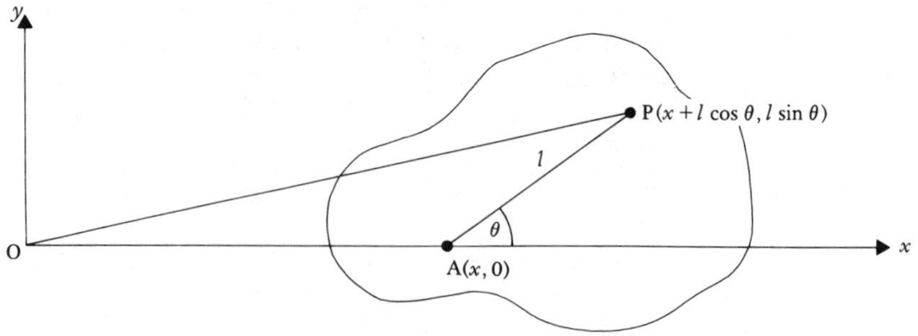

The coordinates of P at any instant are (X, Y) where

$$X = x + l\cos\theta \quad \text{and} \quad Y = l\sin\theta$$

Differentiating with respect to time, we have

$$\dot{X} = \dot{x} - l\sin\theta\,\dot{\theta}$$

$$\dot{Y} = l\cos\theta\,\dot{\theta}$$

The velocity components of P can therefore be used in the form illustrated below.

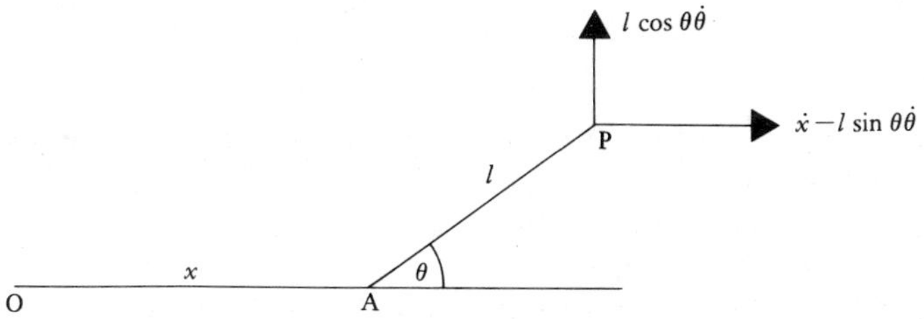

Alternatively, since $\overrightarrow{OP} = \overrightarrow{OA} + \overrightarrow{AP}$, the motion of P is a combination of the motion of A and the motion of P relative to A.

Now, relative to A, P is a point with polar coordinates (l, θ).
The radial and transverse components of the velocity and acceleration of a point with general polar coordinates (r, θ) were derived in Chapter 2.
For the motion of P relative to A, r is of constant value l, so these general results become

	Radial component	Transverse component
Velocity	0	$l\dot{\theta}$
Acceleration	$-l\dot{\theta}^2$	$l\ddot{\theta}$

Thus the motion of P relative to O is made up of the components shown in the diagrams below.

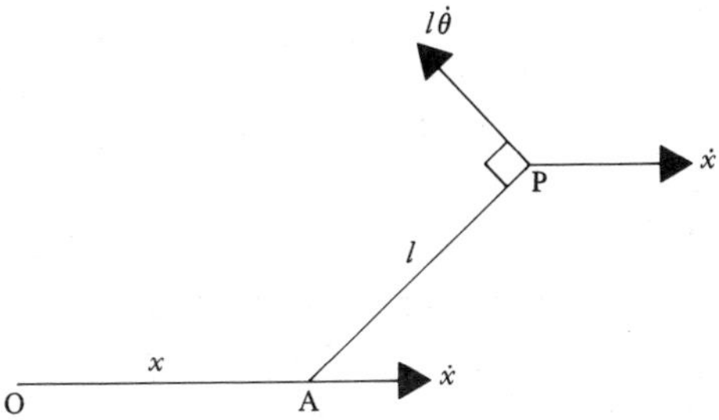

Velocity components

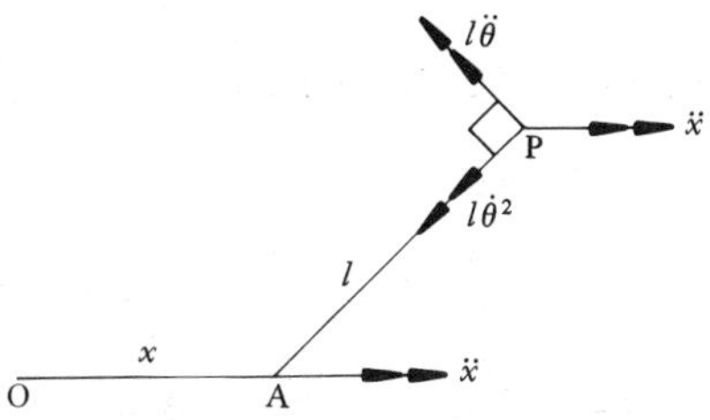

Acceleration components

EXAMPLES 13a

1) A lamina is moving in its own plane so that a point A of the lamina moves with constant acceleration a in a straight line Ox, and the lamina rotates about an axis through A with constant angular velocity ω. Find

(a) the equation relative to A of the locus of the set of points with acceleration of magnitude $a\omega^2$.

(b) in terms of t the equation relative to O of the set of points with speed $2a\omega$, given that the initial speed of A is u.

The angular velocity of the lamina is constant, i.e. $\dot{\theta} = \omega$, so the angular acceleration $\ddot{\theta}$ is zero.

(a) Thus the acceleration components of a point P distant r from A are as shown in the diagram below.

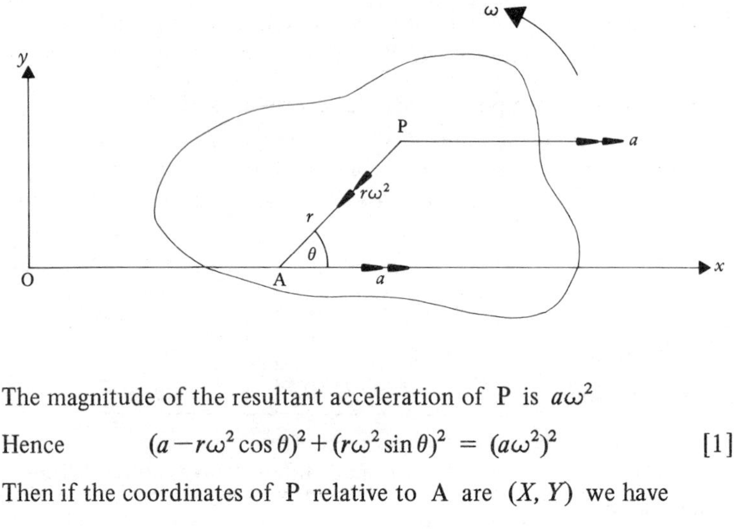

The magnitude of the resultant acceleration of P is $a\omega^2$

Hence $$(a - r\omega^2\cos\theta)^2 + (r\omega^2\sin\theta)^2 = (a\omega^2)^2 \qquad [1]$$

Then if the coordinates of P relative to A are (X, Y) we have

$$X = r\cos\theta \quad \text{and} \quad Y = r\sin\theta$$

Hence equation [1] becomes

$$(a - X\omega^2)^2 + (Y\omega^2)^2 = (a\omega^2)^2$$

or $$\left(X - \frac{a}{\omega^2}\right)^2 + Y^2 = a^2$$

This equation gives the relationship between X and Y and is therefore the equation, relative to A, of the locus of points with acceleration $a\omega^2$.

This locus can be identified as a circle with centre $\left(\frac{a}{\omega^2}, 0\right)$ and radius a.

(b) The velocity of A at time t is $u+at$ so the components of the velocity of P at time t are as shown in the following diagram.

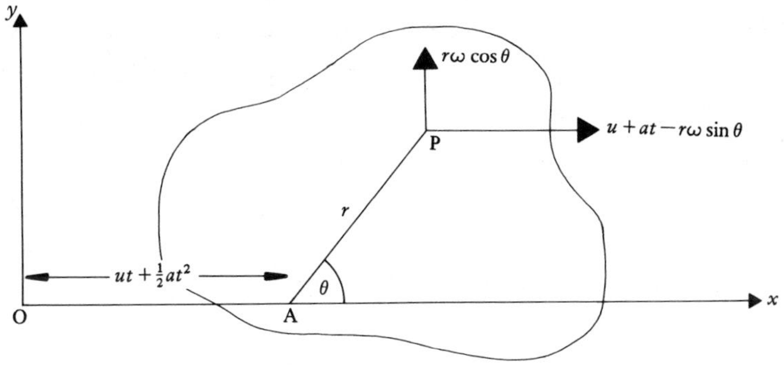

Points with speed $2a\omega$ are such that

$$(u+at-r\omega\sin\theta)^2+(r\omega\cos\theta)^2 = (2a\omega)^2 \qquad [2]$$

Now $\text{OA} = ut+\frac{1}{2}at^2$, so the coordinates (x, y) of P relative to O are

$$x = ut+\tfrac{1}{2}at^2+r\cos\theta$$

$$y = r\sin\theta$$

Hence equation [2] becomes

$$(u+at-\omega y)^2 + (x-ut-\tfrac{1}{2}at^2)^2\omega^2 = (2a\omega)^2$$

i.e.

$$\left(y-\frac{u+at}{\omega}\right)^2+\left(x-\frac{2ut+at^2}{2}\right)^2 = 4a^2$$

This is the equation, relative to O, of the locus of points with speed $2a\omega$ at time t.

(The locus is seen to be a circle with centre $\left(\frac{u+at}{\omega}, \frac{2ut+at^2}{2}\right)$ and radius $2a$.)

Independence of Rotation and Translation

Consider a lamina of mass M, whose centre of mass is at the point G, moving under the action of a set of forces whose resultant is **R**. It was established in Chapter 7 that the acceleration of the point G is the same as that of a particle of mass M moving under the action of **R**.

If the direction of **R** is constant and initially G was either at rest or moving in the direction of **R**, then G subsequently moves in a straight line Ox parallel to **R**.

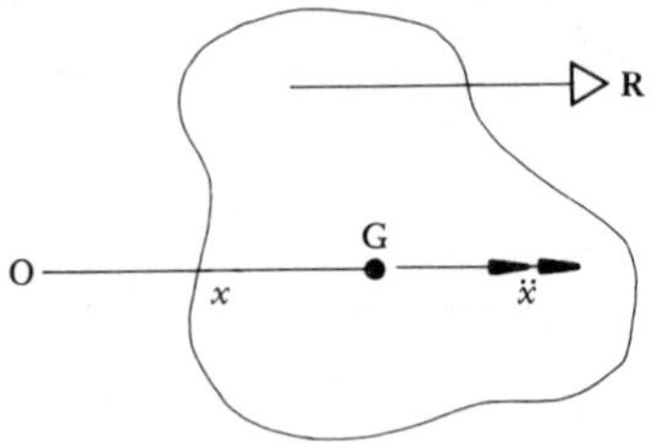

Thus the equation of motion of G is

$$R = M\ddot{x}$$

The acceleration components of any point P have already been established and are shown in diagram (i) below. If, at P, there is a constituent particle of the lamina of mass m, the force acting on this particle is shown as a pair of perpendicular components parallel and perpendicular to GP, in diagram (ii).

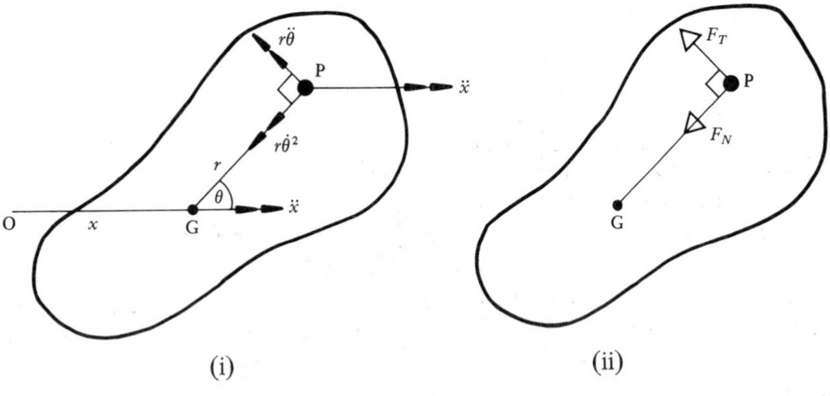

The rotational behaviour of the lamina will now be investigated.

Applying Newton's law to the motion of the particle at P we have, in the direction of F_T,

$$F_T = m(r\ddot{\theta} - \ddot{x}\sin\theta)$$

Hence $$rF_T = mr(r\ddot{\theta} - \ddot{x}\sin\theta)$$

For the whole body therefore,

$$\sum rF_T = \ddot{\theta}\sum mr^2 - \ddot{x}\sum mr\sin\theta \qquad [1]$$

Now $\sum rF_T = C$ where C is the resultant torque about G

$$\sum mr^2 = I_G$$

and $\sum mr\sin\theta = 0$ as G is the centre of mass.

Therefore equation [1] becomes

$$C = I\ddot{\theta}$$

Hence the lamina rotates as though it were moving under the action of the resultant torque about a fixed axis through G.

The motion of a lamina rotating and translating in its own plane can therefore be analysed by considering independently
(a) the linear motion of the centre of mass,
(b) rotation about an axis through the centre of mass.

These results are confirmed if we consider that the given force **R** is equivalent to an equal force **R** acting through G, together with a couple of moment C where $C = Ra$.

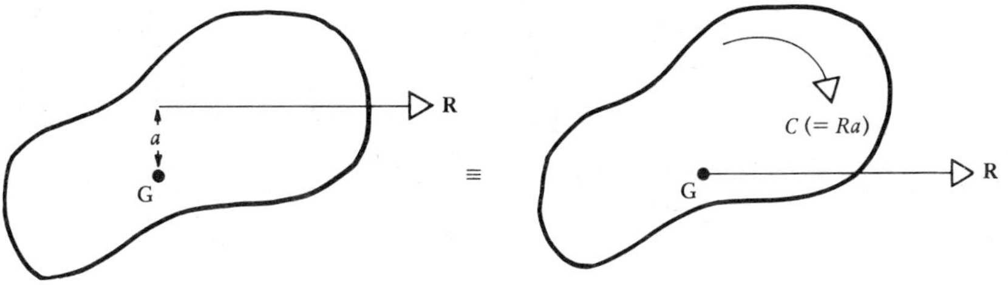

The equivalent system clearly indicates that G can be expected to move in a straight line and that the lamina will rotate about a perpendicular axis through G. If the resultant **R**, of the given set of forces, itself acts through G then $C = 0$. In this case the lamina will move in the same way as a particle of mass M at G. (This property justifies the method used in *Mechanics and Probability* to solve a number of problems involving the motion of rigid bodies, in which the body was treated as a particle.)

EXAMPLES 13a (continued)

2) One end of a light inextensible string is fastened to a point on the rim of a uniform circular disc of mass m and radius a. The string is wrapped several times round the circumference and the other end is then attached to a fixed point A. The disc is held in the vertical plane containing A so that the portion of string which is not in contact with the disc is taut and vertical. If the disc is then released from rest find the acceleration of its centre and the tension in the string.

(As the disc descends, its centre O moves vertically downwards since the only forces acting on the disc are vertical. The disc also rotates as it falls, because the resultant force does not act through O.)

Let the linear velocity and acceleration of O be $\dot{x}$ and $\ddot{x}$ and let the angular velocity and acceleration of the disc be $\dot{\theta}$ and $\ddot{\theta}$.

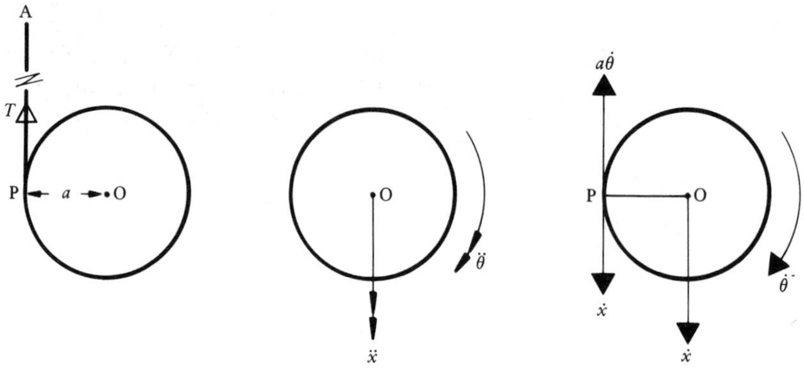

(i) (ii) (iii)

The point P where the string loses contact with the disc has a velocity $\dot{x}-a\dot{\theta}$ vertically downward (diagram (iii)). But this point is momentarily at rest (in contact with a stationary point on the string).

i.e. $$\dot{x} = a\dot{\theta} \quad \Rightarrow \quad \ddot{x} = a\ddot{\theta}$$

For the linear motion of the centre of mass, Newton's law gives

$$mg-T = m\ddot{x} \qquad [1]$$

For the rotation of the disc about an axis through O, using $C = I\ddot{\theta}$ gives

$$Ta = (\tfrac{1}{2}ma^2)\ddot{\theta}$$

But $\ddot{x} = a\ddot{\theta}$ hence $$Ta = \tfrac{1}{2}ma\ddot{x} \qquad [2]$$

From equations [1] and [2] we see that the tension in the string is $\frac{1}{3}mg$ and that the centre of the disc descends vertically with acceleration $\frac{2}{3}g$.

EXERCISE 13a

In questions 1 to 3, a lamina is moving in its own plane so that it rotates about a point C fixed in the lamina, while C moves along a fixed straight line Ox. A point P of the lamina is distant d from C and initially, i.e. when $t = 0$, P is on Ox.

1) The point C has constant speed u and the lamina rotates about C with constant angular speed ω. Using t as the only variable, find the components parallel and perpendicular to Ox, of the velocity of P at any time t.

2) Point C moves with constant speed u and is at O when $t = 0$. The lamina rotates about C with constant angular acceleration α. Find, in terms of t, the locus of the set of points P, relative to Ox and a perpendicular axis Oy, given that P is initially at rest.

3) The lamina rotates about C with constant angular velocity ω and C moves with constant acceleration a. Show that, when $t = \pi/\omega$, the acceleration of P is parallel to Ox and is of magnitude $a + d\omega^2$.

4) A long string is wrapped completely round the rim of a uniform circular ring of mass m, radius a and centre C; the string ends at a point A on the ring. This free end of string is attached to a fixed point and the ring is released from rest with AC horizontal. Find the acceleration of the centre of the ring in the subsequent motion. Find also the tension in the string while it is taut. Describe the motion of the ring after it falls off the end of the string.

5) A yoyo consists of two uniform circular discs each of mass m and radius $4a$ between which is a small uniform solid cylinder of mass $3m$ and radius a. The line through the axis of the cylinder passes through the centre of each disc. A light inextensible string has one end fastened to a point on the cylinder and is wrapped round the cylinder. The yoyo is projected vertically downward with speed $4\sqrt{ag}$ and the string, as it unwinds, becomes vertical. Find the speed of the centre when a length $126a$ of string has unwound.

ROTATION AND TRANSLATION OF A RIGID BODY

A rigid body can be regarded as being made up of a series of parallel laminate sections. Consider such a body, moving so that

its centre of mass is moving in a line,

each of its constituent laminas is rotating in its own plane,

the axis of rotation passes through the centre of mass of each section and hence through the centre of mass of the body.

Then each lamina moves as though it were rotating, under the action of its own resultant torque, about the common axis through the centres of mass.
Also, because the body is rigid, every lamina has the same angular acceleration.

Thus the whole body moves as though it were rotating under the action of the total resultant torque about an axis through its centre of mass.

The motion of a rigid body can therefore be analysed in the same way as the motion of a lamina moving in its own plane.

Kinetic Energy

Consider first a lamina, whose centre of mass is G, moving in its own plane under the action of a set of forces whose resultant has a constant direction. The velocity components of a constituent particle P of the body are shown below.

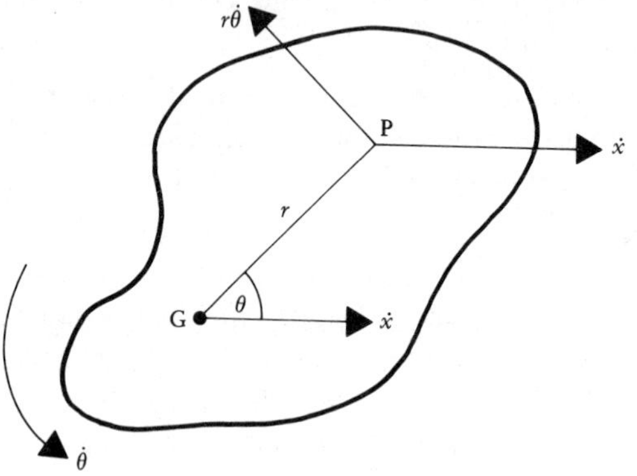

If the mass of the particle is m, its kinetic energy is

$$\tfrac{1}{2}m\{(\dot{x}-r\dot{\theta}\sin\theta)^2+(r\dot{\theta}\cos\theta)^2\}$$

Summing for the whole lamina we have

$$\begin{aligned}\text{K.E.} &= \sum \tfrac{1}{2}m\{\dot{x}^2-2\dot{x}r\dot{\theta}\sin\theta+r^2\dot{\theta}^2\}\\ &= \tfrac{1}{2}\dot{x}^2\sum m-\dot{x}\dot{\theta}\sum mr\sin\theta+\tfrac{1}{2}\dot{\theta}^2\sum mr^2\end{aligned}$$

But $\sum m = M$ where M is the total mass of the lamina

$\sum mr^2 = I$ where I is the moment of inertia of the lamina about the axis through G

and $\sum mr\sin\theta = 0$ (Since G is the centre of mass.)

The kinetic energy of the lamina is therefore given by

$$\text{K.E.} = \tfrac{1}{2}M\dot{x}^2+\tfrac{1}{2}I\dot{\theta}^2$$

This shows that the kinetic energy of a rotating lamina is made up of the linear kinetic energy of a particle of mass M at G and the rotational kinetic energy of the lamina about an axis through G.

By considering, as we did before, a rigid body to comprise a set of parallel laminas, we see that the results derived above can be extended to give the kinetic energy of a rigid body.
Hence the following facts can be used in problems involving the translation and rotation of a rigid body of mass M and centre of mass G.

(a) The centre of mass moves as though it were a particle of mass M acted on by the resultant of all forces affecting the body.
(b) The rotation of the body about an axis through G is independent of the linear movement of G.
(c) The kinetic energy of the body is made up of the linear kinetic energy of a particle of mass M at G and the rotational kinetic energy of the body about an axis through G.

ROLLING WITHOUT SLIPPING

EXAMPLES 13b

1) A uniform solid sphere of radius a and mass M is rolling without slipping on a horizontal surface so that its centre of mass moves in a straight line with constant speed v. Find the kinetic energy of the sphere.

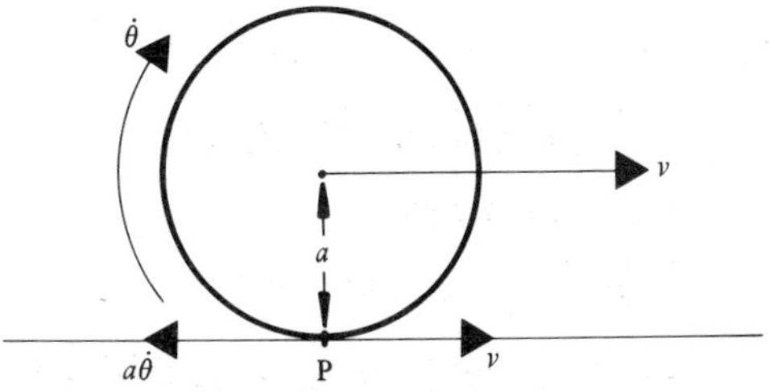

The velocity components of the point P, which is momentarily in contact with the plane, are shown in the diagram.
The disc does not slip on the plane, so P is instantaneously at rest,

i.e. $$v - a\dot{\theta} = 0 \quad \Rightarrow \quad \dot{\theta} = \frac{v}{a}$$

Now the kinetic energy of the sphere is given by

$$\begin{aligned} \text{K.E.} &= \tfrac{1}{2}Mv^2 + \tfrac{1}{2}I\dot{\theta}^2 \\ &= \tfrac{1}{2}Mv^2 + \tfrac{1}{2}(\tfrac{2}{5}Ma^2)\left(\frac{v}{a}\right)^2 \\ &= \tfrac{7}{10}Mv^2 \end{aligned}$$

2) A uniform solid cylinder of mass $2m$ and radius r rolls without slipping down a plane inclined at $30°$ to the horizontal. Find its angular acceleration and the least possible value of the coefficient of friction between the cylinder and the plane.

(A circular object cannot roll down an inclined plane unless the plane is rough because it is the moment of the frictional force which causes rotation about an axis through the centre of mass. If the object rolls without slipping, the point of application of the frictional force is the point of contact between the object and the plane. As this point is always momentarily at rest, no work is done by the frictional force.)

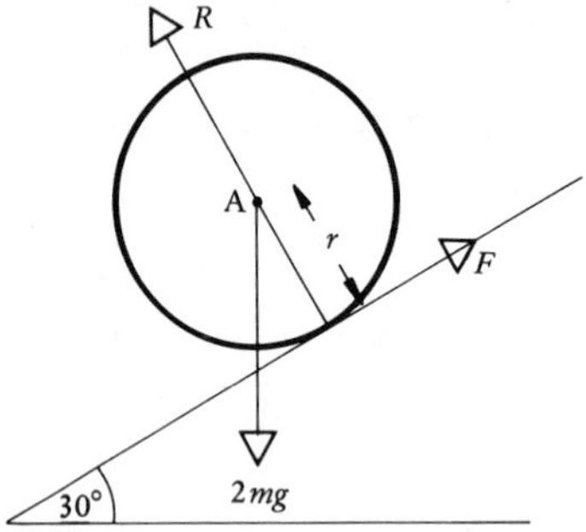

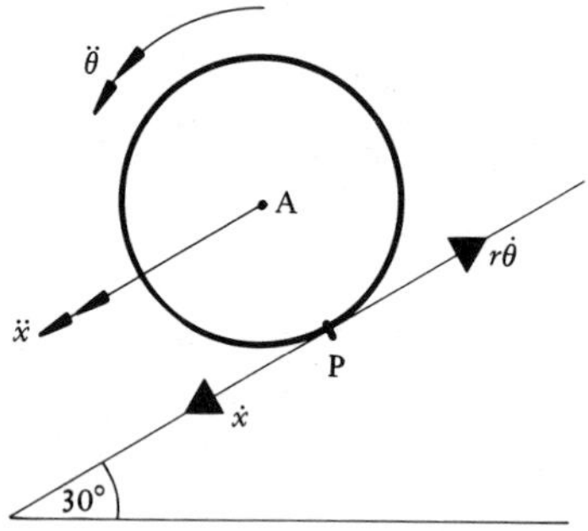

The velocity components of P, the point of contact, are shown in the diagram.

But P is momentarily at rest so

$$\dot{x} - r\dot{\theta} = 0$$

Therefore $$\ddot{x} - r\ddot{\theta} = 0$$

For the linear motion of a particle of mass $2m$ at A we have

$$\swarrow \quad 2mg\sin 30° - F = 2m\ddot{x} \qquad [1]$$

$$\nwarrow \quad R - 2mg\cos 30° = 0 \qquad [2]$$

Considering rotation about an axis through A we have

A $$Fr = I\ddot{\theta}$$

But $$I = 2m(\tfrac{1}{2}r^2) \quad \text{and} \quad r\ddot{\theta} = \ddot{x}$$

So $$Fr = mr^2\left(\frac{\ddot{x}}{r}\right)$$

i.e. $$F = m\ddot{x} \qquad [3]$$

Solving equations [1] and [3] gives

$$F = \tfrac{1}{3}mg$$

and

$$\ddot{x} = \tfrac{1}{3}g$$

Hence the angular acceleration of the disc is $\frac{1}{3}g/r$.

Now $\quad F = \frac{1}{3}mg \quad$ and $\quad R = mg\sqrt{3}$

But $$F \leqslant \mu R$$

i.e. $$\tfrac{1}{3}mg \leqslant \mu mg\sqrt{3}$$

Hence $$\mu \geqslant \frac{\sqrt{3}}{9}$$

3) A ring of mass M and radius a rolls from rest without slipping down a plane inclined at an angle α to the horizontal. Find the angular velocity of the ring when it has travelled a distance d down the plane.

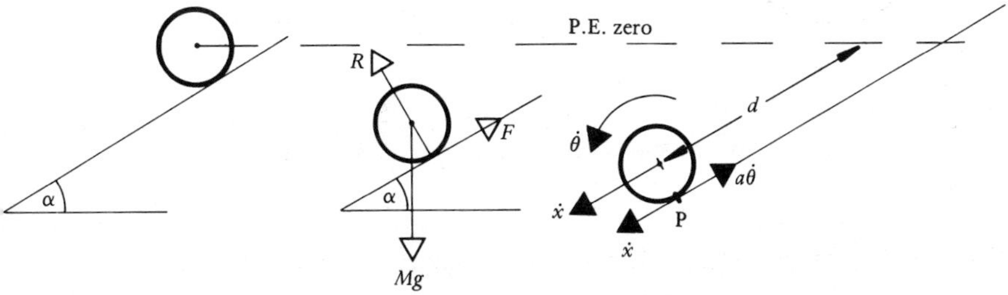

Initial position General position Final position

Considering the point P, as before, we have $\dot{x} = a\dot{\theta}$.

No work is done by R or F since neither force moves its point of application, so there is no change in the total mechanical energy.

The initial mechanical energy is zero.

The final potential energy is $-Mgd\sin\alpha$

The final kinetic energy is $\frac{1}{2}M\dot{x}^2 + \frac{1}{2}I\dot{\theta}^2 = \frac{1}{2}M(a\dot{\theta})^2 + \frac{1}{2}(Ma^2)\dot{\theta}^2$

Hence $$0 = -Mgd\sin\alpha + Ma^2\dot{\theta}^2$$

i.e. $$\dot{\theta}^2 = \frac{gd\sin\alpha}{a^2}$$

Thus the angular velocity of the ring is $\dfrac{1}{a}\sqrt{gd\sin\alpha}$

EXERCISE 13b

1) Find the kinetic energy of a body rolling without slipping on a horizontal plane if
(a) the body is a ring of mass M and radius a and its angular velocity is ω,
(b) the body is a uniform hollow sphere of mass $5M$ and radius a and the centre is moving with velocity v.

2) A uniform solid cylinder of mass 8 kg and diameter 1 m is placed with its axis horizontal on a rough horizontal plane. When a horizontal force of 20 N is applied to the midpoint of, and perpendicular to, the highest generator, the cylinder begins to roll without slipping on the plane. Find the angular acceleration of the cylinder and the range of values of the coefficient of friction between the cylinder and the plane for which this motion is possible.

3) A body rolls without slipping down a plane inclined at an angle α to the horizontal. Find the angular acceleration and the least value of the coefficient of friction if the body is
(a) a uniform solid sphere of radius a,
(b) a uniform hollow cylinder of radius a,
(c) a uniform hollow sphere of radius a.

4) A uniform solid sphere of radius 0.25 m rolls from rest, without slipping, on a plane inclined at 30° to the horizontal. Find the velocity of the centre of the sphere
(a) after 5 seconds,
(b) after moving a distance 1 m down the plane.

5) A body with constant circular cross-section of radius a, has a radius of gyration k about its axis of symmetry. If it rolls without slipping down a plane inclined at an angle α to the horizontal, show that

$$\tan\alpha \leqslant \mu\left(1+\frac{a^2}{k^2}\right)$$

6) A uniform hollow sphere rolls without slipping down a rough plane inclined at an angle α to the horizontal. A uniform solid sphere rolls without slipping down another plane inclined at an angle β to the horizontal. If the accelerations of the centres of the two spheres are equal, show that $21\sin\alpha = 25\sin\beta$.

7) A uniform solid cylinder rolls without slipping down a plane inclined at $\arctan\frac{7}{24}$ to the horizontal. Find its angular velocity when its centre has moved, from rest, through a distance $21a$ and find the time taken to travel this distance. What is the least value of the coefficient of friction between the cylinder and the plane, given that the radius of the sphere is a?

ROLLING AND SLIPPING

When an object such as a disc slips as it rolls on a plane, the point of the disc momentarily in contact with the plane is not at instantaneous rest. Therefore there is no physical relationship between the angular and linear velocities

i.e. $a\dot{\theta} \neq \dot{x}$

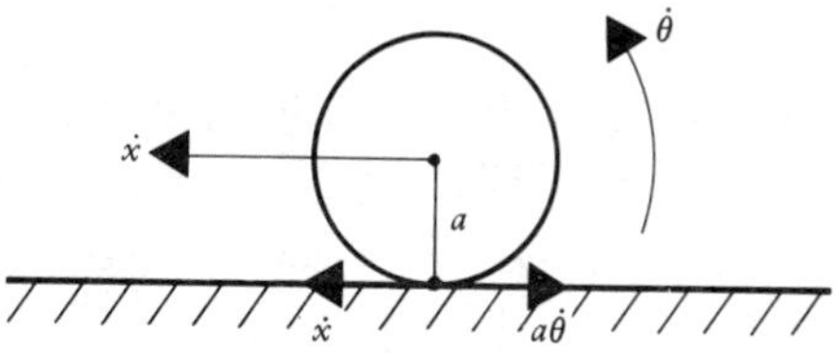

If it is *known* that slipping occurs then it follows that friction is limiting and $F = \mu R$ can be used.
On the other hand, to *show* that slipping occurs as well as rolling, it is necessary to show that the value of $\dot{x} - a\dot{\theta}$ for the point of momentary contact, is not zero. Friction must not be assumed to be limiting until slipping has been established.

EXAMPLES 13c

1) A uniform disc of mass $2m$ and radius a is placed with its rim on a rough plane inclined at $30°$ to the horizontal and is released from rest. If the coefficient of friction between the disc and the plane is $\frac{1}{10}\sqrt{3}$, show that the disc slips down the plane while rolling. Find the time taken for the centre of the disc to move a distance $14a$ down the plane.

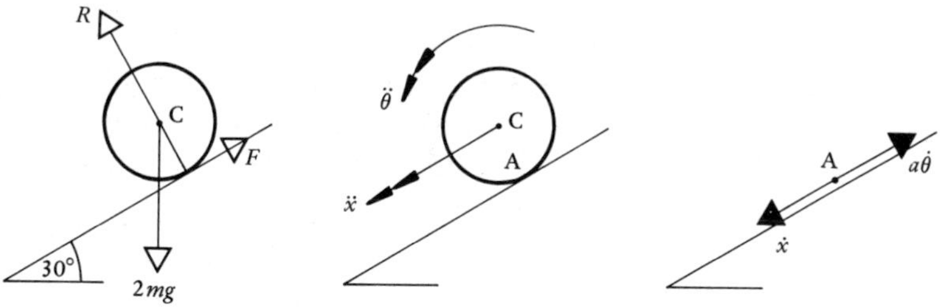

Using Newton's law for translation and rotation gives

$$\swarrow \quad 2mg\sin 30° - F = 2m\ddot{x} \qquad [1]$$

$$↺C \qquad Fa = I_C\ddot{\theta} = \tfrac{1}{2}(2m)a^2\ddot{\theta}$$

$$\Rightarrow \qquad F = ma\ddot{\theta} \qquad [2]$$

We also know that $F \leqslant \mu R$, i.e. $F \leqslant \frac{1}{10}\sqrt{3}(2mg\cos 30°)$

$$\Rightarrow \qquad F \leqslant \tfrac{3}{10}mg \qquad [3]$$

From [1] and [3] we have,

$$2mg\sin 30^\circ - 2m\ddot{x} \leqslant \tfrac{3}{10}mg$$

$$\Rightarrow \qquad \ddot{x} \geqslant \tfrac{7}{20}g$$

From [2] and [3] we have

$$a\ddot{\theta} \leqslant \tfrac{3}{10}g$$

Both $\ddot{x}$ and $a\ddot{\theta}$ are constant, and the disc starts from rest, therefore the equations for constant acceleration give

$$a\dot{\theta} = 0 + a\ddot{\theta}t \quad \Rightarrow \quad a\dot{\theta} \leqslant \tfrac{3}{10}gt$$

and

$$\dot{x} = 0 + \ddot{x}t \quad \Rightarrow \quad \dot{x} \geqslant \tfrac{7}{20}gt$$

Therefore the velocity of the point A down the plane is

$$\dot{x} - a\dot{\theta} \geqslant \tfrac{7}{20}gt - \tfrac{3}{10}gt$$

i.e.

$$\dot{x} - a\dot{\theta} \geqslant \tfrac{1}{20}gt \quad \text{which is positive.}$$

So the disc slides down the plane as it rolls.

Hence friction is limiting and $\ddot{x} = \frac{7}{20}g$.

For the motion of C, we can use $s = ut + \frac{1}{2}at^2$

When $s = 14a$,

$$14a = 0 + \tfrac{1}{2}(\tfrac{7}{20}g)t^2$$

$$\Rightarrow \qquad t^2 = \frac{80a}{g}$$

The time taken for the centre of the disc to move a distance $14a$ down the plane is $4\sqrt{5a/g}$.

2) A uniform solid sphere of mass $5M$ and radius a is placed gently on a rough horizontal plane with which the coefficient of friction is μ. Initially the sphere is rotating about a horizontal diameter with angular velocity ω. Show that the sphere slips on the plane until its centre O has moved a distance $\dfrac{2a^2\omega^2}{49\mu g}$.

Find the angular velocity at the instant that slipping ceases.

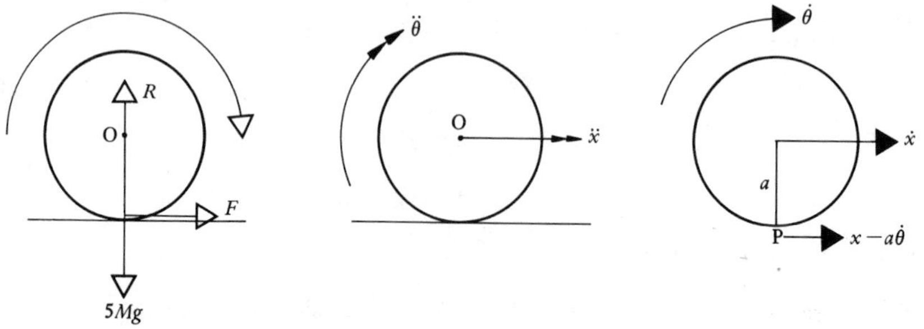

$$\uparrow \quad R = 5Mg$$

$$\leftarrow \quad F = 5M\ddot{x}$$

$$\circlearrowright O \qquad Fa = I_O\ddot{\theta} = \tfrac{2}{5}(5M)a^2\ddot{\theta} = 2Ma^2\ddot{\theta}$$

Also $\qquad F \leqslant \mu R$

Initially $\dot{x} = 0$ therefore $\ddot{x} = \dfrac{F}{5M} \Rightarrow \dot{x} = \dfrac{Ft}{5M}$

Initially $\dot{\theta} = \omega$ therefore $\ddot{\theta} = \dfrac{F}{2Ma} \Rightarrow \dot{\theta} = \omega - \dfrac{F}{2Ma}$

Therefore, at time t, the velocity of P, the instantaneous point of contact with the plane, is

$$\dot{x} - a\dot{\theta} = \frac{Ft}{5M} - \left(a\omega - \frac{Ft}{2M}\right) = \frac{7Ft}{10M} - a\omega$$

When $t = 0$ the speed of P is $-a\omega$, i.e. the sphere initially slips on the plane and will continue to do so until

$$\dot{x} - a\dot{\theta} = 0 \Rightarrow t = \frac{10a\omega M}{7F}$$

Now, while slipping occurs, $F = 5\mu Mg$

Therefore the sphere slips for a time $\dfrac{2a\omega}{7\mu g}$

At this time, for the motion of the centre O, we have

$$u = 0, \quad \ddot{x} = \frac{F}{5M} = \mu g, \quad t = \frac{2a\omega}{7\mu g}$$

Therefore $\qquad s = \tfrac{1}{2}(\mu g)\left(\dfrac{2a\omega}{7\mu g}\right)^2 = \dfrac{2a^2\omega^2}{49\mu g}$

Also $\dot{\theta} = \omega - \dfrac{Ft}{2Ma}$ so after a time $\dfrac{2a\omega}{7\mu g}$ we have

$$\dot{\theta} = \omega - \left(\frac{5\mu Mg}{2Ma}\right)\left(\frac{2a\omega}{7\mu g}\right) = \tfrac{5}{7}\omega$$

Therefore the angular velocity at the moment that slipping ceases is $\frac{5}{7}\omega$.

EXERCISE 13c

1) A uniform solid sphere is released from rest on a rough plane inclined at 45° to the horizontal and with which the coefficient of friction is μ.
If the sphere rolls without slipping show that $\mu \geqslant \frac{2}{7}$ and that the centre of the sphere has an acceleration $\frac{5}{14}g\sqrt{2}$ down the plane.
If $\mu < \frac{2}{7}$ and a is the acceleration of the centre of the sphere down the plane, show that $a > \frac{5}{14}g\sqrt{2}$.

2) A uniform solid cylinder is rotating about its axis, which is horizontal, with angular velocity ω. The cylinder is placed gently on a rough plane, inclined at 30° to the horizontal, in such a way that the cylinder tends to move directly up the plane. If the coefficient of friction is $\frac{1}{3}\sqrt{3}$, show that the cylinder stays in the same place for a time $a\omega/g$ and then moves down the plane with a linear acceleration $g/3$.

3) A uniform circular disc of mass m and radius a, whose plane is vertical, is placed on a rough plane inclined to the horizontal at an angle θ. A couple of moment $4mga\sin\theta$ acts in the plane of the disc in the sense tending to move the disc up the plane. If the coefficient of friction between the disc and the plane is $2\tan\theta$, show that the disc slips and find the acceleration of the centre of the disc.

4) A uniform ring, of mass m and radius a, is rotating with angular velocity ω about a horizontal axis through its centre and perpendicular to the ring. It is placed gently on a rough horizontal plane with which the coefficient of friction is μ. Prove that, at first, the ring slips and find the distance moved before slipping ceases.

5) A uniform sphere, rotating about a horizontal diameter with angular velocity Ω is placed on a rough inclined plane, the sense of rotation being such as to tend to roll the sphere down the plane. The sphere is of mass M and radius a; the plane is inclined at 30° to the horizontal and the coefficient of friction between the sphere and the plane is $\sqrt{3}/14$. Find the time that elapses before the sphere stops slipping down the plane and the angular velocity at this moment.

IMPULSIVE MOTION

The independent analysis of translation and rotation can also be applied to the effect of an impulse on a rigid body which is completely free to move, i.e.,

(a) the moment of the impulse about an axis through the centre of mass is equal to $I\Omega$ where I is the moment of inertia about that axis and Ω is the sudden change in angular velocity,

(b) the impulse is equal to the increase in momentum of a particle of mass M at the centre of mass, where M is the mass of the body.

EXAMPLES 13d

1) A rod AB of mass M and length $2a$ is lying at rest on a smooth horizontal table. An impulse MV is applied to the end A in a horizontal direction perpendicular to AB. Find the speed of A immediately after the impulse.

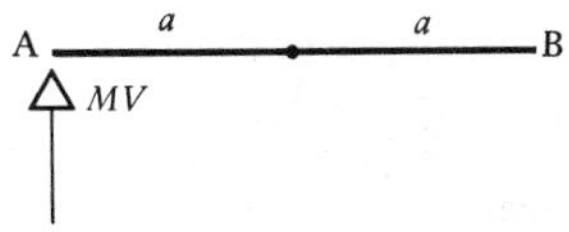

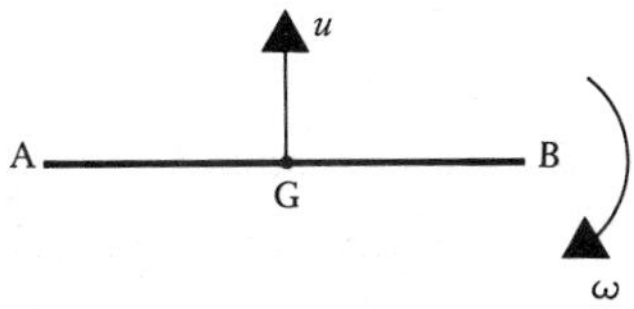

Immediately after the impulse let the velocity of the centre of mass of the rod be u and the angular velocity of the rod be ω.

(**Note** that we use the velocity of G, and not of A.)

Using impulse = increase in linear momentum gives

$$MV = Mu \Rightarrow u = V$$

Using moment of impulse = increase in angular momentum gives

$$MVa = I_G\omega = \tfrac{1}{3}Ma^2\omega$$

$$\Rightarrow \quad a\omega = 3V$$

Now we consider the velocity of A.

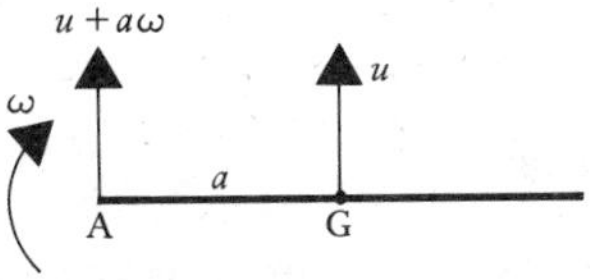

At A, the velocity is $u + a\omega = V + 3V$.
Therefore the instantaneous velocity of A is $4V$.

2) A rod AB of mass $3m$ and length $4a$ is falling freely in a horizontal position, and C is a point distant a from A.
When the speed of the rod is u, the point C collides with a particle of mass m which is moving vertically upwards with speed u. If the impact between the particle and the rod is perfectly elastic find

(a) the velocity of the particle immediately after impact,
(b) the angular velocity of the rod immediately after impact,
(c) the speed of B immediately after impact.

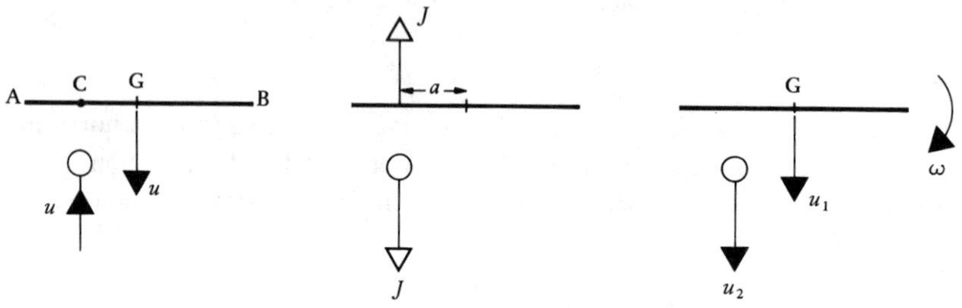

Immediately after impact, let the speed of G be u_1, the speed of the particle u_2 (both downwards) and the angular velocity of the rod ω.

Using impulse = increase in linear momentum gives

for the rod $$J = -3mu_1-(-3mu) \qquad [1]$$

for the particle $$J = mu_2-(-mu) \qquad [2]$$

Using moment of impulse = change in angular momentum gives

↺G $$Ja = I_G\omega = \tfrac{1}{3}(3m)(2a)^2\omega \qquad [3]$$

In order to use the law of restitution we need the speed of the point C, which is $u_1-a\omega$.

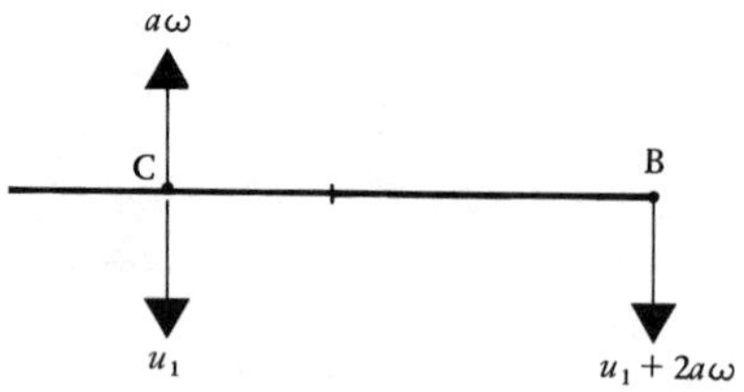

The law of restitution now gives

$$e(u+u) = u_2-(u_1-a\omega)$$

But $e = 1$,

therefore $$2u = u_2-u_1+a\omega \qquad [4]$$

From [1] $$u_1 = u - \frac{J}{3m}$$

From [2] $$u_2 = \frac{J}{m} - u$$

From [3] $$a\omega = \frac{J}{4m}$$

Hence [4] becomes

$$2u = \left(\frac{J}{m} - u\right) - \left(u - \frac{J}{3m}\right) + \frac{J}{4m}$$

$$\Rightarrow \qquad J = \tfrac{48}{19}mu$$

Hence $u_1 = \frac{3}{19}u, \quad u_2 = \frac{29}{19}u, \quad a\omega = \frac{12}{19}u$

(a) The velocity of the particle is $\dfrac{29}{19}u$

(b) The angular velocity of the rod is $\dfrac{12u}{19a}$

(c) The speed of B is $u_1 + 2a\omega = \dfrac{27}{19}u$

3) A uniform ring of mass m, radius a and centre C lies at rest on a smooth horizontal table. A point P on the circumference is struck horizontally and begins to move in a direction that is at 60° to PC. If the magnitude of the blow is $mV\sqrt{7}$, find the initial speed of point P.

(The initial direction of motion of P is *not* in the direction of the blow. The direction of the blow gives the initial direction of motion of the centre of mass of the ring.)

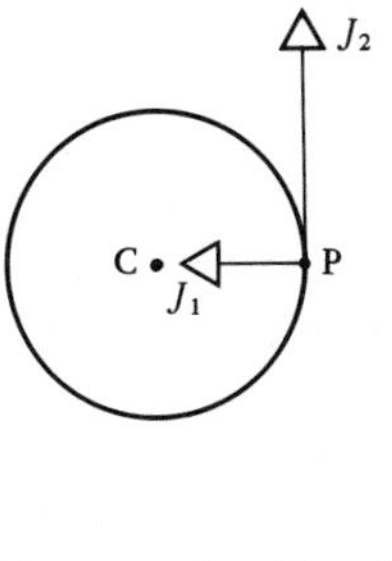

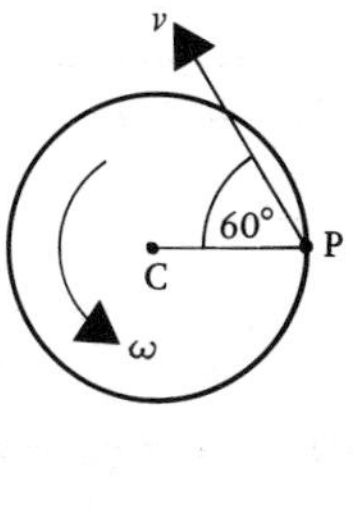

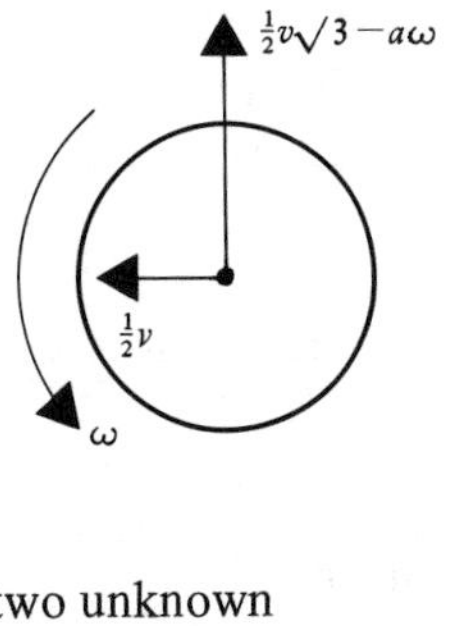

Because the direction of the impulse is not known we use two unknown components, and an unknown speed v for point P.

For the motion of the centre of mass,

$$J_1 = \tfrac{1}{2}mv \qquad [1]$$

$$J_2 = m(\tfrac{1}{2}v\sqrt{3} - a\omega) \qquad [2]$$

For the initial rotation,

$$J_2 a = ma^2\omega \qquad (I_C = ma^2) \qquad [3]$$

From [2] and [3], $J_2 = \tfrac{1}{4}mv\sqrt{3}$

Now $$J_1^2 + J_2^2 = (mV\sqrt{7})^2$$

Therefore $$(\tfrac{1}{2}mv)^2 + (\tfrac{1}{4}mv\sqrt{3})^2 = (mV\sqrt{7})^2$$

$\Rightarrow$ $$\tfrac{7}{16}m^2v^2 = 7m^2V^2$$

$\Rightarrow$ $$v^2 = 16V^2$$

Therefore the speed of P is $4V$.

EXERCISE 13d

1) A uniform rod AB of mass m and length $4a$ is lying on a smooth horizontal plane when it is struck, at a point P distant a from A, by a horizontal blow, of impulse J, perpendicular to AB. Find the initial angular velocity of the rod and the initial speed of B.

2) A uniform disc, resting on a smooth horizontal table, has mass m, radius a and centre C. A particle of mass m, moving across the table with speed v, strikes a point A on the circumference of the disc when travelling at right angles to AC. If the particle adheres to the disc, find the initial velocity of C.

3) A light rod AB, of length $2l$, has a particle attached at each end and lies at rest on a horizontal plane. A horizontal impulse of magnitude mV is applied to A at 45° to AB. Find the initial angular velocity of AB and the initial speed of the centre of mass if
(a) the two particles are of equal mass m,
(b) the particle at A is of mass $2m$ and that at B is of mass m.

4) A particle of mass $2m$ is moving with speed v along a smooth horizontal surface, at right angles to a uniform rod AB of mass m and length $2a$ which lies at rest on the surface.
The particle strikes the rod at the end B and the coefficient of restitution at impact is $\tfrac{1}{2}$. Find the angular velocity with which the rod begins to rotate and the speed of the particle after impact. How far has the particle travelled when the rod has rotated through 180°?

GENERAL PROBLEMS

This chapter has so far dealt with the solution of specific types of problems involving translation and rotation. There is, however, a wide range of problems which are not easy to categorise.

To solve these, all the mechanics principles covered in this book, and in *Mechanics and Probability* are available. In a particular situation the reader must decide which principles are both valid and appropriate. A small selection of general problems is given in the following examples.

EXAMPLES 13e

1) A uniform rod AB of mass $3m$ and length $2a$ has a small smooth ring of mass m attached at A. The ring is threaded on to a horizontal fixed wire. Initially the rod is held horizontally alongside the wire and is released from rest in this position. Find the angular velocity of the rod when AB is first vertical and the speed of B at this instant.

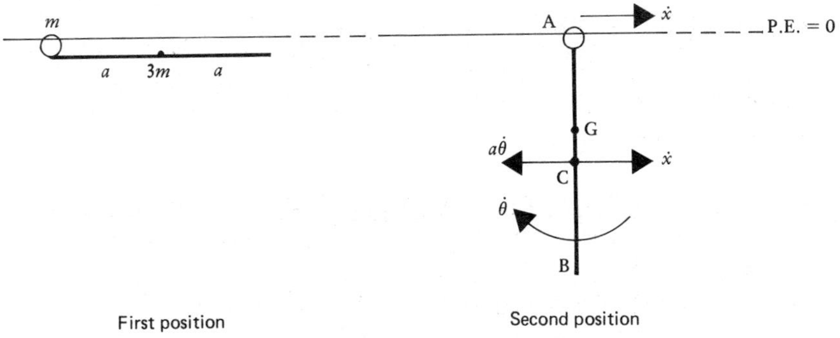

(The reaction R between the ring and the wire does no work so conservation of mechanical energy applies.)

In the first position the total M.E. is zero.
In the second position we have,

for the ring K.E. $= \frac{1}{2}m(\dot{x})^2$

P.E. $= 0$

for the rod K.E. $= \frac{1}{2}(3m)(\dot{x}-a\dot{\theta})^2 + \frac{1}{2}(\frac{1}{3}\{3ma^2\})\dot{\theta}^2$

P.E. $= -3mga$

Conservation of M.E. gives

$$0 = \frac{1}{2}m\dot{x}^2 + \frac{3}{2}m(\dot{x}-a\dot{\theta})^2 + \frac{1}{2}ma^2\dot{\theta}^2 - 3mga \qquad [1]$$

Now considering the motion of the centre of mass of the system, which is at the point G where $AG = \frac{3}{4}a$, we see that, as all the forces acting on the system are vertical, G moves vertically downwards.

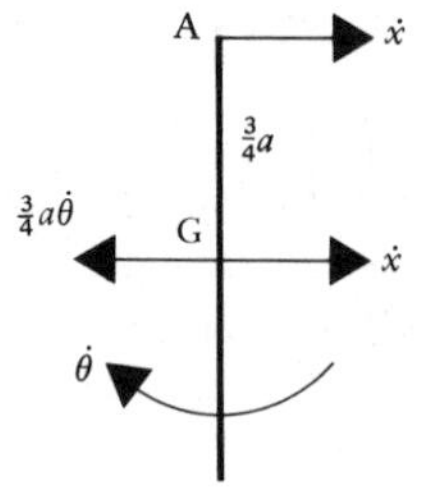

Therefore $\dot{x} - \frac{3}{4}a\dot{\theta} = 0$ [2]

Solving equations [1] and [2] gives

$$(\tfrac{3}{4}a\dot{\theta})^2 + 3(\tfrac{3}{4}a\dot{\theta} - a\dot{\theta})^2 + a^2\dot{\theta}^2 - 6ga = 0$$

$$\Rightarrow \quad \tfrac{28}{16}a^2\dot{\theta}^2 = 6ga$$

$$\Rightarrow \quad \dot{\theta} = 2\sqrt{\frac{6g}{7a}}$$

Hence $\dot{x} = \frac{3}{2}\sqrt{\frac{6}{7}ga}$

For B the speed is $\left|\dot{x} - 2a\dot{\theta}\right| = \left|-\frac{5}{2}\sqrt{\frac{6}{7}ga}\right|$

Therefore, when AB is vertical, the angular velocity of the rod is $2\sqrt{\dfrac{6g}{7a}}$ and the speed of B is $\frac{5}{2}\sqrt{\frac{6}{7}ga}$.

2) A uniform rod AB of mass M and length $2l$ rests in a vertical plane, perpendicular to a smooth vertical wall. The end A is in contact with the wall, B rests on smooth horizontal ground, and AB is at an angle of 30° to the wall when the rod is released from rest. Find the inclination of the rod to the wall when A is about to leave the wall.

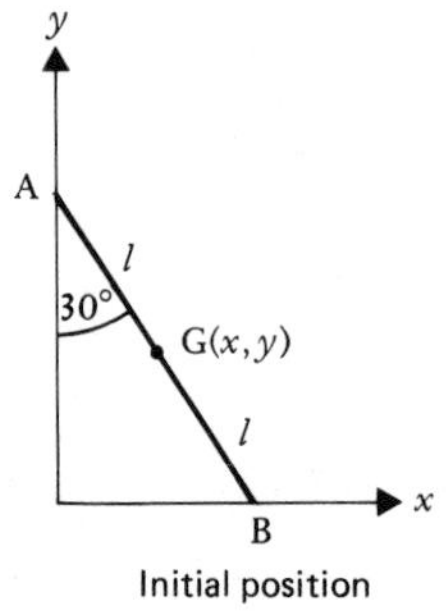

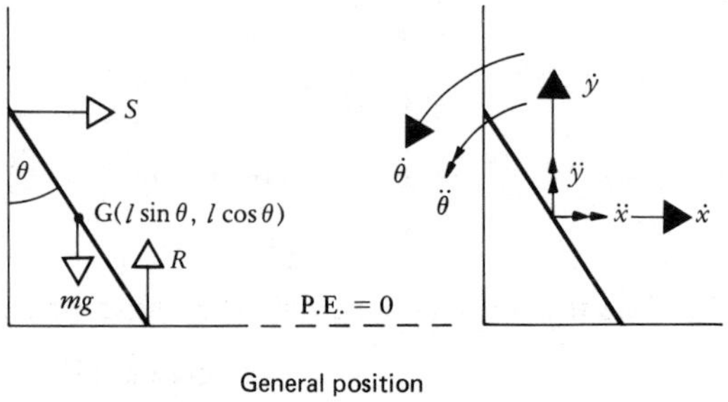

(The rod leaves the wall when $S=0$ so we use Newton's law to find an expression for S. Also, neither R nor S do any work so mechanical energy is conserved.)

The coordinates of the centre of mass G can be expressed in terms of θ, i.e.

$$x = l\sin\theta \qquad y = l\cos\theta$$

$$\Rightarrow \quad \dot{x} = l\cos\theta\,\dot{\theta} \qquad \dot{y} = -l\sin\theta\,\dot{\theta}$$

$$\Rightarrow \quad \ddot{x} = l\cos\theta\,\ddot{\theta} - l\sin\theta\,\dot{\theta}^2 \qquad \ddot{y} = -l\sin\theta\,\ddot{\theta} - l\cos\theta\,\dot{\theta}^2$$

Newton's law gives

$$S = M\ddot{x} = Ml(\cos\theta\,\ddot{\theta} - \sin\theta\,\dot{\theta}^2) \qquad [1]$$

Conservation of mechanical energy gives

$$Mgl\cos 30^\circ = \tfrac{1}{2}M(\dot{x}^2+\dot{y}^2) + \tfrac{1}{2}(I_G)\dot{\theta}^2 + Mgl\cos\theta$$

$$\Rightarrow \quad Mgl(\sqrt{3} - 2\cos\theta) = M(l^2\cos^2\theta\,\dot{\theta}^2 + l^2\sin^2\theta\,\dot{\theta}^2) + \tfrac{1}{3}Ml^2\dot{\theta}^2$$

$$\Rightarrow \quad \tfrac{4}{3}l\dot{\theta}^2 = (\sqrt{3} - 2\cos\theta)g \qquad [2]$$

Differentiating with respect to time gives

$$\tfrac{4}{3}l(2\dot{\theta}\,\ddot{\theta}) = 2g\sin\theta\,\dot{\theta}$$

$$\ddot{\theta} = \frac{3g}{4l}\sin\theta \qquad [3]$$

Substituting from [2] and [3] in [1] gives:

$$S = Ml\left\{\frac{3g}{4l}\sin\theta\cos\theta - \frac{3g}{4l}(\sqrt{3} - 2\cos\theta)\sin\theta\right\}$$

When $S=0$ we have

$$\frac{9g}{4l}\sin\theta\cos\theta = \frac{3\sqrt{3}g}{4l}\sin\theta$$

$$\Rightarrow \quad \text{either} \quad \sin\theta = 0 \quad \text{or} \quad \cos\theta = \tfrac{1}{3}\sqrt{3}$$

Therefore the rod leaves the wall when its inclination to the wall is

$$\arccos\left(\tfrac{1}{3}\sqrt{3}\right)$$

3) A sphere of mass m and radius $2a$ is rolling with angular velocity ω along horizontal ground when it strikes a step, of height a, which is perpendicular to the line of motion of the sphere. If the edge of the step is rough enough to prevent slipping show that the sphere can surmount the step provided that

$\omega > \frac{1}{9}\sqrt{70g/a}$

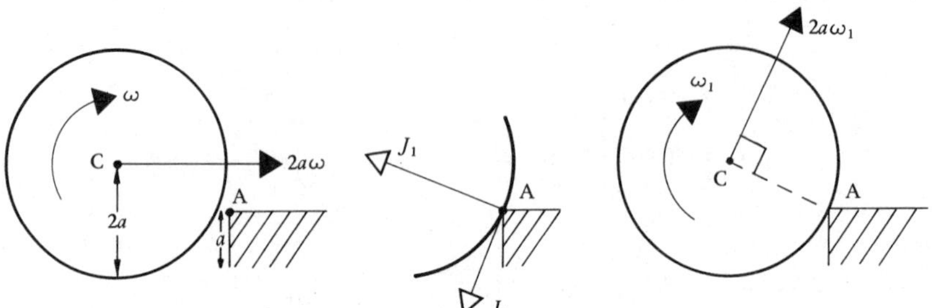

Just before impact   Just after impact

(At the moment of impact with the edge of the step there is a frictional impulse, J_2, in addition to the normal impulse, J_1. These impulses do not, however, change the moment of momentum about an axis along the edge of the step.)

Considering the total moment of momentum about A we have,

$$I_C\omega + (2ma\omega)(a) \qquad \text{just before impact}$$

and $$I_C\omega_1 + (2ma\omega_1)(2a) \qquad \text{just after impact}$$

(**Note** that we consider the angular momentum about an axis through the centre of mass, together with the linear momentum of the centre of mass.)

Using conservation of moment of momentum gives

$$(\tfrac{8}{5}ma^2)\omega + 2ma^2\omega = (\tfrac{8}{5}ma^2)\omega_1 + 4ma^2\omega_1$$

$$\Rightarrow \qquad \omega_1 = \tfrac{9}{14}\omega$$

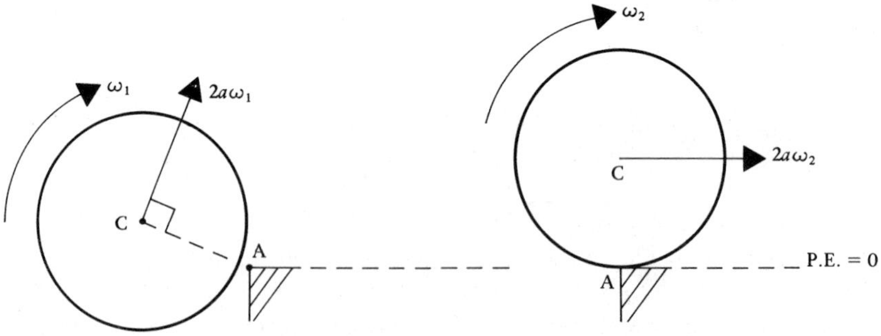

(While the sphere is mounting the step, the resultant contact force at A does no work because the sphere does not slip. Therefore mechanical energy is conserved.)

Conservation of M.E. gives

$$\tfrac{1}{2}I_C\omega_1^2+\tfrac{1}{2}m(2a\omega_1)^2+mga = \tfrac{1}{2}I_C\omega_2^2+\tfrac{1}{2}m(2a\omega_2)^2+2mga$$

$$\Rightarrow \qquad \tfrac{14}{5}ma^2\omega_1^2-mga = \tfrac{14}{5}ma^2\omega_2^2$$

The sphere can surmount the step only if ω_2 has a real value,

i.e. $$\tfrac{14}{5}ma^2\omega_1^2-mga > 0$$

i.e. if $\omega_1^2>\dfrac{5g}{14a} \Rightarrow \left(\dfrac{9}{14}\omega\right)^2>\dfrac{5g}{14a}$

So the sphere can surmount the step if $\omega>\dfrac{1}{9}\sqrt{\dfrac{70g}{a}}$

Some problems of a general nature can be found at the end of the next miscellaneous exercise.

MULTIPLE CHOICE EXERCISE 13

(Instructions for answering these questions are given on page ix.)

TYPE II

1) A constant torque C acts for a time t on a lamina rotating in a horizontal plane about a smooth vertical axis. If the moment of inertia of the lamina about that axis is I and the angular velocity changes from ω_1 to ω_2 in the time t,
(a) the lamina has a constant angular acceleration,
(b) the work done by the torque is equal to $I(\omega_2^2-\omega_1^2)$,
(c) the impulse of the torque is Ct,
(d) $Ct = I(\omega_2-\omega_1)$.

2) A rod AB is free to rotate in a vertical plane about a horizontal axis through A. It is slightly disturbed from rest in its position of unstable equilibrium and when it is next vertical the end B collides with a fixed peg and rebounds. If the rod comes to instantaneous rest when AB is horizontal,
(a) the total mechanical energy of the rod is constant throughout,
(b) the coefficient of restitution between the rod and the peg is $\frac{1}{2}$,
(c) the angular momentum of the rod is constant except for a sudden change at the instant of impact with the peg,
(d) the sudden change in the angular momentum of the rod at the instant of impact is equal to the moment of the impulse at the peg.

3) A uniform solid cylinder rolls without slipping through a distance d down an inclined plane.
(a) The angular acceleration is constant.
(b) The work done by the frictional force F is Fd.
(c) Friction is limiting.
(d) The mechanical energy of the cylinder is constant.

4) A uniform rod AB of mass m and length $2l$ is rotating freely, on a smooth horizontal table, about a vertical axis through A. A particle P of mass m collides normally with the midpoint of the rod, when moving with speed v, and as a result of the impact is brought to rest.
(a) The rod is also brought to rest.
(b) The rod exerts an impulse of magnitude mv on the particle.
(c) The collision is inelastic.
(d) The total moment of momentum is not changed by the impact.

TYPE III

5) (a) A sphere is rolling without slipping on a horizontal plane.
(b) A sphere is moving on a rough horizontal plane.

6) (a) The impulse of a torque C is Ct.
(b) A constant torque C acts for a time t.

7) (a) There is no sudden change in the total moment of momentum of a rotating rod and a moving particle when they collide.
(b) There is a perfectly elastic collision between a rotating rod and a moving particle.

8) (a) A rotating cylinder is slipping on a rough horizontal plane.
(b) The axis of a cylinder rotating on a horizontal plane is stationary.

TYPE IV

9) A torque C acts on a uniform circular disc which can rotate about a diameter. If the disc starts from rest find its angular velocity after t seconds;
(a) the radius of the disc is r,
(b) the mass of the disc is m,
(c) no other force exerts a torque on the disc,
(d) the torque is constant.

10) The end B of a rod AB, hanging vertically from a fixed horizontal axis through A, is struck a blow. If the rod first comes to instantaneous rest after turning through an angle $2\pi/3$ find the impulse of the blow given that
(a) the length of the rod is $2l$,
(b) the rod is uniform,
(c) the axis is smooth,
(d) the mass of the rod is m.

11) A ring rolls down an inclined plane. Find the least value of the coefficient of friction between the ring and the plane if
(a) the plane is inclined at 30° to the horizontal,
(b) the mass of the ring is m,
(c) the ring does not slip on the plane,
(d) friction is limiting.

12) A particle is placed gently on a rotating turntable and does not slip. Find the angular velocity of the turntable immediately afterwards.
(a) The mass of the particle is m.
(b) The turntable is a uniform disc of mass $4m$ and radius r.
(c) The angular velocity of the turntable just before the addition of the particle is ω.
(d) Friction between the particle and the turntable is limiting.

TYPE V

13) The impulse of a torque C acting on a body for a time t is given by $\int C\,dt$ only if no other torque acts on the body.

14) A body cannot roll on a smooth surface.

15) If, throughout a specified time interval, the moment of momentum of a system is constant, the total mechanical energy of the system must also be constant during the same interval.

16) A sphere of mass $5m$ and radius a is moving, on a horizontal plane, with angular velocity ω about a horizontal diameter, and its centre of mass has a linear speed v. The kinetic energy of the sphere is $\frac{5}{2}mv^2 + ma^2\omega^2$.

17) When a rod lying on a horizontal plane is struck by a blow, it will begin to rotate.

MISCELLANEOUS EXERCISE 13

1) A uniform circular hoop, of mass m and radius r, starts from rest and rolls with its plane vertical and without slipping down a line of greatest slope of a fixed plane which is inclined at an angle α to the horizontal. Prove that the hoop rolls down the plane with constant acceleration $\frac{1}{2}g\sin\alpha$. (U of L)

2) A uniform circular disc, of mass m and radius r, starts from rest and rolls, without slipping and with its plane vertical down a line of greatest slope of a fixed plane of inclination α to the horizontal. Find the time taken for the disc to move through a distance x down the plane.
Show that the coefficient of friction between the disc and the plane during this motion is at least $\frac{1}{3}\tan\alpha$. (U of L)

3) A uniform solid cylinder rolls without slipping down a rough plane inclined at an angle α to the horizontal. Find the acceleration of the centre of gravity of the cylinder and the least possible value of the coefficient of friction between the cylinder and the plane.
A uniform solid sphere rolls without slipping down a rough plane inclined at an angle β to the horizontal. The acceleration of the centre of the sphere is the same as the acceleration of the centre of gravity of the cylinder.
Show that $14 \sin \alpha = 15 \sin \beta$. (AEB)

4) A uniform sphere is released from rest on a rough plane inclined at an angle α to the horizontal. The coefficient of friction is μ. Prove that, if $\mu > \frac{2}{7} \tan \alpha$, the sphere will roll down the plane and its centre will have an acceleration $\frac{5}{7}g \sin \alpha$ down the plane.
Prove also that, if $\mu < \frac{2}{7} \tan \alpha$, the acceleration of the centre of the sphere down the plane is greater than $\frac{5}{7}g \sin \alpha$. (O)

5) A closed container in the form of a right circular cylinder, of radius a and height $4a$, is made from thin uniform sheet metal. If the mass of the container is M, show that its moment of inertia about its axis is $\frac{9}{10}Ma^2$.
This container rolls without slipping down a plane inclined at an angle α to the horizontal. Calculate the angular speed of the container when its axis has travelled, from rest, a distance $10a$ down the plane. (AEB)

6) Show that the moment of inertia of a uniform solid cylinder of mass M and radius r about its axis is $\frac{1}{2}Mr^2$.
State the moment of inertia of a thin cylindrical ring, also of mass M and radius r, about its axis.
The solid cylinder and the ring roll from rest, without slipping, down lines of greatest slope of a plane inclined at α to the horizontal. Show that the solid cylinder covers a distance $\frac{1}{12}gt^2 \sin \alpha$ further than the ring in time t.
Find the minimum value of the coefficient of friction between the plane and the rolling bodies. (AEB)

7) Prove that the moment of inertia of a uniform solid sphere, of mass M and radius r, about a diameter is $(2Mr^2)/5$.
The sphere is placed in contact with a rough plane inclined at $30°$ to the horizontal. On being released the sphere rolls, without slipping, down a line of greatest slope of the plane. Show that the centre of the sphere has constant acceleration and that it acquires a speed of 7 metres per second after rolling 7 metres down the plane. What is the smallest possible value of the coefficient of friction between the sphere and the plane? (AEB)

8) A thin uniform spherical shell has mass m and radius a. Show, by integration, that the moment of inertia of the spherical shell about a diameter is $\frac{2}{3}ma^2$.
The shell rolls from rest, without slipping, down a rough plane inclined at an angle α to the horizontal. Show that the centre of the shell has acceleration $(3g\sin\alpha)/5$.
After travelling a distance $5a$ the shell is at the bottom of the inclined plane and begins to roll along a rough horizontal plane.
(a) Find the time taken for it to reach the bottom of the inclined plane.
(b) Explain why the shell now rolls with constant speed along the horizontal plane. (AEB)

9) A uniform sphere of radius a is standing at rest on the rough upper surface of a flat board which is lying on a horizontal table. The board is now moved along the table with a constant acceleration f. Prove that, if the coefficient of friction μ between the sphere and the board is greater than $\frac{2}{7}(f/g)$, the sphere rolls on the board with angular acceleration $\frac{5}{7}(f/a)$.
Find the angular acceleration of the sphere if $\mu < \frac{2}{7}(f/g)$. (U of L)

10) A cylinder of mass M and radius a rolls, without slipping and with its axis horizontal, down a plane inclined at an angle α to the horizontal. In two successive seconds it is observed to make two complete revolutions and three complete revolutions. Find:
(a) the angular acceleration of the cylinder,
(b) the moment of inertia of the cylinder about its axis,
(c) the minimum value of the coefficient of friction between the cylinder and the plane. (AEB)

11) A circular cylinder of radius a has its centre of mass in its axis and has a radius of gyration k about this axis. Prove that, when the cylinder rolls down a plane of inclination α, the acceleration of its centre of mass is

$$\frac{a^2 g\sin\alpha}{a^2+k^2}.$$

If the density of the cylinder at a distance r from its axis is proportional to r, find k in terms of a. Determine the minimum value of the coefficient of friction between the plane and the cylinder. (U of L)

12) A uniform solid reel of mass $2M$ is made from three right circular cylinders of equal lengths, the outer two each being of radius $2a$ and the middle one of radius a, fixed together so that the cylinders have a common axis. Show that the moment of inertia of the reel about its axis is $11Ma^2/3$.
A light inextensible string, wound several times around the perfectly rough central cylinder of the reel, passes over a fixed smooth peg at B and carries a particle of mass M at its free end. The reel stands on a rough plane inclined at $30°$ to the horizontal. The straight portion of the string from B to the reel is parallel to a line of greatest slope of the plane and the portion BC is vertical.

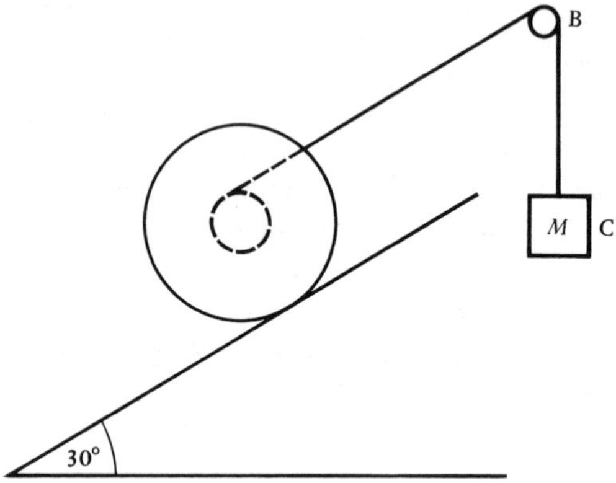

The system moves with the string taut, C falls vertically and the reel, whose axis is horizontal, rolls up the inclined plane without slipping. Calculate
(a) the acceleration of C,
(b) the minimum value of the coefficient of friction between the reel and the inclined plane. (AEB)

13) A uniform solid sphere, initially rotating about a horizontal diameter with angular speed Ω, is gently placed on a rough horizontal plane. The radius of the sphere is a, and the coefficient of friction between the sphere and the plane is μ. Prove that the sphere moves a distance $2a^2\Omega^2/(49\mu g)$ before slipping ceases, and find the angular speed at that instant.
Prove that in the subsequent motion the sphere rolls on the plane.

14) A uniform sphere, of radius a, is rotating about a horizontal diameter with angular speed Ω. The sphere is gently placed on a rough plane which is inclined at an angle α to the horizontal, the sense of rotation being such as to tend to cause the sphere to move up the plane along a line of greatest slope. Show that, if the coefficient of friction is $\tan\alpha$, the centre of the sphere will remain at rest for a time $2a\Omega/(5g\sin\alpha)$, and will then move down the plane with acceleration $(5g\sin\alpha)/7$.

15) A uniform circular disc, of mass m and radius r, is on a rough plane which is inclined to the horizontal at an angle α, the disc being in a vertical plane through a line of greatest slope. A couple of moment $kmgr \sin\alpha$ acts in the plane of the disc, tending to roll the disc up the plane. Given that the coefficient of friction between the disc and the plane is $2\tan\alpha$, find the acceleration of the centre of the disc in each of the cases $k = 1$, $k = 2$, $k = 3$.
Sketch a graph showing how this acceleration varies with k for $0 \leqslant k \leqslant 3$.
(U of L)

16) A uniform rigid rod AB, of mass M and length $2a$, is falling freely without rotation under gravity with AB horizontal. Suddenly the end A is fixed when the speed of the rod is v. Find the angular speed with which the rod begins to rotate. (U of L)

17) A uniform rod AB, of mass m and length $2a$, lies at rest on a smooth horizontal table. A horizontal impulse is applied to the rod at a point P so that, immediately after the impulse, the end A is still stationary.
Find the distance AP.
Show that, when the rod is first perpendicular to its initial direction, the centre of mass of the rod has moved a distance $\pi a/2$. (U of L)

18) Two particles, A, B, each of mass m, are connected by a light rigid rod of length l. The system is at rest on a smooth horizontal table with B fixed to the table. A horizontal impulse I is applied to A in a direction inclined at an angle $\pi/3$ to BA. Find the angular velocity of AB.
Show that, if both A and B are free to move, then initially A moves in a direction inclined at an angle θ, where $\tan\theta = 2\sqrt{3}$, with BA and find the kinetic energy generated by the impulse in this case. (U of L)

19) A uniform circular disc, of mass m, radius a and centre O, lies at rest on a smooth horizontal table. The disc is given a horizontal blow at a point A of its circumference, such that the initial velocity of A is of magnitude V at an angle of 45° to AO. Show that the magnitude of the blow is $mV(\sqrt{5})/3$, and that the kinetic energy given to the disc is $mV^2/3$.
Find the distance travelled by the centre O while the disc makes one revolution. (U of L)

20) Show by integration that the moment of inertia of a uniform solid sphere of mass M and radius a about a diameter is $2Ma^2/5$.
Two such spheres are joined together by a light rod of length $2a$ so that their centres P and Q are at a distance $4a$ apart. When the spheres are at rest on a smooth horizontal table, a horizontal impulse is given to the sphere with centre P along a line through P at right angles to PQ. Given that the initial speed of Q is u, find the magnitude of the impulse. Show that when the rod has turned through a right angle its midpoint will have moved through a distance $11\pi a/10$. (U of L)

21) A uniform rod AB of length l and mass m hangs from a fixed point A at which it is freely hinged. From the end B hangs a uniform disc of radius a and mass M, the rod being freely hinged to the disc at a point B of its circular edge. When the system hangs in equilibrium an impulse I, horizontal and in the same vertical plane as the disc, is given to the rod at a point distant x below A. Find the angular velocities imparted to the rod and the disc. Prove that, if the impulsive reaction at the fixed end A is zero, then $x = 2l(M+m)/(2M+3m)$. (O)

22) A uniform circular disc of radius a whose plane is vertical rolls without slipping and with angular speed ω along a rough horizontal plane. It strikes a thin horizontal inelastic rod fixed at a height $\frac{1}{3}a$ above the plane and perpendicular to the plane of the disc. If the rod is sufficiently rough to prevent slipping, prove that the disc surmounts the rod without losing contact with it if

$$36g < 49a\omega^2 < 54g. \qquad \text{(U of L)}$$

23) A uniform solid sphere is held at rest on a rough plane inclined at an angle α to the horizontal. The sphere is released and rolls without slipping down the plane. Find the acceleration of the centre of the sphere. Find also the least possible value of the coefficient of friction between the sphere and the plane.
The sphere, which has radius a, is next placed on a rough horizontal table. It is then struck a horizontal blow in a vertical plane containing the centre of the sphere. The line of action of the blow is at a height h above the table. Given that the sphere starts to roll without slipping, show that $5h = 7a$. (U of L)

24) State the moment of inertia of a uniform circular hoop of radius a and mass m about an axis through its centre, perpendicular to the plane of the hoop. The hoop rolls without slipping in a vertical plane on a rough horizontal floor. If the speed of the centre of the hoop is v, show that the actual speed of a point P on the hoop is

$$\{2v^2(1-\sin\alpha)\}^{1/2},$$

where the line joining P to the centre of the hoop makes an angle α with the horizontal, as shown below.
A particle of mass m is attached to the inner surface of the hoop at a point Q. The hoop is now placed vertically on the rough horizontal floor with OQ horizontal. The hoop starts to roll from rest, without slipping, on the floor. Show that, when the hoop has turned through an angle θ,

$$a(\dot{\theta})^2\,[2-\sin\theta] = g\sin\theta.$$

When the hoop has turned through a right angle find
(a) its angular velocity,
(b) its angular acceleration.

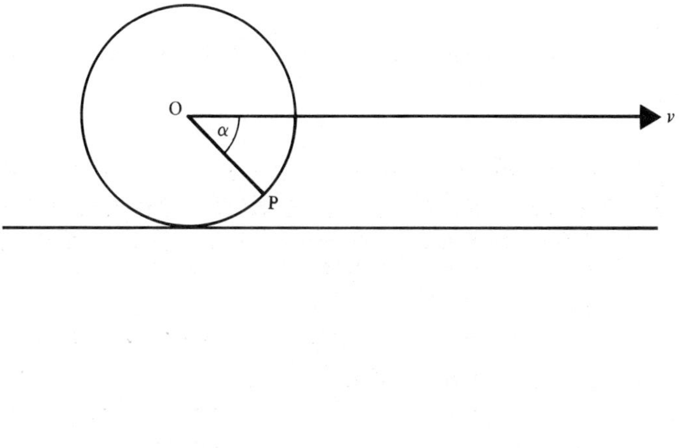

(AEB)

25) A light ring is fixed to a rigid rod, of uniform density and length $4a$, at a distance a from one end. The ring is free to slide along a smooth horizontal wire and the rod can move in the vertical plane containing the wire. The rod is held alongside the wire and released from rest. Describe the path of the centre of mass of the rod in the subsequent motion.
Show that, if the rod has rotated through an angle θ at time t, then

$$a\left(\frac{d\theta}{dt}\right)^2 = (6g\sin\theta)/(4+3\cos^2\theta).$$

Find the speed of the ring when $\theta = \pi/4$. (U of L)

26) A uniform rod AB, of length $2a$ and mass m, is held at rest with the end A in contact with a smooth vertical wall and the end B in contact with smooth horizontal ground. The inclination of the rod to the vertical is α, where $0 < \alpha < \frac{\pi}{2}$.

The rod is released from rest when in a vertical plane perpendicular to the wall. Show that the end A leaves the wall when its inclination θ to the wall satisfies

$$\cos\theta = \tfrac{2}{3}\cos\alpha. \qquad \text{(U of L)}$$

27) A uniform thin circular disc has mass $4m$ and radius a. A particle of mass m is attached to a point A on the disc which is at a distance $\frac{3}{4}a$ from the centre O of the disc. The disc is released from rest when standing in a vertical plane with OA horizontal and rolls, without slipping, along a horizontal plane. Find the speeds of O and A when A is vertically below O and show that in this position the angular speed of the disc is $\sqrt{\left(\frac{24g}{97a}\right)}$.

When OA has turned through an angle $30°$ from the horizontal, the angular speed of the disc is ω. Find
(a) the speed of A in terms of a and ω,
(b) ω in terms of a and g. (AEB)

28) Four equal uniform rods, each of mass M and length $2a$, form a rigid square frame ABCD. The frame is free to rotate in a horizontal plane about a fixed vertical axis through its centre of mass and perpendicular to the plane of the frame. A mouse, of mass $M/9$, is at E, the midpoint of AB, and frame and mouse are initially at rest. If the mouse now runs along the frame from E towards B, show that, when the mouse is at a point P on EB, where $EP = x$, and the frame has rotated through an angle θ, then the components of the velocity of the mouse parallel to AB and CD are $\dot{x} - a\dot{\theta}$ and $x\dot{\theta}$ respectively. Hence, by using the conservation of moment of momentum, show that

$$\dot{\theta} = a\dot{x}/(49a^2 + x^2).$$

Deduce that, when the mouse reaches B, the frame will have turned through an angle ϕ where $7\tan 7\phi = 1$. (U of L)

29) A particle P, of mass $m/10$, is fixed to the surface of a uniform solid sphere, of centre O, mass m and radius a. The sphere is placed on a rough horizontal table and initially the system is held at rest, with OP inclined to the upward vertical at $\pi/3$. The sphere is released and rolls without slipping. Show that, if OP makes an angle θ with the upward vertical at time t after release, the kinetic energy of P is

$$ma^2\left(\frac{d\theta}{dt}\right)^2(1+\cos\theta)/10.$$

Obtain, in terms of m, a, θ and $\frac{d\theta}{dt}$, an expression for the kinetic energy of the sphere and hence, or otherwise, show that

$$g(1-2\cos\theta) = 2a\left(\frac{d\theta}{dt}\right)^2(8+\cos\theta).$$

Find the speed of O and the speed of P when OP is first horizontal. (U of L)

30) A uniform thin ring has mass m and radius a. State the moment of inertia of the ring about a diameter.
Such a smooth uniform ring is free to rotate about its vertical diameter which is fixed, and a small bead of mass $m/2$ is free to slide on the ring. Initially the ring is spinning, with angular velocity Ω, with the bead at the lowest point of the ring. If the bead is slightly displaced, prove that in the subsequent motion the angular velocity of the ring is

$$\frac{\Omega}{1+\sin^2\theta},$$

where θ is the angle between the downward vertical and the radius from the centre of the ring to the bead.
Show that when $\dot{\theta} = 0$, θ satisfies the relation

$$\frac{a\Omega^2\sin^2\theta}{1+\sin^2\theta} = 2g(1-\cos\theta).$$

In the case when $\Omega^2 = 7g/a$, deduce that when the bead reaches its highest point on the ring, the angular velocity of the ring is $4\Omega/7$. (JMB)

31) A uniform rod, of length $2a$ and mass m, stands vertically on a rough horizontal plane. The rod is given a small displacement from the vertical and is then released from rest. Show that, when the rod has turned through an angle θ, its angular velocity $\dot{\theta}$, is given by $2a\dot{\theta}^2 = 3g(1-\cos\theta)$, assuming that slipping has not by then occurred at the lower end of the rod. Find the horizontal and vertical components of the reaction between the rod and the plane at this instant.
If slipping between the rod and the plane begins when $\cos\theta = 4/5$, show that the coefficient of friction between the rod and the plane is $18/49$. (U of L)

32) A shell of mass $2m$ is fired vertically upwards with a velocity v from a point on a horizontal plane. When the shell comes instantaneously to rest an internal explosion splits it into two equal fragments of mass m. Given that the explosion supplies kinetic energy mu^2 to the system, show that the greatest possible distance between the points where the fragments hit the ground is

$\dfrac{2uv}{g}$ if $u \leqslant v$ or $\dfrac{(u^2+v^2)}{g}$ if $u \geqslant v$. (AEB)

33) A uniform rod AB of mass $2m$ is freely jointed at B to a second uniform rod BC of mass m. The rods lie on a smooth horizontal table at right angles to each other when an impulse I is applied at A in a direction parallel to BC. Find the velocities of the centres of mass of each rod immediately after the impulse is applied. Prove that the total kinetic energy generated by the impulse is $\dfrac{5I^2}{6m}$ and find the impulsive reaction of the hinge at B on both rods. (AEB)

34) A fixed solid smooth circular cylinder, whose radius and height are both equal to l, has its axis vertical and a smooth, thin, uniform rod OA, of length l and mass m, is placed on its upper plane surface so that the axis of the cylinder passes through O. The end A is attached to a light elastic string of natural length l and modulus $mg/3$, the other end of this string being attached to a fixed point B on the circumference of the lower plane surface of the cylinder.

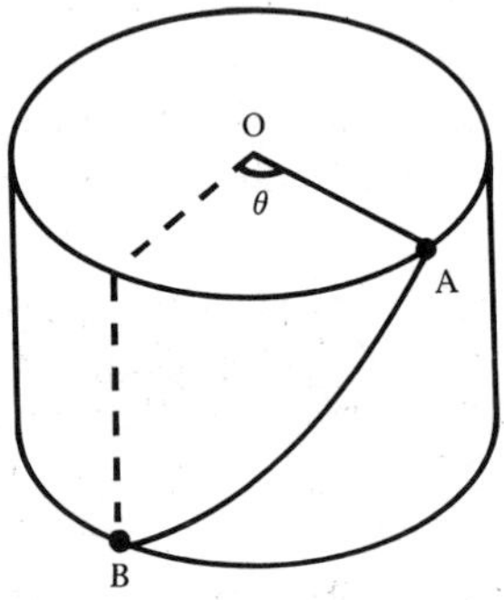

The rod, which can rotate smoothly about the axis of the cylinder, is displaced through an angle α from its equilibrium position and released from rest. Assuming that the string lies along the shortest possible curve joining A and B and lying entirely in the curved surface of the cylinder, show that θ, the displacement of the rod from its equilibrium position, satisfies the equation

$$l\,[(1+\theta^2)+(1+\theta^2)^{1/2}]\ddot{\theta} = -g\theta^3.$$

(It may be helpful to note that the portion of the curved surface of a cylinder between any two generators becomes a rectangle when rolled flat.)
Find an expression, as a definite integral, for the time that elapses before the rod first reaches its equilibrium position. (JMB)

CHAPTER 14

DIFFERENTIAL EQUATIONS

FORMATION OF DIFFERENTIAL EQUATIONS

There are many physical situations where the relationship between two variables, initially defined in words, can be interpreted as a differential equation.

The reader has already encountered some of these cases in mechanics. For example, if the motion of a particle moving along Ox is described as, 'such that the rate of increase of its displacement from O is inversely proportional to that displacement' we can write

$$\frac{dx}{dt} = \frac{k}{x}$$

This differential equation can then be solved if the initial values of x and t are known.

Mechanics is only one of many areas in which such differential equations occur naturally and it is not necessary to understand the background of the problem in order to form the appropriate equation. The solution depends upon the type of differential equation produced. In *The Core Course for A-level* and *Further Pure Mathematics* methods are given for solving a variety of differential equations and the reader is advised to revise these methods at this stage.

EXAMPLES 14a

1) Find a differential equation that describes each of the following situations. In each case state the number of extra facts that are needed in order to find a particular solution.

(a) Twenty grams of a powder are to be dissolved in a small quantity of water. The rate at which the powder dissolves is inversely proportional to the weight, ω grams, already in solution.

(b) For a certain shrub the number of leaves falling off at any time is proportional to the number, n, of leaves then on the shrub and the number of new leaves sprouting is proportional to $\sqrt{n}$.

(c) In a greenhouse, greenfly are eating the leaves of a plant at a rate of one-tenth of a leaf per greenfly per time unit. Pesticide is sprayed on to the leaves and kills four greenfly per leaf per time unit. At time t, there are n leaves and m greenfly. Form a differential equation in which n and t are the only variables.

(a) The amount of powder left undissolved at time t is $(20-\omega)$ grams, therefore

$$\frac{d}{dt}(20-\omega) = \frac{k}{\omega}$$

Two extra facts are needed; one to find k and one to find the constant of integration. (There is only one constant of integration for a first order differential equation.)

(b) The current number of leaves decreases at a rate λn and increases at a rate $\mu\sqrt{n}$, therefore

$$\frac{dn}{dt} = \mu\sqrt{n}-\lambda n$$

Three further facts are needed to find λ, μ and the one constant of integration.

(c) The number of greenfly killed per time unit is $4n$ therefore

$$\frac{dm}{dt} = -4n \qquad [1]$$

The number of leaves eaten per time unit is $\frac{1}{10}m$ therefore

$$\frac{dn}{dt} = -\tfrac{1}{10}m \qquad [2]$$

Differentiating equation [2] with respect to time gives

$$\frac{d^2n}{dt^2} = -\frac{1}{10}\frac{dm}{dt} = -\frac{1}{10}(-4n)$$

$$\Rightarrow \qquad \frac{d^2n}{dt^2} = \frac{2}{5}n$$

This second order differential equation requires two extra facts to evaluate the two constants of integration.

2) Write down a differential equation which describes the following chemical reaction. The number of molecules of a chemical A is initially 2500 and n is the number that have been converted to another chemical B after t seconds. The rate at which the number of converted molecules increases is proportional to the square root of the number not yet converted. If the rate of conversion is 80 molecules per second when $n = 900$ find the number of A molecules left after 20 seconds.

The number of A molecules present at time t is $(2500-n)$, therefore

$$\frac{dn}{dt} = k\sqrt{2500-n}$$

Now $\frac{dn}{dt} = 80$ when $n = 900$, hence

$$80 = k\sqrt{1600} \quad \Rightarrow \quad k = 2$$

Therefore $\frac{dn}{dt} = 2\sqrt{2500-n}$

Now if N is the number of A molecules converted in 20 seconds, separating the variables gives

$$\int_0^N \frac{dn}{\sqrt{2500-n}} = \int_0^{20} 2dt$$

$$\Rightarrow \qquad -2\left[\sqrt{2500-n}\right]_0^N = 40$$

$$\Rightarrow \qquad -\sqrt{2500-N}+50 = 20$$

$$\Rightarrow \qquad N = 1600$$

Therefore the number of A molecules left is 900.

3) The position vector of a moving particle P at time t is given by $\mathbf{r} = 2\sin t\mathbf{i} + 5\mathbf{j} + (4t + 2\cos t)\mathbf{k}$.

Show that this equation satisfies the relation $\mathbf{V} \times \dot{\mathbf{r}} = \mathbf{A} - \ddot{\mathbf{r}}$

for certain values of the constant vectors $\mathbf{V}$ and $\mathbf{A}$ and find these values.

$$\mathbf{r} = 2\sin t\mathbf{i} + 5\mathbf{j} + (4t + 2\cos t)\mathbf{k}$$

$$\Rightarrow \quad \dot{\mathbf{r}} = 2\cos t\mathbf{i} + (4 - 2\sin t)\mathbf{k}$$

$$\Rightarrow \quad \ddot{\mathbf{r}} = -2\sin t\mathbf{i} - 2\cos t\mathbf{k}$$

If these equations satisfy $\mathbf{V} \times \dot{\mathbf{r}} = \mathbf{A} - \ddot{\mathbf{r}}$ and $\mathbf{V} = a\mathbf{i} + b\mathbf{j} + c\mathbf{k}$, $\mathbf{A} = f\mathbf{i} + g\mathbf{j} + h\mathbf{k}$, then

$$b(4 - 2\sin t)\mathbf{i} + (2c\cos t - 4a + 2a\sin t)\mathbf{j} - 2b\cos t\mathbf{k}$$

$$= (f + 2\sin t)\mathbf{i} + g\mathbf{j} + (h + 2\cos t)\mathbf{k}$$

Hence

$$4b - 2b\sin t = f + 2\sin t \quad \Rightarrow \quad b = -1, \quad f = -4$$

$$2c\cos t - 4a + 2a\sin t = g \quad \Rightarrow \quad a = c = 0, \quad g = 0$$

$$-2b\cos t = h + 2\cos t \quad \Rightarrow \quad h = 0$$

Therefore the given equation does satisfy the given relation provided that

$$\mathbf{V} = -\mathbf{j} \quad \text{and} \quad \mathbf{A} = -4\mathbf{i}$$

(**Note.** In a question worded like this one, the non-differential relationship between the variables is intended to be *used* to form the differential relationship. A solution of the differential equation is not intended.)

EXERCISE 14a

1) During the initial stages of growth of yeast cells in a certain culture, the number of cells present increases in proportion to the number already formed. Given that n is the number of cells at a particular time t seconds, and treating n as a continuous variable, write down a differential equation satisfied by n. Given that there are 2 million cells present when $t = 0$ and 4 million cells present when $t = 1$, find the number of cells present when $t = 4$. (U of L)

2) Initially a tank contains 25 litres of water. Beginning at time $t = 0$, brine of salt concentration 0.2 kg per litre flows into the tank at a rate of 1 litre per minute. The mixture is kept uniform by stirring, and leaves the tank at a rate of 1.5 litres per minute. Given that there is x kg of salt in the tank at time t minutes, show that $\dfrac{dx}{dt} = \dfrac{1}{5} - \dfrac{3x}{50 - t}$, $(0 \leqslant t < 50)$.

Solve this equation to find the value of x at a time t minutes, where $t < 50$, and hence show that the maximum value of x is $10/(3\sqrt{3})$. (U of L)

3) Newton's law of cooling states that the rate of decrease of temperature of a body is proportional to the difference of the temperature T of the body and the temperature T_0, where $T_0 < T$, of the surrounding space. Write down the differential equation satisfied by T and find its general solution.
A body takes 5 minutes to cool from 80°C to 50°C, the temperature of the surrounding space being 20°C. Find the time taken for the body to cool a further 10°C, giving your answer to the nearest second. (U of L)

4) A fertilized egg initially contains an embryo of mass m_0 together with a mass $100m_0$ of nutrient, all of which is available as food for the embryo. At time t the embryo has mass m and the mass of nutrient which has been consumed is $5(m-m_0)$. Show that, when three-quarters of the nutrient has been consumed, $m = 16m_0$.
The rate of increase of the mass of the embryo is a constant μ multiplied by the product of the mass of the embryo and the mass of the remaining nutrient. Show that

$$\frac{dm}{dt} = 5\mu m(21m_0 - m).$$

(U of L)

5) In a chemical reaction, a mass $3m$ of a compound A can decompose into a mass m of an element B and a mass $2m$ of an element C, the rate of decomposition being equal to k times the mass of A present, where k is a constant. At the same time a mass m of B and a mass $2m$ of C can recombine into a mass $3m$ of A, the rate of combination being equal to $2k$ times the product of the masses of B and C present. Show that, if the masses of A, B, C present at time t are $1-3y$, y, $2y$ respectively, then

$$3\frac{dy}{dt} = k(1-4y)(1+y).$$

Given that initially only compound A is present, find y in terms of k and t and deduce that y is always less than $1/4$. (U of L)

6) The population x of a colony of bacteria increases at a rate equal to the product of the number x of bacteria present at time t and the capability C of the environment to support the number present at time t. The capability C is measured by the excess of the maximum number b of bacteria that can be supported by the environment over the number of bacteria actually present. Write down the differential equation governing the growth of bacteria.
Solve this differential equation, expressing x in terms of t, given that the population at time $t = 0$ is p, where $p < b$. State what happens to the population after the passage of a large interval of time. (U of L)

7) In the manufacture of wine, a jar initially contains 20 litres of a mixture which consists of a liquid into which 5 kg of sugar have been dissolved. At any given time the sugar is converted into alcohol at a rate of $4c$ kg/day, where c kg/litre is the concentration of unconverted sugar in the mixture. Assuming that there is no change in the volume of the mixture, show that, if y kg of sugar remain after t days, then

$$5\frac{dy}{dt} = -y.$$

Hence find the time taken for the quantity of sugar in the mixture to fall to 0.005 kg.
A second jar also initially contains 5 kg of sugar in 20 litres of mixture but, as it is not sealed correctly, the liquid evaporates at a constant rate of 100 ml/day. Show that, if x kg of sugar remain after t days, then

$$\frac{dx}{dt} = -\frac{40x}{200-t}.$$

Hence find, to the nearest day, the time taken for the sugar content to fall to 0.005 kg in the second jar.
(You may assume that $(0.8414)^{40} = 0.001$) (U of L)

8) An army of red ants encounters an army of black ants. Each red ant then proceeds to kill α black ants per unit time while each black ant kills 4α red ants per unit time, where α is a positive constant. The number of live red ants and the number of live black ants at time t are x and y respectively. Assuming that x and y may be considered as continuous variables, show that

$$\frac{d^2x}{dt^2} = 4\alpha^2 x.$$

Solve this equation for x, and hence determine x and y, given that $x = 5N$ and $y = 2N$ when $t = 0$.
Show that your solutions are not realistic for $\alpha t > \frac{1}{2}\ln 3$ and find the time at which the army of black ants is reduced to half of its original size. (U of L)

9) At any instant the birth rate of a colony of ants is proportional to N, the number of ants present, and the death rate is proportional to N^2. Write down a differential equation for N, and show that the solution takes the form

$$N = \frac{A}{1+Be-\lambda^t},$$

where A, B, λ are constants and t is the number of weeks which have elapsed since the population was first examined.
Find the values of A, B and λ given that initially there are 20 000 ants, that after a very long time the ant population becomes 100 000, and, if there were no deaths and the birth rate was unchanged, the population would double after two weeks. (U of L)

10) Newton's law of cooling states that the rate at which a body cools is proportional to the difference between the temperature of the body and that of the medium in which the body is situated. If θ is the temperature of the body and α is the constant temperature of the surrounding medium, show that $\theta - \alpha$ must be proportional to e^{-kt}, where t is the time and k is a positive constant.

At time $t = 0$ a body with a temperature of 80°C is placed in a medium whose temperature is maintained at 50°C. At $t = 300$ s the body has cooled to a temperature of 70°C. Find, correct to two decimal places, the temperature of the body when $t = 600$ s.
Find also, to the nearest second, the time at which the temperature of the body is 60°C. (U of L)

11) At any instant, a spherical asteroid in space is gaining mass because of two effects:

(a) mass from the solar wind is condensing on to it at a rate which is proportional to the surface area of the asteroid,

(b) the gravitational field of the asteroid attracts mass on to the asteroid, the rate being proportional to the mass of the asteroid at that instant.

Assuming that the rate of increase of mass of the asteroid is given by the sum of the rates given in (a) and (b) above and that the asteroid remains spherical and of constant density, show that the differential equation satisfied by the radius r of the asteroid at time t can be written in the form

$$\frac{dr}{dt} = \frac{r + r_0}{t_0},$$

r_0 and t_0 being two constants.

Given that $r = r_0$ when $t = t_0$, find r in terms of r_0, t_0 and t.
Find also the time at which $r = 3r_0$. (U of L)

12) A particle, of mass m, moves such that its velocity at any instant is given by $\mathbf{r} = \boldsymbol{\Omega} \times \mathbf{r}$, where $\boldsymbol{\Omega}$ is a constant vector and $\mathbf{r}$ is its position vector. Given that $\boldsymbol{\Omega} = \omega\mathbf{k}$ (ω constant) and $\mathbf{r} = x\mathbf{i} + y\mathbf{j} + z\mathbf{k}$, show that

(a) $\ddot{\mathbf{r}} = -\omega^2(x\mathbf{i} + y\mathbf{j})$

(b) $\boldsymbol{\Omega}$ is perpendicular to $\dot{\mathbf{r}}$.

Write down the resultant force acting on the particle and find the moment of this force about the origin.
Verify that $\mathbf{r} = a\cos\omega t\mathbf{i} + a\sin\omega t\mathbf{j} + b\mathbf{k}$, where a and b are constants, gives the position vector for a possible motion of the particle and show that, in this case, the speed of the particle is constant. (AEB)

13) At time t, the position vector $\mathbf{r}(t)$ of a moving particle satisfies the relation

$$\ddot{\mathbf{r}} = \mathbf{E} + \dot{\mathbf{r}} \times \mathbf{B},$$

where $\mathbf{E}$ and $\mathbf{B}$ are constant vectors. Show that

$$\mathbf{r} = \cos t\,\mathbf{i} + (\sin t + 3t)\mathbf{k}$$

satisfies this relation provided $\mathbf{E}$ and $\mathbf{B}$ take certain values which are to be determined. (JMB)

VECTOR DIFFERENTIAL EQUATIONS

Many of the techniques used in the solution of differential equations can also be applied when some of the variables are vector quantities, *provided that any integration or differentiation is carried out with respect to a scalar quantity.*

Some examples have already occurred in this book, e.g. if a particle has a velocity $3t\mathbf{i} - 4t^2\mathbf{j}$ at time t then we can write

$$\frac{d\mathbf{r}}{dt} = 3t\mathbf{i} - 4t^2\mathbf{j}$$

This is a first order linear differential equation which can be solved by integrating with respect to the scalar, t.

The solution of further vector differential equations is given in the following examples.

EXAMPLES 14b

1) Solve the equation $\ddot{\mathbf{r}} - 4\dot{\mathbf{r}} + 3\mathbf{r} = \mathbf{0}$ given that $\mathbf{r} = \mathbf{i}$ and $\dot{\mathbf{r}} = \mathbf{i} + 2\mathbf{j}$ when $t = 0$.

This equation can be written

$$\frac{d^2\mathbf{r}}{dt^2} - 4\frac{d\mathbf{r}}{dt} + 3\mathbf{r} = \mathbf{0}$$

and is easily identified as a second order linear differential equation with numerical coefficients.

The solution can therefore be found by using the associated quadratic equation

$$u^2-4u+3 = 0 \quad \Rightarrow \quad u = 3 \text{ or } 1$$

Hence $$\mathbf{r} = \mathbf{A}e^{3t}+\mathbf{B}e^{t}$$

(**Note** that the constants of integration, **A** and **B**, are vectors.)

If $\mathbf{r}=\mathbf{i}$ when $t=0$ we get

$$\mathbf{i} = \mathbf{A}+\mathbf{B}$$

If $\dot{\mathbf{r}}=\mathbf{i}+2\mathbf{j}$ when $t=0$, then

$$\mathbf{i}+2\mathbf{j} = 3\mathbf{A}+\mathbf{B}$$

These two equations give $\mathbf{A}=\mathbf{j}$ and $\mathbf{B}=\mathbf{i}-\mathbf{j}$.

Hence $$\mathbf{r} = e^{t}\mathbf{i}+(e^{3t}-e^{t})\mathbf{j}$$

2) If $\mathbf{r}=\mathbf{i}+\mathbf{j}$ and $\dfrac{d\mathbf{r}}{dt}=\mathbf{i}-\mathbf{j}$ when $t=0$, use the equation $\dfrac{d^2\mathbf{r}}{dt^2}+n^2\mathbf{r}=\mathbf{0}$ to find an expression for **r**.

Writing the given equation in the form

$$\ddot{\mathbf{r}} = -n^2\mathbf{r}$$

shows it to be similar to the equation of single harmonic motion $(\ddot{x}=-n^2x)$ and the solution can be quoted.

i.e. $$\mathbf{r} = \mathbf{A}\cos nt+\mathbf{B}\sin nt$$

$\Rightarrow$ $$\dot{\mathbf{r}} = -\mathbf{A}n\sin nt+\mathbf{B}n\cos nt$$

When $t=0$, $\mathbf{r}=\mathbf{i}+\mathbf{j}$ and $\dot{\mathbf{r}}=\mathbf{i}-\mathbf{j}$, therefore

$$\mathbf{i}+\mathbf{j} = \mathbf{A}$$

and $$\mathbf{i}-\mathbf{j} = n\mathbf{B}$$

Hence $$\mathbf{r} = \left(\cos nt+\frac{1}{n}\sin nt\right)\mathbf{i}+\left(\cos nt-\frac{1}{n}\sin nt\right)\mathbf{j}$$

3) The velocity **v** of a particle P satisfies the equation $\mathbf{v}=\dfrac{d\mathbf{v}}{dt}$. Given that $\mathbf{v}=2\mathbf{i}+\mathbf{j}$ and $\mathbf{r}=\mathbf{j}$ when $t=0$, find an expression for the position vector, **r**, of P at time t.

The form of this equation must be changed before it can be solved, because we cannot integrate with respect to the vector **v**.

Using $\mathbf{v} = \frac{d\mathbf{r}}{dt}$ gives

$$\frac{d\mathbf{r}}{dt} = \frac{d^2\mathbf{r}}{dt^2} \quad \text{or} \quad \ddot{\mathbf{r}} - \dot{\mathbf{r}} = \mathbf{0}$$

This is a second order linear differential equation with numerical coefficients. The associated quadratic equation, $u^2 - u = 0$, has roots 0 and 1.

Therefore the solution of the differential equation is

$$\mathbf{r} = \mathbf{A} + \mathbf{B}e^t$$

$$\Rightarrow \quad \mathbf{v} = \mathbf{B}e^t$$

Now $\mathbf{r} = \mathbf{j}$ and $\mathbf{v} = 2\mathbf{i} + \mathbf{j}$ when $t = 0$, therefore

$$\mathbf{j} = \mathbf{A} + \mathbf{B} \quad \text{and} \quad 2\mathbf{i} + \mathbf{j} = \mathbf{B}$$

Hence $$\mathbf{r} = -2\mathbf{i} + (2\mathbf{i} + \mathbf{j})e^t$$

or $$\mathbf{r} = (2e^t - 2)\mathbf{i} + e^t\mathbf{j}$$

Note. The method used in the following example can be used as the first step in an alternative solution to this problem. The reader may be interested to compare the two methods.

4) If $\mathbf{p}$ is a vector that varies with the time t, find an expression for $\mathbf{p}$ given that $\frac{d\mathbf{p}}{dt} = 3\mathbf{p}$ and, when $t = 0$, $\mathbf{p} \times \mathbf{i} = 4\mathbf{j}$ and $\mathbf{p}.\mathbf{i} = 5$

Considering $\frac{d\mathbf{p}}{dt} = 3\mathbf{p}$ as a linear differential equation with an associated 'quadratic' equation $u - 3 = 0$ gives

$$\mathbf{p} = \mathbf{A}e^{3t}$$

$$\Rightarrow \quad \dot{\mathbf{p}} = 3\mathbf{A}e^{3t}$$

When $t = 0$, $\mathbf{p} = \mathbf{A}$ and $\dot{\mathbf{p}} = 3\mathbf{A}$.

Taking $\mathbf{A} = a\mathbf{i} + b\mathbf{j} + c\mathbf{k}$ then, when $t = 0$,

$$\mathbf{A} \times \mathbf{i} = 4\mathbf{j} \quad \Rightarrow \quad c\mathbf{j} - b\mathbf{k} = 4\mathbf{j}$$

$$\Rightarrow \quad c = 4 \quad \text{and} \quad b = 0$$

$$3\mathbf{A}.\mathbf{i} = 5 \quad \Rightarrow \quad a = \tfrac{5}{3}$$

Therefore $$\mathbf{p} = (\tfrac{5}{3}\mathbf{i} + 4\mathbf{k})e^{3t}$$

5) At time t, the position vector $\mathbf{r}$ of a moving point P is such that

$$\frac{d^2\mathbf{r}}{dt^2}+4\mathbf{r} = (3\sin t)\mathbf{i}$$

If, when $t=0$, $\mathbf{r}=\mathbf{j}$ and $\mathbf{v}=\mathbf{i}$, find the Cartesian equation of the locus of P.

First we find the C.F. (complementary function) of the given second order linear differential equation

$$\frac{d^2\mathbf{r}}{dt^2}+4\mathbf{r} = \mathbf{0}$$

C.F. is $\mathbf{r} = \mathbf{A}\cos 2t + \mathbf{B}\sin 2t$.

Now to find the P.I. (particular integral) we use, as a trial solution,

$$\mathbf{r} = \mathbf{C}\sin t + \mathbf{D}\cos t$$

$$\Rightarrow \quad \dot{\mathbf{r}} = \mathbf{C}\cos t - \mathbf{D}\sin t$$

$$\Rightarrow \quad \ddot{\mathbf{r}} = -\mathbf{C}\sin t - \mathbf{D}\cos t$$

Using these in the given equation gives

$$-\mathbf{C}\sin t - \mathbf{D}\cos t + 4(\mathbf{C}\sin t + \mathbf{D}\cos t) = (3\sin t)\mathbf{i}$$

Hence $\quad 3\mathbf{C}\sin t = (3\sin t)\mathbf{i} \quad \Rightarrow \quad \mathbf{C} = \mathbf{i}$

and $\quad 3\mathbf{D}\cos t = \mathbf{0} \quad \Rightarrow \quad \mathbf{D} = \mathbf{0}$

Therefore the complete solution for $\mathbf{r}$ is

$$\mathbf{r} = \mathbf{A}\cos 2t + \mathbf{B}\sin 2t + \mathbf{i}\sin t$$

Now $\mathbf{r}=\mathbf{j}$ when $t=0$, therefore $\mathbf{j}=\mathbf{A}$.

Also $\quad \dot{\mathbf{r}} = -2\mathbf{A}\sin 2t + 2\mathbf{B}\cos 2t + \mathbf{i}\cos t$

Using $\dot{\mathbf{r}}=\mathbf{v}=\mathbf{i}$ when $t=0$ gives

$$\mathbf{i} = 2\mathbf{B}+\mathbf{i} \quad \Rightarrow \quad \mathbf{B} = \mathbf{0}$$

Hence, at time t,

$$\mathbf{r} = (\sin t)\mathbf{i} + (\cos 2t)\mathbf{j}$$

i.e. for P, $x = \sin t$ and $y = \cos 2t$

Using $\cos 2t = 1-2\sin^2 t$ gives $y = 1-2x^2$ and this is the Cartesian equation of the locus of P.

EXERCISE 14b

In questions 1 to 7 solve the given equation.

1) $\dfrac{d^2\mathbf{r}}{dt^2}-2\dfrac{d\mathbf{r}}{dt}+\mathbf{r}=\mathbf{0};\quad \mathbf{r}=\mathbf{i}-\mathbf{j}$ and $\dfrac{d\mathbf{r}}{dt}=\mathbf{i}+\mathbf{j}$ when $t=0$.

2) $\dfrac{d^2\mathbf{r}}{dt^2}+2\dfrac{d\mathbf{r}}{dt}+\mathbf{r}=\mathbf{0};\quad \mathbf{r}=\mathbf{i}$ and $\dfrac{d\mathbf{r}}{dt}=\mathbf{i}+\mathbf{j}+\mathbf{k}$ when $t=0$.

3) $\dfrac{d^2\mathbf{r}}{dt^2}+\dfrac{d\mathbf{r}}{dt}+\mathbf{r}=\mathbf{0};\quad \mathbf{r}=-2\mathbf{i}$ and $\dfrac{d\mathbf{r}}{dt}=\mathbf{i}+\sqrt{3}\mathbf{j}$ when $t=0$.

4) $\dfrac{d^2\mathbf{r}}{dt^2}+2\dfrac{d\mathbf{r}}{dt}=\mathbf{0};\quad \mathbf{r}=\mathbf{0}$ and $\dot{\mathbf{r}}=4\mathbf{i}$ when $t=0$.

5) $\ddot{\mathbf{r}}+2\mathbf{r}=4\mathbf{j}\cos t;\quad \mathbf{r}=\mathbf{0}$ and $\dot{\mathbf{r}}=2\sqrt{2}\mathbf{i}-\sqrt{2}\mathbf{j}$ when $t=0$.

6) $\ddot{\mathbf{r}}+\mathbf{r}=10e^{2t}\mathbf{i};\quad \mathbf{r}=\mathbf{i}$ and $\dot{\mathbf{r}}=2\mathbf{j}$ when $t=0$.

7) $\ddot{\mathbf{r}}+\lambda^2\mathbf{r}=\mathbf{a}\sin 2\lambda t$ where $\mathbf{a}$ is a constant vector and λ is a positive scalar constant; $\mathbf{r}=3\mathbf{a}$ and $\dot{\mathbf{r}}=\mathbf{a}/3\lambda$ when $t=0$.

8) A point P moves so that its position vector at time t is given by $\ddot{\mathbf{r}}+9\mathbf{r}=(8\sin t)\mathbf{i}$. When $t=0$, P is at the origin with velocity $\mathbf{i}+3\mathbf{j}$. Show that the Cartesian equation of the path of P is $y=3x-4x^3$.

9) Integrate the vector equation

$$\frac{d^2\mathbf{r}}{dt^2}+n^2\mathbf{r}=\mathbf{0}$$

to find $\mathbf{r}$, given that $\mathbf{r}=(\mathbf{i}+\mathbf{j})a$ and $\dfrac{d\mathbf{r}}{dt}=(\mathbf{i}-2\mathbf{k})b$ when $t=0$.

(U of L)p

10) Solve the equations
(a) $\ddot{\mathbf{r}}+5\dot{\mathbf{r}}+6\mathbf{r}=\mathbf{0}$, given that $\mathbf{r}=\mathbf{i}$, $\dot{\mathbf{r}}=\mathbf{j}$, at $t=0$,
(b) $\ddot{\mathbf{r}}+2n\dot{\mathbf{r}}+n^2\mathbf{r}=\mathbf{0}$, where n is constant, given that $\mathbf{r}=\mathbf{0}$, $\dot{\mathbf{r}}=\mathbf{V}$ at $t=0$,
(c) $\ddot{\mathbf{r}}+\omega^2\mathbf{r}=\mathbf{a}\cos 2\omega t$, where $\mathbf{a}$ is a constant vector and ω is a positive constant.

(U of L)

11) (a) At time t, the velocity $\mathbf{v}$ of a particle P satisfies the equation $\dfrac{d\mathbf{v}}{dt} + 2\mathbf{v} = \mathbf{0}$. At time $t = 0$, the position vector and velocity of P are $(\mathbf{i}+\mathbf{j})$ and $4(-\mathbf{i}+\mathbf{j})$ respectively.
Show that at time t the position vector of P is

$$\mathbf{r} = (-\mathbf{i}+3\mathbf{j}) + 2(\mathbf{i}-\mathbf{j})e^{-2t}.$$

(b) A particle moves in the x-y plane such that its position vector $\mathbf{r}$ satisfies the differential equation

$$\frac{d^2\mathbf{r}}{dt^2} + 4\mathbf{r} = \mathbf{0}.$$

Given that $\mathbf{r} = 3\mathbf{i}$ when $t = 0$ and $\mathbf{r} = \mathbf{j}$ when $t = \pi/4$, show that the particle describes the ellipse

$$x^2 + 9y^2 = 9.$$

(U of L)

12) (a) Solve the vector differential equation

$$\frac{d^2\mathbf{r}}{d\theta^2} = \frac{d\mathbf{r}}{d\theta},$$

given that $\dfrac{d\mathbf{r}}{d\theta} = \mathbf{i}$ and $\mathbf{r} = \mathbf{j}$ when $\theta = 0$.

(b) Find the general solution of the vector differential equation

$$\frac{d^2\mathbf{r}}{d\theta^2} - 2\frac{d\mathbf{r}}{d\theta} + 10\mathbf{r} = \mathbf{0}.$$

Given that $\mathbf{r} = \mathbf{j}$ when $\theta = 0$, find $\mathbf{r}$ when $\theta = \pi$. (U of L)

13) (a) Solve the differential equation $d\mathbf{r}/dt = 4\mathbf{r}$, given that, when $t = 0$, $\mathbf{r}.\mathbf{i} = 1$ and $\mathbf{r} \times \mathbf{i} = \mathbf{j} + \mathbf{k}$.

(b) Find the solution of the differential equation

$$d^2\mathbf{r}/dt^2 + 2d\mathbf{r}/dt + 2\mathbf{r} = \mathbf{0}$$

such that $\mathbf{r} = \mathbf{i} + \mathbf{k}$ when $t = 0$, and $d\mathbf{r}/dt = \mathbf{0}$ when $t = \pi/2$.

(c) At time t the position vector $\mathbf{r}$ of the point P satisfies the differential equation

$$d^2\mathbf{r}/dt^2 + 4\mathbf{r} = 3\mathbf{i}\sin t.$$

When $t = 0$, P passes through the point $\mathbf{r} = \mathbf{j}$ with velocity $\mathbf{i}$.
Find the equation of the locus of P. (U of L)

CHAPTER 15

SHEARING FORCE AND BENDING MOMENT

INTERNAL STRESSES IN A RIGID BODY

Hitherto in our study of the equilibrium of rigid bodies we have confined ourselves to the effect of external forces. We must now give some consideration to the internal forces which act between adjacent parts of a rigid body. A beam, or girder, is a particularly important body to study in this way because of its many uses in the construction industry.

In general the forces acting on a beam give rise to stresses within the beam which can cause it to stretch, break, bend or twist. To deal fully with these effects requires complex analysis beyond the scope of this book. Consequently we will confine ourselves to the study of a horizontal girder to which coplanar vertical forces only are applied and which is rigid enough for the distortion that those forces produce to be ignored.

SHEARING FORCE AND BENDING MOMENT

Consider first a uniform light rigid girder AB which rests horizontally on two supports, one at each end. A concentrated load of weight $2W$ is carried at the midpoint C.

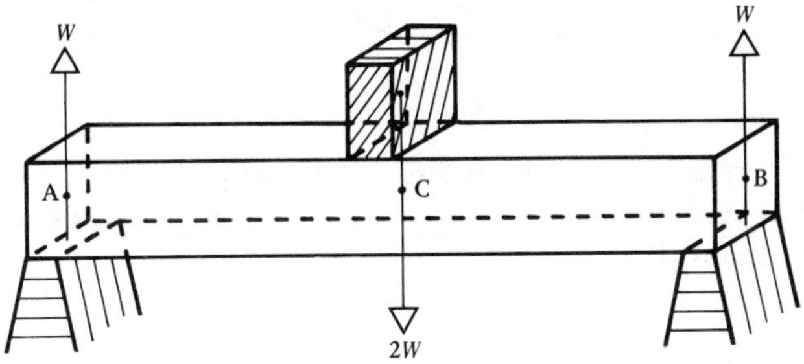

If we were to cut vertically through the girder at a point P and remove the part PB, then the section AP would collapse.
To prevent this from happening, an upward force S and an anticlockwise torque $\mathfrak{M}$ have to be applied at P so that the equilibrium of the section AP is restored.

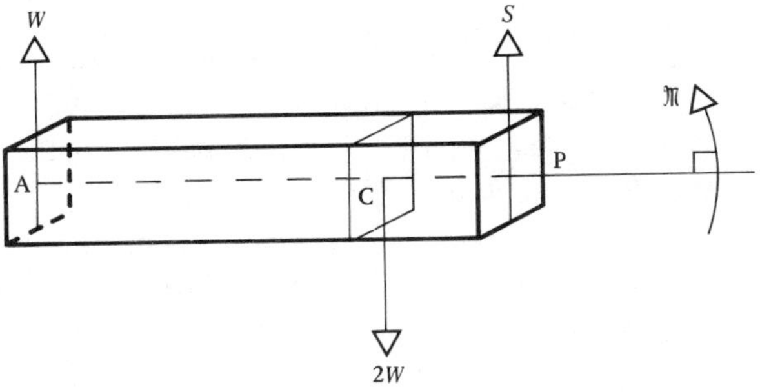

We therefore conclude that, because the girder was in equilibrium before it was divided at P, the section PB must have been responsible for exerting on the section AP a force equal to S and a torque equal to $\mathfrak{M}$.

But internal forces act in equal and opposite pairs, so the section AP must also have exerted on the section PB a force equal and opposite to S and a torque equal and opposite to $\mathfrak{M}$. (See diagram.)

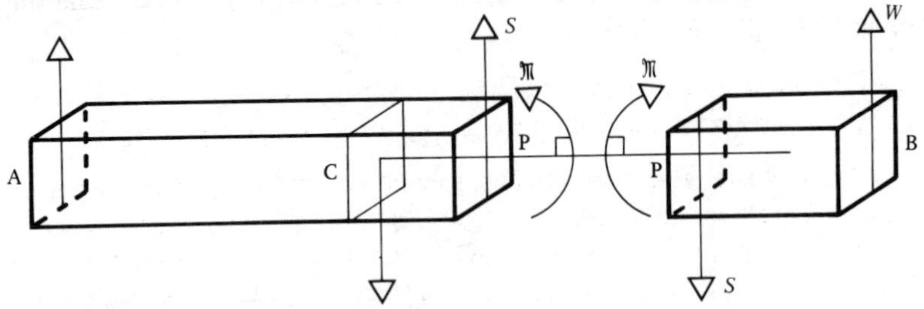

The force S is called the *shearing force* at P. The pair of equal and opposite shearing forces at that point tends to make the beam break across its vertical cross-section at P (i.e. to *shear*).

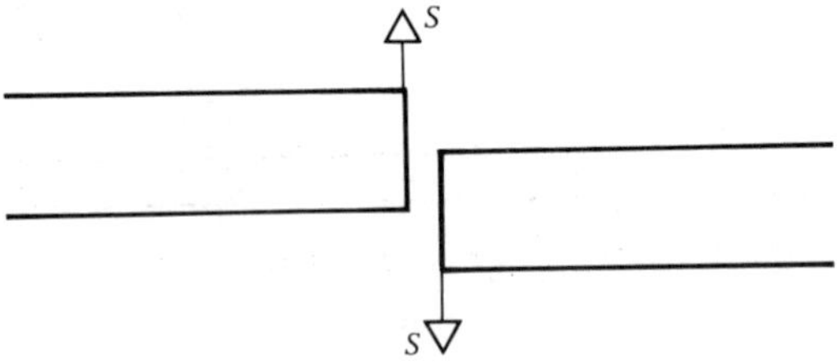

The torque $\mathfrak{M}$ is called the *bending moment* at P. Its effect on the beam is to produce a tendency to bend in a vertical plane.

SIGN CONVENTION

The sense in which S and $\mathfrak{M}$ are taken to be positive is a purely arbitrary choice. In this book, when considering the left-hand portion of the beam the upward direction is taken as positive for the shearing force while the positive sense for bending moment is anticlockwise.

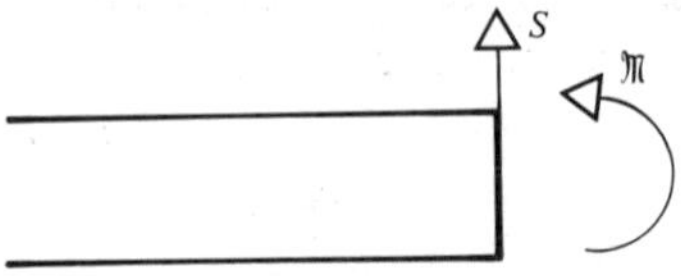

The numerical values of S and $\mathfrak{M}$ can, of course, be calculated by considering the equilibrium of the right-hand section of the beam, but it must be remembered that the bending moment and shearing force act in opposite directions on the two portions of the beam that meet at P.

CONCENTRATED AND DISTRIBUTED FORCES

The effect of some of the forces which act on beams is concentrated at a point (or on an area which is so small compared with the area of the beam that it can be regarded as a point).

Such a force is called a *concentrated* or *point force.* An example of a force of this type is the reaction exerted on a beam by a knife-edge support.

Other forces are such that their effect is spread out over an appreciable length of the beam; these are called *distributed* forces. The commonest example of this type is the weight of a uniform heavy beam. Although for overall analysis the total weight of the beam can be taken to act through its midpoint, when studying the internal stresses within the beam it must be appreciated that the weight of each element or section of the beam acts through the centre of gravity of that portion.

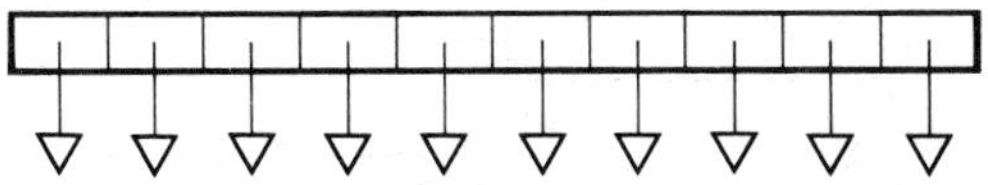

In the diagrams illustrating the examples that follow, concentrated forces will be marked in the usual way, but distributed forces will be indicated by shading along the beam and by a broken line marking the overall force, e.g. a uniform heavy beam of weight W will be shown as:

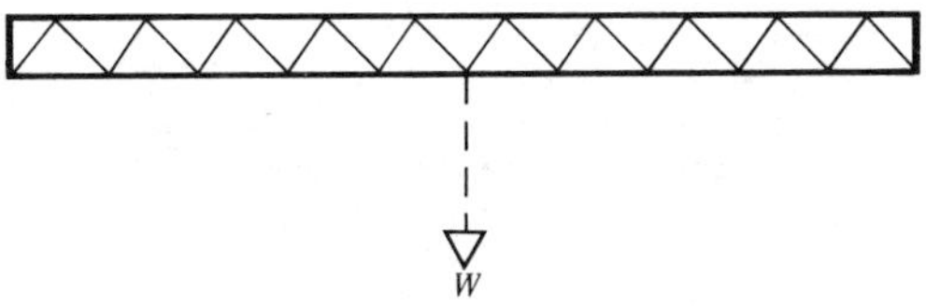

The internal stresses caused by concentrated forces differ somewhat from those caused by distributed forces so we shall consider these two cases separately.

CALCULATION OF SHEARING FORCE AND BENDING MOMENT

Case 1. A Light Beam with a Concentrated Load

Consider a light beam AB of length $2a$, resting horizontally on two end supports and carrying a concentrated load $2W$ at the midpoint, C.

From symmetry we see that each end support exerts an upward force W on the beam.

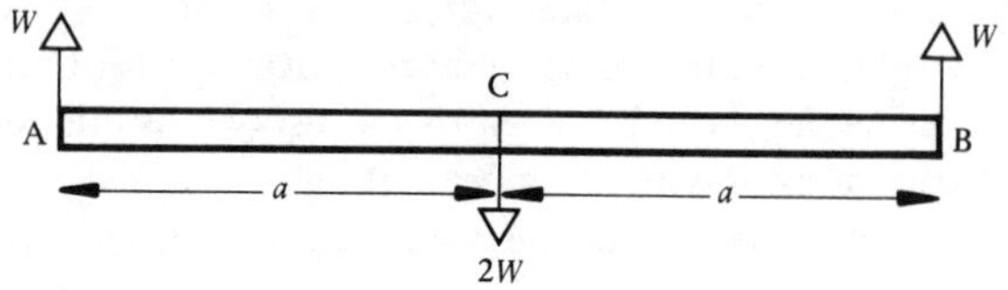

Let P be any point on the beam such that $AP = x$. Then, if S and $\mathfrak{M}$ are the shearing force and bending moment at P, we have:

(a) $0<x<a$ i.e. when P is between A and C.

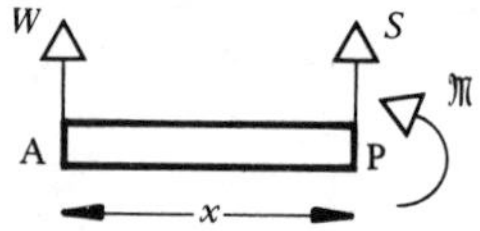

Resolving vertically and taking moments about a horizontal axis through P:

$$\uparrow \qquad W+S = 0$$

$$\text{P)} \qquad \mathfrak{M} - Wx = 0$$

Hence $S = -W$ and $\mathfrak{M} = Wx$.

(b) $a<x<2a$ i.e. when P is between C and B.

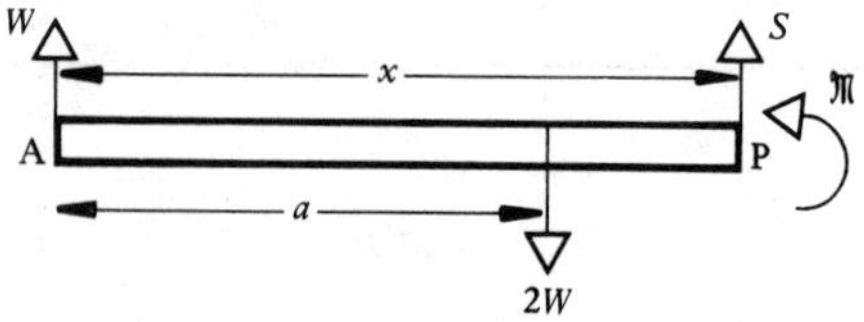

$$\uparrow \qquad W + S - 2W = 0$$

$$\text{P)} \qquad \mathfrak{M} + 2W(x-a) - Wx = 0$$

Hence $S = W$ and $\mathfrak{M} = W(2a-x)$.

From these results we see that the expressions for S and $\mathfrak{M}$ are *different* for AC and CB.

Between A and C $\begin{cases} S \text{ is constant} \quad (S = -W), \\ \mathfrak{M} \text{ increases uniformly with } x. \end{cases}$

Between C and B $\begin{cases} S \text{ is constant} \quad (S = +W), \\ \mathfrak{M} \text{ decreases uniformly with } x. \end{cases}$

At C there is a sudden change in the value of S from $-W$ to W. The magnitude of this sudden change (discontinuity) is equal to the magnitude of the concentrated load at C.

At A and B, $\mathfrak{M} = 0$

Both for $0 < x < a$ and $a < x < 2a$ (but *not* when $x = a$ where there is a discontinuity) $\dfrac{d\mathfrak{M}}{dx} = -S$.

At C there is a sudden change in the gradient of $\mathfrak{M}$.

These properties can be illustrated graphically as follows:

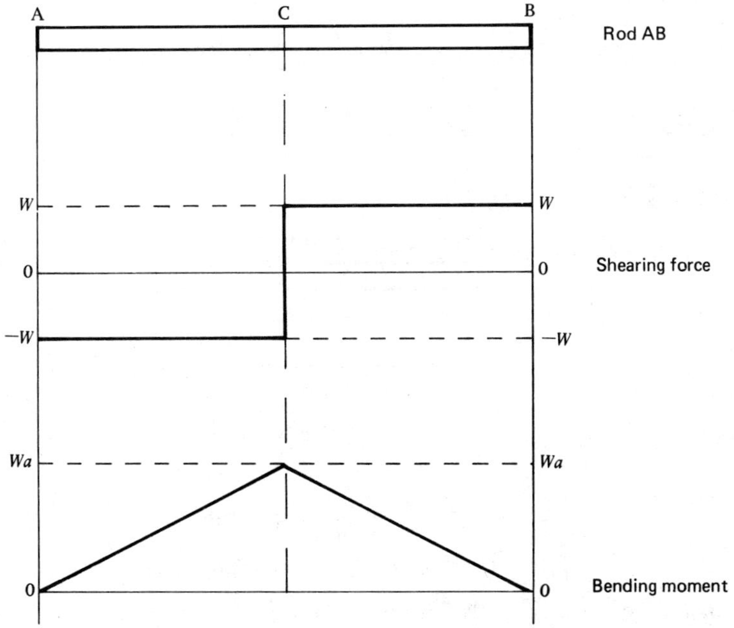

Case 2. A Uniform Heavy Beam

Again we shall consider a beam AB of length $2a$ resting horizontally on two end supports but this time the weight of the beam, $2W$, is distributed uniformly along the length of the beam. The weight per unit length of the beam is therefore $\frac{W}{a}$.

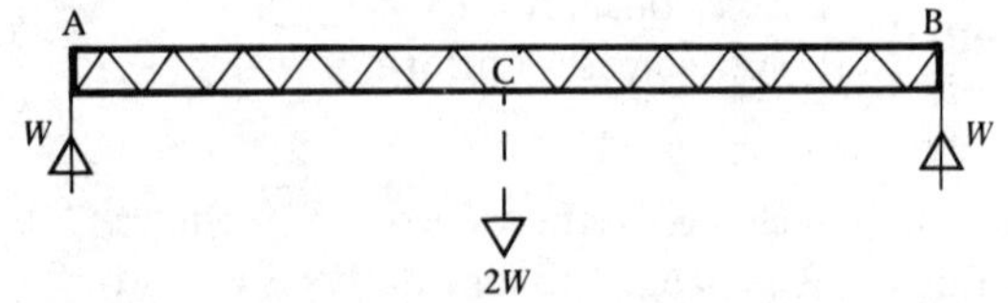

If P is any point such that $AP = x$ then the weight of the section AP is $\frac{Wx}{a}$.

Considering the equilibrium of the section AP, we have:

(a) $0 < x < a$

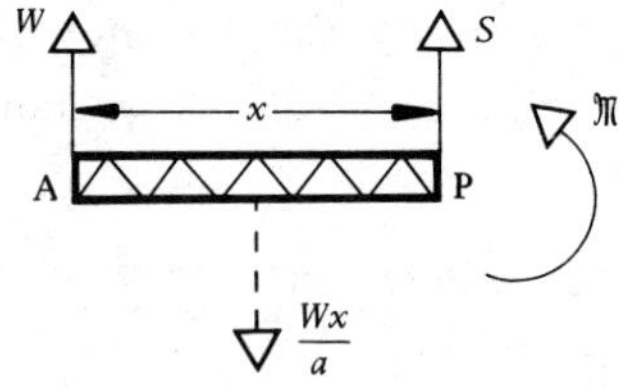

$$\uparrow \qquad W + S - \frac{Wx}{a} = 0$$

$$\text{P} \curvearrowright \qquad \mathfrak{M} + \frac{Wx}{a}\left(\frac{x}{2}\right) - Wx = 0$$

Hence $S = \frac{W}{a}(x - a)$ and $\mathfrak{M} = \frac{Wx}{2a}(2a - x)$.

(b) $a < x < 2a$

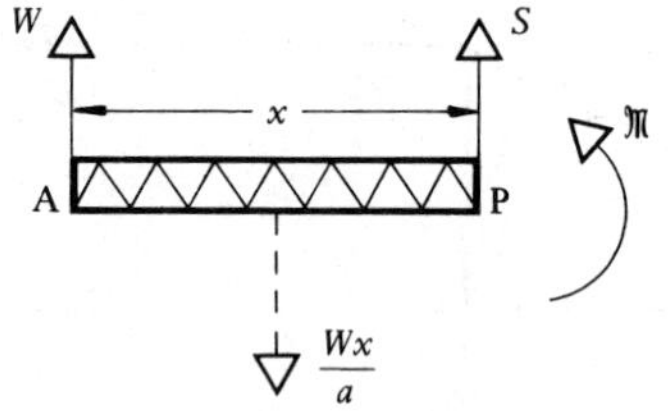

$$\uparrow \qquad W + S - \frac{Wx}{a} = 0$$

$$\text{P} \curvearrowright \qquad \mathfrak{M} + \frac{Wx}{a}\left(\frac{x}{2}\right) - Wx = 0$$

So again $S = \frac{W}{a}(x - a)$ and $\mathfrak{M} = \frac{Wx}{2a}(2a - x)$.

This time the characteristic properties are as follows.

The expressions for S and $\mathfrak{M}$ are *the same* for AC and CB so it was not necessary to consider two separate sections.

S increases uniformly with x throughout.
$\mathfrak{M}$ is a quadratic function of x throughout.

At A, $S = -W$ and $\mathfrak{M} = 0$.
At B, $S = W$ and $\mathfrak{M} = 0$.
At C, $S = 0$.

For all values of x, $\dfrac{d\mathfrak{M}}{dx} = -S$.

There is no discontinuity in the value of S, or the gradient of $\mathfrak{M}$.

The graphical illustrations of these properties are shown below.

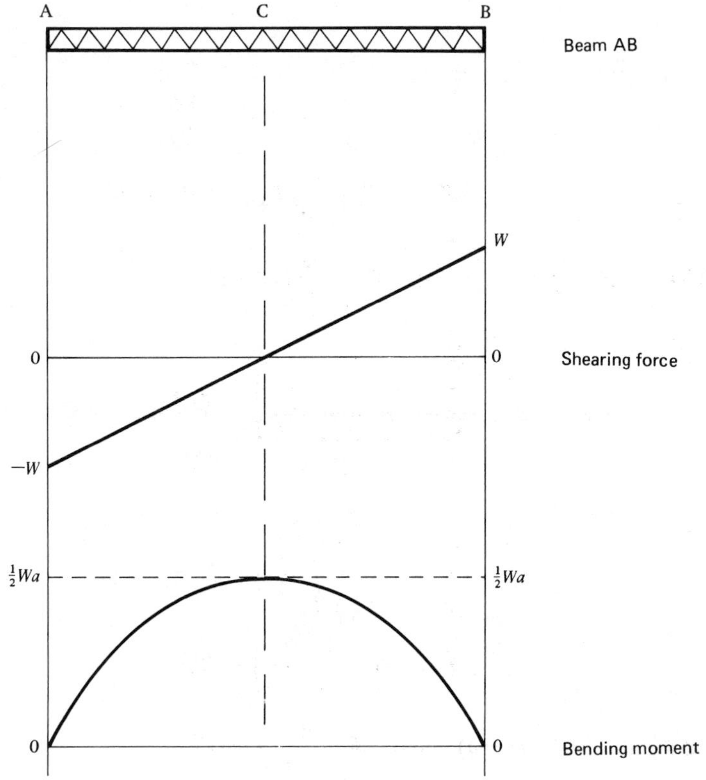

RELATIONSHIP BETWEEN BENDING MOMENT AND SHEARING FORCE

In the two particular cases we have so far studied, observation of the functions derived for $\mathcal{M}$ and S shows that $\frac{d\mathcal{M}}{dx} = -S$.

This, in fact, is a property of all uniform beams and it can be proved by considering a small section of a heavy beam whose weight per unit length is w. If, at a point P where $AP = x$, the shearing force and bending moment are S and $\mathcal{M}$, then at an adjacent point Q, where $AQ = x + \delta x$, the shearing force and bending moment are $S + \delta S$ and $\mathcal{M} + \delta\mathcal{M}$. Provided that no concentrated forces act between P and Q the forces which act on the section PQ are as follows:

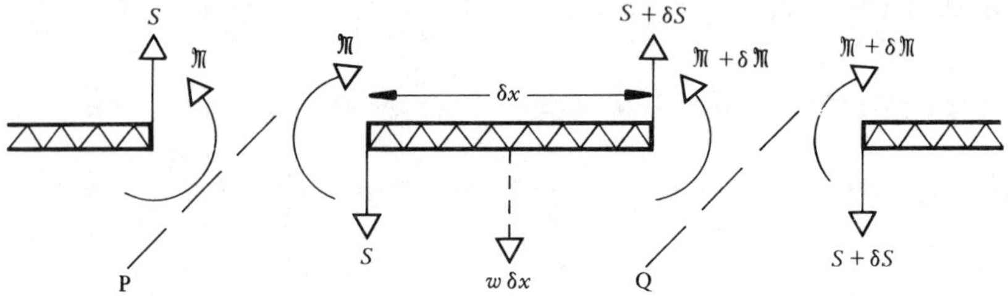

Taking moments about a horizontal axis through Q we have,

$$(\mathcal{M} + \delta\mathcal{M}) + S\delta x + w\delta x\left(\frac{\delta x}{2}\right) - \mathcal{M} = 0 \qquad [1]$$

i.e.
$$\delta\mathcal{M} = -S\,\delta x - \tfrac{1}{2}w(\delta x)^2$$

Hence
$$\frac{\delta\mathcal{M}}{\delta x} = -S - \tfrac{1}{2}w\delta x$$

As $\delta x \to 0$
$$\frac{d\mathcal{M}}{dx} = -S$$

If we now consider a light beam in the same way, equation [1] becomes

$$(\mathcal{M} + \delta\mathcal{M}) + S\,\delta x - \mathcal{M} = 0$$

and again
$$\frac{d\mathcal{M}}{dx} = -S$$

Thus, for all uniform beams, $\frac{d\mathcal{M}}{dx} = -S$

Note. If a different convention is used for the positive direction of $\mathcal{M}$ and S, this relationship may appear in the form $\frac{d}{dx} = +S$.

Now consider a section of a uniform beam bounded by $x = a$ and $x = b$.

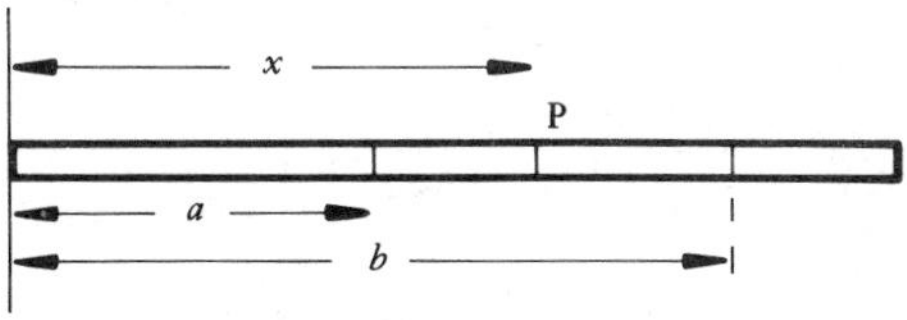

If no concentrated forces are applied within this section then at a general point P,

$$\frac{d\mathfrak{M}}{dx} = -S$$

Hence
$$\int_{x=a}^{x=b} d\mathfrak{M} = -\int_{x=a}^{x=b} S\,dx$$

i.e.
$$\mathfrak{M}_{x=a} - \mathfrak{M}_{x=b} = \int_a^b S\,dx$$

But $\int_a^b S\,dx$ is the area under the shearing force graph between $x = a$ and $x = b$.

> Thus the change in bending moment over a uniform section is equal in magnitude to the corresponding area under the shearing force diagram.

EXAMPLES 15a

1) A light girder ABCD $(AB = BC = CD = a)$ rests horizontally on supports at A and C and carries concentrated loads $2W$ and W at B and D respectively. Draw diagrams showing the variation in shearing force and bending moment along the girder and calculate the greatest bending moment.

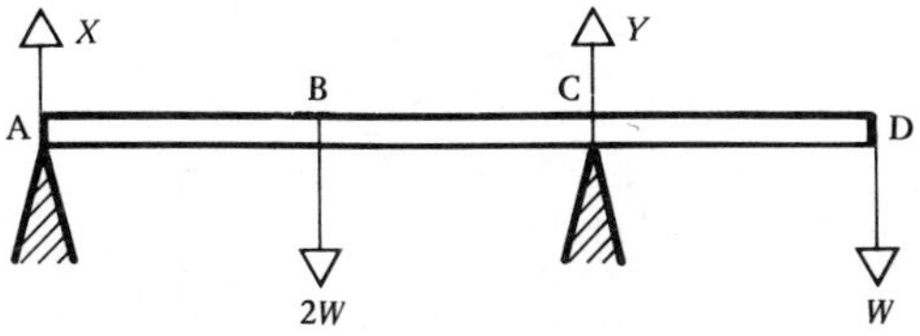

(This girder is not symmetrically supported or loaded so we must first calculate the values of the supporting forces X and Y.)

Considering the equilibrium of the girder under the action of the external forces only, and taking moments about horizontal axes through A and C we have:

$$\text{A)} \qquad 2Wa + 3Wa - 2Ya = 0$$

$$\text{C)} \qquad Wa + 2Xa - 2Wa = 0$$

Hence $\qquad X = \frac{1}{2}W \quad \text{and} \quad Y = \frac{5}{2}W$

Now consider the shearing force S and the bending moment $\mathfrak{M}$ at a point P on the beam, where $AP = x$.

(There are concentrated forces (and hence discontinuities) at B and C so it is necessary to consider separately the sections for which $0 < x < a$, $a < x < 2a$ and $2a < x < 3a$.)

(a) $0 < x < a$

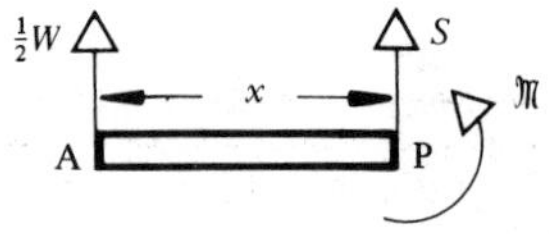

$$\uparrow \qquad S + \tfrac{1}{2}W = 0$$

$$\text{P)} \qquad \mathfrak{M} - \tfrac{1}{2}Wx = 0$$

Hence $S = -\frac{1}{2}W$ and $\mathfrak{M} = \frac{1}{2}Wx$.

(b) $a < x < 2a$

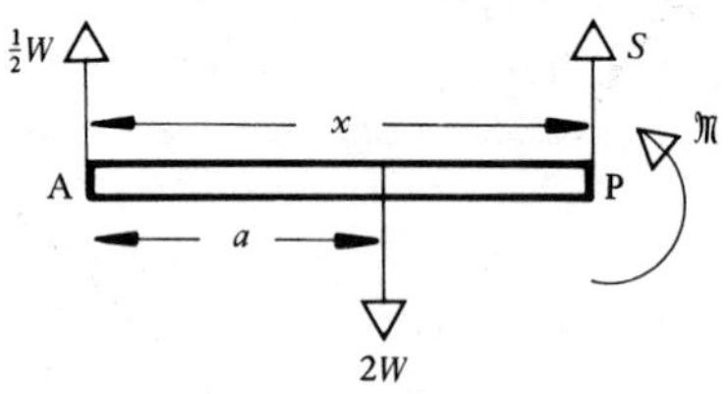

$$\uparrow \qquad S + \tfrac{1}{2}W - 2W = 0$$

$$\text{P)} \qquad \mathfrak{M} - \tfrac{1}{2}Wx + 2W(x-a) = 0$$

Hence $S = \frac{3}{2}W$ and $\mathfrak{M} = W(4a-3x)/2$.

(c) $2a < x < 3a$

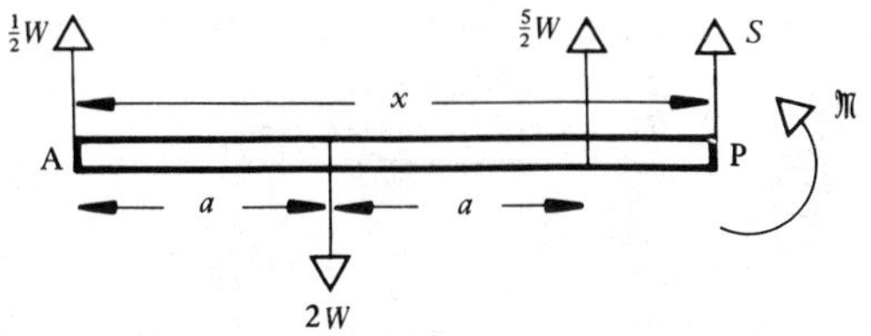

$$\uparrow \qquad \tfrac{1}{2}W + \tfrac{5}{2}W - 2W + S = 0$$

$$\overset{\frown}{A} \qquad \mathfrak{M} + Sx + 5Wa - 2Wa = 0$$

Hence $S = -W$ and $\mathfrak{M} = W(x - 3a)$.

The values of $\mathfrak{M}$ at A, B, C and D are 0, $\frac{1}{2}Wa$, $-Wa$ and 0 respectively. The shearing force and bending moment diagrams can now be drawn.

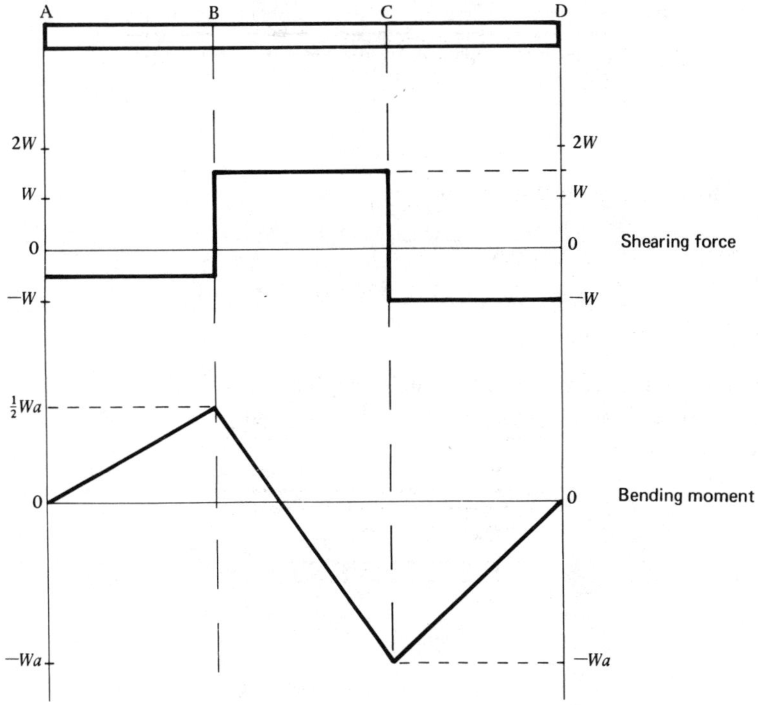

The greatest bending moment (i.e. the greatest *magnitude* of $\mathfrak{M}$ regardless of sign) occurs at C and is of magnitude Wa.

Note: The expressions for S and $\mathfrak{M}$ in the section CD of the beam can, if preferred, be derived as follows:

(c) $2a < x < 3a$

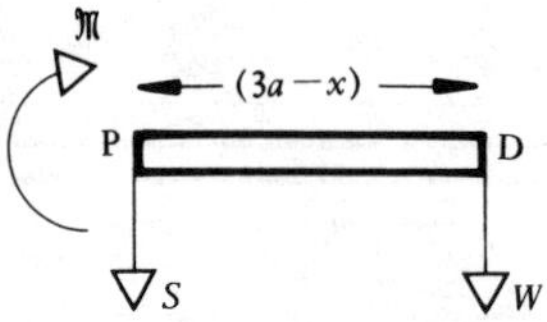

$$\downarrow \qquad S + W = 0$$

$$\text{P}\circlearrowright \qquad \mathfrak{M} + W(3a - x) = 0$$

Thus $S = -W$ and $\mathfrak{M} = W(x - 3a)$ as before.

2) A light beam AB of length $2a$ is supported horizontally at A and B. A load of weight $4W$ is uniformly distributed between A and C where $AC = a$. Find expressions for the bending moment and shearing force of the beam, indicating the point at which bending is most likely to occur.

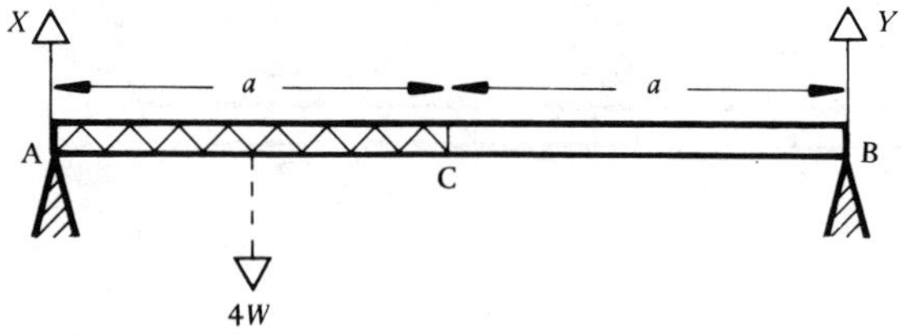

Considering external forces only:

$$\text{A}\circlearrowleft \qquad 2aY - 4W(\tfrac{1}{2}a) = 0$$

$$\text{B}\circlearrowright \qquad 2aX - 4W(\tfrac{3}{2}a) = 0$$

Hence $X = 3W$ and $Y = W$.

Now using $4W/a$ as the weight per unit length along AC and considering the shearing force S and bending moment $\mathfrak{M}$ at a point P where $AP = x$, we have,

(a) $0 < x < a$

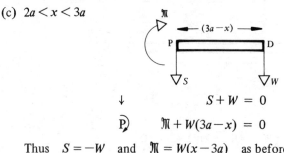

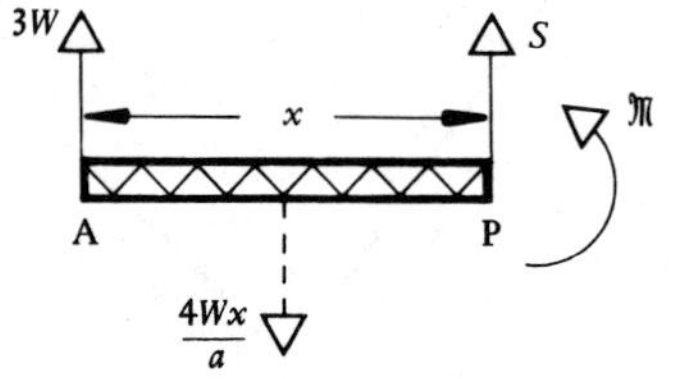

$$\uparrow \qquad 3W + S - 4Wx/a = 0$$

$$\text{P}\circlearrowleft \qquad \mathfrak{M} + \frac{4Wx}{a}(\tfrac{1}{2}x) - 3Wx = 0$$

Hence $S = W(4x - 3a)/a$ and $\mathfrak{M} = Wx(3a - 2x)/a$

When $x = 0$ $S = -3W$ and $\mathfrak{M} = 0$.

When $x = a$ $S = W$ and $\mathfrak{M} = Wa$.

(b) $a < x < 2a$

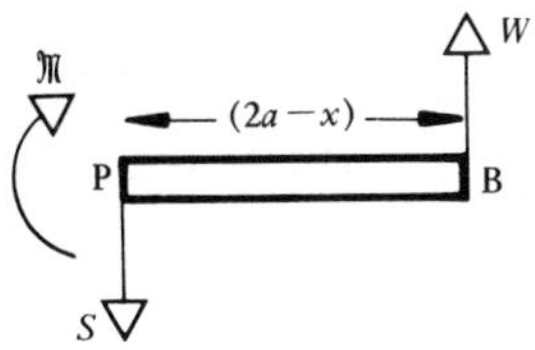

(using the right-hand section)

$\uparrow$ $W - S = 0$

P↻ $\mathfrak{M} - W(2a - x) = 0$

Hence $S = W$ and $\mathfrak{M} = W(2a - x)$

When $x = 2a$, $\mathfrak{M} = 0$.

Now for $0 < x < a$, $\mathfrak{M}$ is a quadratic function of x so its graph is a parabola whose vertex is at a point where $\dfrac{d\mathfrak{M}}{dx} = 0$.

But $\dfrac{d\mathfrak{M}}{dx} = -S$, so the vertex of the bending moment parabola is at the point where $x = \frac{3}{4}a$. For this value of x, $\mathfrak{M} = \frac{9}{8}Wa$ and this is the greatest value of $\mathfrak{M}$ between A and C.

For $a < x < 2a$, $\mathfrak{M}$ is a linear function of x with extreme values Wa and 0.

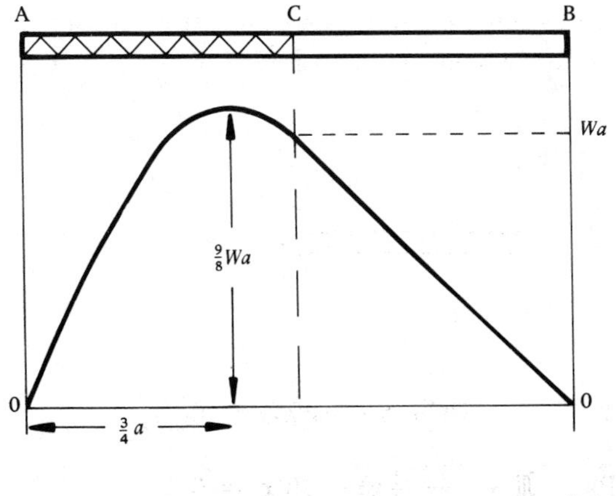

Bending moment

So the beam is most likely to bend at a point distant $\frac{3}{4}a$ from A.

3) A uniform heavy beam AB is held in a horizontal position by a clamp at the end A. Sketch the bending moment and shearing force diagrams.

A beam which is fixed by a clamp at one end is called a *cantilever.*
In order to support the beam, the clamp must exert on it both an upward force and a torque, whose magnitudes must be evaluated before the internal stress analysis can begin.
Let the weight and length of the rod be W and l respectively, and let F and C be the force and couple exerted by the clamp.

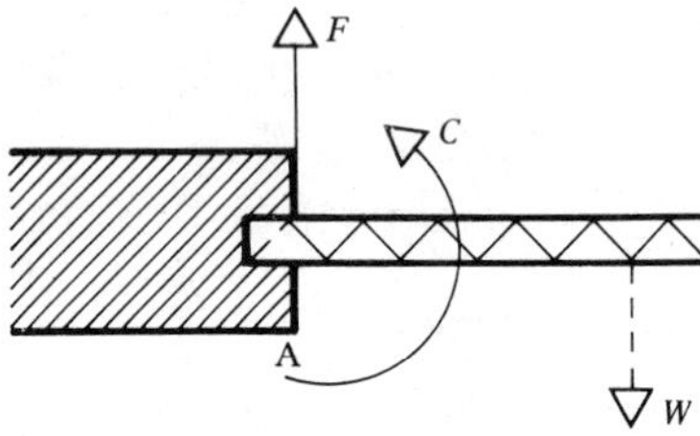

Considering the equilibrium of the beam

$$\uparrow \qquad F - W = 0$$

$$\text{A}\circlearrowright \qquad C - \tfrac{1}{2}Wl = 0$$

Now consider the shearing force and bending moment at a point P where AP $= x$.

As there is no concentrated force acting between A and B, the expressions for S and $\mathfrak{M}$ will not change at any intermediate point.

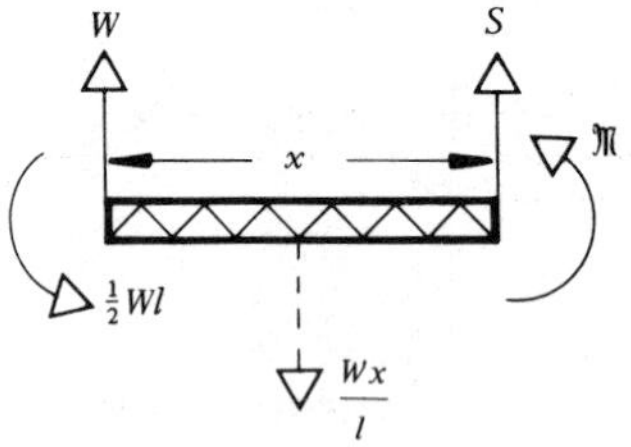

$$\uparrow \qquad S + W - Wx/l = 0$$

$$\text{P}\circlearrowright \qquad \tfrac{1}{2}Wl + \mathfrak{M} - Wx + W\frac{x^2}{2l} = 0$$

Hence $\qquad S = \dfrac{W}{l}(x-l) \quad$ and $\quad \mathfrak{M} = \dfrac{W}{2l}(2lx-x^2-l^2)$

i.e. $\qquad S = -\dfrac{W}{l}(l-x) \quad$ and $\quad \mathfrak{M} = \dfrac{W}{2l}(l-x)^2$

The shearing force and bending moment diagrams therefore become:

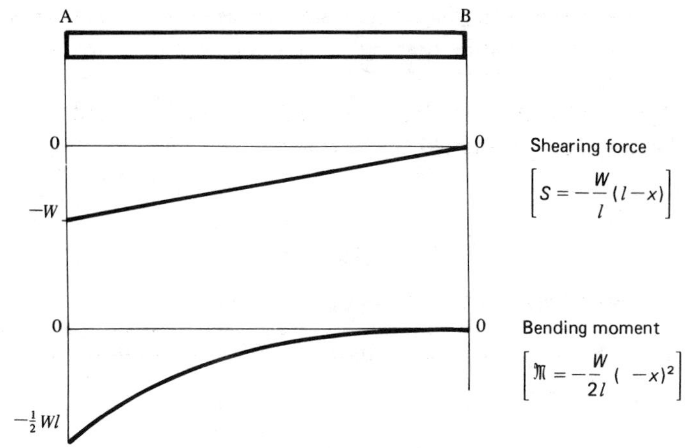

4) A rigid beam of length 10 m rests horizontally on two supports at points distant 2 m from each end, and carries a uniformly distributed load of 1000 N per m along its complete length. Draw shearing force and bending moment diagrams for this beam.
If the supports are to remain equidistant from the ends of the beam, calculate their position so that the magnitudes of the greatest positive and negative bending moments shall be the same.

(The reactions at the two supports are equal since the beam is mechanically symmetrical.)

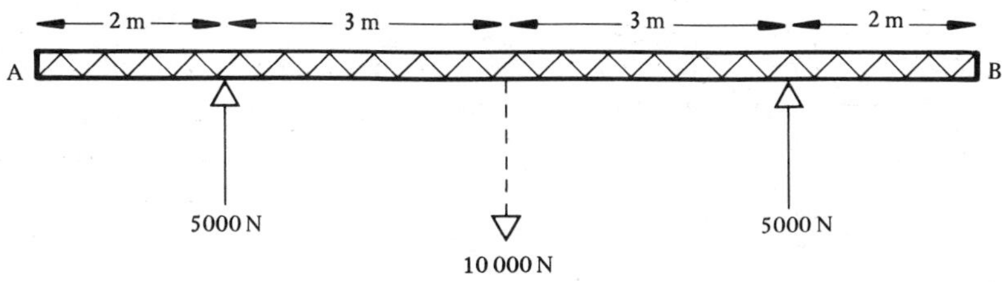

Consider the shearing force and bending moment at a point P distance x from A. Since the beam is symmetrical we need deal with only half of it. We therefore consider two separate ranges of values of x: $0 < x < 2$ and $2 < x < 5$.

(a) $0<x<2$

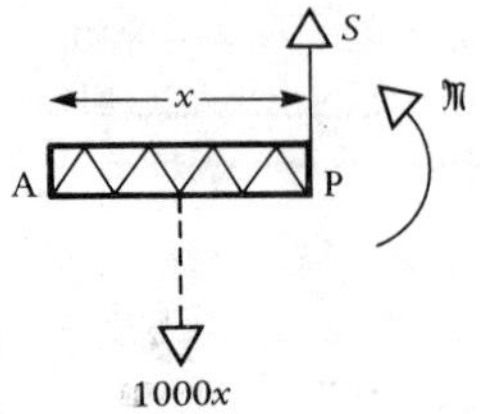

$$\uparrow \qquad S-1000x = 0$$

$$\overset{\frown}{P} \qquad \mathfrak{M} + 1000x(\tfrac{1}{2}x) = 0$$

Hence $\mathfrak{M} = -500x^2$ (a parabola with vertex where $x = 0$).

When $x = 0$ $S = 0$ and $\mathfrak{M} = 0$.

When $x = 2$ $S = 2000$ N and $\mathfrak{M} = -2000$ N m.

(b) $2<x<5$

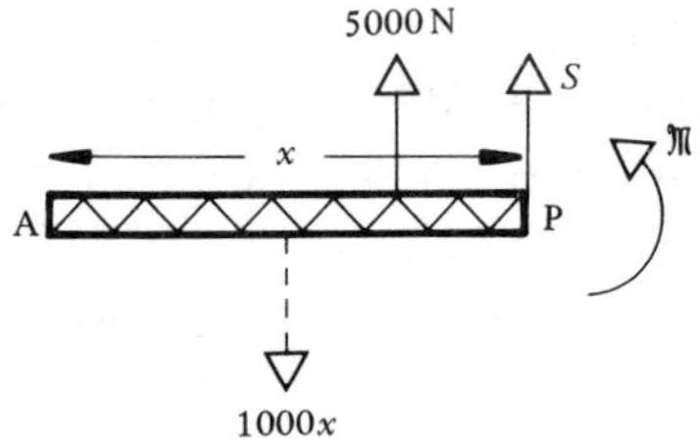

$$\uparrow \qquad S - 1000x + 5000 = 0$$

$$\overset{\frown}{P} \qquad \mathfrak{M} - 5000(x-2) + 500x^2 = 0$$

(The vertex of this bending moment parabola is at the point where $\dfrac{d\mathfrak{M}}{dx} = -S = 0$; i.e. when $x = 5$ m.)

When $x = 2$ $S = -3000$ N and $\mathfrak{M} = -2000$ N m.

When $x = 5$ $S = 0$ and $\mathfrak{M} = 2500$ N m.

So the bending moment and shearing force diagrams are:

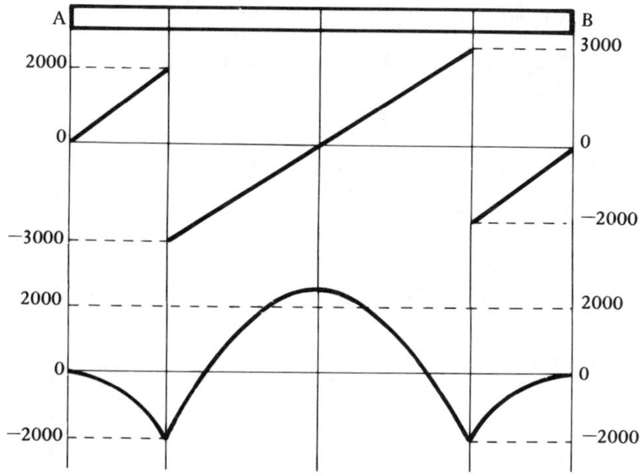

Now let the supports be moved to points distant d from each end so that the magnitudes of the greatest positive and the greatest negative bending moments are equal.

The expressions for S and $\mathfrak{M}$, and the corresponding graphs become:

$0 < x < d$ $\quad S = 1000x$

$\mathfrak{M} = -500x^2$

$d < x < 5$ $\quad S = 1000x - 5000$

$\mathfrak{M} = 5000(x-d) - 500x^2$

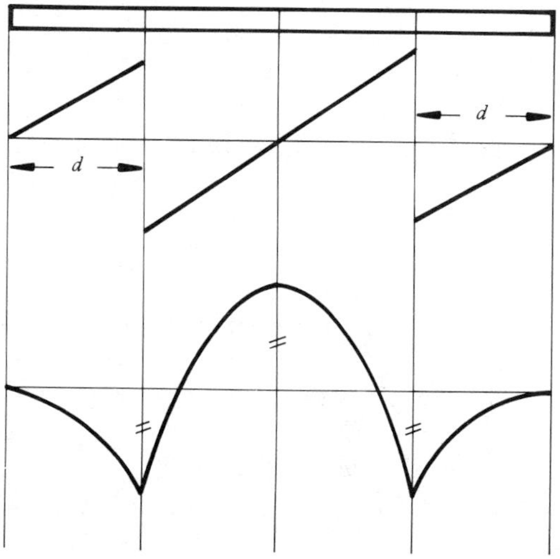

The greatest positive value of $\mathfrak{M}$ occurs at the midpoint.

When $x = 5$ $\quad \mathfrak{M} = 5000(5-d) - 12\,500$

$= 12\,500 - 5000d$

The greatest negative value of $\mathfrak{M}$ occurs at the supports.

When $x = d$ $\quad \mathfrak{M} = -500d^2$

If the magnitudes of these bending moments are equal,

$$12\,500 - 5000d = 500d^2$$

i.e. $$d^2 + 10d - 25 = 0$$

Hence $$d = 5(\sqrt{2}-1)$$

Thus the supports should be placed at a distance $5(\sqrt{2}-1)$ m from each end.

Note. It is not always necessary to find expressions for shearing force and/or bending moment for every section of a beam. Each section can be dealt with independently.

Suppose, for instance, that in Example 1 on p. 335 we were interested only in the bending moment at a point E, midway between B and C. The supporting force at A must first be found, then we can go directly to E as follows:

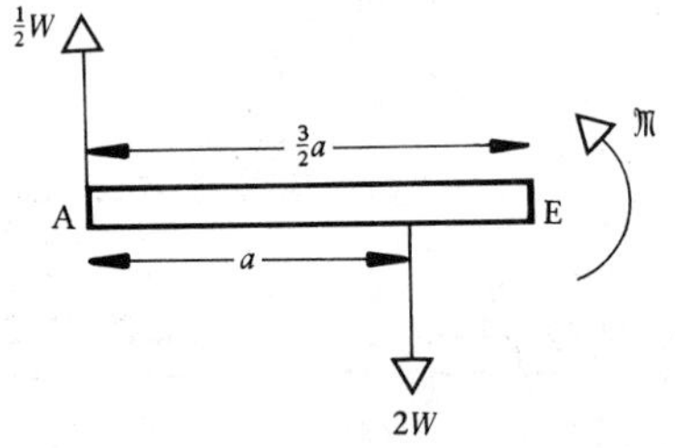

E↺ $$\mathfrak{M} = (\tfrac{1}{2}w)(\tfrac{3}{2}a) - (2w)(\tfrac{1}{2}a) = -\tfrac{1}{4}wa$$

SUMMARY

For a uniform horizontal beam under the action of coplanar vertical forces only:

1. At the junction P of two sections of the beam, each section exerts on the other a force S and a torque $\mathfrak{M}$.
   S is vertical and is called the *shearing force* at P.
   $\mathfrak{M}$ is in a vertical plane parallel to the axis of the beam and is called the *bending moment* at P.

2. The sign convention used in this book is: considering the left-hand section of the beam, S is positive upwards and $\mathfrak{M}$ is positive in the anticlockwise sense.

3. Over a uniform section of the beam:
   (a) if the beam is light, S is constant and $\mathfrak{M}$ is a linear function of x (the distance of P from the left-hand end of the beam);
   (b) if the beam is heavy, S is a linear function and $\mathfrak{M}$ a quadratic function of x.

4. At a point of application of a concentrated load there is a discontinuity in the value of S and in the gradient of $\mathfrak{M}$.

5. At any point other than the point of application of a concentrated load,
$$\frac{d\mathfrak{M}}{dx} = -S.$$

6. The change in the value of $\mathfrak{M}$ over a uniform section of the beam is equal in magnitude to the area under the corresponding section of the shearing force graph.

EXERCISE 15a

In each of the following examples calculate (where necessary) the forces and/or torque supporting the beam. Find expressions for the shearing force S and the bending moment $\mathfrak{M}$ at any point on the beam and illustrate your results graphically.

1)
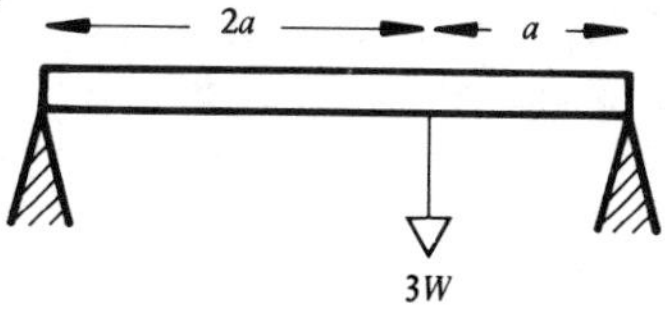

2)
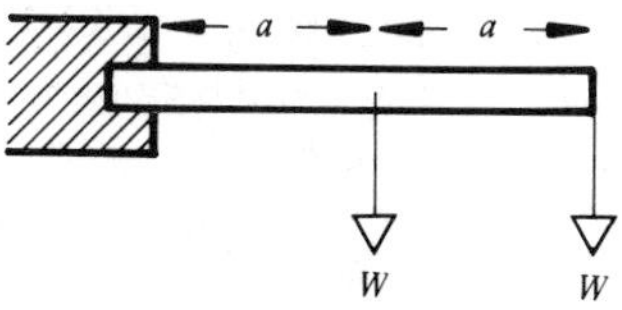

3)
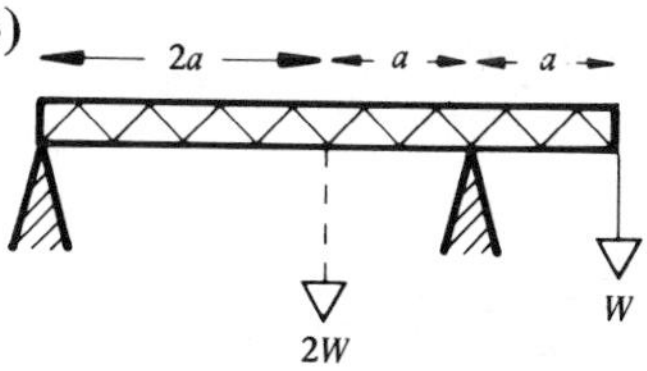

4)
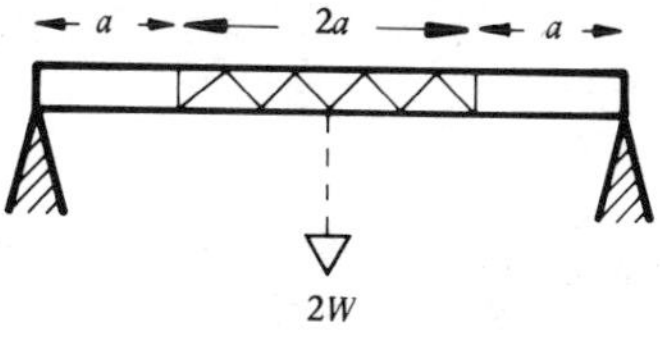

5)
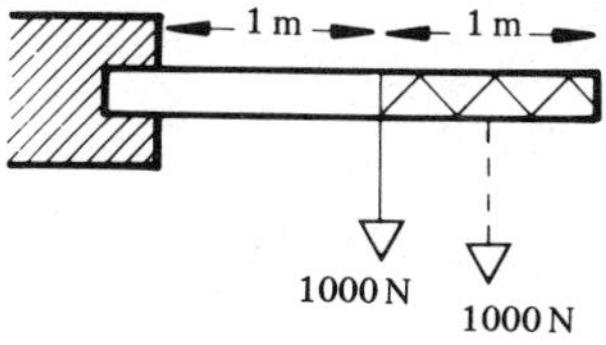

6)
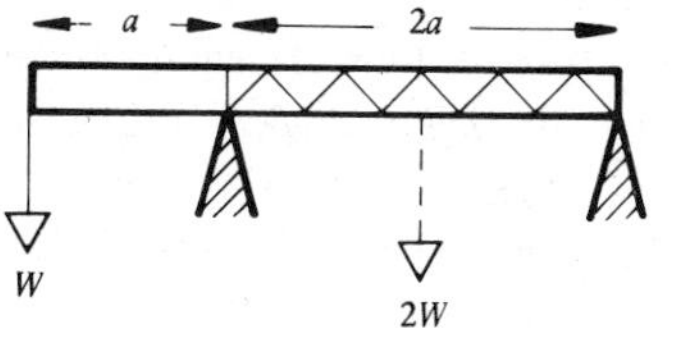

7)

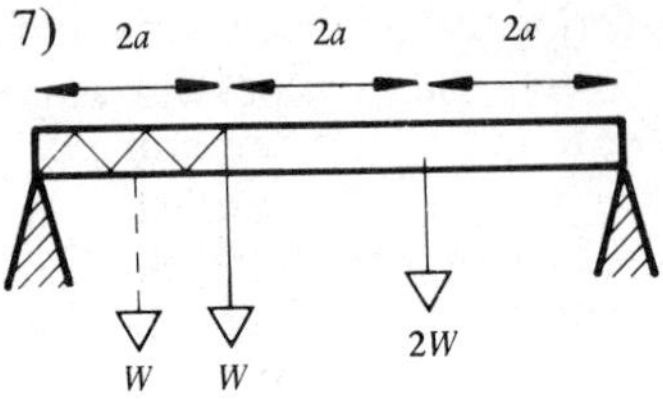

8)

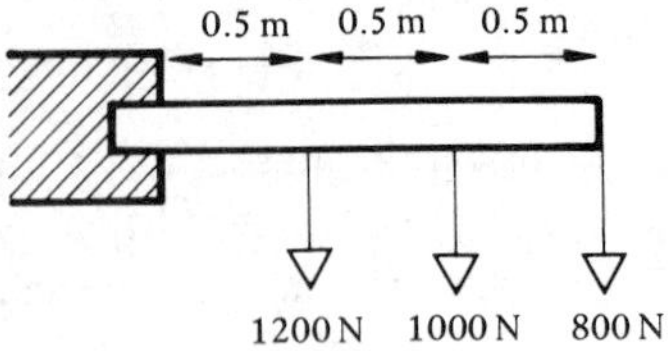

MISCELLANEOUS EXERCISE 15

1) A light beam AB, which is 20 m long, is supported at two points C and D which are 5 m and 15 m respectively from A. The beam carries uniformly distributed loads of 1 newton per metre along AC and DB together with a concentrated load of 20 N at the mid-point of the beam. Find the reactions at the supports C and D and sketch the shearing force and bending moment diagrams.
The concentrated load of 20 N is now removed and distributed uniformly along the section CD of the beam. Sketch the appropriate shearing force and bending moment diagrams for this case. Also find the maximum bending moment stating the positions at which it occurs. (AEB)

2) A light rigid beam ABCD, in which $AB = BC = CD = 4$ m, is simply supported in a horizontal position at B and D. Between A and B the beam carries a uniformly distributed load of 500 newtons per metre. Sketch the shearing force and bending moment diagrams when an additional load is placed on the beam
(a) in the form of an isolated load of 1000 N attached at C,
(b) in the form of a uniformly distributed load of 250 newtons per metre between C and D.
Compare the magnitudes and positions of the greatest bending moment to which the beam is subjected in the two cases. (AEB)

3) A straight rigid beam ACDB, 4 metres long, rests horizontally on two supports at A and D and carries a load of 20 newtons at C where $AC = DB = 1$ metre. Sketch the shearing force and bending moment diagrams when:
(a) the beam is of negligible weight,
(b) the beam weighs 10 newtons per metre, and the weight is uniformly distributed.
In each case, calculate the forces at the supports and find the maximum bending moment. (AEB)

4) On a light rigid beam ACDB of length $4l$ the points C and D are such that $AC = DB = l$. A weight $2W$ is attached at C and a uniformly distributed load of $4W$ is spread over the beam from C to B. The beam is held at rest in a horizontal position by means of two vertical strings attached at A and D. Draw the shearing force and bending moment diagrams for this loading of the beam and state the maximum bending moment to which the beam is subjected and where this occurs.
A weight is now attached at B. Find its magnitude if it is just sufficient to reduce the tension in one string to zero. Draw the shearing force and bending moment diagrams in this case. (AEB)

5) A uniform beam ACDB, of weight 800 newtons and length 5 metres, rests horizontally on two supports at A and D and carries a load of 400 newtons at C, where $AC = DB = 1$ metre. Sketch the shearing force and the bending moment diagrams and find the greatest bending moment.
When an additional load is attached at B the bending moment at C is zero. Calculate the magnitude of this additional load and the new bending moment at D. (AEB)

6) A uniform horizontal beam ABC is of length $4a$ and weight $4W$. It is supported in a horizontal position by vertical supports at B and C where $AB = a$. A particle of weight $2W$ is suspended from the midpoint of the rod. Show that the reactions at the supports at B and C are $4W$ and $2W$ in magnitude, respectively.
Find the shearing force and bending moment for each point of the rod and sketch graphs to display your results. (Accurate graphs are *not* required.)
Find the points of the rod which have greatest and least bending moment. (SUJB)

7) A uniform beam is supported in a horizontal position. Derive the relationships between bending moment, shearing force and the weight of the beam.
A uniform beam ABC, of weight 160 N and length 16 m, rests horizontally on two supports, one at the end A and the other at B which is 6 m from the other end C. A load of 40 N is attached at the end C. Sketch the shearing force and bending moment diagrams, showing the position and the magnitude of the maximum bending moment. (AEB)

8) A uniform rigid beam AB of length 10 m and of weight 500 N per m rests horizontally on two supports P and Q, where $AP = 1$ m and $QB = 3$ m. A weight of 800 N is suspended from the midpoint of AB. Draw the shearing force and bending moment diagrams, and find the distances from A of all sections of the beam where the bending moment is zero.

9) Prove that the change in bending moment between two points on a span is equal to the area of the shearing force diagram over that portion of the span.
A rigid beam of length 6 m is supported at two points 1 m from each end and carries a uniformly distributed load of 400 N per m over its whole length. Draw the shearing force and bending moment diagrams for this beam.
If the supports are to remain equidistant from the ends of the beam calculate their position so that the maximum positive and negative bending moments shall be the same.

10) A uniform beam AB of length $5a$ and weight w per unit length is supported in a horizontal position by vertical forces at C and D, where $AC = DB = a$, and a load of $3wa$ is suspended from the midpoint E of the beam. Copy and complete the following table which shows the shearing force S and the bending moment $\mathfrak{M}$ for different distances x from A.

Distance from A	S	$\mathfrak{M}$
$0 \leqslant x < a$	wx	
$a < x < 5a/2$		
$5a/2 < x < 4a$		$-w(x^2 - 2ax - 7a^2)/2$
$4a < x \leqslant 5a$		

Sketch the shearing force and bending moment diagrams separately, showing each stage clearly. (AEB)

11) A uniform beam AB, of weight 108 N, is 12 m long and rests horizontally on two supports at the ends A and B. The beam carries a load of 36 N at the point 3 m from A. Find the reactions at the supports and sketch the shearing force and bending moment diagrams.
A further load of 36 N is now placed on the beam at the point 9 m from A. Sketch the resulting shearing force and bending moment diagrams.
For each of the above cases, find the maximum bending moment and the position at which this maximum occurs. (AEB)

12) A rigid light horizontal beam ACDB of length 12 m is simply supported at C and D, where $AC = DB = 3$ m, and carries a uniformly distributed load of 250 N per m over its whole length AB. Draw the shearing force and bending moment diagrams.
An extra load added at E, where $CE = 5$ m, has the effect of making the support at D carry twice the load of that at C. Draw the shearing force and bending moment diagrams for this new loading and find the greatest values of the shearing force and the bending moment.

13) A light rigid beam ACDB, in which $AC = CD = DB = 3\ m$, is simply supported, in a horizontal position, at the end B and at C. Between A and C the beam carries a uniformly distributed load of $180\ N\ m^{-1}$. Draw the shearing force and bending moment diagrams for the beam in each of the following cases:

(a) when an additional point load of 400 N is attached at D,

(b) when this additional load of 400 N is spread uniformly from the midpoint of CD to the midpoint of DB.

14) A uniform bridge span, of length 200 m and weight $6 \times 10^6\ N$, is freely supported at its ends. Draw shearing force and bending moment diagrams. A uniform train, of length 50 m and weight $3 \times 10^6\ N$, runs on to the span and stops just on it. Draw, on the same figure as before, the new bending moment and shearing force diagrams. Calculate:

(a) the point at which the shearing force is unchanged,

(b) the ratio of the new bending moment to the old at the centre point of the span.

CHAPTER 16

DISCRETE RANDOM VARIABLES

The work in this chapter assumes knowledge of the principles of probability covered in Chapter 18 of *Mechanics and Probability*. We strongly recommend that the reader revises that chapter before studying this one.
A knowledge of elementary descriptive statistics, including the drawing of histograms, the calculation of the mean, median and mode, is also assumed.

DISCRETE AND CONTINUOUS VARIABLES

Any quantity that can take different values is a variable. For example
real numbers whose squares are less than 9,
the heights of blades of grass,
the suits of playing cards,
the numbers of people waiting at a bus stop,
are all variables.

From the third example we see that a variable does not necessarily have numerical values; the suit of a playing card is either a club, a diamond, a heart or a spade.

A discrete variable can take only exact, and therefore separate, numerical values, or assume distinct and separate attributes.
The second two of the examples above are discrete variables; in the first of these the variable can be only a club, a diamond, a heart or a spade; in the second of these examples the values of the variable are limited to $0, 1, 2, 3, \dots$

A continuous variable is not restricted to separate and distinct values; it can take any value within a specified range. In the first example, the variable can take any value in the interval $[-3, 3]$. In the second example, the height of a blade of grass can be any value greater than zero.

Notation

A variable is denoted by a capital letter, e.g. X, and the values that it can take by a small letter, e.g. x.

For example, if the variable X is 'the number of heads obtained when two coins are tossed',

then X can take the values $0, 1, 2,$ only
and we can define x by $x = 0, 1, 2$.

We can describe the variable more briefly as

$$X = x \quad \text{for} \quad x = 0, 1, 2$$

EXERCISE 16a

1) State whether the following are discrete or continuous variables
(a) the number of sixes that can be scored when a die is thrown ten times,
(b) the heights of trees,
(c) the numbers of leaves on trees,
(d) the speeds of cars passing a police radar control,
(e) the brightness of lights,
(f) the colours of snooker balls.

2) If X is the variable, 'the number of sixes that can be obtained when three dice are tossed', define x.

3) If X is the variable, 'the number at which a ball on a roulette wheel can settle', define x.

4) If X is the variable, 'the possible score when one die is thrown', define x.

A DISCRETE RANDOM VARIABLE

Consider the discrete variable X, 'the number of heads possible when two coins are tossed'.
X can take the value 0, 1 or 2, i.e. it is possible to get either no heads, or one head or two heads.

These events are mutually exclusive, i.e.

$$P(X=0 \cup X=1 \cup X=2) = P(X=0) + P(X=1) + P(X=2) \qquad [1]$$

We have listed all the possible events so one of them must happen,

i.e. $$P(X=0 \cup X=1 \cup X=2) = 1 \qquad [2]$$

When a variable satisfies these two conditions it is called a *random variable.*

Showing that the variable satisfies these conditions is simplified if equations [1] and [2] are combined to give the single condition

$$P(X=0) + P(X=1) + P(X=2) = 1$$

In general if X is a variable that can take values $x_1, x_2, \ldots, x_n$ and if $X=x_1$, $X=x_2$, ..., $X=x_n$ (a) are mutually exclusive events, (b) comprise *all* the possible events so that one of them must happen, then the equations resulting from these two conditions can be combined to give the single condition

$$P(X=x_1) + P(X=x_2) + \ldots + P(X=x_n) = 1$$

Hence

X is a discrete random variable $\Longleftrightarrow \sum_{i=1}^{n} P(X=x_i) = 1$

where $x_1, x_2, \ldots, x_n$ are the values that X can take.

Consider again the variable X, 'the number of heads possible when two coins are tossed'.

Now $X=x$ for $x=0, 1, 2,$

$$P(X=0) = P(0\text{H}, 2\text{T}) = (\tfrac{1}{2})^2 = \tfrac{1}{4}$$

$$P(X=1) = P(1\text{H}, 1\text{T}) = 2(\tfrac{1}{2})(\tfrac{1}{2}) = \tfrac{1}{2}$$

$$P(X=2) = P(2\text{H}, 0\text{T}) = (\tfrac{1}{2})^2 = \tfrac{1}{4}$$

The sum of these probabilities is 1, hence in this case X is a discrete random variable.

EXAMPLE 16b

The probability that a tennis player gets his serve in on any one attempt is 0.8. If X is the variable, 'the number of attempts required to get a serve in', prove that X is a random variable.

Either the first serve is good, in which case $X = 1$, or the first serve is out and the second serve is good, in which case $X = 2, \ldots$ and so on to infinity. Therefore X can take the values $1, 2, 3, \ldots$

$$\begin{aligned} P(X=1) &= 0.8 \\ P(X=2) &= (0.2)(0.8) \\ P(X=3) &= (0.2)^2(0.8) \end{aligned}$$

and so on

Therefore

$$\sum_{i=1}^{\infty} P(X=x_i) = 0.8+(0.2)(0.8)+(0.2)^2(0.8)+\ldots$$

$$= \sum_{n=0}^{\infty} (0.8)(0.2)^n$$

This is a G.P. where $a = 0.8$ and $r = 0.2$. The sum to infinity of a G.P. is given by $\dfrac{a}{1-r}$

Therefore $$\sum_{n=0}^{\infty} (0.8)(0.2)^n = \frac{0.8}{1-0.2} = 1$$

i.e. $$\sum_{i=1}^{\infty} P(X=x_i) = 1$$

Therefore X is a random variable.

EXERCISE 16b

In each of the following questions prove that X is a random variable.

1) X is the variable, 'the number of sixes scored when two dice are tossed'.

2) X is the variable, 'the number of heads obtained when three coins are tossed'.

3) X is the variable, 'the number of aces drawn when four cards are drawn (with replacement) from a pack of playing cards'.

4) A bag contains 5 red discs and 3 yellow discs. X is the variable, 'the number of red discs drawn when two discs are removed (without replacement) from the bag'.

5) X is the variable, 'the number of times a coin has to be tossed before a head is obtained'.

6) X is the variable, 'the number of times a die has to be thrown before a six is obtained'.

7) X is the variable, 'the score obtained when a die is tossed'.

8) X is the variable, 'the colour of a pen drawn at random from a box containing 1 red, 1 blue, 1 green and 1 black pen'.

9) X is the variable, 'the sum of the scores obtained when two ordinary dice are thrown'.

10) X is the variable, 'the sum of the scores obtained when two tetrahedral dice are thrown'. Assume that the faces of each die are numbered 1, 2, 3 and 4 and that the score is the face on which the die lands.

PROBABILITY DISTRIBUTIONS

Consider the random variable X, 'the number of heads possible when three coins are tossed'.
X can take the values 0, 1, 2, 3 and the probability that X takes each of these values can be calculated as follows

$$P(X=0) = P(0H, 3T) = (\tfrac{1}{2})^3 = \tfrac{1}{8}$$

$$P(X=1) = P(1H, 2T) = 3(\tfrac{1}{2})(\tfrac{1}{2})^2 = \tfrac{3}{8}$$

$$P(X=2) = P(2H, 1T) = 3(\tfrac{1}{2})^2(\tfrac{1}{2}) = \tfrac{3}{8}$$

$$P(X=3) = P(3H, 0T) = (\tfrac{1}{2})^3 = \tfrac{1}{8}$$

This is a complete list of the probabilities for X assuming each of its possible values and it is called the *probability distribution* of X.

The probability distribution of X is more conveniently displayed in a table as follows,

x	0	1	2	3
$P(X=x)$	$\frac{1}{8}$	$\frac{3}{8}$	$\frac{3}{8}$	$\frac{1}{8}$

and can be illustrated by a histogram.

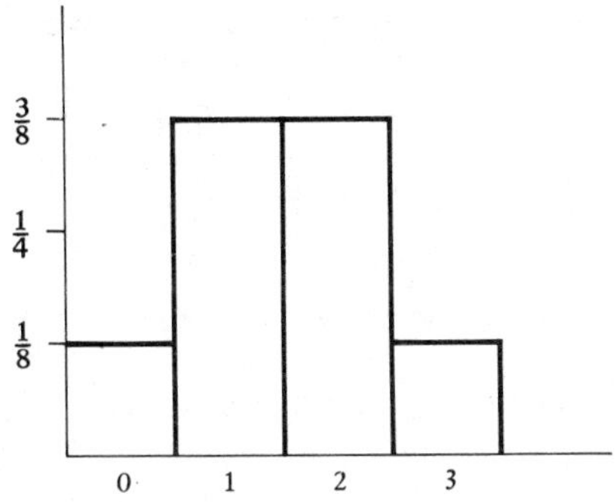

(A histogram is a bar chart where the *area* of each bar is proportional to the quantity represented by that bar.)

Probability Density Functions

In the example above, the probability that X takes each one of the values 0, 1, 2 and 3 was worked out individually. In this case, however, it is possible to find a function of x which gives each of these probabilities.
Consider $P(X=x)$ where

$$
\begin{aligned}
P(X=x) &= P(x \text{ heads}, 3-x \text{ tails}) \\
&= P(x \text{ heads}) \times P(3-x \text{ tails}) \times (\text{no. of combinations of } x \text{ heads and } 3-x \text{ tails}) \\
&= (\tfrac{1}{2})^x(\tfrac{1}{2})^{(3-x)}({}^xC_3) \\
&= {}^xC_3(\tfrac{1}{2})^3
\end{aligned}
$$

The expression is a function of x,
i.e. $P(X=x)=f(x)$ where $f(x)={}^xC_3(\frac{1}{2})^3$ for $x=0,1,2,3$.

Now $f(x)$ gives the rule for calculating $P(X=x)$ and it is called the *probability density function* (p.d.f.) of X.

For any discrete random variable, the function of x that gives the values of $P(X=x)$ is called the probability density function of x.

It is important to realise however, that it is not always possible to give one simple function that gives all the values of $P(X=x)$. Consider, for example, two tetrahedral dice with their faces numbered 1, 2, 3 and 4.

If X is the random variable, 'the sum of the scores when the two dice are thrown' then X can take the values 2, 3, 4, 5, 6, 7 and 8.

The probability distribution of X is given in the following table.

x	2	3	4	5	6	7	8
$P(X=x)$	$\frac{1}{16}$	$\frac{2}{16}$	$\frac{3}{16}$	$\frac{4}{16}$	$\frac{3}{16}$	$\frac{2}{16}$	$\frac{1}{16}$

Now for $x = 2, 3, 4$ and 5, $P(X=x)$ is given by $\dfrac{x-1}{16}$

but for $x = 6, 7$ and 8, $P(X=x)$ is given by $\dfrac{9-x}{16}$

Therefore the p.d.f. of X is a two-part function given by $f(x)$ where

$$f(x) = \begin{cases} \dfrac{x-1}{16} & \text{for } x = 2, 3, 4, 5 \\ \dfrac{9-x}{16} & \text{for } x = 6, 7, 8 \end{cases}$$

EXAMPLES 16c

1) The p.d.f. of a random variable X is given by $f(x) = \frac{1}{6}x$ for $x = 1, 2, 3$.

(a) Write out the probability distribution of X and illustrate it with a histogram.

(b) Find $P(X<3)$.

(a) $P(X=x) = \frac{1}{6}x$ for $x = 1, 2, 3$ therefore the probability distribution of X is

x	1	2	3
$P(X=x)$	$\frac{1}{6}$	$\frac{1}{3}$	$\frac{1}{2}$

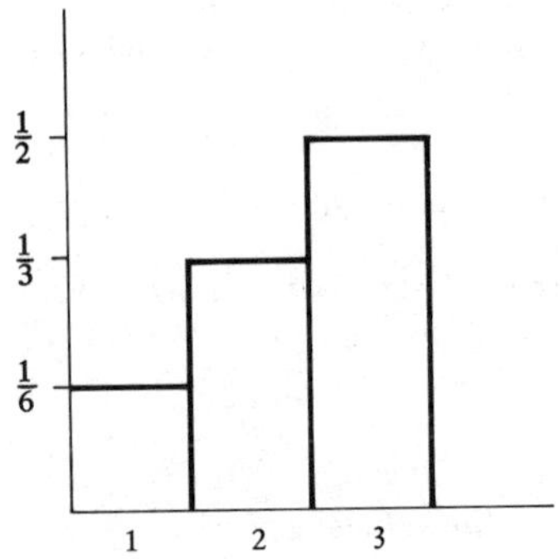

(b)
$$P(X<3) = 1-P(X=3) = 1-\tfrac{1}{2} = \tfrac{1}{2}$$

Note that $\Sigma P(X=x) = \frac{1}{6}+\frac{1}{3}+\frac{1}{2} = 1$, confirming that X is a random variable.

Note also that the area of the histogram is 1.

2) The probability density function of a random variable X is given by $f(x) = c(x+2)$ for $x = 0, 1, 2, 3$ and 4.
Find the value of c and hence find $P(X>1)$.

As X is a random variable, $\sum_{i=0}^{4} P(X=x) = 1$ where $P(X=x) = c(x+2)$

Now
$$P(X=0) = f(0) = 2c$$
$$P(X=1) = f(1) = 3c$$
$$P(X=2) = f(2) = 4c$$
$$P(X=3) = f(3) = 5c$$
$$P(X=4) = f(4) = 6c$$

Therefore $\sum_{i=0}^{4} P(X=x) = 20c$

$$\Rightarrow \quad 20c = 1$$
$$\Rightarrow \quad c = \tfrac{1}{20}$$

Hence
$$f(x) = \frac{x+2}{20}$$

Now
$$P(X>1) = 1 - P(X=0 \cup X=1) = 1 - (\tfrac{2}{20} + \tfrac{3}{20}) = \tfrac{3}{4}$$

3) The probability density function of a random variable X is given by $f(x) = (\frac{1}{5})(\frac{4}{5})^x$ for $x = 0, 1, 2, 3, \ldots$ Find $P(X \geqslant 3)$.

$$P(X \geqslant 3) = 1 - P(X=0 \cup X=1 \cup X=2)$$

Now
$$P(X=0) = f(0) = \tfrac{1}{5} = 0.2$$
$$P(X=1) = f(1) = (\tfrac{1}{5})(\tfrac{4}{5}) = 0.16$$
$$P(X=2) = f(2) = (\tfrac{1}{5})(\tfrac{4}{5})^2 = 0.128$$

Therefore
$$P(X \geqslant 3) = 1-(0.2+0.16+0.128) = 0.512$$

EXERCISE 16c

1) X is the random variable, 'the number of heads obtained when two coins are tossed'. Write out the probability distribution of X, and illustrate with a histogram.

2) The probability density function of a random variable X is given by $f(x) = \frac{1}{14}x^2$ for $x = 1, 2, 3$. Write out the probability distribution of X and find $P(X < 3)$.

3) The probability density function of a random variable Y is given by $f(y) = \frac{1}{5}$ for $y = 0, 1, 2, 3, 4$. Write out the probability distribution of Y and find $P(Y > 1)$.

4) The probability density function of a random variable X is $f(x) = c(4-x)$ for $x = 1, 2, 3$. Find the value of c.

5) The probability density function of a random variable Y is $f(y) = y^2(k-y)$ for $y = 1, 2, 3$. Find the value of k.

6) X is the random variable, 'the sum of the scores obtained when two dice are tossed'. Write out the probability distribution table for X, and hence find the probability density function of X as a function of x.

7) X is the random variable, 'the score obtained when an ordinary die is tossed'. Find the probability density function of X, and illustrate the probability distribution with a histogram.

8) Four straws each have one end painted black. These straws together with sixteen unpainted, but otherwise identical, straws are placed in a can so that the black ends cannot be seen. One straw is removed from the can and then replaced. If the straw has a black end, a prize is given. One person plays this game until he wins a prize.
If X is the random variable, 'the number of attempts needed to win a prize', find $P(X = 1)$, $P(X = 2)$ and $P(X = 3)$.
Find, as a function of x, the probability density function of X.

THE BINOMIAL PROBABILITY DISTRIBUTION

Consider the discrete random variable X, 'the number of sixes obtained when a die is tossed 4 times'.
X can take the values $0, 1, 2, 3, 4$ and we can construct a probability distribution table for X as follows.

x	0	1	2	3	4
$P(X=x)$	$(\frac{5}{6})^4$	${}^4C_1(\frac{1}{6})(\frac{5}{6})^3$	${}^4C_2(\frac{1}{6})^2(\frac{5}{6})^2$	${}^4C_3(\frac{1}{6})^3(\frac{5}{6})$	$(\frac{1}{6})^4$

The entries in the bottom line of this table are the terms in the expansion of the binomial expression $(\frac{5}{6}+\frac{1}{6})^4$. We say that X has a binomial probability distribution.

The properties in this example that give X a binomial probability distribution are

(a) X is a discrete variable,

(b) the event (in this case, obtaining a six) either happens (a success) or it does not happen,

(c) on every attempt or trial (one throw of the die) the probability, p, that the event happens is constant ($p = \frac{1}{6}$ in this case),

(d) there is a fixed number, n, of trials ($n = 4$ in this case).

These properties can be condensed as follows.

X has a binomial distribution if

> X is the number of successes in n independent trials where the probability of success in any one trial is p (and the probability of failure is $1-p$).

Any random variable with these properties also has a binomial probability distribution as we can see if we look at the general case.

Consider an event A which can either happen or not happen (i.e. A and $\bar{A}$ are mutually exclusive), such that, at any one trial, the probability that A happens is constant and equal to p, say. If n independent trials are made then the probability that A happens x times out of the n trials is given by

$$\text{P}(A \text{ happens } x \text{ times}) \times \text{P}(\bar{A} \text{ happens } n-x \text{ times}) \times (\text{the no. of combinations of } xA\text{s and } (n-x)\bar{A}\text{s})$$

$$= (p)^x(1-p)^{n-x}(^n\text{C}_x)$$

$$= {}^n\text{C}_x p^x q^{n-x} \qquad \text{where} \quad q = 1-p$$

This is the $(x+1)$th term in the expansion of $(q+p)^n$, therefore if X is the discrete random variable, 'the number of times that A happens in n independent trials', then X has a binomial probability distribution and the p.d.f. of X is given by

$$\text{f}(x) = {}^n\text{C}_x p^x q^{n-x} \quad \text{for} \quad x = 0, 1, \ldots n \quad \text{where} \quad q = 1-p$$

The values of n and p vary with different experiments; they are called the *parameters* of the distribution.

Note that $\sum_{x=0}^{n} \text{P}(X=x) = (p+q)^n = 1$ since $q = 1-p$.

This confirms that X is a *random* variable.

Summing up:

If X has a binomial probability distribution with parameters n and p, then the p.d.f. of X is given by

$$\begin{aligned} P(X = x) &= (x+1)\text{th term in the expansion of } (q+p)^n \\ &= {}^nC_x p^x q^{n-x} \quad \text{for } x = 0, 1, \ldots n \quad \text{where } q = 1-p \end{aligned}$$

EXAMPLES 16d

1) Determine which of the following variables have a binomial probability distribution and, for those that do, give the values of the parameters:

(a) the number of faulty components in a batch of ten chosen at random from a production line for which it is known that 10% of the components will be faulty,

(b) the sum of the scores when two dice are tossed six times,

(c) the time of a particular journey on six consecutive days.

(a) This does follow a binomial probability distribution because
the number of faulty components is a discrete variable,
the component is either faulty or not faulty,
for any one component $P(\text{faulty}) = 0.1$,
there are 10 trials (components picked at random).
In this case $p = 0.1$ and $n = 10$.

(b) The condition that there is one event which can either happen or not happen, does not apply in this case. Therefore this is not a binomial distribution.

(c) This is not a binomial distribution because the variable (time) is continuous.

2) It is known that 20% of a population of laboratory mice are infertile.

(a) If 5 mice are chosen at random, find the probability that exactly 2 of them are infertile.

(b) Find the least number of mice that need to be chosen for the probability of choosing at least one infertile mouse to be greater than 0.99.

(a) Let X be the variable, 'number of infertile mice in a batch of 5 mice'. X has a binomial probability distribution, with $p = 0.2$ and $n = 5$.

$$\begin{aligned} P(X = 2) &= {}^5C_2(0.2)^2(0.8)^3 \\ &= 0.205 \quad \text{correct to 3 s.f.} \end{aligned}$$

Therefore the probability of getting exactly two infertile mice is 0.205 to 3 s.f.

(b) Let X be 'the number of infertile mice in a batch of n mice'.
X has a binomial probability distribution with $p = 0.2$ and n unknown.

$$P(X \geqslant 1) > 0.99$$

$$P(X \geqslant 1) = 1 - P(X = 0)$$

$$= 1 - (0.8)^n \qquad \text{(i.e. } 1 - {}^nC_0(0.2)^0(0.8)^n\text{)}$$

i.e. $$1 - (0.8)^n > 0.99$$

$$(0.8)^n < 0.01$$

$$n \log 0.8 < \log 0.01$$

$$n > \frac{\log 0.01}{\log 0.8} \qquad \text{(log 0.8 is negative)}$$

$$n > 20.6 \qquad \text{(to 3 s.f.)}$$

n is an integer, so $n \geqslant 21$.

Therefore at least 21 mice must be picked to ensure that the probability that at least one is infertile is greater than 0.99.

3) A cubical die has its faces numbered 2, 3, 4, 5, 6, 6. Find the probability that, in five tosses of the die, more than three sixes are obtained.

On any one throw $P(\text{a six}) = \frac{1}{3}$.

If X is the random variable, 'the number of sixes obtained in 5 tosses', then X has a binomial probability distribution.

Therefore $$P(X > 3) = P(X = 4 \text{ or } X = 5)$$

$$= {}^5C_4(\tfrac{1}{3})^4(\tfrac{2}{3}) + (\tfrac{1}{3})^5$$

$$= 0.0453 \quad \text{to 3 s.f.}$$

Hence the probability of throwing more than three sixes is 0.0453 to 3 s.f.

EXERCISE 16d

1) A random variable X has a binomial probability distribution with parameters n and p. For each of the following values of n and p, construct the probability distribution and illustrate it with a sketch histogram.

(a) $n = 4, \quad p = \frac{1}{5}$
(b) $n = 4, \quad p = \frac{1}{3}$
(c) $n = 4, \quad p = \frac{1}{2}$
(d) $n = 4, \quad p = \frac{4}{5}$
(e) $n = 3, \quad p = \frac{2}{5}$
(f) $n = 5, \quad p = \frac{2}{5}$
(g) $n = 7, \quad p = \frac{2}{5}$
(h) $n = 10, \quad p = \frac{2}{5}$

2) A random variable X has a binomial probability distribution with $n = 5$ and $p = 0.6$. Find $P(X = 0)$ and $P(X < 5)$.

3) A random variable X has a binomial probability distribution with $n = 8$ and $p = 0.3$. Find $P(X = 4)$ and $P(X > 6)$.

4) Determine which of the following variables have a binomial probability distribution.
(a) The number of bad peaches in a batch of five taken at random from a lorry load of peaches of which 10% are bad.
(b) The possible score obtained when a die is thrown.
(c) The number of heads possible when a coin is tossed six times.
(d) The heights of blades of grass picked at random from a grazing field.
(e) The number of cars passing through a police control in one hour when the average is 20 cars per hour.
(f) The number of substandard items in a box of ten items when the items are taken at random from a container that contains 5% substandard items.

5) A coin is tossed five times. Find the probability of getting exactly 3 heads.

6) 30% of a large batch of loaves are substandard. If ten loaves are selected at random find the probability that at least one loaf is substandard.

7) Assuming that the sex of any child born to a particular couple is equally likely to be male or female, find the probability that if they have 4 children they are all girls.

8) A coin is biased so that the probability of getting a head is twice that of getting a tail. If the coin is tossed 10 times, find the probability of getting exactly 5 tails.

9) It is known that 15% of the apples produced by an archard are blemished. The apples are packed in boxes of ten. Find the probability that a box contains more than two blemished apples.

10) In a large population of aphids, 10% are type A and the remainder are type B. If five aphids are selected at random, what is the probability that they are all of type B?
What is the smallest number of aphids that must be selected from the population if the probability that this sample contains at least one type A aphid is to be greater than 0.9?

11) A drawer contains a large number of socks of various colours, 50% of them being white.
What is the least number of socks that have to be taken out of the box to ensure that the probability of taking at least 2 white socks is greater than 0.9?

THE DISCRETE RECTANGULAR PROBABILITY DISTRIBUTION

We looked at the binomial probability distribution in some detail because it arises in many practical situations. We now investigate two other probability distributions that arise quite often.

Consider the discrete random variable X, 'the score possible when an ordinary die is thrown'.
Now $X = x$ for $x = 1, 2, 3, 4, 5, 6$
and $P(X = x) = \frac{1}{6}$ for all these values of x because the probability of obtaining any one score is the same as that of obtaining any other score.

This histogram illustrating this distribution has a distinctive rectangular shape:

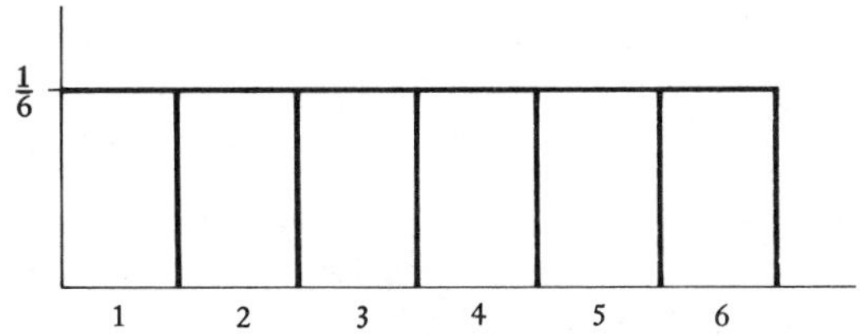

Any probability distribution in which the discrete random variable is equally likely to take any of the possible values, will give a rectangular histogram. Such a distribution is called a rectangular distribution.

> If X is a discrete random variable which is *equally likely* to take any of its possible values $x_1, x_2 \dots x_n$ then X has a rectangular distribution and the p.d.f. of X is given by
>
> $$f(x) = \frac{1}{n} \quad \text{for} \quad x = x_1, x_2, \dots x_n$$

THE GEOMETRIC PROBABILITY DISTRIBUTION

Consider the discrete random variable X, 'the number of throws needed to score a double at darts', given that the probability of scoring a double at any one throw is 0.2.

If it takes x throws to get a double then the first $x - 1$ throws result in failure (i.e. not getting a double).

Therefore $P(X = x) = (0.8)^{x-1}(0.2)$ for $x = 1, 2, 3, \dots$

Now $(0.8)^{x-1}(0.2)$ is the xth term of a G.P. with $a = 0.2$ and $r = 0.8$, so we say that X has a geometric probability distribution.

In any experiment where the probability of success is constant, and equal to p, say,
and where X is the random variable, 'the number of times the experiment is repeated to achieve a success',
then the p.d.f. of X is given by

$$\begin{aligned} f(x) &= P(X=x) \\ &= P(\text{1st } x-1 \text{ trials resulting in failure and the } x\text{th trial resulting in success}) \\ &= (1-p)^{x-1}(p) \end{aligned}$$

i.e. $$f(x) = p(1-p)^{x-1} \quad \text{for} \quad x = 1, 2, 3, \ldots$$

This is the xth term of a G.P. with first term p and common ratio $1-p$. Therefore X has a geometric probability distribution.

EXERCISE 16e

1) A fair coin is tossed and X is the random variable
(a) the number of times the coin has to be tossed to get a head,
(b) the result of tossing the coin once,
(c) the number of heads possible in three tosses.
In each case name the probability distribution.

2) An ordinary pack of playing cards is cut and X is the random variable
(a) the face value of the card showing (take a jack as 11, a queen as 12, a king as 13 and an ace as 1),
(b) the number of aces possible in 10 cuts of the pack,
(c) the number of times the pack has to be cut to get an ace.
In each case name the probability distribution and find $P(X=3)$.

THE EXPECTED VALUE OF X (MEAN VALUE OF X)

In *Mechanics and Probability* we defined expectation as the average result from a series of experiments. We also defined the expected number of times that an event A occurs in N experiments as $NP(A)$ assuming that each experiment is independent.

If X is a random variable then the expected value of X is denoted by $E(X)$ and means the value, on average, that X takes.

For example, if X is 'the number of heads possible when 4 coins are tossed', then $E(X)$ is the expected number of heads from one toss of the 4 coins. It is fairly obvious that the average outcome of this experiment is 2 heads, i.e. $E(X) = 2$.

We can confirm this by finding the total number of heads expected when the 4 coins are tossed N times, say, and then calculating the average number of heads expected per toss.

The probability distribution of X is given by

x	0	1	2	3	4
$P(X=x)$	$\frac{1}{16}$	$\frac{4}{16}$	$\frac{6}{16}$	$\frac{4}{16}$	$\frac{1}{16}$

In N tosses we expect

$$\text{no heads } \frac{N}{16} \text{ times, } \quad 1 \text{ head } \frac{4N}{16} \text{ times, } \quad \ldots, \quad 4 \text{ heads } \frac{N}{16} \text{ times}$$

therefore the total number of heads expected in N tosses is

$$(0)\left(\frac{N}{16}\right)+(1)\left(\frac{4N}{16}\right)+(2)\left(\frac{6N}{16}\right)+(3)\left(\frac{4N}{16}\right)+(4)\left(\frac{N}{16}\right) = 2N$$

Therefore the average number of heads per toss is $\dfrac{2N}{N} = 2$

Note that the result is independent of N.

It is also worth noting that, in this case, the value of $E(X)$ is one of the values that X can take, but this is not always the case. Consider the random variable, Y 'the number of heads possible when three fair coins are tossed'. Common sense tells us that $E(Y) = 1.5$, i.e. we would get an average of $1\frac{1}{2}$ heads per toss.

The histograms for the two examples considered illustrate that, in both cases,
the distribution is symmetric,
the value of x at the line of symmetry is the expected value.

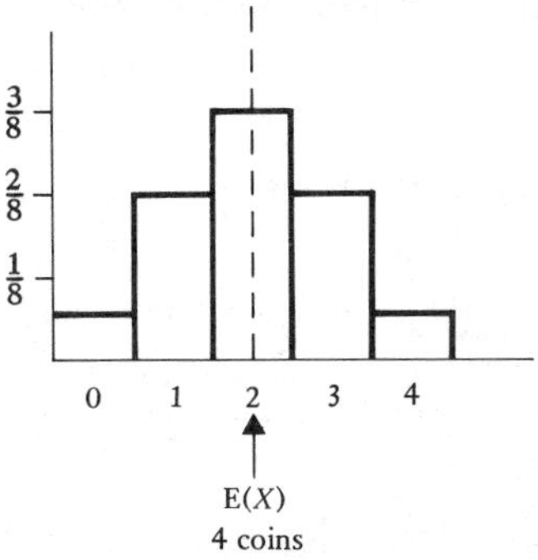

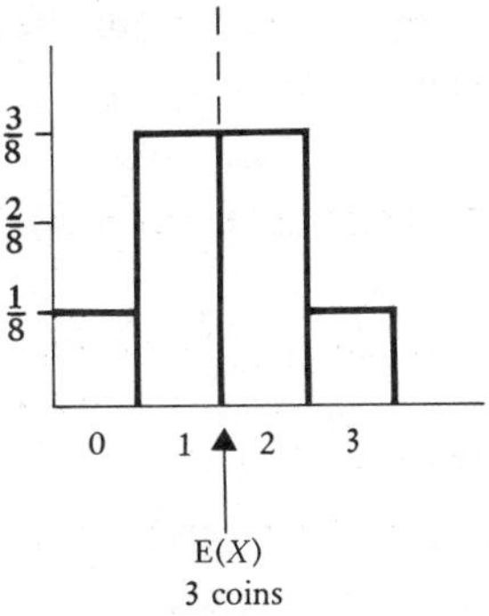

For any symmetric probability distribution, $E(X)$ is the value of x about which the distribution is symmetric.

(It is not necessary to draw the histogram to find this value of x because symmetry is obvious from the probability distribution.)

When the distribution is not symmetric, the value of $E(X)$ can be found by the method that we used to find the expected number of heads when four coins are tossed.

If $E(X)$ has a p.d.f. given by $f(x)$ for $x = x_1, x_2, \dots, x_n$ then in N experiments we expect

$X = x_1$ to occur $Nf(x_1)$ times giving $Nx_1f(x_1)$

$X = x_2$ to occur $Nf(x_2)$ times giving $Nx_2f(x_2)$

and so on,

giving a total value of $Nx_1f(x_1) + Nx_2f(x_2) + \dots + Nx_nf(x_n)$

$$\text{Therefore} \qquad E(X) = \frac{N[x_1f(x_1) + x_2f(x_2) + \dots + x_nf(x_n)]}{N}$$

$$= x_1f(x_1) + x_2f(x_2) + \dots + x_nf(x_n)$$

i.e. if X is a discrete random variable whose p.d.f. is given by $f(x)$ for $x = x_1, x_2, \dots x_n$ then

$$E(X) = \Sigma x f(x)$$

Note that the expected value of X is sometimes called the *mean value* of X and can be denoted by μ, i.e. $\mu = E(X)$.

For a given p.d.f., the calculation of $E(X)$ involves the summation of a number series. When only a few terms are involved, their sum can be found by simple addition but for a larger number of terms, the techniques for summing number series may be needed.

EXAMPLE 16f

The probability density function of a discrete random variable X is given by $f(x) = \frac{1}{20}(6 - x)$ for $x = 0, 1, 2, 3, 4, 5$. Find the mean value of X.

Using $E(X) = \Sigma x f(x)$ gives

$$E(X) = \sum_{x=0}^{5} \frac{x}{20}(6 - x)$$

$$= \tfrac{3}{10} \sum_{x=0}^{5} x - \tfrac{1}{20} \sum_{x=0}^{5} x^2$$

Using $\sum_{r=1}^{n} r = \frac{n}{2}(n+1)$ and $\sum_{r=1}^{n} r^2 = \frac{n}{6}(n+1)(2n+1)$ gives

$$\begin{aligned} E(X) &= (\tfrac{3}{10})(\tfrac{5}{2})(6)-(\tfrac{1}{20})(\tfrac{5}{6})(6)(11) \\ &= 1.75 \end{aligned}$$

Therefore the mean value of X is 1.75.

THE MEAN OF THE BINOMIAL DISTRIBUTION

If X has a binomial probability distribution with parameters n and p, then X is 'the number of times an event A can happen in n independent trials'.

In one trial, $P(A) = p$
Therefore in n trials we expect A to happen np times, i.e. $E(X) = np$.

This is a quotable result, i.e.

if X has a binomial probability distribution with parameters n and p then

$$E(X) = np$$

For example, if X is 'the number of bad apples possible in a sample of 10 apples taken from a large store of apples' when it is known that 12% of the apples in the store are bad,
then X has a binomial distribution where $n = 10$ and $p = \frac{3}{25}$

Therefore $E(X) = np = 1.2$
i.e. the expected number of bad apples in the sample is 1.2.

EXERCISE 16f

1) The probability distribution of a random variable X is

(a)

x	0	1	2	3	4
$P(X = x)$	$\frac{1}{2}$	$\frac{1}{3}$	$\frac{1}{12}$	$\frac{1}{24}$	$\frac{1}{24}$

(b)

x	0	1	2	3	4
$P(X = x)$	$\frac{1}{12}$	$\frac{1}{4}$	$\frac{1}{3}$	$\frac{1}{4}$	$\frac{1}{12}$

(c)

x	0	1	2	3	4	5
$P(X=x)$	0.09	0.1	0.21	0.4	0.15	0.05

(d)

x	0	1	2	3	4	5
$P(X=x)$	0.05	0.15	0.3	0.3	0.15	0.05

Find $E(X)$ in each case.

2) The probability density function of a discrete random variable X is given by $f(x) = \frac{1}{6}(3-x)$ for $x = 0, 1, 2$. Find the mean value of X.

3) The probability density function of a random variable Y is $f(y)$ where

$$f(y) = \frac{1}{14}y^2 \quad \text{for} \quad y = 1, 2, 3$$

Find $E(Y)$.

4) It is known that 8% of the items from the output of a certain production line are substandard. Ten items are selected at random from this production line. Find the expected number of substandard items in the sample.

5) A box contains a large number of ball bearings of identical shape, 20% of which are nylon and the rest are metal. Fifteen ball bearings are removed at random from the box. Find the expected number of metal ball bearings.

6) An ordinary die is thrown and X is the random variable, 'the score thrown'. Find $E(X)$.

7) A tennis player has a probability of 0.8 of getting a serve in on any one attempt. If X is the random variable, 'the number of serves needed to get one good serve', find the mean value of X.

FUNCTIONS OF A RANDOM VARIABLE

We have looked at the mean of a probability distribution and in the next section we consider the spread of a distribution about its mean value. However before we can do that we need to give a meaning to 'a function of X'.

Suppose that X is 'the possible score where an ordinary die is thrown', then, for example, X^2 would be 'the square of the possible score'.

Now X takes the values $1, 2, 3, 4, 5, 6$ whereas X^2 takes the values $1, 4, 9, 16, 25, 36$.

However the probability that X^2 should take any one of its possible values is the *same* as the probability that X should take the corresponding value,

i.e. $$P(X^2 = x^2) = P(X = x)$$

and this is true of *any* function of X.

If we now consider the expected value of X^2 then we have

$$\begin{aligned} E(X^2) &= \Sigma x^2 P(X^2 = x^2) \\ &= \Sigma x^2 P(X = x) \end{aligned}$$

This result is true in general.

If $g(X)$ is any function of X
then the expected value of $g(X)$ is given by

$$E[g(X)] = \Sigma g(x) P(X = x) \quad \text{for} \quad x = x_1, \ldots, x_n$$

or $$E[g(X)] = \Sigma g(x) f(x) \quad \text{where} \quad f(x) \text{ is the p.d.f. of } X.$$

Properties of the Operator E

Using the definition given above we can show that E has the following properties.

1. If a is a constant then

$$\begin{aligned} E(a) &= \Sigma a\, f(x) \\ &= a \Sigma f(x) \end{aligned}$$

But $\Sigma f(x) = 1$ since $f(x)$ is the p.d.f. of a random variable.

Hence $$E(a) = a$$

2. $$\begin{aligned} E(aX) &= \Sigma ax\, f(x) \\ &= a \Sigma x f(x) \\ &= a E(X) \end{aligned}$$

i.e. $$E(aX) = a E(X)$$

3. If g and h are two functions then

$$\begin{aligned} E[g(X)+h(X)] &= \Sigma\,[g(x)+h(x)]\,f(x) \\ &= \Sigma\,g(x)\,f(x)+\Sigma\,h(x)\,f(x) \\ &= E\,[g(X)]\,+E\,[h(X)] \end{aligned}$$

i.e.

$$E[g(X)+h(X)] = E[g(X)]+E(h(X)]$$

EXAMPLE 16g

The random variable X has the following probability distribution.

x	0	1	2	3	4
$P(X=x)$	0.2	0.3	0.35	0.1	0.05

Find (a) $E(X)$ (b) $E(X^2)$ (c) $E(X^2+2X)$

(a)
$$\begin{aligned} E(X) &= \sum_{x=0}^{4} x\,P(X=x) \\ &= (0)(0.2)+(1)(0.3)+(2)(0.35)+(3)(0.1)+(4)(0.05) \\ &= 1.5 \end{aligned}$$

(b)
$$\begin{aligned} E(X^2) &= \sum_{x=0}^{4} x^2 P(X=x) \\ &= (0)(0.2)+(1)(0.3)+(4)(0.35)+(9)(0.1)+(16)(0.05) \\ &= 3.4 \end{aligned}$$

(c)
$$\begin{aligned} E(X^2+2X) &= E(X^2)+E(2X) \\ &= E(X^2)+2E(X) \\ &= 3.4+(2)(1.5) \\ &= 6.4 \end{aligned}$$

EXERCISE 16g

1) The random variable X has the probability distribution tabulated below. Find (a) $E(X)$ (b) $E(X^2)$ (c) $E(3X^2-5)$

x	1	2	3	4	5
$P(X=x)$	0.1	0.3	0.3	0.2	0.1

2) The random variable X has a p.d.f. given by $f(x) = \frac{8}{5}x(1-x)$ for $x = \frac{1}{4}, \frac{1}{2}, \frac{3}{4}$.
Make a probability distribution table and use it to find
(a) $E(X)$ (b) $E(X^2)$ (c) $E(2X^2-3X)$

VARIANCE

It is useful to have a measure of the dispersion of a distribution about its mean value. One way of doing this is to find the average value of the deviations from the mean, i.e. to find the average of $X-\mu$. The disadvantage of this is that, as μ is somewhere in the middle of a distribution, some values of $X-\mu$ will be negative and some will be positive so they will tend to cancel each other out. Finding the squares of the deviations from the mean, eliminates this problem and the average values of these squares is called the variance of the distribution,
i.e. $E[(X-\mu)^2]$ is called the *variance* of X.

The variance of X is denoted by $\text{Var}(X)$, hence

$$\text{Var}(X) = E[(X-\mu)^2] \quad \text{where} \quad \mu = E(X)$$

The square root of the variance is called the *standard deviation* (or, sometimes, the root mean square deviation).

The standard deviation of a probability distribution is denoted by σ, so

$$\sigma = \sqrt{[\text{Var}(X)]} \quad \text{or} \quad \sigma^2 = \text{Var}(X)$$

The variance can be found directly from the definition given above, but the calculation is easier if we use the properties of E to write the formula in another form:

$$\begin{aligned} \text{Var}(X) &= E[(X-\mu)^2] \\ &= E[X^2-2\mu X+\mu^2] \\ &= E(X^2)-2\mu E(X)+\mu^2 \\ &= E(X^2)-\mu^2 \qquad (E(X) = \mu) \end{aligned}$$

i.e.

$$\text{Var}(X) = E(X^2)-\mu^2 \quad \text{where} \quad \mu = E(X)$$

Note that $\mu^2 = [E(X)]^2$ and $[E(X)]^2$ is written $E^2(X)$ so that the formula may also be written $\text{Var}(X) = E(X^2) - E^2(X)$.

EXAMPLES 16h

1) The discrete random variable X has a probability distribution given by

(a)

x	1	2	3	4	5
$P(X=x)$	$\frac{1}{10}$	$\frac{3}{20}$	$\frac{1}{2}$	$\frac{3}{20}$	$\frac{1}{10}$

(b) $P(X=x)=\frac{1}{5}, \quad x=1,2,3,4,5.$

Find the variance in each case.

(a) (When a probability distribution is given in a table, adding another row for the values of x^2 can help with the calculation of $E(X^2)$.)

x	1	2	3	4	5
$P(X=x)$	$\frac{1}{10}$	$\frac{3}{20}$	$\frac{1}{2}$	$\frac{3}{20}$	$\frac{1}{10}$
x^2	1	4	9	16	25

From symmetry $E(X)=3$

$$\begin{aligned} E(X^2) &= \sum_{x=1}^{5} x^2 P(X=x) \\ &= (1)(\tfrac{1}{10})+(4)(\tfrac{3}{20})+(9)(\tfrac{1}{2})+(16)(\tfrac{3}{20})+(25)(\tfrac{1}{10}) \\ &= 10.1 \\ \text{Var}(X) &= E(X^2)-E^2(X) \\ &= 10.1-9 = 1.1 \end{aligned}$$

(b) $E(X) = 3$ from symmetry

$$\begin{aligned} E(X^2) &= \sum_{x=1}^{5} x^2(\tfrac{1}{5}) = \tfrac{1}{5}\sum_{x=1}^{5} x^2 = \tfrac{1}{5}(55) = 11 \\ \text{Var}(X) &= E(X^2)-E^2(X) \\ &= 11-9 = 2 \end{aligned}$$

Note that X takes the same values in both distributions, but in (a) the further away from the mean we are, the smaller is $P(X=x)$; whereas in (b) $P(X=x)$ is the same for all x, so we would expect the variance in (a) to be smaller than the variance in (b).

The Variance of the Binomial Distribution

If X has a binomial probability distribution with parameters n and p, then $\mathrm{P}(X=x)={}^n\mathrm{C}_x p^x q^{n-x}$ where $q=1-p$, and $\mathrm{E}(X)=np$.

$$\mathrm{E}(X^2) = \sum_{x=0}^{n} (x^2)\,{}^n\mathrm{C}_x p^x q^{n-x}$$

Now

$$(x^2)\,{}^n\mathrm{C}_x p^x q^{n-x} = (x^2)\left(\frac{n!}{(n-x)!\,x!}\right)p^x q^{n-x}$$

$$= np\left[(x)\,\frac{(n-1)!}{(n-x)!\,(x-1)!}\,p^{x-1}\,q^{n-x}\right]$$

$$= np\left[(1+x-1)\,\frac{(n-1)!}{(n-x)!\,(x-1)!}\,p^{x-1}\,q^{n-x}\right]$$

$$= np\left[\frac{(n-1)!}{(n-x)!\,(x-1)!}\,p^{x-1}\,q^{n-x}+\frac{(n-1)!}{(n-x)!\,(x-2)!}\,p^{x-1}\,q^{n-x}\right]$$

$$= np\left[{}^{n-1}\mathrm{C}_{x-1}\,p^{x-1}\,q^{n-x}+p(n-1)\,{}^{n-2}\mathrm{C}_{x-2}\,p^{x-2}q^{n-x}\right]$$

$\therefore\quad \mathrm{E}(X^2)=np(p+q)^{n-1}+(np)(p\{n-1\})(p+q)^{n-2}$ but $p+q=1$

so $$\mathrm{E}(X^2) = np+n^2p^2-np^2$$

Hence
$$\begin{aligned}\mathrm{Var}\,(X) &= \mathrm{E}(X^2)-\mathrm{E}^2(X)\\ &= np+n^2p^2-np^2-n^2p^2\\ &= np\,(1-p)\end{aligned}$$

i.e. if X has a binomial probability distribution then

$$\mathrm{Var}\,(X) = np\,(1-p).$$

EXAMPLES 16h (continued)

2) A random variable X has a binomial probability distribution with parameters n and p. The mean value is 3 and the variance is 1.2. Find the values of n and p and hence calculate, correct to three significant figures, $\mathrm{P}(X<4)$.

X has a binomial distribution with parameters n and p, therefore $E(X) = np$ and $Var(X) = np(1-p)$.

Hence $$np = 3 \qquad [1]$$

and $$np(1-p) = 1.2 \qquad [2]$$

[1] and [2] give $p = 0.6$ and $n = 5$

Therefore the p.d.f. of X is ${}^5C_x(0.6)^x(0.4)^{5-x}$ for $x = 0, 1, 2, 3, 4, 5$

Now

$$P(X<4) = 1 - P(X=4 \text{ or } X=5) = 1 - [P(X=4) + P(X=5)]$$

$$P(X=4) = {}^5C_4(0.6)^4(0.4) = 0.2592$$

$$P(X=5) = (0.6)^5 = 0.077\,76$$

Therefore $P(X<4) = 0.663$ correct to 3 s.f.

3) A box contains n pieces of ribbon of lengths $1, 2, 3, \ldots, n$ cm. A piece of ribbon is removed at random and the probability that it is of length l cm is cl. If L is the random variable, 'the length of the ribbon removed', find
(a) the value of c,
(b) the expected value of L,
(c) the standard deviation of L.

(a) $$P(L=l) = cl$$

Therefore $$\sum_{l=1}^{n} cl = 1$$

$\Rightarrow$ $$c(1+2+\ldots+n) = 1$$

$\Rightarrow$ $$c\left(\frac{n}{2}(n+1)\right) = 1$$

$\Rightarrow$ $$c = \frac{2}{n(n+1)}$$

(b)
$$E(L) = \sum_{l=1}^{n} l\,P(L = l)$$
$$= \sum_{l=1}^{n} \frac{2l^2}{n(n+1)}$$
$$= \frac{2}{n(n+1)}[1^2 + 2^2 + \ldots + n^2]$$
$$= \frac{2}{n(n+1)}\left[\frac{n}{6}(n+1)(2n+1)\right]$$
$$= \tfrac{1}{3}(2n+1)$$

i.e. the expected value of L is $\frac{1}{3}(2n+1)$

(c)
$$\text{Var}(L) = E(L^2) - E^2(L)$$
$$E(L^2) = \sum_{l=1}^{n} l^2\,P(L = l)$$
$$= \sum_{l=1}^{n} \frac{2l^3}{n(n+1)}$$
$$= \frac{2}{n(n+1)} \sum_{l=1}^{n} l^3$$
$$= \frac{2}{n(n+1)}\left[\frac{n^2}{4}(n+1)^2\right]$$
$$= \tfrac{1}{2}n(n+1)$$

Therefore $\text{Var}(L) = \frac{1}{2}n(n+1) - \frac{1}{9}(2n+1)^2$
$= \frac{1}{18}(n^2 + n - 2)$

Hence the standard deviation is $\frac{1}{3}[\frac{1}{2}(n^2 + n - 2)]^{1/2}$

4) X is a discrete random variable and a is a constant. Prove that
$\text{Var}(aX) = a^2\,\text{Var}(X)$.

By definition
$$\text{Var}(aX) = E(a^2X^2) - E^2(aX)$$
$$= a^2E(X^2) - [a\,E(X)]^2$$
$$= a^2[E(X^2) - E^2(X)]$$
$$= a^2\,\text{Var}(X)$$

Therefore $\text{Var}(aX) = a^2\,\text{Var}(X)$.

EXERCISE 16h

1) A discrete random variable X has the following probability distribution.

x	1	2	3	4
$P(X=x)$	0.1	0.2	0.6	0.1

Find (a) $E(X)$ (b) $Var(X)$.

2) A discrete random variable X has a p.d.f. given by $f(x) = \frac{1}{10}x$ for $x = 1, 2, 3, 4$.
Find the mean and variance of X.

3) A discrete random variable X has the following probability distribution.

x	0	1	2	3	4
$P(X=x)$	0.12	0.15	0.4	0.25	0.08

Find the mean and standard deviation of the distribution.

4) A random variable X has a binomial probability distribution with mean 2 and variance $1\frac{1}{3}$. Find the values of the parameters n and p and hence find $P(X<2)$, giving your answer correct to 3 s.f.

5) A random variable X has a rectangular probability distribution for which $P(X=x) = \frac{1}{4}$ for $x = 1, 2, 3, 4$. Find $E(X)$ and $Var(X)$.

6) A random variable X has a rectangular probability distribution for which $P(X=x) = \dfrac{1}{n}$ for integer values of x from $x = 1$ to $x = n$. Find the mean and variance of the distribution.

7) A random variable X has a geometric probability distribution where X is 'the number of attempts required to achieve success'. The probability of success on any one attempt is 0.2, so the probability of success after x attempts, $P(X=x)$, is $(0.8)^{x-1}(0.2)$. Find the mean and variance of X.

8) X is a discrete random variable with a geometric probability distribution such that $P(X=x) = p(1-p)^{x-1}$ for $x = 1, 2, 3, \ldots$
Find, in terms of p, the mean and variance of X.

9) A machine manufactures washers and 20% of the production is substandard. A random sample of 10 washers is selected. Find the mean and standard deviation of the number of substandard washers in the sample.

10) Prove that (a) $Var(a) = 0$,

(b) $Var(a+X) = Var(X)$

where a is a constant.

11) A fair coin is tossed repeatedly until a head appears. Find the expected number of tosses required and the variance.

12) A box contains a large number of discs of which 8% are red and the rest are blue. Discs are removed one at a time until a red disc is taken out. Find the expected number of discs that are removed and the variance.

SUMMARY

Basic Definitions

If X is a discrete random variable that can take the values $x_1, x_2, \ldots, x_n$, and whose p.d.f. is $f(x)$, then

$$\sum_{i=1}^{n} P(X = x_i) = 1$$

$$\mu = E(X) = \sum_{i=1}^{n} x_i P(X = x_i) = \sum x f(x)$$

$$E[g(X)] = \sum g(x) f(x)$$

$$\sigma^2 = \mathrm{Var}(X) = E[(X-\mu)^2] = E(X^2) - E^2(X)$$

The Binomial Probability Distribution

If X is the number of successes in n trials, where the probability of success in one trial is p, then X has a binomial probability distribution for which

$$f(x) = {}^nC_x\, p^x (1-p)^{n-x}$$

$$E(X) = np$$

$$\mathrm{Var}(X) = np(1-p)$$

The Discrete Rectangular Distribution

If X is such that $f(x)$ is constant, then X has a rectangular probability distribution.

The Geometric Probability Distribution

If X is the number of trials necessary to achieve a success and if the probability of success on any one trial is p, then X has a geometric probability distribution for which

$$f(x) = p(1-p)^{x-1}$$

MULTIPLE CHOICE EXERCISE 16

TYPE I

1) Three fair dice are tossed. The probability of getting at least 1 six is
(a) $\frac{1}{216}$ (b) $\frac{125}{216}$ (c) $\frac{91}{216}$ (d) $\frac{25}{216}$.

2) The probability distribution of a random variable X is

x	0	1	2
$P(X=x)$	$\frac{1}{2}$	$\frac{3}{8}$	$\frac{1}{8}$

The mean value of X is
(a) $\frac{9}{8}$ (b) $\frac{5}{8}$ (c) $\frac{5}{16}$ (d) $\frac{5}{24}$.

3) In a by-election 20% of the electorate voted for Mr X. If 5 voters are chosen at random from this electorate, what is the probability that 20% of the sample voted for Mr X?
(a) 1 (b) $\frac{1}{5}$ (c) $\frac{1}{625}$ (d) $\frac{256}{625}$.

4) Five unbiased coins are tossed. The expected number of tails is
(a) 5 (b) 2.5 (c) 2 (d) 3.

5) A tetrahedral die has its faces numbered 1, 2, 3, 4 and when it is tossed, the score is the number on the face on which it lands. Assuming the die to be unbiased, the expected score is
(a) $2\frac{1}{2}$ (b) 2 (c) 3 (d) none of these.

6) R is a discrete random variable whose p.d.f. is given by $f(r)=\frac{1}{3}$ for $r=2,3,4$ and $f(r)=0$ for other values of r. $E(R^2)$ is
(a) 3 (b) $\frac{29}{3}$ (c) 9 (d) 29.

TYPE II

7) X is a discrete random variable that takes values x_1, x_2, x_3 only.
(a) $P(X=x_1 \cup X=x_2 \cup X=x_3)=1$.
(b) $x_1=1$, $x_2=2$, $x_3=3$.
(c) $x_1+x_2+x_3=1$.

8) X has a binomial probability distribution for which $n=6$ and $p=\frac{1}{2}$.
(a) $E(X)=3$.
(b) X has a p.d.f. given by $f(x)={}^6C_x\frac{1}{64}$ for $x=1,2,\ldots,6$.
(c) The standard deviation of the distribution is $\frac{3}{4}$.

9) X is a variable such that $P(X=x) = \frac{1}{4}x$ for $x=0,1,2,3$, only.
(a) X is a random variable.
(b) X is continuous.
(c) $P(X<3)=\frac{3}{4}$.

TYPE III

10) (a) X is a discrete random variable.
(b) $\Sigma P(X = x) = 1$ for all values of x that X can assume.

11) X is a discrete random variable.
(a) $E(X) = 4$ and $E(X^2) = 17.5$.
(b) $\text{Var}(X) = 1.5$.

12) X is a discrete random variable.
(a) $E(X) = 0$.
(b) X can assume only the value zero.

13) X is a discrete random variable and a is a constant.
(a) $a = 0$.
(b) $E(aX) = 0$.

14) X is a discrete random variable that can assume values x_1, x_2, x_3 only.
(a) $\text{Var}(X) = 0$.
(b) $E(X) = 0$.

MISCELLANEOUS EXERCISE 16

This exercise contains some questions on conditional probability.

1) The probability of a hurdler knocking down any particular hurdle in a race is 1/5. Find a numerical expression for the probability that, in a race over 8 hurdles, she will knock down less than 3 hurdles. (U of L)

2) In a certain population 20% are of Rhesus negative blood group. Find, to 2 decimal places, the probability of there being more than one of this blood group in a random sample of six people from this population. (U of L)

3) In a certain class 30% of the students failed Mathematics, 15% failed Chemistry, and 10% failed both Mathematics and Chemistry. A student is selected at random.
(a) Given that he failed Chemistry, find the probability that he failed Mathematics.
(b) Given that he passed Mathematics, find the probability that he failed Chemistry.
(c) Find also the probability that he failed one and only one of the two subjects. (U of L)

4) A population consists of 100 men and 900 women and it is known that exactly 10 men and 30 women have red hair. A person is chosen at random from the population. Find the probability that the selected person
(a) is male,
(b) has red hair,
(c) is male given that the person has red hair.
In a further selection, 5 people are chosen at random and with replacement. Find, to 3 decimal places, the probability that exactly two of the people chosen have red hair. Find, also to 3 decimal places, an expression for the probability that at least two have red hair. (U of L)

5) (a) Two buses A and B are scheduled to arrive at a town centre bus station at noon. The probability that bus A will be late is 0.2. The probability that bus B will be late is 0.28. The time keepings are not independent and the conditional probability that B is late given that A is late is 0.9. Calculate the probability that neither bus will be late on a particular day.
Calculate also the conditional probability that bus A is late, given that bus B is late.
(b) Two dice are thrown together.
(i) In one throw of the two dice, the sum of the two scores was 6. Calculate the probability that the numbers showing are 2 and 4.
(ii) In each of seven throws of the two dice the sum of the scores is to be recorded. Calculate the probability that 11 will be recorded exactly twice.
[You may assume that $\left(\frac{17}{18}\right)^5 = 0.751.$] (U of L)

6) A crossword is published in *The Times* each day of the week, except Sunday. A man is able to complete, on average, 8 out of 10 of the crossword puzzles.
(a) Find the expected value and the standard deviation of the number of completed crosswords in a given week.
(b) Show that the probability that he will complete at least 5 in a given week is 0.655 (to 3 significant figures).
(c) Given that he completes the puzzle on Monday, find to three significant figures the probability that he will complete at least 4 in the rest of the week. (C)

7) Of the valves made by a new process, 10% are found to be defective. The valves are packed into boxes each containing 12 valves. Find the probability that a box will contain
(a) exactly one defective valve,
(b) not more than two defective valves.
Find also the probability that, of 100 such boxes, less than two will contain more than two defective valves. (U of L)

8) One car in four fails an inspection test. State the mean and variance for the distribution of the number of cars that fail the test when random samples of 100 cars are inspected.
A sample of 10 cars is selected at random. Find an expression for the probability that at least two cars fail this inspection test. (U of L)

9) The output of a factory is 500 units a day and is provided by three machines A, B and C, which produce respectively 200, 175 and 125 units a day. Over a long period it is found that the percentage of defective output is 5% from machine A, 4% from machine B and 4% from machine C.
(a) Find the probability that a unit chosen at random from the output of the factory is defective.
(b) If a unit is chosen from the output of the factory and is found to be defective, find the probability that it came from machine A.
(c) On three different days a unit is chosen at random from the output of the factory. Find the probability that exactly one of these units will be defective.
(d) If one unit is chosen at random from the output of each machine, find the probability that exactly one unit will be defective. (WJEC)

10) A factory manufactures items with a probability of 0.06 for each item that is not of standard quality. A second factory manufactures items with a probability of 0.02 for each item that is not of standard quality. Nine hundred items from the first factory are randomly stored with one hundred items from the second factory. A customer purchases one item from this stock. Find the probability that the item
(a) was produced by the second factory,
(b) was produced by the first factory and is not of standard quality,
(c) is not of standard quality,
(d) was produced by the second factory and is of standard quality.
If, instead, the customer purchases 4 items, find an approximate value, to 3 significant figures, for the probability that exactly two of these items are not of standard quality. (U of L)

11) A discrete random variable Y has a uniform distribution over the set of values $\{1, 2, 3, \ldots n\}$. Find $\mathrm{E}(Y)$.
Evaluate $\mathrm{Var}(Y)$ when $n = 5$. (C)

12) The probability that any electric bulb produced by a certain factory is defective is 0.01. A shipment of 10 000 bulbs from this factory is sent to a wholesaler. Find (a) the expected number of defective bulbs, (b) the standard deviation of the number defective. (U of L)

13) A seed merchant mixes a large number of seeds of red, yellow and white wallflowers together in the proportion 4 red to 2 yellow to 1 white, and then makes up packets of 20 seeds for sale. Find, correct to 3 s.f. the probability that a packet contains more than 4 white seeds. Find the mean and standard deviation of the number of white seeds in the packets.

14) A firm makes pocket calculators, and the production line is such that the probability that any one calculator operates correctly is 0.998. The calculators are then packed into boxes of ten for distribution. Find the probability that a box contains one or more faulty calculators. Find also the mean and variance of the number of faulty calculators in a box.

15) Two unbiased coins are spun together. State the probabilities p_0, p_1, p_2 of getting no heads, one head or two heads respectively. Obtain the mean and the variance of the number of heads. (U of L)

16) The picture cards, including the aces, are removed from a pack of playing cards. The remaining cards are shuffled and X is the number on the top card. Find the mean and variance of X, and $\mathrm{P}(X \leqslant 4)$.

17) A fair six-sided die has two of its faces numbered 2, three of its faces numbered 3 and the remaining face numbered 4. The die is tossed twice and the sum of the scores is denoted by X. Make a probability distribution table for X and use it to find the mean and standard deviation of X.

18) A multiple choice examination has 30 questions. For each question the candidate has to select one of 4 possible answers, only 1 of which is correct. If a candidate ticks at random what is the probability that he gets a particular answer correct? Find the mean and variance of the number of questions that he gets correct.

19) Two people, A and B, draw a card alternately from a well shuffled pack of ordinary playing cards, look at it and then replace it. The first person to draw an ace wins.

If A cuts first, find

(a) the probability that B wins with his first cut,

(b) the probability that A wins with his third cut,

(c) the probability that B wins.

If X is the total number of cuts needed for either A or B to win, find the p.d.f. of X and $\mathrm{E}(X)$.

20) A random number generator is set up to produce five figure integers where any of the figures, including the first, can be zero. Find the probability of getting

(a) zero,

(b) a number ending in zero.

Find also the mean number of zeros in a number.

21) For a particular species of animal, the probability of any one baby being male is p, independent of the sex of other babies there may be in the litter. What is the probability that at least one baby of a litter of given size R is female?
The size R of the litter is a random variable such that the probability that $R = r$, where r can take the values $1, 2, 3, \ldots$ (but not 0), is

$$k\alpha^r/r!$$

where k and α are positive constants. Show that $k = (e^\alpha - 1)^{-1}$, and find, in terms of α and p, the probability that a litter will contain at least one female, simplifying your answer as much as possible. (O and C)

22) (a) Starting from the definition

$$\text{Var}(X) = \text{E}[(X-\mu)^2]$$

for any random variable X having mean μ, derive the result

$$\text{Var}(X) = \text{E}[X(X-1)] + \mu - \mu^2$$

(b) The random variable X takes all *non-negative* integer values, with

$$\text{P}(X = x) = \frac{e^{-\theta}\theta^x}{x!} \quad \text{for} \quad x = 0, 1, 2, \ldots, \quad \theta > 0$$

Show that the mean of X is θ, and using the result of part (a) or otherwise obtain the variance of X.

(c) The random variable Y takes all *positive* integer values, with

$$\text{P}(Y = y) = k\frac{e^{-\theta}\theta^y}{y!} \quad \text{for} \quad y = 1, 2, \ldots, \quad \theta > 0$$

Find the constant k (in terms of θ) and determine $\text{E}[Y]$. (MEI)

CHAPTER 17

CONTINUOUS RANDOM VARIABLES

CONTINUOUS VARIABLES

A continuous variable is not limited to discrete values but can take any value within a specified range.

Continuous Random Variables

If X is a continuous variable that can take values in the range from x_1 up to x_n, then X is a *random* variable if

$$P(x_1 \leqslant X < x_n) = 1 \quad \text{and} \quad P(X < x_1) = 0, \quad P(X \geqslant x_n) = 0$$

PROBABILITY DISTRIBUTIONS

Consider a continuous random variable X that can take values only in the range $0 \leqslant X < 2$.
Now consider the mutually exclusive ranges

$$0 \leqslant X < \tfrac{1}{2}, \quad \tfrac{1}{2} \leqslant X < 1, \quad 1 \leqslant X < 1\tfrac{1}{2}, \quad 1\tfrac{1}{2} \leqslant X < 2$$

If the probabilities that X takes values in each of these ranges are known then we have a probability distribution for these ranges which can be illustrated by a histogram.

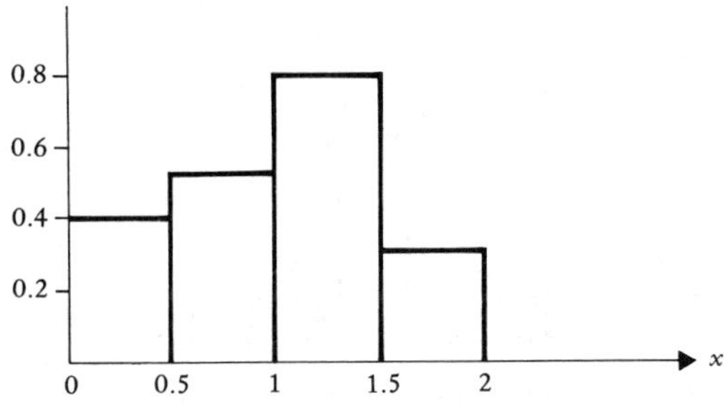

It is the *area* of each bar, and not the height, that represents the probability associated with that bar, so the vertical scale is not a probability scale.

In this case $P(0 \leqslant X < \frac{1}{2})$ is represented by the area of the first bar,

i.e. $$P(0 \leqslant X < \tfrac{1}{2}) = (0.4)(0.5) = 0.2$$

Similarly $P(0 \leqslant X < 1)$ can be found by adding together the areas of the first two bars,

i.e. $$\begin{aligned} P(0 \leqslant X < 1) &= (0.4)(0.5)+(0.5)(0.5) \\ &= 0.45 \end{aligned}$$

Now X is a random variable, so the sum of the probabilities is 1, hence the total area of the histogram is 1. Therefore, showing that the total area of a probability histogram is 1 proves that the variable is random.

EXERCISE 17a

1)

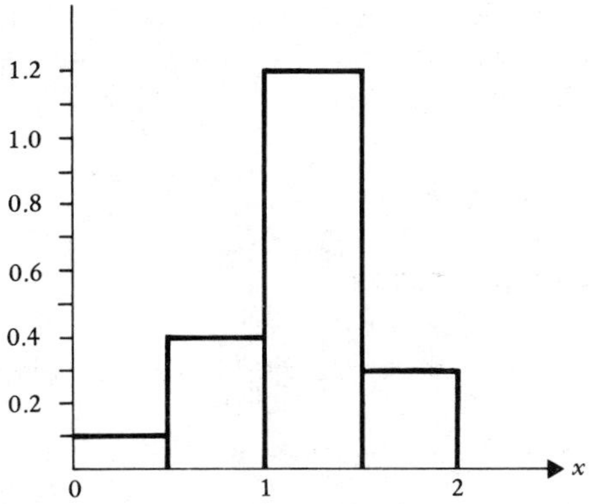

This histogram illustrates the probabilities associated with a continuous variable X.
Confirm that X is a random variable and find $P(0 \leqslant X < 1.5)$.

2)

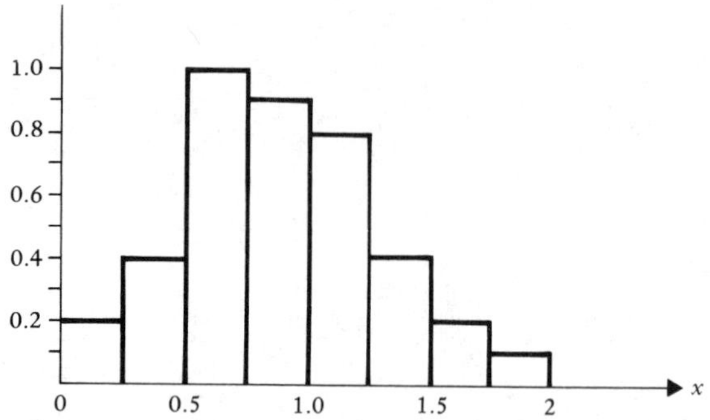

This histogram represents the probabilities associated with a continuous variable X.
Show that X is a random variable and find
(a) $P(0 \leqslant X < 0.5)$ (b) $P(0.75 \leqslant X < 1.5)$.

PROBABILITY DENSITY FUNCTIONS

A histogram illustrating probabilities for given ranges of X can be used to find the probability that X lies in one of those ranges, or in any sum of those ranges but it cannot be used to find the probability that X lies in any other range.
We cannot, for example, find $P(0.1 \leqslant X < 0.2)$ from any histogram given so far in this chapter.

The thinner the bars of the histogram are, the greater is the choice of ranges for which probabilities can be found.

Complete flexibility is obtained if each 'bar' is 'infinitely thin', giving an infinite number of bars for the range of values that X can assume. In this case the top boundary of the histogram is a continuous line without any 'steps'.

e.g.

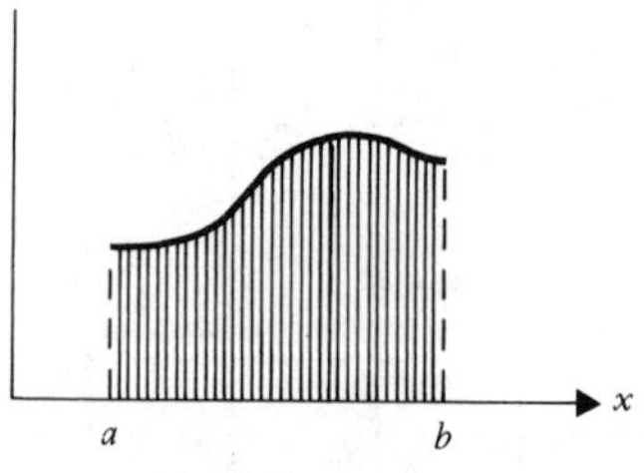

The probability that X lies in *any* chosen range, within the overall range, can then be found from the appropriate area under the curve.

The function of x that defines the top boundary of the histogram is called the *probability density function* of X.

Consider a continuous random variable X for which $a \leqslant X < b$ and whose p.d.f. is $f(x)$.

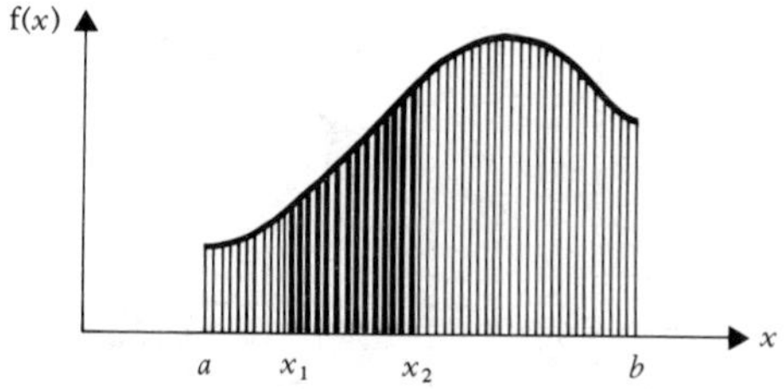

If x_1 and x_2 are two values in the range $a \leqslant x < b$ then $P(x_1 \leqslant X < x_2)$ is represented by the sum of areas of the 'bars' between x_1 and x_2, i.e. by the area under $y = f(x)$ from x_1 to x_2.

Hence
$$P(x_1 \leqslant X < x_2) = \int_{x_1}^{x_2} f(x)\,dx$$

Note that the area represented by $\int_{x_1}^{x_2} f(x)\,dx$ is the same whether or not the boundary lines are included.
Therefore

$$P(x_1 \leqslant X < x_2) = P(x_1 \leqslant X \leqslant x_2) = P(x_1 < X \leqslant x_2) = P(x_1 < X < x_2)$$

Now X is a random variable $\Rightarrow$ $P(a \leqslant X < b) = 1$

Hence
$$\int_a^b f(x)\,dx = 1$$

Note that although a probability can be found from an integral, it is often quicker to find the probability as an area from the sketch of $y = f(x)$.

It is also worth noting that, for a discrete random variable, where $f(x)$ is not continuous and is defined only for isolated values of x, a value of the p.d.f. gives a probability, i.e. $f(x_i) = P(X = x_i)$.

In the case of a continuous random variable however, $f(x)$ is continuous and particular values of $f(x)$ do *not* give probabilities.

Probabilities are always positive, therefore in both cases the p.d.f. must be positive, i.e. $f(x) \geqslant 0$.

EXAMPLES 17b

1) A continuous random variable X has a p.d.f. given by

$$f(x) = kx \quad \text{for} \quad 1 \leqslant x < 3$$
$$f(x) = 0 \quad \text{elsewhere.}$$

Find the value of k and the probability that X should take a value between 1.5 and 2.

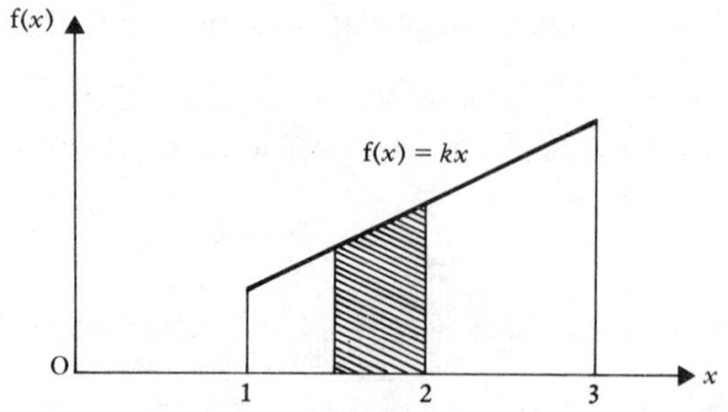

As X is a continuous random variable the area of the largest trapezium is 1

$$\Rightarrow \qquad \tfrac{1}{2}(k+3k)(2) = 1$$

$$\Rightarrow \qquad k = \tfrac{1}{4}$$

therefore $\qquad f(x) = \tfrac{1}{4}x$

Now $P(1.5 \leqslant X < 2)$ is the area of the shaded trapezium

$$\Rightarrow \qquad P(1.5 \leqslant X < 2) = \tfrac{1}{2}\left(\tfrac{1.5}{4} + \tfrac{2}{4}\right)(0.5)$$

$$= 0.219 \quad \text{to } 3 \text{ s.f.}$$

2) A continuous random variable X has a probability density function $f(x)$ given by

$$f(x) = k(1-x^2) \quad \text{for} \quad -1 \leqslant x < 0$$

$$f(x) = k(1-x) \quad \text{for} \quad 0 \leqslant x < 1$$

$$f(x) = 0 \quad \text{for} \quad x < -1 \quad \text{and} \quad x \geqslant 1$$

Find the value of k.

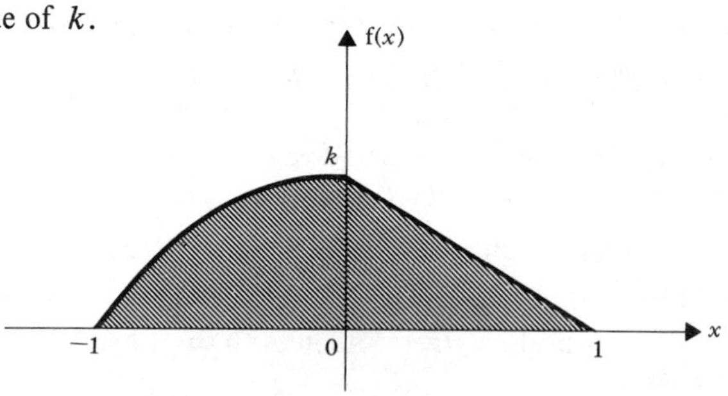

As X is a random variable the area between $y = f(x)$ and the x axis is 1.

$$\Rightarrow \qquad \int_{-1}^{0} k(1-x^2)\,dx + \tfrac{1}{2}k = 1$$

$$\Rightarrow \qquad \left[kx - \tfrac{1}{3}kx^3\right]_{-1}^{0} + \tfrac{1}{2}k = 1$$

$$\Rightarrow \qquad k = \tfrac{6}{7}$$

The Median

If X is a continuous random variable that can take values in the range $x_0 \leqslant X < x_n$, then the median, M, is the value of x for which $P(x_0 \leqslant X < x) = \frac{1}{2}$, i.e.

if M is the median of a continuous probability distribution then

$$\int_{x_0}^{M} f(x)\,dx = \tfrac{1}{2}$$

The Mode

The mode of a continuous random variable is the value of x for which $f(x)$ has its greatest value.

EXAMPLES 17b (continued)

3) A continuous random variable X has a probability density function given by

$$f(x) = \tfrac{1}{9}(x-3)^2 \quad \text{for} \quad 0 \leqslant x < 3$$

$$f(x) = 0 \quad \text{elsewhere.}$$

Find the median and the mode of X.

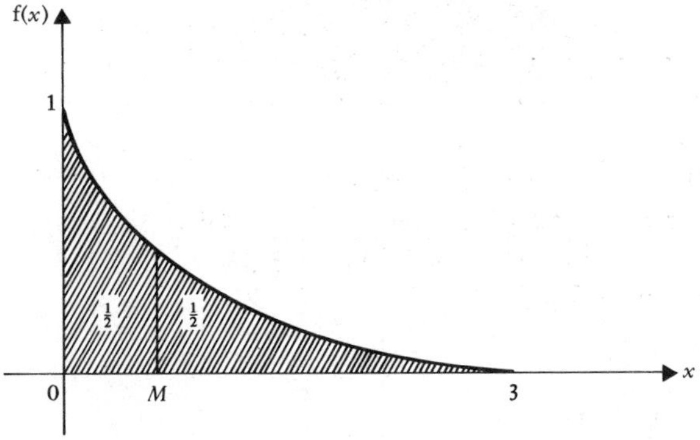

If M is the median then

$$\int_0^M \tfrac{1}{9}(x-3)^2 = \tfrac{1}{2}$$

$$\Rightarrow \qquad \tfrac{1}{9}\left[\tfrac{1}{3}(x-3)^3\right]_0^M = \tfrac{1}{2}$$

$$\Rightarrow \qquad (M-3)^3+27 = \tfrac{27}{2}$$

$$\Rightarrow \qquad M-3 = \sqrt[3]{(-\tfrac{27}{2})} = -2.381$$

$$\Rightarrow \qquad M = 0.619 \quad \text{to 3 s.f.}$$

The mode is where $f(x)$ has its greatest value and this is when $x=0$. Therefore the mode of X is 0.

EXERCISE 17b

1) A random variable X has a p.d.f. defined by

$$f(x) = kx \quad \text{for} \quad 0 \leqslant x < 2$$

$$f(x) = 0 \quad \text{for} \quad x < 0 \quad \text{and} \quad x \geqslant 2.$$

Find
(a) the value of k,
(b) $P(0.5 \leqslant X < 1)$,
(c) the median.

2) A continuous random variable X has a p.d.f. $f(x)$ given by $f(x)=k$ for $-1 \leqslant x < 1$ and $f(x)=0$ elsewhere.
(a) Find the value of k.
(b) Sketch $y=f(x)$.
(c) Find $P(X \geqslant 0)$.

3) A continuous random variable L has a probability density function defined by

$$f(l) = \tfrac{3}{16}(4-l^2) \quad \text{for} \quad 0 \leqslant l < 2$$

$$f(l) = 0 \quad \text{elsewhere.}$$

Find
(a) $P(l < 1)$,
(b) $P(l > 1.5)$,
(c) the mode of L.

4) A continuous random variable R has a probability density function $f(r)$ where

$$f(r) = 1-\tfrac{1}{2}r \quad \text{for} \quad 0 \leqslant r < k$$

$$f(r) = 0 \quad \text{for} \quad r < 0 \quad \text{and} \quad r \geqslant k.$$

Find the value of the constant k.

5) A continuous random variable X has a probability density function $f(x)$ where

$$\begin{aligned} f(x) &= kx && \text{for} \quad 0 \leqslant x < 1 \\ f(x) &= -\tfrac{1}{2}k(x-3) && \text{for} \quad 1 \leqslant x < 3 \\ f(x) &= 0 && \text{for} \quad x < 0 \quad \text{and} \quad x \geqslant 3. \end{aligned}$$

Find the value of k and sketch $y = f(x)$.
What is the mode of X?

6) A continuous random variable X has a probability density function $f(x)$ defined by

$$f(x) = \begin{cases} A\,e^{-x} & \text{for} \quad 0 \leqslant x < 2 \\ 0 & \text{for} \quad x < 0 \quad \text{and} \quad x \geqslant 2. \end{cases}$$

Find
(a) the value of A (b) $P(X < 1)$.

7) The probability density function of a random variable X is defined by

$$\begin{aligned} f(x) &= Ax(1-x^2) && \text{for} \quad 0 \leqslant x < 1 \\ f(x) &= 0 && \text{elsewhere.} \end{aligned}$$

Find the value of the constant A.
If M is the median of X show that $2M^4 - 4M^2 + 1 = 0$.

8) A random variable X has a probability density function $f(x)$ defined by the graph.

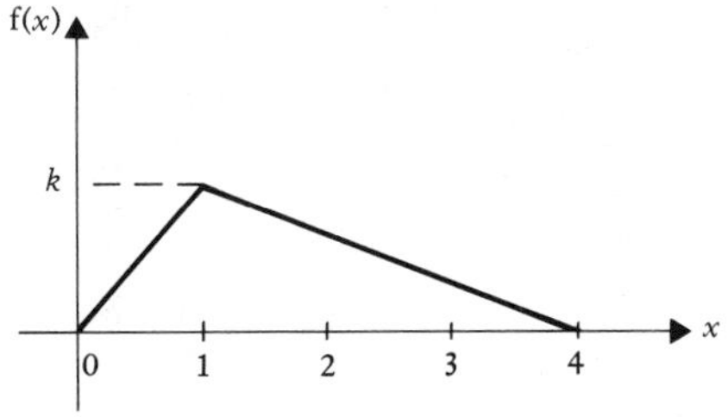

Find the value of k and hence define $f(x)$ algebraically.

9) A continuous random variable X has a probability density function $f(x)$ where

$$\begin{aligned} f(x) &= \frac{1}{x^2} && \text{for} \quad x \geqslant 1 \\ f(x) &= 0 && \text{for} \quad x < 1 \end{aligned}$$

Sketch the curve $y = f(x)$ and find
(a) $P(1 \leqslant X < 2)$ (b) $P(X > 4)$.

10) A continuous random variable X has a probability density function defined by

$$f(x) = k\,e^{-x} \quad \text{for} \quad x \geqslant 0$$
$$f(x) = 0 \quad \text{for} \quad x < 0.$$

Find the value of k and the median of X.

THE RECTANGULAR DISTRIBUTION

Consider a continuous random variable X that can take values only in the range $a \leqslant X < b$ and such that the probability that X lies in any one particular range is the same as the probability that X lies in any other range of the same width,

i.e. $$P\begin{pmatrix}X \text{ lies in a range} \\ \text{of width } l\end{pmatrix} = P\begin{pmatrix}X \text{ lies in any other} \\ \text{range of width } l\end{pmatrix}$$

If $f(x)$ is the p.d.f. of X, this means that on the graph of $f(x)$, the areas of equal width 'bars' must be the same. This can only be so if $f(x)$ is constant, i.e. if $f(x) = k$, and the area under the graph of $f(x)$ is a rectangular shape. In this case X is said to have a rectangular, or uniform, probability distribution.

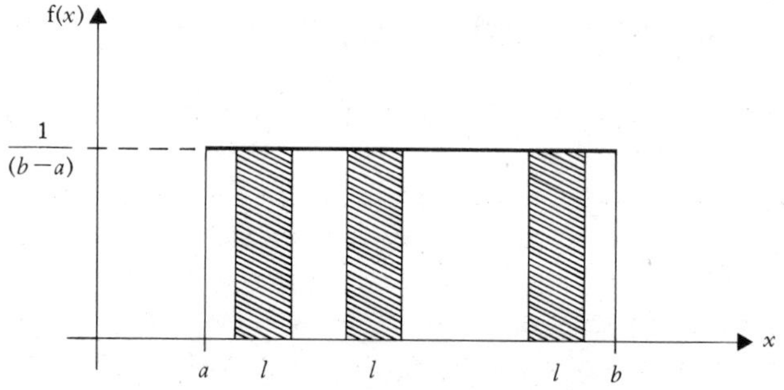

Now $P(a \leqslant X < b) = 1$, so the rectangle enclosed by $y = f(x)$, $x = a$, $x = b$ and the x axis has unit area.

The 'width' of the rectangle is $b - a$ so the 'height' must be $\dfrac{1}{b-a}$,

i.e. $$f(x) = \frac{1}{b-a}$$

A continuous random variable X has a rectangular distribution when the p.d.f. of X is given by

$$f(x) = \frac{1}{b-a} \quad \text{for} \quad a \leqslant x < b$$

$$f(x) = 0 \quad \text{elsewhere.}$$

EXAMPLES 17c

1) A random variable X has a p.d.f. given by

$$f(x) = \tfrac{1}{8} \quad \text{for} \quad -2 \leqslant x < k$$
$$f(x) = 0 \quad \text{elsewhere.}$$

Find the value of k and $P(3 \leqslant x < 3.5)$.

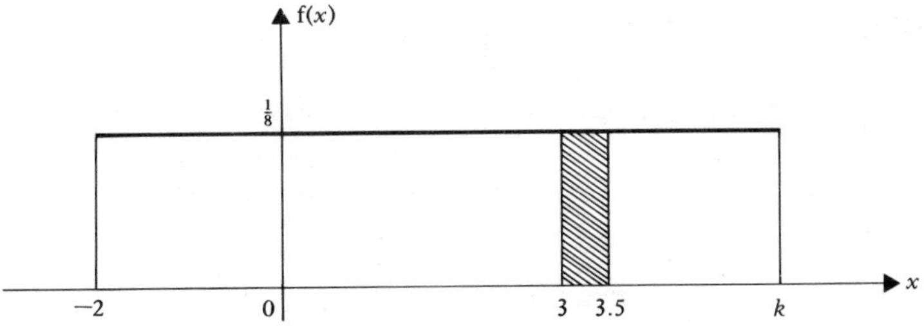

X has a rectangular distribution so the area under $f(x)$ is rectangular in shape and is equal to 1

$\Rightarrow$ $$\tfrac{1}{8}[k-(-2)] = 1$$

$\Rightarrow$ $$k = 6$$

$P(3 \leqslant x < 3.5)$ is represented by the shaded area in the diagram

$\Rightarrow$ $$P(3 \leqslant x < 3.5) = \tfrac{1}{8}(0.5) = \tfrac{1}{16}$$

2) A flywheel in a car has a spot of paint marking a point on its circumference. The radius of the flywheel is 20 cm. A small viewing hole is provided, through which it is possible to see 3 cm of the circumference of the flywheel. Assuming that the mark on the flywheel is as likely to be in any one position as in any other, find the probability that, when the engine is off, the mark is visible from the viewing hole.

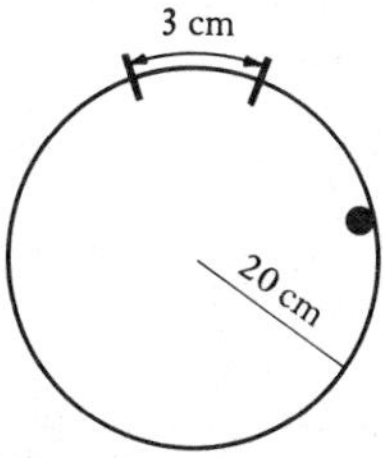

This is a rectangular distribution.
The circumference of the flywheel is 40π

Therefore $P(\text{spot of paint is visible}) = \dfrac{3}{40\pi}$

$$= 0.0239 \quad \text{to 3 s.f.}$$

3) A random variable X has a rectangular distribution in the range $0 \leqslant x < 4$. Find the probability density function of the random variable Y where $Y = X^2$.

For X we know that $\int_0^4 \frac{1}{4}\,dx = 1$.

For Y we need to find $f(y)$ and values y_1 and y_2 such that

$$\int_{y_1}^{y_2} f(y)\,dy = 1$$

Now $Y = X^2 \Rightarrow y = x^2$

Therefore $x = 0 \Rightarrow y_1 = 0$ and $x = 4 \Rightarrow y_2 = 16$.

Then $$\int_0^4 \frac{1}{4}\,dx = 1 \Rightarrow \int_0^{16} \frac{1}{4}\left(\frac{dx}{dy}\right) dy = 1$$

But $y = x^2 \Rightarrow \dfrac{dy}{dx} = 2x$

Therefore $$\int_0^{16} \frac{1}{4}\left(\frac{1}{2x}\right) dy = 1$$

$$\Rightarrow \qquad \int_0^{16} \tfrac{1}{8}y^{-1/2}\,dy = 1$$

Comparing with $\int_{y_1}^{y_2} f(y)\,dy = 1$ we see that the p.d.f. of Y is given by

$f(y) = \frac{1}{8}y^{-1/2}$ for $0 \leqslant y < 16$.

(As a check we can evaluate $\int_0^{16} \frac{1}{8}y^{-1/2}\,dy = \left[\frac{1}{4}y^{1/2}\right]_0^{16} = \frac{1}{4}\sqrt{16} = 1$.)

EXERCISE 17c

1) A continuous random variable X has a rectangular distribution over the range $0 \leqslant x < 6$. Find the p.d.f. of X and $P(X > 5)$.

2) A random variable X has a rectangular distribution over the whole interval $-2 \leqslant x < 3$. Find the p.d.f. of X and $P(X < 0)$.

3) The p.d.f. of a random variable R is $f(r)$ where $f(r) = \frac{1}{10}$ for $2 \leqslant r < 12$. Find $P(R < 6)$.

4) The p.d.f. of a continuous random variable Y is $f(y)$ where $f(y) = k$ for $-2 \leqslant y < 4$. Find k.

5) The p.d.f. of a continuous random variable A is $f(a)$ where $f(a) = \frac{1}{12}$ for $k \leqslant a < 8$. Find k and $P(A < 6)$.

6) A continuous random variable X is uniformly distributed over the range $2 \leqslant x < 6$. Find the p.d.f. of X^2.

7) A continuous random variable X is uniformly distributed over the interval $0 \leqslant x < 6$. Find the p.d.f. of the continuous random variable A where $A = X(8 - X)$.

EXPECTATION, E(X)

The expected or mean value of a continuous random variable X is the value, on average, that we expect X to take and it is denoted by $E(X)$.

If a probability distribution has a symmetrical shape then the value of x at the line of symmetry is obviously the mean value.
For example, if X has a rectangular distribution over the range $a \leqslant x < b$, the distribution is symmetrical about $x = \frac{1}{2}(a+b)$, so $E(X) = \frac{1}{2}(a+b)$.

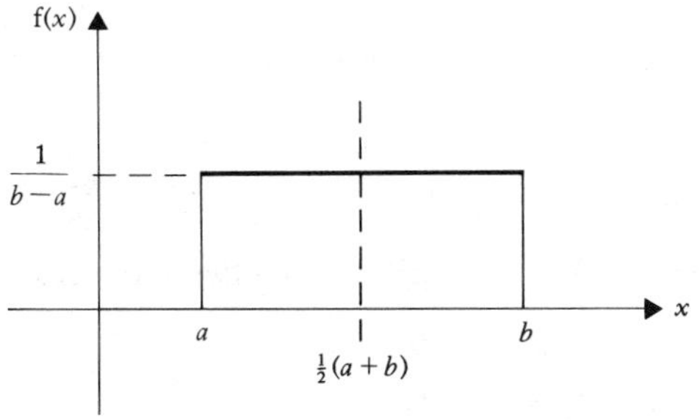

Similarly if X has a p.d.f. given by $f(x) = \frac{3}{2}(1-x)(x-3)$ for $1 \leqslant x < 3$ then $y = f(x)$ is symmetrical about $x = 2$.
Therefore $E(X) = 2$.

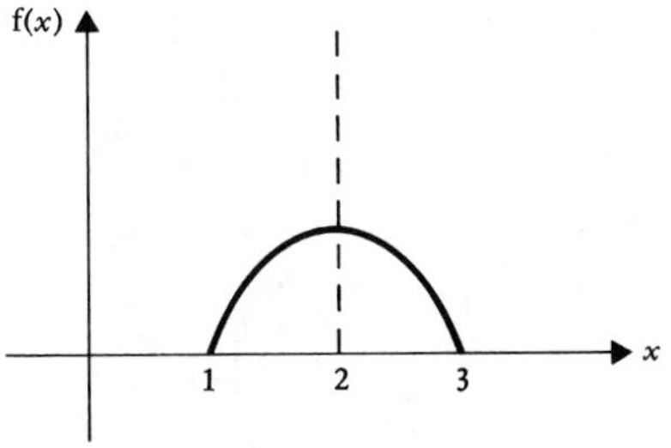

We will now consider the general case when $y = f(x)$ has no vertical line of symmetry.

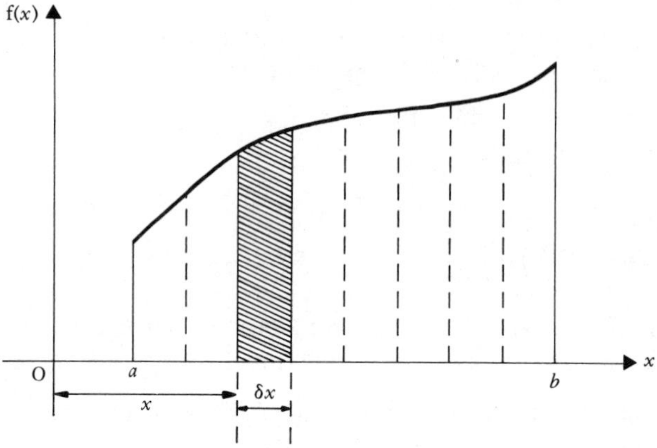

Consider a continuous random variable X whose p.d.f. is $f(x)$ for $a \leqslant x < b$.

If the range $a \leqslant x < b$ is divided into smaller ranges, each of width δx then for a typical range,

$$P(x \leqslant X < x + \delta x) \simeq f(x)\,\delta x$$

Hence the expected value of X arising from that range is approximately $x\,f(x)\,\delta x$.

Summing for all the ranges we have

$$E(X) \simeq \sum_{x=a}^{x=b} x\,f(x)\,\delta x$$

This approximation improves as δx gets smaller,

therefore $$E(X) = \lim_{\delta x \to 0} \sum_{x=a}^{x=b} x\,f(x)\,\delta x$$

i.e. $$E(X) = \int_a^b x\,f(x)\,dx$$

For example if X has a p.d.f. given by $f(x) = 1 - \frac{1}{2}x$ for $0 \leqslant x < 2$ then

$$\begin{aligned} E(X) &= \int_0^2 x(1 - \tfrac{1}{2}x)\, dx \\ &= \tfrac{2}{3} \end{aligned}$$

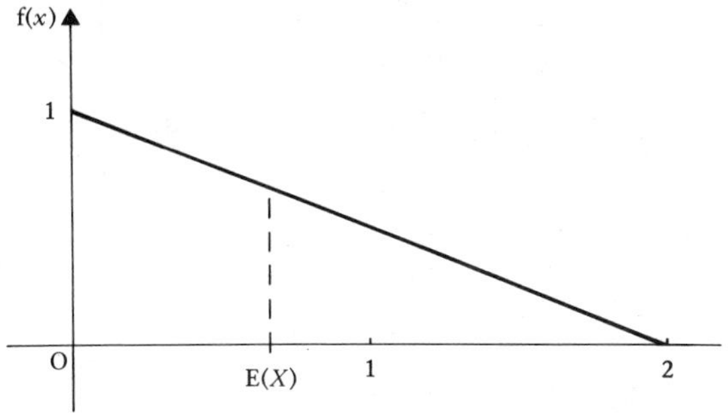

The Expectation of a Function of X

As is the case with a discrete random variable, we need to define the expected value of a function of X to facilitate the calculation of variance.

If X is a continuous random variable whose p.d.f. is $f(x)$ for $a \leqslant x < b$ then $E\,g(X)$ is defined by

$$E\,g(X) = \int_a^b g(x)\, f(x)\, dx$$

For example, if X has a p.d.f. given by $f(x) = 1 - \frac{1}{2}x$ for $0 \leqslant x < 2$ then $E(X^2)$ is given by

$$\begin{aligned} E(X^2) &= \int_0^2 x^2(1 - \tfrac{1}{2}x)\, dx \\ &= \tfrac{2}{3} \end{aligned}$$

Using the definition above it can be shown that the operator E has the same properties when applied to a continuous variable as it has when applied to a discrete variable.

EXERCISE 17d

1) A continuous random variable X has a p.d.f. defined by

(a) $f(x) = \frac{1}{9}x^2$ for $0 \leqslant x < 3$
$f(x) = 0$ elsewhere,

(b) $f(x) = \frac{1}{2}x$ for $0 \leqslant x < 2$
$f(x) = 0$ elsewhere,

(c) $f(x) = \frac{1}{2}$ for $-1 \leqslant x < 1$
$f(x) = 0$ elsewhere.

In each case find $E(X)$ and $E(X^2)$.

2) A continuous random variable R has a p.d.f. defined by

(a) $f(r) = k(9 - r^2)$ for $-3 \leqslant r < 3$
$f(r) = 0$ elsewhere,

(b) $f(r) = kr^3$ for $0 \leqslant r < 1$
$f(r) = 0$ elsewhere,

(c) $f(r) = k(r+1)$ for $-1 \leqslant r < 0$
$f(r) = k(1-r)$ for $0 \leqslant r < 1$
$f(r) = 0$ elsewhere.

In each case find the value of k and $E(X)$.

3) The length of a side of a square is represented by the continuous random variable X where X is uniformly distributed in the range $1 \leqslant x < 2$.
Find the p.d.f. of X, $E(X)$ and $E(X^2)$.
If A is the random variable representing the area of the square, find the p.d.f. of A in the form $f(a)$ and find $E(A)$.
Compare the values of $E(A)$ and $E(X^2)$.

VARIANCE, VAR (X)

The variance of a probability distribution is defined on page 384. It is the average value of the squares of the deviations of X from the mean value of X, i.e.

$$\text{Var}(X) = E[(X - \mu)^2]$$

As before, using the properties of the operator E, this formula can be written in the form

$$\text{Var}(X) = E(X^2) - \mu^2$$

Either version of the formula may be used to find $\text{Var}(X)$, although the second form usually leads to a simpler calculation.

EXAMPLES 17e

1) A continuous random variable X has a p.d.f. given by $f(x)$ where

$$f(x) = 6x(1-x) \quad \text{for} \quad 0 \leqslant x < 1$$
$$f(x) = 0 \qquad \text{elsewhere.}$$

Find the mean, μ, and variance, σ^2, of X.
Find also, correct to 3 s.f. $P(|X-\mu| < \sigma)$.

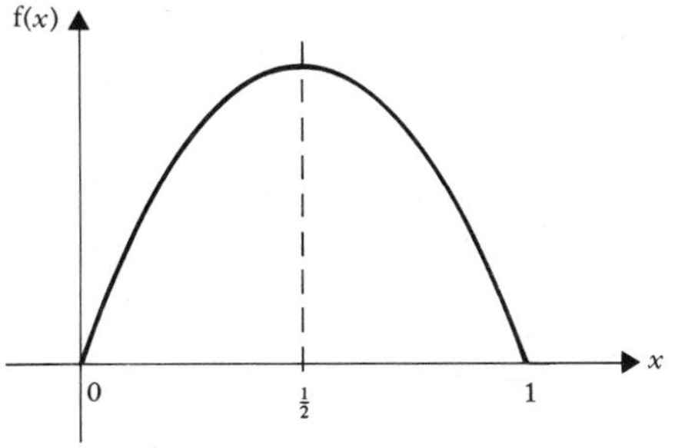

The distribution is symmetrical about $x = \frac{1}{2}$

Therefore $\mu = E(X) = \frac{1}{2}$

Using $\text{Var}(X) = E(X^2) - \mu^2$ we have

$$E(X^2) = \int_0^1 (x^2)(6x)(1-x)\,dx$$
$$= \left[\tfrac{6}{4}x^4 - \tfrac{6}{5}x^5\right]_0^1$$
$$= \tfrac{3}{10}$$

and $E(X) = \frac{1}{2} \Rightarrow \mu^2 = \frac{1}{4}$

Therefore $\sigma^2 = \text{Var}(X) = \frac{3}{10} - \frac{1}{4}$

$$= \tfrac{1}{20}$$

Now $P(|X-\mu| < \sigma) = P(-\sigma < X-\mu < \sigma)$

$$= P(\mu - \sigma < X < \mu + \sigma)$$

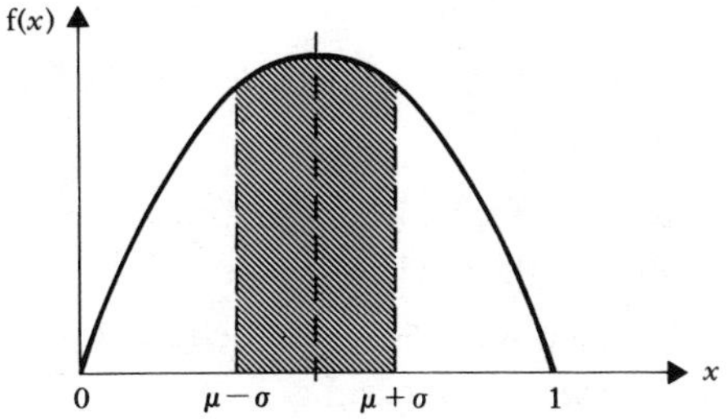

Now $\sigma^2 = \frac{1}{20} \Rightarrow \sigma = 0.2236$ to 4 s.f. and $\mu = \frac{1}{2}$

Therefore

$$P(|X-\mu| < \sigma) = P(0.2764 < X < 0.7236)$$
$$= \int_{0.2764}^{0.7236} 6x(1-x)\,dx$$
$$= \left[3x^2 - 2x^3\right]_{0.2764}^{0.7236}$$
$$= 0.8130 - 0.1870$$
$$= 0.626 \text{ to 3 s.f.}$$

2) The continuous random variable X has a p.d.f. given by

$$f(x) = (x+1) \quad \text{for} \quad -1 \leqslant x < 0$$
$$f(x) = e^{-2x} \quad \text{for} \quad x \geqslant 0$$
$$f(x) = 0 \quad \text{elsewhere.}$$

Find the mean value of X and show that the median is 0.

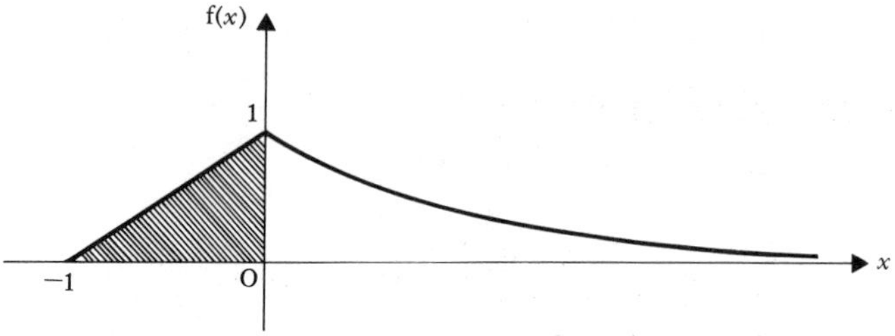

There is no symmetry.

Hence

$$\mu = \int_{-1}^{0} x(x+1)\,dx + \int_0^\infty x\,e^{-2x}\,dx$$
$$= \left[\tfrac{1}{3}x^3 - \tfrac{1}{2}x^2\right]_{-1}^{0} + \left[-\tfrac{1}{2}x\,e^{-2x}\right]_0^\infty - \int_0^\infty -\tfrac{1}{2}e^{-2x}\,dx$$

(integrating by parts)

$$= \tfrac{1}{6} + 0 + \left[-\tfrac{1}{4}e^{-2x}\right]_0^\infty$$

($x\,e^{-2x} \to 0$ as $x \to \infty$)

$$= \tfrac{1}{6} + (+\tfrac{1}{4}) = \tfrac{5}{12}$$

Therefore the mean of X is $\frac{5}{12}$

The median is the value of x for which $P(-1 \leqslant X < x) = \frac{1}{2}$

From the sketch, the area of the shaded triangle is $\frac{1}{2}$,

therefore $P(-1 \leqslant X < 0) = \frac{1}{2}$,

hence the median is 0.

EXERCISE 17e

1) A continuous random variable X has a p.d.f. defined by

(a) $f(x) = \frac{2}{9}x$ for $0 \leqslant x < 3$
$f(x) = 0$ for $x < 0$ and $x \geqslant 3$,

(b) $f(x) = \dfrac{3}{x^4}$ for $x \geqslant 1$
$f(x) = 0$ elsewhere.

In each case find $\text{Var}(X)$.

2) A continuous random variable X has a p.d.f. given by

(a) $f(x) = kx$ for $1 \leqslant x < 2$
$f(x) = 0$ elsewhere,

(b) $f(x) = ke^{-x}$ for $x \geqslant 1$
$f(x) = 0$ for $x < 1$.

In each case find the value of k and the mean and standard deviation.

MISCELLANEOUS EXERCISE 17

1) Verify that

$$\phi(x) = \begin{cases} \dfrac{3}{x^4}, & \text{for } x \geqslant 1 \\ 0, & \text{for } x < 1 \end{cases}$$

is a possible probability density function for a random variable X. Find that value of a which is such that

$$P(X < a) = 0.75$$ (U of L)

2) A random variable X has a probability density function given by

$$f(x) = a\,e^{-x/3} \quad \text{for } x \geqslant 0$$
$$f(x) = 0 \quad \text{for } x < 0$$

Show that the value of a is $\frac{1}{3}$ and find the probabilities $P(0 \leqslant X \leqslant 1)$ and $P(X \geqslant 2)$. (U of L)

3) The probability density function $f(x)$ for a random variable X is given by

$$f(x) = 0, \quad x < 0$$
$$f(x) = C(3x - x^2), \quad 0 \leqslant x \leqslant 3$$
$$f(x) = 0, \quad x > 3$$

Determine the value of the constant C and find the mean value of X. (U of L)

4) A probability density function of a random variable X is defined by

$$f(x) = \begin{cases} 0, & x < 0 \\ kx, & 0 \leqslant x \leqslant 1 \\ 0, & x > 1 \end{cases}$$

where k is a positive constant. Evaluate k and obtain the mean value of the random variable. (U of L)

5) A random variable X has a p.d.f. given by

$$f(x) = k(3-x)^2 \quad \text{for} \quad 0 \leqslant x < 3$$

Show that $k = \frac{1}{9}$ and find $E(X)$ and $\text{Var}(X)$.

6) A random variable X has probability density function f given by

$$f(x) = \begin{cases} 0, & x < 0 \\ \lambda(1-\cos x), & 0 \leqslant x \leqslant \pi/2 \\ \lambda \sin x, & \pi/2 < x \leqslant \pi \\ 0, & \pi < x \end{cases}$$

where λ is a constant. Calculate λ and the mean value of X. (U of L)

7) A continuous random variable has a p.d.f. given by

$$f(x) = c \tan^{-1} x \quad \text{for} \quad 0 \leqslant x < 1$$
$$f(x) = 0 \quad \text{elsewhere.}$$

Find the value of c and $E(X)$.

8) A probability density function f of a random variable X is defined by

$$f(x) = \begin{cases} 0, & x < 0 \\ \lambda, & 0 \leqslant x < 1 \\ \lambda/x, & 1 \leqslant x < 3 \\ 0, & 3 \leqslant x \end{cases}$$

where λ is a positive constant. Sketch the graph of f and prove that

$$\lambda = (1 + \ln 3)^{-1}.$$ (U of L)

9) Derive the mean and variance of the random variable X whose p.d.f. is given by

$$f(x) = a\,e^{-ax} \quad \text{for} \quad x \geqslant 0$$
$$f(x) = 0 \quad \text{elsewhere}$$

(This is called the exponential probability distribution.)

10) A random variable X takes values x such that $-1 \leqslant x \leqslant 1$ and its probability density function $f(x)$ is given by

$$f(x) = k(x^2+3x+2) \quad \text{when} \quad -1 \leqslant x \leqslant 0$$
$$f(x) = k(x^2-3x+2) \quad \text{when} \quad 0 \leqslant x \leqslant 1,$$
$$f(x) = 0 \quad \text{otherwise}$$

where k is a positive constant. Show that $k = 3/5$.
Determine the probability that X takes a value between $-\frac{1}{2}$ and $+\frac{1}{2}$.
(U of L)

11) The probability density function f of a continuous random variable X is given by

$$f(x) = kx^2(2-x) \quad \text{for } 0 \leqslant x \leqslant 2$$
$$f(x) = 0 \quad \text{elsewhere}$$

where k is a constant.
(a) Evaluate k.
(b) Draw a sketch of $f(x)$, giving the x-coordinate of the maximum point.
(c) Calculate $P(1 \leqslant X \leqslant 2)$.
(d) Find the mean and variance of X. (U of L)

12) A random variable X has a p.d.f. given by

$$f(x) = \tfrac{3}{2}(1-x^2) \quad \text{for} \quad 0 \leqslant x < 1$$
$$f(x) = 0 \quad \text{elsewhere.}$$

Find the mean and standard deviation of X and show that $P(\mu-\sigma \leqslant X < \mu+\sigma) = 0.66$ approximately.

13) Find the probability distribution (probability density function) $f(x)$ such that $f(x)$ is a linear function of x in the interval $1 \leqslant x \leqslant 2$, $f(x) = 0$ outside this interval and $f(2) = 0$. (U of L)

14) The continuous random variable X, $a \leqslant x \leqslant b$, has probability density function $f(x)$ with mean value μ. Show that

$$\int_a^b (x-\mu)^2 f(x)\,dx = \int_a^b x^2 f(x)\,dx - \mu^2$$

The probability density function of the random variable X is given by

$$f(x) = \lambda x^2(1-x) \quad \text{for} \quad 0 \leqslant x \leqslant 1$$
$$f(x) = 0 \quad \text{for} \quad x < 0 \quad \text{and} \quad x > 1$$

Find (a) the value of λ,
(b) the mean and the standard deviation of X.
Given that the median of X is M, show that

$$6M^4 - 8M^3 + 1 = 0$$

(U of L)

15) A random variable X has a rectangular distribution over the interval $-1 \leqslant x < 1$. Find $E(X)$ and $\text{Var}(X)$. The random variable Y is such that $Y = X^2$. Show that $0 \leqslant y \leqslant 1$ and find the p.d.f. of Y.

16) State the conditions which must be satisfied by any probability density function.
The function $f(x)$, where

$$f(x) = \begin{cases} c\,[a-(x-b)^2] & \text{for } -1 \leqslant x \leqslant 1 \\ 0 & \text{otherwise} \end{cases}$$

where c, a and b are constants and $c > 0$, is a probability density function with zero mean. Show that $b = 0$, find the set of possible values of a and express c in terms of a.
Find also the smallest possible value of the variance of the distribution. (U of L)

17) The random variable X has a rectangular distribution whose probability density function $f(x)$ is given by

$$f(x) = \begin{cases} \frac{1}{6} & \text{for } -2 \leqslant x \leqslant 4, \\ 0 & \text{otherwise.} \end{cases}$$

Sketch the probability density function of X and hence, or otherwise, find the probabilities that
(a) $X \leqslant 2$,
(b) $|X| \leqslant 2$,
(c) $|X| \leqslant x$ for $0 \leqslant x \leqslant 2$,
(d) $|X| \leqslant x$ for $2 < x \leqslant 4$,
(e) $|X| \leqslant x$ for $x > 4$.
Hence obtain, and sketch, the probability density function of $|X|$.
Thus determine the mean of $|X|$. (MEI)

18) The continuous random variable X has probability density function

$$f(x) = \begin{cases} \lambda e^{-\lambda x} & x \geqslant 0 \\ 0 & \text{elsewhere,} \end{cases}$$

where $\lambda > 0$. Obtain the mean and variance of X.
Show that $P(X > a + b \mid X > a)$ is equal to $P(X > b)$, for a and b both greater than zero.
A random variable having this p.d.f. is often postulated as a good model for the length of life of items such as certain types of electrical components, light bulbs, and so on. Suppose a certain organisation uses a large number of components whose life-length is indeed described by this p.d.f. What does your last result above lead you to conclude about a 'preventive maintenance' policy of replacing such a component by a new one at pre-specified intervals even if the old one is still working satisfactorily? (MEI)

19) The random variable X has probability density function (p.d.f.) given by

$$f(x) = \begin{cases} \dfrac{1}{\lambda}e^{-x/\lambda} & x \geqslant 0, \quad \text{where} \quad \lambda > 0 \\ 0 & \text{otherwise} \end{cases}$$

Show that the mean of this distribution is λ.
A manufacturer of hi-fi equipment buys a certain type of integrated circuit from Superduper Enterprises at a cost of 50 p each. The time to failure of these circuits can be taken as a random variable with the above p.d.f. and mean 700 hours. It costs the manufacturer £1.50 to honour his guarantee whenever a circuit of this type in his equipment lasts for 400 hours or less. What is the expected cost to the manufacturer of using one such circuit? (You may assume that at most one claim with respect to this circuit may occur in the guarantee period.)
A rival company, Buzz Electronics, offers to supply the hi-fi manufacturer with these circuits at a cost to be negotiated, where, once again, the time to failure can be taken to be a random variable with the above distribution, but this time with mean life 400 hours. By considering the expected costs, calculate the maximum amount the manufacturer may pay Buzz Electronics for a circuit, for this new circuit to be a cheaper buy. (AEB)

20) The number of people who attend an open air swimming pool on a fine day may be considered to be a continuous random variable X having probability density function

$$f(x) = k_1 x \quad 0 \leqslant x \leqslant 500$$
$$f(x) = 0 \quad \text{otherwise}$$

and on a wet day, as a continuous random variable Y having probability density function

$$g(y) = k_2(500 - y) \quad 0 \leqslant y \leqslant 500$$
$$g(y) = 0 \quad \text{otherwise}$$

(a) Find k_1 and k_2, and the expected number of people who attend the swimming pool on both wet and fine days.
(b) Prove that $P(X < N) = P(Y > 500 - N)$.
(c) The running costs of the pool are the same on wet or fine days and amount to £50 per day. The entrance fee is 20 p on wet days and 40 p on fine days. If the pool is open every day of the week, find the number of fine days needed in a week for the expected profit of the pool to be positive.
(d) If the probabilities that a day will be fine or wet in the summer are 0.55 and 0.45 respectively, find the probability that more than 300 people will attend the pool on any one day in the summer. (WJEC)

CHAPTER 18

THE POISSON AND NORMAL DISTRIBUTIONS

In this chapter we investigate two distributions, one discrete and one continuous, which are important because they act as good models for many naturally occuring phenomena.

THE POISSON PROBABILITY DISTRIBUTION

Consider the following variables,

the number of flaws per metre of cloth,

the number of accidents per month on a particular stretch of road,

the number of incoming calls per day at a particular house,

the number of misprints per page of a book,

the number of diseased cells per slide made from plant material.

In each case

(a) the variable is a number of events, so it is discrete and takes only positive integral values, i.e. $0, 1, 2, 3, \dots,$

(b) the number of events is considered per unit of a quantity (the quantity is anything that can be divided into units, e.g. time, mass, distance, pages) and there must be no obvious maximum number of events per unit quantity,

(c) the event occurs randomly.

If, in addition to these properties, the event is rare, i.e. the probability of the event occurring at any given instant is low, then experimentally it is found that the probability of x occurrences of the event per unit quantity is given by

$e^{-\lambda}\dfrac{\lambda^x}{x!}$, where λ is a constant.

This probability distribution is called *the Poisson distribution.*

We will now show that a variable X with such a probability distribution is a *random* variable.

If $\quad P(X=x) = e^{-\lambda}\dfrac{\lambda^x}{x!} \quad$ for $\quad x = 0, 1, 2, 3, \ldots$

then

$$\sum_{x=0}^{\infty} P(X=x) = e^{-\lambda}\left[1+\lambda+\frac{\lambda^2}{2!}+\frac{\lambda^3}{3!}+\ldots\right]$$
$$= e^{-\lambda}\left[e^{\lambda}\right]$$
$$= 1$$

Therefore X is a discrete random variable.

Hence if a discrete random variable X has a Poisson distribution, the p.d.f. of X is given by

$$f(x) = e^{-\lambda}\frac{\lambda^x}{x!} \quad \text{for} \quad x = 0, 1, 2, 3, \ldots$$

The constant λ is called the parameter of the distribution.

The Mean of the Poisson Distribution

If the p.d.f. of X is given by $\quad f(x) = e^{-\lambda}\dfrac{\lambda^x}{x!} \quad$ for $\quad x = 0, 1, \ldots \quad$ then

$$E(X) = \sum_{x=0}^{\infty} x\, e^{-\lambda}\frac{\lambda^x}{x!}$$
$$= \left[(0)(e^{-\lambda}) + (1)(\lambda)(e^{-\lambda}) + (2)\left(\frac{\lambda^2}{2!}\right)(e^{-\lambda}) + (3)\left(\frac{\lambda^3}{3!}\right)(e^{-\lambda}) + \ldots\right]$$
$$= e^{-\lambda}\left[\lambda+\lambda^2+\frac{\lambda^3}{2!}+\frac{\lambda^4}{3!}+\ldots\right]$$
$$= \lambda e^{-\lambda}\left[1+\lambda+\frac{\lambda^2}{2!}+\frac{\lambda^3}{3!}+\ldots\right]$$
$$= \lambda e^{-\lambda} e^{\lambda}$$
$$= \lambda$$

Therefore the mean of the Poisson distribution is λ.

The Variance of the Poisson Distribution

If the p.d.f. of X is $e^{-\lambda}\dfrac{\lambda^x}{x!}$ for $x = 0, 1, 2, \ldots$

then $\text{Var}(X) = \text{E}(X^2) - \mu^2$

Now $\mu = \text{E}(X) = \lambda$

and
$$\begin{aligned}
\text{E}(X^2) &= \sum_{x=0}^{\infty} x^2 \frac{\lambda^x}{x!} e^{-\lambda} \\
&= e^{-\lambda}\left[(0)(1) + (1)(\lambda) + (4)\left(\frac{\lambda^2}{2!}\right) + (9)\left(\frac{\lambda^3}{3!}\right) + \ldots\right] \\
&= \lambda e^{-\lambda}\left[1 + 2\lambda + \frac{3\lambda^2}{2!} + \frac{4\lambda^3}{3!} + \ldots\right] \\
&= \lambda e^{-\lambda}\frac{d}{d\lambda}\left[\lambda + \lambda^2 + \frac{\lambda^3}{2!} + \frac{\lambda^4}{3!} + \ldots\right] \\
&= \lambda e^{-\lambda}\frac{d}{d\lambda}\left[\lambda e^{\lambda}\right] \\
&= \lambda e^{-\lambda}[e^{\lambda} + \lambda e^{\lambda}] \\
&= \lambda + \lambda^2
\end{aligned}$$

Therefore $\text{Var}(X) = (\lambda + \lambda^2) - \lambda^2 = \lambda$

Hence the variance of the Poisson distribution is λ.

Summarising,

> if X is a discrete random variable with a Poisson probability distribution then the p.d.f. of X is given by
>
> $$f(x) = \frac{\lambda^x}{x!}e^{-\lambda} \quad \text{for} \quad x = 0, 1, 2, 3, \ldots$$
>
> where λ is both the mean *and* variance of X.

The Poisson distribution can be used if X is stated to be a Poisson variable, but when this is not the case we must satisfy ourselves that X fulfils the necessary conditions,
i.e. X is the number of occurrences of a randomly occurring event in a given interval (i.e. the unit quantity) when the event itself is fairly unlikely, and there is no maximum number of occurrences. The value of λ, i.e. the mean number of occurrences in a given interval, must also be known before the Poisson distribution can be used.

EXAMPLES 18a

1) The number of flaws in cloth woven on a loom averages 2 per piece (where a piece is a 25 m length of this cloth). Assuming that the number of flaws per piece has a Poisson distribution find, to 3 d.p., the probabilities that there are 0, 1, 2, 3, 4, 5 and 6 flaws in one piece and hence find the probability that there are more than 6 flaws in any one piece.

Let X be the number of flaws per piece.
X has a Poisson distribution with mean 2.

Therefore $\mathrm{P}(X=x) = \dfrac{2^x}{x!}\mathrm{e}^{-2}$ for $x = 0, 1, 2, \ldots$

giving $\mathrm{P}(X=0) = \mathrm{e}^{-2} = 0.135$

$\mathrm{P}(X=1) = 2\mathrm{e}^{-2} = 0.271$

$\mathrm{P}(X=2) = \dfrac{2^2}{2!}\mathrm{e}^{-2} = 0.271$

$\mathrm{P}(X=3) = \dfrac{2^3}{3!}\mathrm{e}^{-2} = \tfrac{2}{3}\mathrm{P}(X=2) = 0.180$

$\mathrm{P}(X=4) = \dfrac{2^4}{4!}\mathrm{e}^{-2} = \tfrac{2}{4}\mathrm{P}(X=3) = 0.090$

$\mathrm{P}(X=5) = \dfrac{2^5}{5!}\mathrm{e}^{-2} = \tfrac{2}{5}\mathrm{P}(X=4) = 0.036$

$\mathrm{P}(X=6) = \dfrac{2^6}{6!}\mathrm{e}^{-2} = \tfrac{2}{6}\mathrm{P}(X=5) = 0.012$

Hence $\mathrm{P}(X > 6) = 1 - \mathrm{P}(X \leqslant 6)$

$= 1 - [0.135 + 0.271 + \ldots + 0.012]$

$= 0.005$

The Shape of the Poisson Distribution

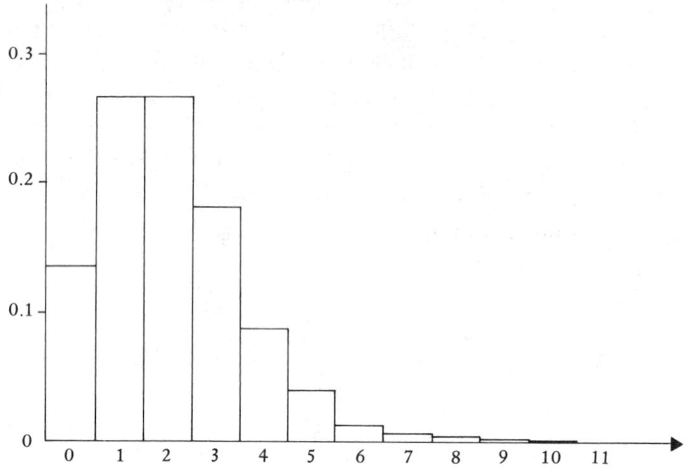

The histogram illustrates the probabilities from Example 18a, No. 1. It shows the typical shape of the Poisson distribution, i.e. it is skewed towards the left and tails off towards zero to the right. (There is no fixed upper value for X, but the probabilities associated with values of X greater than 7 get progressively smaller, so the tops of the bars approach the horizontal axis.)

There is no fixed maximum value for X in any Poisson distribution; hence the histogram will always have a similar shape with a peak around the mean value and a tail-off towards the right. The diagrams below show the shape of the histogram for some different values of λ.

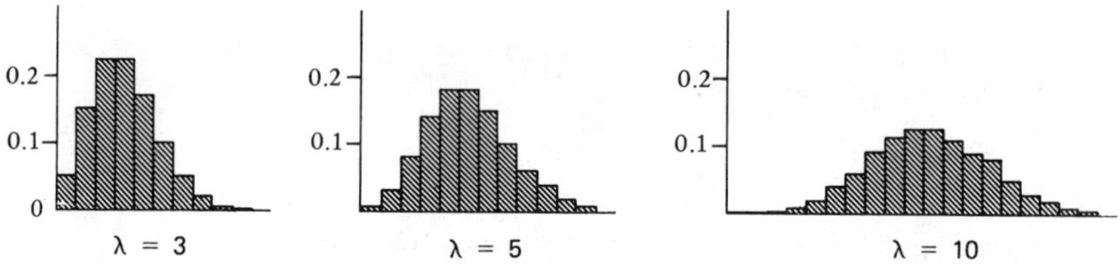

EXAMPLES 18a (continued)

2) The number of misprints per page of a book is a Poisson variable with a mean value of 0.1 misprints per page. Find, to 3 d.p., the probability of
(a) one page containing more than one misprint,
(b) one chapter of 20 pages containing more than one misprint.

(a) Let X be the number of misprints per page,

then $\quad P(X=x) = \dfrac{(0.1)^x}{x!}e^{-0.1}$

Hence
$$\begin{aligned} P(X > 1) &= 1-[P(X=0)+P(X = 1)] \\ &= 1-e^{-0.1}(1+0.1) \\ &= 0.005 \quad \text{to 3 d.p.} \end{aligned}$$

(b) (The unit interval is now 20 pages so the variable is the number of misprints per 20 pages of the book. The average is 0.1 per page, giving an average of 20×0.1 per 20 pages.)

Let Y be the number of misprints per 20 pages

then $P(Y=y) = \dfrac{2^y}{y!}e^{-2}$

Therefore
$$\begin{aligned} P(Y > 1) &= 1-[P(Y = 0)+P(Y = 1)] \\ &= 1-e^{-2}[1+2] \\ &= 0.594 \quad \text{to 3 d.p.} \end{aligned}$$

3) The number of people asking for emergency dental treatment per day at surgery A is a Poisson variate with a mean value of 2. The number of requests for emergency treatment per day at surgery B is an independent Poisson variate with a mean value of 3. Find, to 3 s.f., the probability that
(a) both surgeries have exactly two requests for emergency treatment,
(b) less than two people in total ask for emergency treatment at either or both surgeries.

Find an expression for the probability that both surgeries have the same number of requests for emergency treatment.

If X is the number of requests for emergency treatment at surgery A, then X has a Poisson distribution with $\lambda = 2$,

so the p.d.f. of X is $\dfrac{2^x}{x!}e^{-2}$ for $x = 0, 1, 2, \ldots$

Let Y be the number of requests for emergency treatment at surgery B, then Y has a Poisson distribution with $\lambda = 3$,

so the p.d.f. of Y is $\dfrac{3^y}{y!}e^{-3}$ for $y = 0, 1, 2, \ldots$

(a)
$$P(X = 2) = \frac{2^2}{2!}e^{-2} = 0.270\,67$$

$$P(Y = 2) = \frac{3^2}{2!}e^{-3} = 0.224\,04$$

$$\begin{aligned} P(X = 2 \text{ and } Y = 2) &= P(X = 2) \times P(Y = 2) \\ &= 0.061 \quad \text{to 3 s.f.} \end{aligned}$$

(b) P(less than 2 requests per day at both surgeries)

$$= P(X=0 \text{ and } Y=0 \quad \text{or} \quad X=0 \text{ and } Y=1 \quad \text{or} \quad X=1 \text{ and } Y=0)$$

$$= P(X = 0)P(Y = 0) + P(X = 0)P(Y = 1) + P(X = 1)P(Y = 0)$$

$$= (e^{-2})(e^{-3}) + (e^{-2})(3e^{-3}) + (2e^{-2})(e^{-3})$$

$$= 6e^{-5} = 0.0404 \quad \text{to 3 s.f.}$$

If both surgeries have the same number of requests in a day, then both of them have either no requests, or one request, or two, or three ...

i.e. $$P(X = Y) = P(X = 0)P(Y = 0) + P(X = 1)P(Y = 1) + \dots$$

$$= (1)(1) + (2e^{-2})(3e^{-3}) + \left(\frac{2^2}{2!}e^{-2}\right)\left(\frac{3^2}{2!}e^{-3}\right) + \dots$$

$$= 1 + 6e^{-5} + \frac{6^2 e^{-5}}{(2!)^2} + \frac{6^3 e^{-5}}{(3!)^2} + \dots$$

$$= e^{-5} \sum_{r=0}^{\infty} \frac{6^r}{(r!)^2}$$

4) A man travels to work by bus. He always arrives at the bus stop at the same time and if he has to wait more than five minutes for a bus he is late for work. The number of buses arriving at the bus stop is a Poisson variate with a mean of 2 buses every five minutes.

(a) Find the probability that, on any one day, he will be late for work.

(b) Find the probability that he will be late for work at least once in a five-day working week.

(a) Let X be the number of buses arriving at the stop in a 5 minute period

then $P(X=x) = \dfrac{2^x}{x!}e^{-2}$

The man is late for work if $X=0$

$$P(X = 0) = e^{-2} = 0.135 \quad \text{to 3 s.f.}$$

(b) In a five-day week the man could be late on 0, 1, 2, 3, 4 or 5 days.

The variable, 'the number of days the man is late' is *not* the variable in (a), *nor* does it have a Poisson distribution.

If Y is the number of days he is late, then Y has a binomial distribution with $n = 5$ and $p = 0.135$

Therefore
$$\begin{aligned} P(Y \geqslant 1) &= 1 - P(Y = 0) \\ &= 1 - (1 - 0.135)^5 \\ &= 0.516 \quad \text{to 3 s.f.} \end{aligned}$$

EXERCISE 18a

1) The random variable X has a Poisson distribution with mean value 4. Find (a) $P(X = 2)$, (b) $P(X < 2)$, (c) $P(X > 2)$.

2) The mean number of red blood cells per prepared slide is 4. If the number of red blood cells on a slide is a Poisson variate find, to 3 d.p., the probability that there are (a) no red blood cells on a slide, (b) exactly 6 red blood cells on a slide.

3) The number of people entering a railway station between the hours of 7.30 and 9.30 a.m. on a weekday is a Poisson variate and averages two people per minute. Find the probability that
(a) five people enter in one minute,
(b) more than four people enter in one minute,
(c) less than four people enter in a two-minute period.

4) On a particular stretch of motorway there is an average of 0.1 accidents per seven-day week. If the number of accidents has a Poisson distribution, find the probability that, on this stretch of motorway, there are
(a) no accidents in one week,
(b) no accidents in one day,
(c) more than 1 accident in a four-week period.

5) The average number of flaws in carpet produced on a particular loom is 0.01 per metre. Assuming that the flaws are randomly spread, find the probability that
(a) there are no flaws in 1 metre,
(b) there are no flaws in a 20-metre length,
(c) there are more than 2 flaws in a 40-metre length.

6) A shop sells two brands of baked beans, brand A and brand B. The daily demand for brand A is a Poisson distribution with mean 3 and the daily demand for brand B has a Poisson distribution with mean 2. Find the probability that, on a particular day

(a) one tin of brand A is asked for,
(b) one tin of brand B is asked for,
(c) one tin of brand A and one tin of brand B are asked for,
(d) the demand is for just 1 tin of baked beans.

7) There are two fire stations in Trutown: station A and station B. The number of calls at station A is a Poisson variate with a mean of 3 calls a day and the number of calls at station B is a Poisson variate with a mean of 4 calls a day. Find the probability that, on a particular day

(a) exactly two calls are received by each station,
(b) exactly two calls are received in total by either or both stations,
(c) less than two calls are received by either or both stations.

8) Slides are prepared from plant material from two sources, A and B. The number of cells with two nuclei is randomly distributed and for slides made from source A, the mean number of double nucleus cells is 1 per slide, while slides made from source B have a mean of 0.5 double nucleus cells per slide. Find the probability that

(a) a slide made from source A has 2 double nucleus cells,
(b) a slide made from source B has 1 double nucleus cell,
(c) a slide chosen at random has exactly 1 double nucleus cell if equal numbers of cells are prepared from each source.

9) A car hire company has three limousines available each day for hire. The demand for these cars is a Poisson variate with a mean of 1 demand for a limousine per day. Find the probability that

(a) the company cannot meet the demand for limousines on any one day,
(b) the company cannot meet the demand for limousines on exactly one day in a five-day working week.

10) In a large office building it is found that the number of light bulbs failing a day is a Poisson variate with a mean of two failures a day. Find the probability that

(a) more than two light bulbs fail on any one day,
(b) more than five light bulbs fail during a period of five consecutive days,
(c) more than two light bulbs fail on each day of a five-day working week.

Using the Poisson Distribution as an Approximation for the Binomial Distribution

The typical Poisson distribution is skewed to the left and tails off towards zero to the right.

Now consider a binomial distribution with parameters n and p. The shape of this distribution depends on the values of n and p. If $p \simeq 0.5$, the distribution is roughly symmetrical. If p is large the distribution is skewed to the right. If p is small the distribution is skewed to the left, and in this case, if n is large, there will be a tail to the right and the histogram looks something like this:

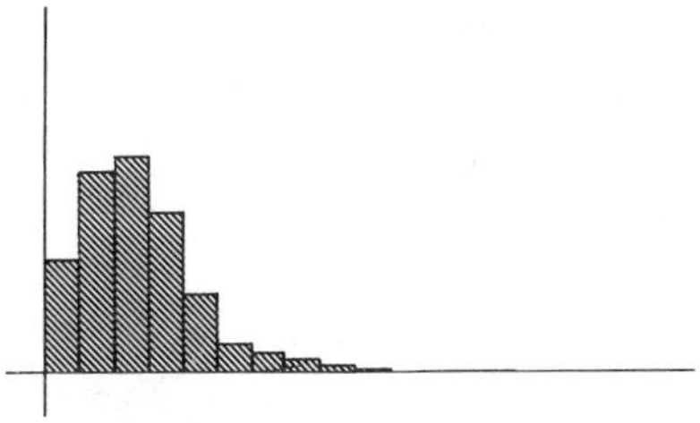

The similarity between this histogram and the shape of the Poisson distribution is obvious.

A further similarity between the distributions is that they both give probabilities for the number of times that an event can happen. (The main difference between them is that for a binomial distribution there is a fixed and finite number of times the event may happen but this is not the case in a Poisson distribution.)

A clear comparison can be made by comparing histograms of the binomial distribution with $n = 100$ and $p = 0.02$ and the Poisson distribution with the same mean value, i.e. $\lambda = np = 2$.

Binomial Distribution

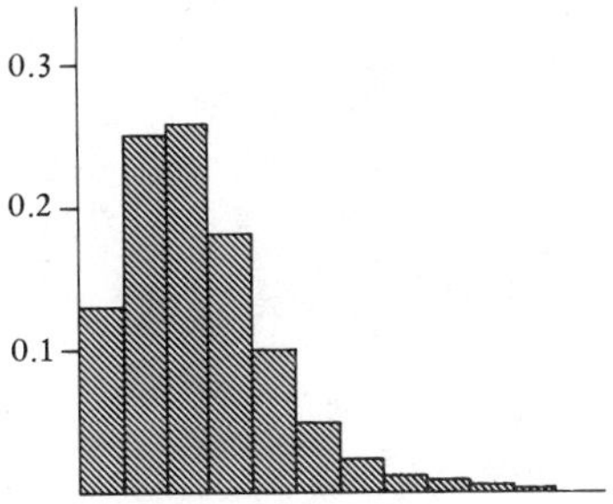

$n = 100, \ p = 0.02$

Poisson Distribution

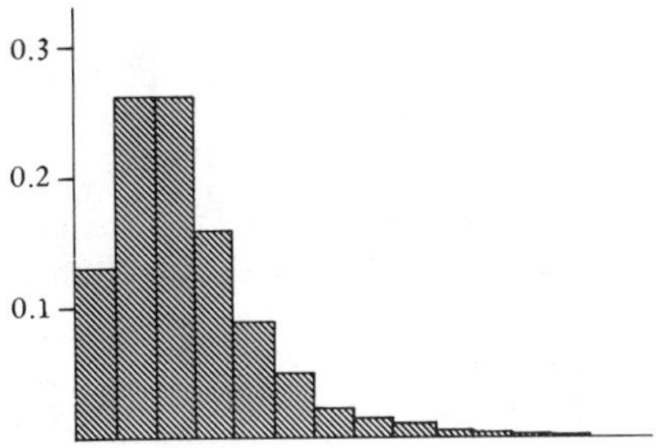

$\lambda = np = 2$

The table below shows the values (to 5 d.p.) of the probabilities for the two distributions.

	Binomial $P(X=x) = {}^{100}C_x(0.02)^x(0.98)^{100-x}$	Poisson $P(X=x) = \frac{2^x}{x!}e^{-2}$
$P(X=0)$	0.132 62	0.135 34
$P(X=1)$	0.270 65	0.270 67
$P(X=2)$	0.273 41	0.270 67
$P(X=3)$	0.182 28	0.180 45
$P(X=4)$	0.090 21	0.090 22
$P(X=5)$	0.035 35	0.036 09
$P(X=6)$	0.011 42	0.012 03
$P(X=7)$	0.003 13	0.003 44
⋮	⋮	⋮

This shows that the probabilities calculated from the Poisson distribution differ very little from those calculated from the binomial distribution.

Therefore, in general

if X has a Binomial distribution with parameters n and p, then if n is large and p is small the Poisson distribution with $\lambda = np$ can be used as an approximation for the Binomial distribution.

If $n > 50$ and $p < 0.15$ the approximation is reasonably good. The approximation improves as n increases and p decreases.

The advantage in using the Poisson distribution as an approximation for the binomial distribution is that the calculations are much simpler.

For example, if X has a binomial distribution with parameters $n = 500$ and $p = 0.01$

then
$$P(X = 3) = {}^{500}C_3(0.01)^3(0.99)^{497}$$
$$= \frac{(500)(499)(498)}{3 \times 2}(0.01)^3(0.99)^{497}$$
$$= 0.1402 \quad \text{to 4 d.p.}$$

Whereas using the Poisson distribution with $\lambda = np = 5$ gives

$$P(X = 3) \simeq \frac{5^3}{3!}e^{-5}$$
$$= 0.1404 \quad \text{to 4 d.p.}$$

We can prove mathematically that as $n \to \infty$ and $p \to 0$, the Binomial distribution tends to the Poisson distribution.

If X has a binomial distribution with parameters n and p then

$$P(X = x) = \frac{n!}{(n-x)!x!}p^x q^{n-x}$$

Let $\lambda = np \Rightarrow p = \frac{\lambda}{n}$ and $q = 1-p = 1-\frac{\lambda}{n}$

Then

$$P(X = x) = \frac{n!}{(n-x)!x!}\left(\frac{\lambda}{n}\right)^x\left(1-\frac{\lambda}{n}\right)^{n-x}$$

$$= \frac{n!}{(n-x)!n^x}\left(\frac{\lambda^x}{x!}\right)\frac{\left(1-\frac{\lambda}{n}\right)^n}{\left(1-\frac{\lambda}{n}\right)^x}$$

$$= \left(1-\frac{\lambda}{n}\right)^n\left(\frac{\lambda^x}{x!}\right)\left(\frac{n!}{(n-x)!n^x}\right)\frac{1}{\left(1-\frac{\lambda}{n}\right)^x}$$

Now $\left(1-\frac{\lambda}{n}\right)^n = 1-\lambda+\frac{n(n-1)}{n^2}\frac{\lambda^2}{2!}-\frac{n(n-1)(n-2)}{n^3}\frac{\lambda^3}{3!}+ \quad \dots \quad +\lambda^n$

Therefore, as $n \to \infty$, $\left(1-\frac{\lambda}{n}\right)^n \to 1-\lambda+\frac{\lambda^2}{2!}-\frac{\lambda^3}{3!}+ \quad \dots \quad = e^{-\lambda}$

Also, as $n \to \infty$, $\frac{\lambda}{n} \to 0$ therefore $\left(1-\frac{\lambda}{n}\right)^x \to 1$ and $\frac{n!}{(n-x)!n^x} \to 1$

Therefore

$$\lim_{n\to\infty} P(X = x) = e^{-\lambda}\frac{\lambda^x}{x!}$$

which is the p.d.f. of the Poisson distribution.

EXAMPLES 18b

1) Two dice are thrown 100 times. Find the probability of getting exactly 3 double sixes.

This is a Binomial distribution with $n = 100$ and $p = \frac{1}{36}$. As n is large and p is small we can use the Poisson distribution with $\lambda = np = \frac{100}{36}$ to give an approximate value for $P(X = 3)$.

i.e. $$P(X = 3) \simeq \frac{(25/9)^3}{3!}e^{-25/9} = 0.222 \quad \text{to 3 d.p.}$$

2) In a large population, one person in a hundred has a rhesus negative blood group. If a random sample of 300 people is taken from this population find approximately the probability that there will be at least 5 people with a rhesus negative blood group.
How many people must a sample contain so that the probability of including at least one person with rhesus negative blood is greater than 0.9?

If X is the random variable, then X has a Binomial distribution with $n = 300$ and $p = 0.01$

Using as an approximation the Poisson distribution with $\lambda = 3$, gives

$$\begin{aligned} P(X \geqslant 5) &= 1 - [P(X = 0) + P(X = 1) + P(X = 2) + P(X = 3) + P(X = 4)] \\ &\simeq 1 - e^{-3}\left[1 + 3 + \frac{3^2}{2!} + \frac{3^3}{3!} + \frac{3^4}{4!}\right] \\ &= 1 - e^{-3}[16.375] \\ &= 0.185 \quad \text{to 3 d.p.} \end{aligned}$$

Let the size of the sample be n, then

$$P(X \geqslant 1) = 1 - P(X = 0) < 0.9$$

Using the Poisson distribution with $\lambda = \frac{n}{100}(np)$ gives

$$\begin{aligned} 1 - e^{-\frac{n}{100}} &< 0.9 \\ e^{-\frac{n}{100}} &> 0.1 \\ \frac{n}{100} &> 2.303 \\ n &> 230.3 \end{aligned}$$

Hence the sample must contain 231 people.

EXERCISE 18b

1) A random variable X has a Binomial distribution with $n = 1000$ and $p = 0.01$.
(a) Use the Binomial distribution to find $P(X = 2)$.
(b) Use the Poisson distribution to find $P(X = 2)$.

2) X is a random variable with a Binomial distribution where
(a) $n = 500$, $p = 0.8$,
(b) $n = 200$, $p = 0.1$,
(c) $n = 5$, $p = 0.04$,
(d) $n = 20$, $p = 0.5$.
In each case state whether it is reasonable to use the Poisson distribution to find probabilities associated with X. Give reasons for your decisions.

3) Eggs are distributed to a supermarket chain from a central depot. On average one egg in 120 eggs is cracked. The eggs are boxed and packed into lorries so that each lorry has a load of 2400 eggs. Find the probability that one lorry load of eggs contains
(a) no cracked eggs,
(b) 10 cracked eggs.

4) A police check on lorries entering the country from a port finds that on average one lorry in fifty contravenes the regulations. Five hundred lorries a week go through this check. Find the probability that in any one week there will be
(a) 3 lorries that contravene regulations,
(b) 10 lorries that contravene regulations.

5) A sample of seeds is taken from a large quantity of seeds of which 1% will not germinate. Find the probability that all the seeds in the sample will germinate if the sample contains
(a) 10 seeds,
(b) 500 seeds.
How many seeds need a sample contain if the probability that at least one seed will not germinate is to be greater than 0.9?

6) The number of cars without an up-to-date tax disc passing a police check is a Poisson variable with a mean of two such cars per hour.
Find the probability that, in one hour, one car without an up-to-date tax disc passes the police check.
If two hundred cars pass the check during this hour, find the probability that exactly three are without a current tax disc.

7) A large population of laboratory mice are such that 95% of them are pure white. The remainder have some colour.
If a batch of 500 of these mice is delivered to one laboratory, find the probability that exactly 10 of them are not pure white.
What is the probability that a group of 10 of the mice, picked at random from the large population, are all white?

8) Two cards are dealt from an ordinary pack of playing cards. What is the probability that they are both aces? If this is repeated 100 times, what is the probability of being dealt two aces on at least four occasions?

THE NORMAL DISTRIBUTION

It has been found experimentally that many naturally occurring continuous variables, such as people's heights and masses, have a probability distribution whose p.d.f. is given by

$$f(x) = \frac{1}{\sigma\sqrt{(2\pi)}} e^{-(x-\mu)^2/2\sigma^2} \quad \text{for all values of } x$$

where μ is the mean and σ^2 is the variance of the distribution.
This distribution is called the *normal probability distribution.*

Some of the properties of the distribution can be deduced from the p.d.f.

(a) The curve $y = f(x)$ is symmetrical about $x = \mu$

(b) When $x = \mu$, $f(x)$ has a maximum value of $\dfrac{1}{\sigma\sqrt{(2\pi)}}$

Deriving the remaining properties requires the evaluation of $\int f(x)\,dx$, which is difficult, so these properties are stated without proof.

(c) $\displaystyle\int_{-\infty}^{\infty} f(x)\,dx = 1$ so a variable with this distribution is a random variable.

(d) $\displaystyle\int_{-3\sigma}^{3\sigma} f(x)\,dx \simeq 0.998$, i.e. $P(\mu - 3\sigma \leqslant X < \mu + 3\sigma) \simeq 0.998$

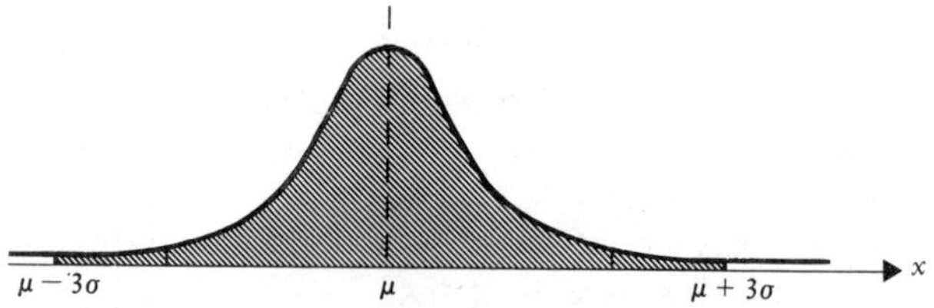

The distribution is bell shaped and it is symmetrical about its mean value. 99.8% of the distribution lies in a range of three times the standard deviation on either side of the mean value. This means that for practical purposes, the spread of the distribution is 6 standard deviations.

The exact shape of the curve depends on the values of μ and σ.

Now the probability that X lies between a and b is given by

$$\int_a^b \frac{1}{\sigma\sqrt{(2\pi)}} e^{-(x-\mu)^2/2\sigma^2}\,dx$$

and, as already stated, this integral is very difficult to evaluate.

Hence to be able to use the normal distribution we have to rely on tables that give the values of the probabilities. It is clearly not practical to provide tables for all possible values of μ and σ so a set of values for the probabilities associated with a normally distributed random variable with mean 0 and variance 1 is used.

The probabilities associated with any other normal distribution can be found from this table of values but first we will investigate the normal distribution whose mean is 0 and whose variance is 1.

The Standard Normal Distribution

The random variable which is normally distributed with a mean value of 0 and variance 1 is called the *standard* normal variable and we will call it Z. The p.d.f. of Z is denoted by $\phi(x)$.

Putting $\mu = 0$ and $\sigma^2 = 1$ in $f(x)$ gives

$$\phi(x) = \frac{1}{\sqrt{(2\pi)}} e^{-x^2/2} \quad \text{for} \quad -\infty < x < \infty$$

The properties of the general normal distribution apply, so the curve $y = \phi(x)$ is symmetrical about $x = 0$, and 99.8% of the area under the curve lies between $x = -3$ and $x = 3$.

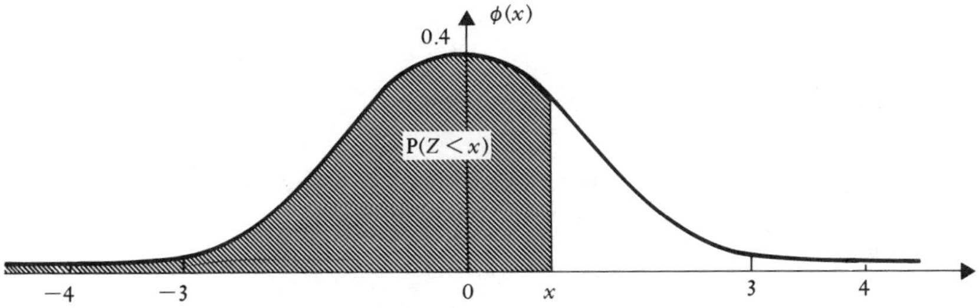

Now $P(Z < x) = \int_{-\infty}^{x} \phi(x)\, dx$.

This integral is denoted by $\Phi(x)$.
It is values of $\Phi(x)$, for differing values of x, that are given in most statistical tables. (Statistical tables do vary however; some give values for $P(Z > x)$, so a check must be made on which function is tabulated. Further, some tables give +ve and −ve values of x while some give only +ve values.)

In the examples that follow, the calculation is based on tables which give $P(Z < x)$, i.e. $\Phi(x)$, only for positive values of x, i.e. tables which give the area on the *left-hand* side of x.

EXAMPLES 18c

1) Z is the standard normal variable. Use tables to find
(a) $P(Z < 1.5)$, (b) $P(Z > 1.5)$,
(c) $P(Z < -1.5)$, (d) $P(Z > -1.5)$.

(a)

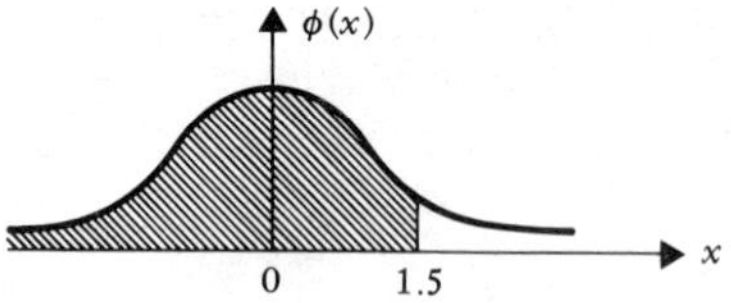

From the tables $P(Z < 1.5) = 0.9332$

(b)

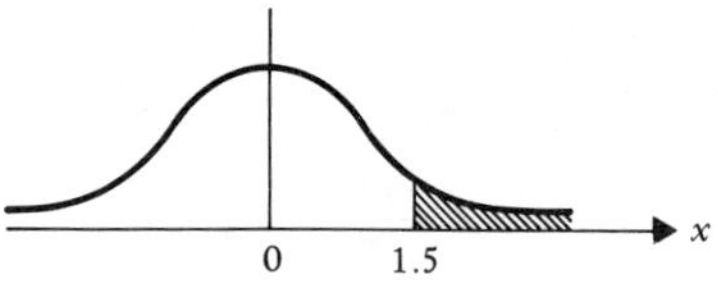

$P(Z > 1.5) = 1 - P(Z < 1.5) = 1 - 0.9332 = 0.0668$

(c)

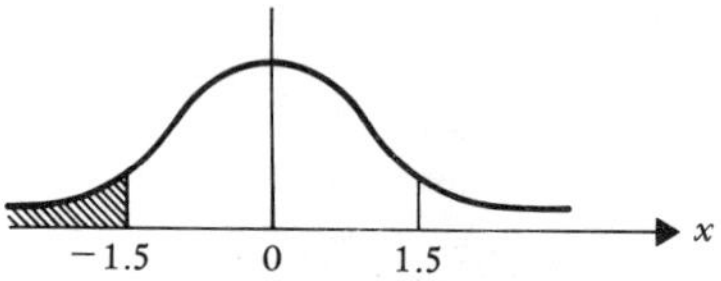

The tables give values of $P(Z < x)$ only for positive values of x. To find $P(Z < x)$ for negative values of x we make use of the symmetric shape of the curve.

Therefore $P(Z < -1.5) = P(Z > 1.5) = 0.0668$

(d)

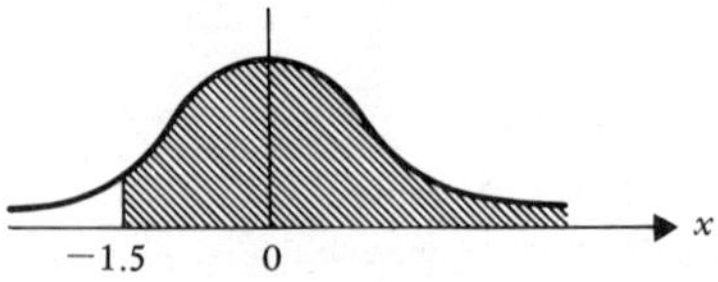

Again making use of the symmetry of the curve we have

$$P(Z > -1.5) = P(Z < 1.5)$$
$$= 0.9332$$

2) If Z is the standard normal variable use tables to find $P(-1.3 < Z < 0.5)$.

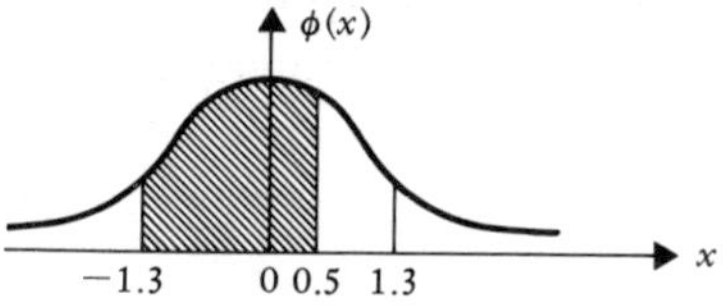

$$P(-1.3 < Z < 0.5) = P(Z < 0.5) - P(Z < -1.3)$$

To find $P(Z < -1.3)$ we use symmetry,

i.e. $$P(Z < -1.3) = P(Z > 1.3) = 1 - P(Z < 1.3)$$

From tables, $$P(Z < 1.3) = 0.9032$$

$\Rightarrow$ $$P(Z > 1.3) = 1 - 0.9032 = 0.0968$$

Hence $$P(Z < -1.3) = 0.0968$$

From tables, $$P(Z < 0.5) = 0.6915$$

Therefore $$P(-1.3 < Z < 0.5) = 0.6915 - 0.0968$$
$$= 0.5947$$

EXERCISE 18c

Z is the standard normal variable. Use tables to find the following probabilities.

1) (a) $P(Z < 0.6)$, (b) $P(Z < 0.05)$, (c) $P(Z < 3.9)$.

2) (a) $P(Z > -1.05)$, (b) $P(Z > -0.22)$, (c) $P(Z > -0.87)$.

3) (a) $P(Z > 1.4)$, (b) $P(Z > 1.82)$, (c) $P(Z > 0.74)$.

4) (a) $P(Z < -1.63)$, (b) $P(Z < -0.93)$, (c) $P(Z < 2.05)$.

5) (a) $P(1.2 < Z < 1.9)$, (b) $P(0.37 < Z < 1.5)$,
(c) $P(1.2 < Z < 2.95)$, (d) $P(-0.9 < Z < -0.5)$,
(e) $P(-1.35 < Z < -0.82)$, (f) $P(-2.6 < Z < -1.03)$,
(g) $P(-0.8 < Z < 1)$, (h) $P(-0.3 < Z < 0.3)$,
(i) $P(-1.6 < Z < 0.7)$, (j) $P(-1.25 < Z < 2.7)$.

6) Find the value of a such that
(a) $P(Z < a) = 0.8413$, (b) $P(Z > a) = 0.7734$,
(c) $P(Z < a) = 0.0668$.

STANDARDISING A NORMAL VARIABLE

In order to use the tables for *any* normal variable X, we need a function which transforms X to the standard normal variable Z.

Consider the function $f(X) = \dfrac{X-\mu}{\sigma}$ where μ and σ^2 are the mean and variance of X.

Now $f(X)$ takes different values from X, but the probabilities associated with corresponding values are the *same*, i.e. $P(X < n) = P\left(f(X) < \dfrac{n-\mu}{\sigma}\right)$.

Using the known properties of E and Var we have

$$E\big(f(X)\big) = E\left(\frac{X-\mu}{\sigma}\right) = \frac{1}{\sigma}\Big[E(X)-\mu\Big] = \frac{1}{\sigma}\Big[\mu-\mu\Big] = 0$$

and $$\mathrm{Var}\big(f(X)\big) = \mathrm{Var}\left(\frac{X-\mu}{\sigma}\right) = \frac{1}{\sigma^2}[\mathrm{Var}(X)+\mathrm{Var}(\mu)] = \frac{\sigma^2}{\sigma^2} = 1$$

Since $f(X)$ has mean 0 and variance 1, it is the *standard normal* variable, i.e. $f(X) = Z$.

Hence

> if X has a normal distribution with mean μ and variance σ^2 then
>
> $$Z = \frac{X-\mu}{\sigma} \quad \text{is the standard normal variable}$$
>
> and $$P(X < n) = P\left(Z < \frac{n-\mu}{\sigma}\right)$$

Using the formula to transform X to Z is called *standardising* X. The following examples show how the tables for probabilities associated with Z can be used to find probabilities associated with X.

EXAMPLES 18d

1) The random variable X has a normal probability distribution with mean 50 and variance 100. Find $P(X < 70)$.

First we standardise X, using $\mu = 50$ and $\sigma^2 = 100$

$$\Rightarrow \qquad Z = \frac{X-50}{10}$$

Then
$$X < 70 \;\Rightarrow\; Z < \frac{70-50}{10}$$
$$\Rightarrow\; Z < 2$$

Therefore $P(X < 70) = P(Z < 2)$.

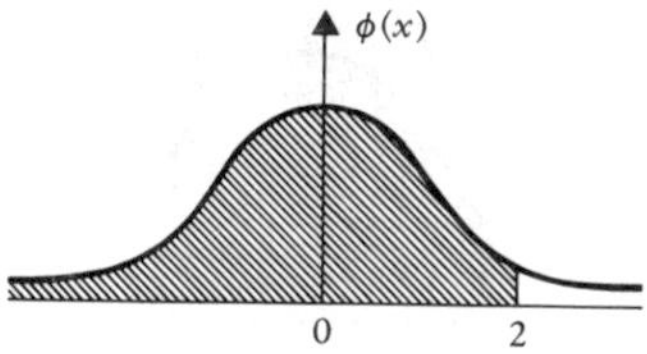

From the tables, $P(Z < 2) = 0.977\,25$

Therefore $P(X < 70) = 0.977\,25$

2) The heights of adult males in a certain population are normally distributed with a mean of 150 cm and a standard deviation of 4 cm. Find the probability that a man picked at random from this population has a height greater than 160 cm.

Let X be the heights of adult males, then X has a normal distribution with mean 150 and standard deviation 4.

Using
$$Z = \frac{X-\mu}{\sigma}$$

gives
$$Z = \frac{X-150}{4}$$

Therefore $P(X > 160) = P\left(Z > \dfrac{160-150}{4}\right) = P(Z > 2.5)$

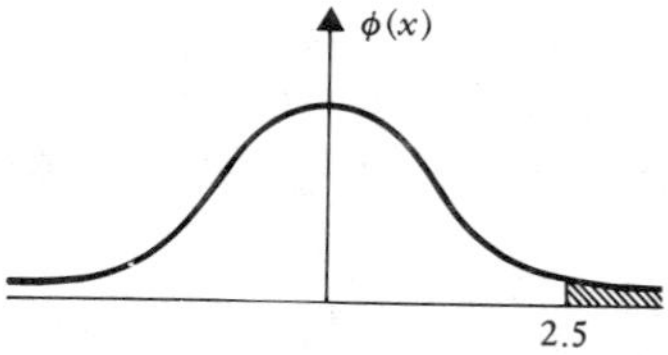

$$P(Z > 2.5) = 1-P(Z < 2.5) = 1-0.993\,79$$
$$= 0.006\,21$$

Therefore $P(X > 160) = 0.006\,21$

3) The weights of potatoes grown in a particular field are normally distributed with a mean of 80 g and a standard deviation of 5 g. If the lightest 5% and the heaviest 5% of the potatoes are discarded, find to 3 s.f. the range of weights of the remaining potatoes.

Let X be the weights of the potatoes.

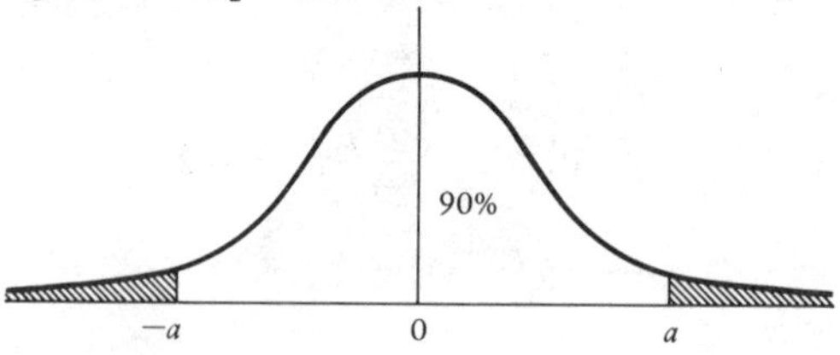

If the heaviest 5% and the lightest 5% of the distribution are discarded, this leaves 90% in the middle.
Working with Z, we need to find the value of a such that

$$P(-a < Z < a) = 0.9$$

Including the lightest 5% gives 95% of the distribution

$$\Rightarrow \qquad P(Z < a) = 0.95$$

From the tables, a is between 1.64 and 1.65, i.e. $a \simeq 1.645$
Therefore the lower 95% of the distribution corresponds to values of Z less then 1.645
Hence from symmetry, the middle 90% of the distribution corresponds to values of Z between -1.645 and 1.645

From these values of Z, corresponding values of X can be found using

$$Z = \frac{X-80}{5} \quad \Rightarrow \quad X = 5Z + 80$$

When $\qquad Z = -1.645, \quad X = 71.775$

When $\qquad Z = 1.645, \quad X = 88.225$

Therefore the remaining potatoes have weights between 71.8 g and 88.2 g to 3 s.f.

4) The times taken by the seeds in a batch to germinate are normally distributed If 20% of the seeds take more than 6 days to germinate and if 10% of the seeds germinate in less than 4 days find the mean time for germination and the standard deviation.

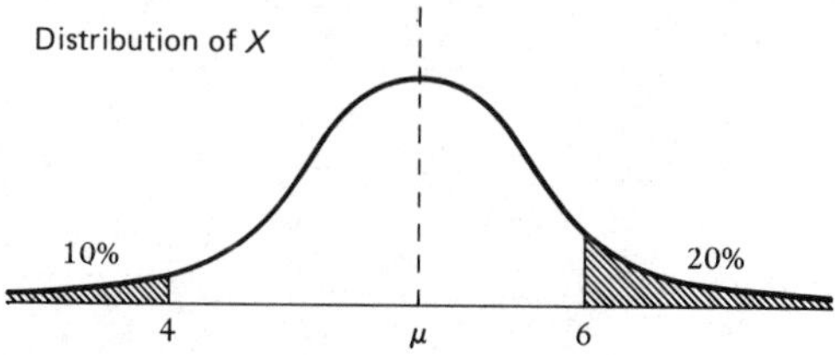

Let X be the time it takes one seed to germinate.

Then $$P(X > 6) = 0.2$$

and $$P(X < 4) = 0.1$$

If μ and σ^2 are the mean and variance, then using the standard normal variable Z gives

$$P\left(Z > \frac{6-\mu}{\sigma}\right) = 0.2 \qquad [1]$$

$$P\left(Z < \frac{4-\mu}{\sigma}\right) = 0.1 \qquad [2]$$

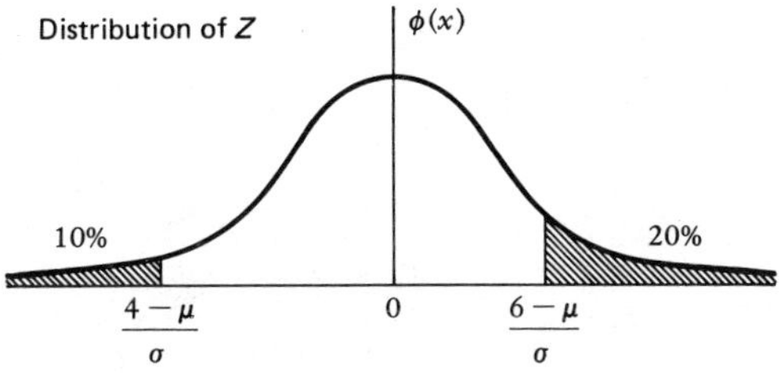

To use the tables we need Z less than a positive number, so equation [1] is written in the form

$$P\left(Z < \frac{6-\mu}{\sigma}\right) = 0.8 \qquad [3]$$

Now $\frac{4-\mu}{\sigma}$ is less than the mean value, so it is negative.

From symmetry $P\left(Z < \frac{4-\mu}{\sigma}\right) = P\left(Z > \frac{\mu-4}{\sigma}\right) = 0.1$

Hence equation [2] becomes

$$P\left(Z < \frac{\mu-4}{\sigma}\right) = 0.9 \qquad [4]$$

From the tables, taking the nearest value of Z giving these probabilities,

[3] $\Rightarrow$ $\frac{6-\mu}{\sigma} = 0.84$

and [4] $\Rightarrow$ $\frac{\mu-4}{\sigma} = 1.28$

$\sigma = 0.943$

$\mu = 5.208$

Therefore the mean time is 5.2 days and the standard deviation is 0.94 days (to 2 s.f.).

5) The heights of a large population of ten-year-olds are normally distributed with a mean of 140 cm and a standard deviation of 6 cm. If five children are picked at random from this population find the probability that they are all less than 145 cm tall.

Let Y be the number of children in the sample that are less than 145 cm tall.

Then Y can be $0, 1, 2, 3, 4$ or 5, i.e. Y has a *Binomial* distribution with $n = 5$ and p as yet unknown.

Now p is the probability that a child picked at random from the population of ten-year-olds is less than 145 cm tall, and this we can find from the normal distribution.

If X is the height of a child, then

$$\begin{aligned} p = \mathrm{P}(X < 145) &= \mathrm{P}\left(Z < \frac{145-140}{6}\right) \\ &= \mathrm{P}(Z < \tfrac{5}{6}) \\ &= \mathrm{P}(Z < 0.833) \\ &= 0.7976 \end{aligned}$$

Therefore Y has a binomial distribution with $n = 5$ and $p = 0.7976$

Hence
$$\begin{aligned} \mathrm{P}(Y = 5) &= (0.7976)^5 \\ &= 0.323 \quad \text{to 3 d.p.} \end{aligned}$$

EXERCISE 18d

1) X is a normal variable with mean 100 and standard deviation 10. Find
(a) $\mathrm{P}(X < 120)$, (b) $\mathrm{P}(X > 105)$, (c) $\mathrm{P}(X > 90)$.

2) X is a normal variable with mean 50 and standard deviation 2. Find the values of X in which
(a) the lowest 70% of the distribution lies,
(b) the highest 80% of the distribution lies,
(c) the middle 50% of the distribution lies.

3) X is a normal variable with mean μ and standard deviation σ. Find μ and σ if
(a) $\mathrm{P}(X < 60) = 0.8$ and $\mathrm{P}(X < 70) = 0.9$,
(b) $\mathrm{P}(X < 10) = 0.7$ and $\mathrm{P}(X < 5) = 0.2$,
(c) $\mathrm{P}(X < 2) = 0.1$ and $\mathrm{P}(X < 1) = 0.05$.

4) After three years' growth, the heights of some spruce seedlings are normally distributed with a mean of 130 cm and standard deviation 3 cm. Find the probability that a randomly chosen seedling has a height
(a) less than 140 cm,
(b) greater than 125 cm,
(c) between 127 cm and 133 cm.

5) In a certain town it is found that the weights of newborn babies are normally distributed with a mean of 3.5 kg and a standard deviation of 0.5 kg. Find the probability that a baby, chosen at random, has a weight
(a) greater than 4 kg,
(b) greater than 3 kg,
(c) less than 2 kg,
(d) between 3.25 kg and 3.75 kg.

6) The masses of eggs from a particular farm are normally distributed with a mean of 60 g and standard deviation of 8 g. If the lightest 5% of the eggs are discarded as being too light for sale, find the range of masses of eggs considered too small for sale.

7) The weights of apples from a particular orchard are normally distributed with a mean of 100 g and a standard deviation of 10 g. The largest 5% and the smallest 10% of these apples are discarded as unfit for sale. Find the range of weights of apples that are sold.

8) The lengths of cucumbers from a certain greenhouse are normally distributed. If 10% of the cucumbers are longer than 35 cm and 10% are shorter than 25 cm find the mean and standard deviation of the lengths.

9) The weights of a population of mice are normally distributed. If the middle 80% of the distribution of weights lies between 45 g and 60 g, find the mean weight and the standard deviation.

10) The times taken by individual members of a large group of people to complete a set task, are normally distributed with a mean of 10 minutes and a standard deviation of 2 minutes. Find the probability that three people chosen at random from this group will all complete the task in less than 8 minutes.

11) The weights and heights of a group of people are each independently and normally distributed, with means of 70 kg and 150 cm and standard deviations of 8 kg and 5 cm.
Find the probability that a person chosen at random will have
(a) a weight less than 75 kg,
(b) a height less than 160 cm,
(c) both a weight less than 75 kg and a height less than 160 cm.
If two people are picked at random find the probability that both of them have a weight less than 75 kg and a height less than 160 cm.

12) A large population has two characteristics A and B that are independently and normally distributed. For A the mean is 2 and the variance is 4. For B the mean is 3 and the variance is 9. Find the probability that a randomly chosen member of this population has
(a) a reading of 2 to the nearest whole number for A,
(b) a reading of 3 to the nearest whole number for B,
(c) a reading of 2 for A and 3 for B, both to the nearest whole number.

Using the Normal Distribution for Discrete Variables

Some characteristics of a naturally occurring population are quantified by discrete variables; for example, IQ is measured by integers, performance in an examination is measured by discrete marks.

If such a discrete distribution is illustrated with a histogram, the shape of the histogram may well be very similar to the shape of the normal distribution with the same mean and standard deviation. If this is the case we say that the discrete variable has *approximately* a normal distribution, and we can then use the normal distribution to analyse the discrete distribution.

EXAMPLES 18e

1) In a certain examination, the marks are found to be approximately normally distributed with a mean of 65 and a standard deviation of 10.
Find the probability that an individual picked at random, has a mark from 60 to 69 inclusive.
If people with the highest 5% of marks are awarded a distinction, find the lowest mark required to obtain a distinction.

Let X be a possible examination mark then, treating X as a continuous variable, X is normally distributed with $\mu = 65$ and $\sigma = 10$.

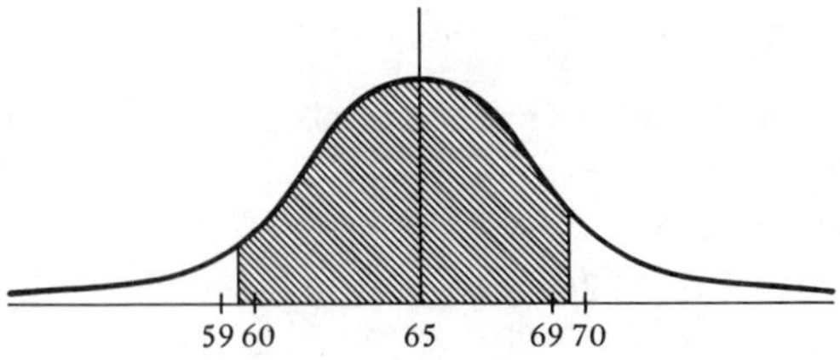

When an individual gets a mark from 60 to 69 inclusive then the range of values of X includes 60 but not 59 and includes 69 but not 70.
We therefore find $P(59.5 < X < 69.5)$.

Now $$\text{P}(59.5 < X < 69.5) = \text{P}\left(\frac{59.5-65}{10} < Z < \frac{69.5-65}{10}\right)$$
$$= \text{P}(-0.55 < Z < 0.45)$$

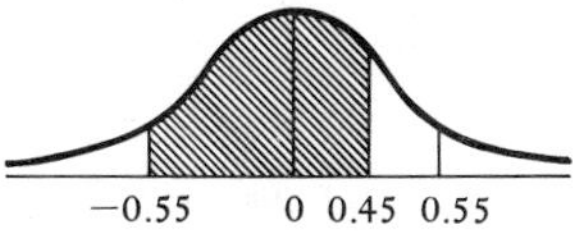

$$\text{P}(-0.55 < Z < 0.45) = \text{P}(Z < 0.45) - [1-\text{P}(Z < 0.55)]$$
$$= 0.6736-(1-0.7088)$$
$$= 0.3824$$

Therefore the probability that an individual gets a mark from 60 to 69 inclusive is approximately 0.382

If the top 5% are awarded a distinction then 95% are not, so we need to find x such that

$$\text{P}(X < x) = 0.95$$

$$\Rightarrow \qquad \text{P}\left(Z < \frac{x-65}{10}\right) = 0.95$$

$$\Rightarrow \qquad \frac{x-65}{10} = 1.645$$

$$\Rightarrow \qquad x = 81.45$$

Therefore a mark greater than 81.45 is needed, i.e. 82 is the lowest mark needed to get a distinction.

Using the Normal Distribution as an Approximation for the Binomial Distribution

The normal distribution is symmetrical and covers an infinite range of values of the variable.
The binomial distribution is symmetrical when $p = \frac{1}{2}$. If, also, n is large (greater than 20 say) then the shape of the binomial distribution is approximately the same as the normal distribution with the same mean and variance.

Hence

> if X has a binomial distribution with $p \simeq 0.5$ and n large, then the normal distribution with $\mu = np$ and $\sigma = \sqrt{(npq)}$ may be used as an approximation for the distribution of X.

The advantage of using the normal distribution in such circumstances is that the calculation involved is much simpler. The approximation is good if $n > 20$ and $0.2 < p < 0.8$ say.

EXAMPLES 18e (continued)

2) A coin is tossed 500 times. Find the probability of getting more than 280 heads.

If X is the number of heads, then X has a binomial distribution with $n = 500$ and $p = \frac{1}{2}$.
(Using the binomial distribution to find $P(X > 280)$ is a daunting prospect, so we use the normal approximation.)

Now X is approximately normally distributed with $\mu = np = 250$ and $\sigma = \sqrt{(npq)} = \sqrt{125} = 11.18$.

Now X is discrete, i.e. it takes the value 280 or 281 but no value between these. The normal distribution, however, is continuous so we use the value midway between 280 and 281, i.e. 280.5 (this is called a *continuity correction*).

Then $$P(X < 280.5) \simeq P\left(Z < \frac{280.5 - 250}{11.18}\right)$$

$$= P(Z < 2.728)$$

$$= 0.996\,81$$

Therefore $$P(X > 280.5) \simeq 1 - 0.996\,81$$

$$= 0.003\,19$$

Hence the probability of getting more than 280 heads is approximately 0.003.

3) In a large population, the probability that an individual has characteristic A is 0.8. Find the smallest number of people that must be chosen from this population so that the probability that more than 70% of them have characteristic A is greater than 0.9.

Let X be the number of people possessing characteristic A in a sample of size n.

Then X has a binomial distribution with $p = 0.8$ and n unknown.

We need to find n such that

$$P\left(X > \frac{7n}{10}\right) > 0.9$$

Assuming that n is reasonably large, we can use the normal distribution with $\mu = 0.8n$ and $\sigma = \sqrt{(0.16n)}$.

Hence $$P\left(Z > \frac{0.7n - 0.8n}{0.4\sqrt{n}}\right) > 0.9$$

$\Rightarrow$ $$P(Z > -0.25\sqrt{n}) > 0.9$$

$\Rightarrow$ $$P(Z < 0.25\sqrt{n}) > 0.9$$

From tables, $$0.25\sqrt{n} > 1.282$$

$$\sqrt{n} > 5.128$$

$$n > 26.3$$

Therefore the sample must contain at least 27 people.

(**Note.** The calculated value of n is just about large enough to justify using the normal distribution.)

The Normal Distribution as an Approximation to the Poisson Distribution

The Poisson distribution is always skewed towards the left, but from the diagrams on p.410, we see that, as λ increases in value the skewness becomes less pronounced. It can be shown that if λ is large (greater than 20 say) then the normal distribution can be used as an approximation for the Poisson distribution.

EXAMPLES 18e (continued)

4) A corner shop stocks fresh milk. The number of bottles of milk sold each day is a Poisson variate with a mean of 25 bottles a day. How many bottles of milk should the shopkeeper stock each day so that the probability that he can meet the demand is greater than 0.95?

Let X be the number of bottles sold in a day, then X has a Poisson distribution with mean 25 and variance 25.

We need to find n such that

$$P(X \leqslant n) > 0.95$$

As $\lambda = 25$, we can use the normal approximation with $\mu = 25$ and $\sigma^2 = 25$, i.e. $\sigma = 5$.

Hence $$P(X \leqslant n) \simeq P\left(Z < \frac{n-25}{5}\right)$$

Now $$P\left(Z < \frac{n-25}{5}\right) > 0.95$$

$$\Rightarrow \qquad \frac{n-25}{5} > 1.645$$

$$\Rightarrow \qquad n > 33.225$$

Therefore the shopkeeper needs to stock 34 bottles.

(**Note** that if the Poisson distribution is used, n must be found from $e^{-25}\left[1+25+\frac{25^2}{2!}+\ldots+\frac{25^n}{n!}\right] > 0.95$ and this calculation is difficult.)

EXERCISE 18e

1) The marks in an examination are approximately normally distributed with mean 55 and standard deviation 10. Find the probability that an individual, picked at random, gets a mark greater than 70.

2) The numbers of eggs laid by individual hens in a poultry farm in a year are approximately normally distributed with a mean of 300 and standard deviation of 20. Find the probability that one hen, chosen at random, lays more than 356 eggs in a year.

3) The marks awarded in a public examination are approximately normally distributed with mean 70 and standard deviation 15. If the top 10% are given a merit award and the lowest 20% fail, find the lowest mark for which a merit award is given and the lowest mark for which a pass is awarded.

4) A box contains a very large number of balloons, 40% of which are red. Twenty-five balloons are removed at random from the box. Find the probability that

(a) more than half of them are red,

(b) the number of red balloons is from 10 to 15 inclusive.

5) The marks in a test are approximately normally distributed with mean 6 and variance 1. Find the probability that an individual, chosen at random, gets a mark of 8.

6) What is the probability of getting a score of 4 or 5 when an ordinary die is tossed? In 50 tosses of the die what is the probability of getting a score of 4 or 5 exactly 20 times?

7) A variety of flower comes in two colours, white and red. A box containing seeds of these flowers has 40% white flowers seeds. How many seeds must be removed from the box so that the probability of getting at least 50 seeds of white flowers is greater than 0.9?

8) A multiple choice test consists of 50 questions each of which has three possible answers, only one of which is correct. If a candidate guesses the answers at random, what is the probability that he will pass if the pass mark is 20?

9) In a particular town it is found that on average 51 boys are born for every 49 girls born. What is the probability that a birth is a male? What is the probability that in 500 births there are fewer boys than girls?

10) The number of people asking for treatment at a certain casualty department is a Poisson variate with a mean of 36 per 24 hours. If the department can cope with 40 people in 24 hours what is the probability that it cannot cope?

11) A box contains a large number of red, white and yellow balls in the proportion 1 red to 6 white to 8 yellow balls. Use a suitable approximation to find the probability that
(a) a sample of 50 contains more than 5 red balls,
(b) a sample of 50 contains less than 25 yellow balls.

12) The number of telephone lines that are engaged at any one moment averages 25 during a 'telephone-in' radio show. How many lines should be available if the probability that a caller gets through is to be greater than 0.7?

SUMMARY

1. If X has a Poisson distribution then the p.d.f. of X is given by

$$P(X = x) = e^{-\lambda}\frac{\lambda^x}{x!} \quad \text{for} \quad x = 0, 1, 2, 3, \ldots$$

λ is the mean *and* the variance of X.

2. If Z has a *standard* normal distribution then Z has mean 0 and variance 1.
The p.d.f. of Z is given by

$$\phi(x) = \frac{1}{\sqrt{(2\pi)}}e^{-x^2/2}$$

$P(Z < x)$ is given in tables for $x \geqslant 0$ (or sometimes $P(Z > x)$ for $x \geqslant 0$).

3. If X has a normal distribution with mean μ and variance σ^2 then

$$P(X < n) = P\left(Z < \frac{n-\mu}{\sigma}\right)$$

4. The Poisson distribution may be used as an approximation for the binomial distribution if n is large and p is small.

5. The normal distribution may be used as an approximation for the binomial distribution if n is reasonably large and $p \simeq 0.5$.

6. The normal distribution may be used as an approximation for the Poisson distribution if λ is large.

MISCELLANEOUS EXERCISE 18

1) A variable X is normally distributed with mean 2.4 and variance 1.44. Find
(i) $P(X < 0)$, (ii) $P(-0.6 < X < 4.2)$, (iii) $P(-1.2 < X < 1.8)$.
Find y, in the cases
(a) $P(2.592 < X < y) = 0.011$,
(b) $P(0.3 < X < y) = 0.44$.
[Give your answers to 3 significant figures.] (U of L)

2) The masses of 1000 students are approximately normally distributed with mean 70 kg and standard deviation 10 kg.
Find, to 2 significant figures, the probability that a student chosen at random has mass
(a) less than or equal to 55 kg,
(b) from 65 to 76 kg,
(c) greater than 84 kg. (U of L)p

3) Given that the lifetimes in hours of electric light bulbs are normally distributed with mean 2040 and standard deviation 60, estimate, to two decimal places, the proportion of bulbs which can be expected to have a lifetime exceeding 1960 hours. (U of L)

4) A trial consists of selecting a card at random from a pack of 52 playing cards and then replacing it. Obtain estimates, to two significant figures, of the probabilities that, in 104 such trials,
(a) the ace of spades is selected at least three times,
(b) at least twenty of the cards selected are spades. (U of L)

5) Given that 5 per cent of a population are left-handed, use the Poisson distribution to estimate the probability that a random sample of 100 people contains 2 or more left-handed people. (U of L)

6) (a) The random variable X has a Poisson distribution and is such that $\text{P}(X = 2) = 3\text{P}(X = 4)$. Find, correct to three decimal places, the values of (i) $\text{P}(X = 0)$, (ii) $\text{P}(X \leqslant 4)$.

(b) The number of characters that are mistyped by a copy typist in any assignment has a Poisson distribution, the average number of mistyped characters per page being 0.8. In an assignment of 80 pages calculate, to three decimal places

(i) the probability that the first page will contain exactly two mistyped characters,

(ii) the probability that the first mistyped character will appear on the third page,

(iii) an approximate value for the probability that the total number of mistyped characters in the 80 pages will be at most 50.

7) (a) The marks gained at a certain examination are approximately normally distributed with mean 60 and standard deviation 15. The top 10% of the students receive grade A whilst the bottom 12% fail. Find to the nearest integer

(i) the minimum mark required to obtain a grade A,

(ii) the maximum mark consistent with a failure.

(b) It is known that 1% of the electric bulbs made in a certain factory are defective. Use the Poisson distribution to estimate to three significant figures, the probability that two or more bulbs are defective in a sample of 200. (U of L)

8) Write down an expression for the probability of exactly k successes given by the Binomial distribution, where p is the probability of success in each of the n independent trials.

Calculate the mean and variance of the following Binomial distributions

(a) $n = 25$, $p = 0.4$,

(b) $n = 100$, $p = 0.02$.

In each case choose which of the Poisson distribution and the Normal distribution you consider to give the better approximation to the Binomial distribution. Use your selected distribution to estimate the probability of more than two successes in each case. Give all your answers to three decimal places. (U of L)

9) (a) Assume that characteristics R and S are independent and are approximately normally distributed with parameters

$$\mu_R = 29.6, \quad \sigma_R = 4.8, \quad \mu_S = 15.9, \quad \sigma_S = 4.1$$

Find, to 3 decimal places, the probability that an individual sampled at random will have a reading of
(i) from 20 to 32 for characteristic R,
(ii) less than 20 for characteristic S,
(iii) less than 20 for each characteristic.
Given that two individuals are sampled at random, find to 2 significant figures, the probability that they both have a reading of less than 20 in each characteristic.

(b) Derive the mean and variance of the exponential distribution, parameter $\alpha > 0$, defined by the probability density function

$$\begin{aligned} f(x) &= \alpha e^{-\alpha x} \quad \text{for } x \geqslant 0 \\ f(x) &= 0 \quad \text{otherwise} \end{aligned}$$

(U of L)

10) State the mean of the probability distribution defined by

$$P(X = r) = \frac{e^{-\mu}\mu^r}{r!} (r = 0, 1, 2, \ldots)$$

where μ is a constant. Prove that the variance of this distribution is equal to the mean.
In a lottery the probability of a ticket being a prize-winning one is $1/100$. Use the Poisson approximation to the binomial distribution to estimate the probability of winning
(a) exactly one prize, (b) at least 2 prizes,
with 50 tickets.
Three people each buy 50 tickets. Estimate the probability that only one of them will fail to win at least one prize. (U of L)

11) (a) The probabilities that a company will require $0, 1, 2, 3, \ldots$ articles of a certain kind in any one month are given by a Poisson distribution with mean 2.
Find the least number of articles that the company should have in stock at the beginning of the month so that the probability of having no stock left at the end of the month is less than 0.02.

(b) In a fairground rifle-range the probability of hitting any one of a number of moving targets is $\frac{1}{3}$. Find the probability that from 5 shots
(i) three targets are hit,
(ii) at least two targets are missed. (U of L)

12) (a) The random variable X has the Poisson distribution defined by

$$P(X = r) = \frac{e^{-\mu}\mu^r}{r!}, \quad r = 0, 1, 2, \ldots$$

Prove that the mean and variance of X are both equal to μ.
The number of green line buses that arrive at a particular bus stop during any period of 10 minutes may be assumed to have a Poisson distribution with mean 1. Find the probability, to 3 decimal places, that

(i) no green line bus will arrive at the bus stop in a given 10 minute interval,

(ii) at least 2 green line buses will arrive at the bus stop in a given 10 minute interval.

(b) The weights of apples are distributed normally with mean 116 g and standard deviation 25 g. The apples are divided into three grades I, II and III. Grade I apples weigh more than 136 g, grade III apples weigh less than 70 g. All the remaining apples are of grade II. Calculate, to one decimal place, the percentage of apples in each grade.

(U of L)

13) (a) The heights of soldiers are known to be normally distributed with mean 1.81 m and standard deviation 5 cm. The heights of a random sample of 1000 soldiers are measured to the nearest one centimetre. Find the expected number of soldiers in the sample who will have *measured* heights

(i) greater than or equal to 1.87 m,

(ii) from 1.72 m to 1.80 m, inclusive.

(b) During a certain morning the number of people entering a supermarket per minute has a Poisson distribution with mean 4. Find, to 2 significant figures in each case, the probabilities that

(i) during a particular minute in the morning no one will enter the supermarket,

(ii) during a particular minute in the morning exactly 3 people will enter the supermarket,

(iii) during a 2 minute period in the morning no one will enter the supermarket.

(U of L)

14) Samples of 40 are taken of mass produced articles of which 1% are defective. Using

(a) the binomial distribution,

(b) the Poisson distribution,

estimate, to three decimal places, the probability that such a sample contains less than 2 defective articles.

Using the Poisson distribution find also the probability that, of 60 such samples, more than one contains 2 or more defective articles.

[You may assume that $(0.99)^{40} = 0.6693$.] (U of L)

15) A large number of students sat an examination. The marks obtained can be regarded as a continuous variable which is normally distributed. Given that 14% of the students obtained less than 30 marks, and $24\frac{1}{2}\%$ obtained more than 50 marks, find the mean and variance of the mark distribution, giving your answers to one decimal place. (U of L)

16) The concentration by volume of methane at a point on the centre line of a jet of natural gas mixing with air is distributed approximately Normally with mean 20% and standard deviation 7%. Find the probabilities that the concentration

(a) exceeds 30%;

(b) is between 5% and 15%.

In another similar jet, the mean concentration is 18% and the standard deviation is 5%. Find the probability that in at least one of the jets the concentration is between 5% and 15%. (MEI)

17) Derive the mean and variance of the Poisson distribution defined by

$$\mathrm{P}(r) = \frac{\mathrm{e}^{-\lambda}\lambda^r}{r!}, \quad r = 0, 1, 2, \ldots$$

where λ is constant.

Given that the probability that an individual will suffer a reaction from an injection of a given drug is 0.002, find the probability that, out of 3000 individuals injected,

(a) exactly 3,

(b) more than 2,

will suffer a reaction.

Find the probability that, in each of 3 groups of 1000 individuals, exactly 1 individual will suffer a reaction. (U of L)

18) (a) Find the probability that, of 10 000 digits each chosen at random from the digits 0 to 9, (i) the digit 7 appears at most 1030 times, (ii) the digit 7 appears at most 980 times.

(b) It is known that 120 misprints are distributed randomly in a book of 600 pages. Given that the number of misprints per page has a Poisson distribution, find the probability that a given page contains (i) exactly two misprints, (ii) one or more misprints. (U of L)

19) (a) The number of calls received by an office during each minute of a working day has a Poisson distribution with mean 4. Find, to 3 decimal places, the probability that more than 4 calls will be received during any particular minute.

Find also, to 2 significant figures, the probability that no call will be received in an unbroken period of 2 minutes chosen at random.

(b) Manufactured articles are known to have lengths which are normally distributed with mean 155 cm and standard deviation 20 cm. Calculate, to 2 significant figures, the probability that a randomly chosen article will have a length
(i) less than 100 cm,
(ii) from 150 cm to 175 cm. (U of L)

20) The number of calls per day to the fire brigade at town A is distributed as a Poisson variate with mean 1, and the number of calls per day to the fire brigade at town B is independently distributed as a Poisson variate with mean 2.
Find the probability, to three decimal places, that on a particular day
(a) both fire brigades have exactly two calls,
(b) the fire brigade from town A receives no calls but the brigade from town B receives at least one call,
(c) between them the two fire brigades have less than 3 calls.
Find an expression, which need not be evaluated, for the probability that on a particular day
(d) both fire brigades have the same number of calls. (U of L)

21) A multiple-choice examination consists of 20 questions, for each of which the candidate is required to tick as correct one of 3 possible answers. Exactly one answer to each question is correct. A correct answer gets 1 mark and a wrong answer gets 0 marks. Consider a candidate who has complete ignorance about every question and therefore ticks at random. What is the probability that he gets a particular answer correct? Calculate the mean and variance of the number of questions he answers correctly.
The examiners wish to ensure that not more than 1% of completely ignorant candidates pass the examination. Use the Normal approximation to the Binomial, working throughout to 3 decimal places, to establish the pass mark that meets this requirement. (MEI)

22) Books on the top shelf at a library are directly accessible only to a person having a reachable height of at least 250 cm. It may be assumed that the reachable heights of adult male readers at the library are normally distributed with a mean of 264 cm and a standard deviation of 8 cm, and that those of adult female readers are normally distributed with a mean of 254 cm and a standard deviation of 5 cm.
(a) Find, correct to three significant figures, the proportion of adult male readers and the proportion of adult female readers who are able to reach books on the top shelf.

(b) Given that 40 per cent of all adult readers at the library are male, find the proportion, correct to three significant figures, of all adult readers who are able to reach books on the top shelf.

(c) The library decides to lower the top shelf so that 95 per cent of all adult female readers will be able to reach books there. Find the corresponding percentage, correct to the nearest integer, of all adult male readers who will then be able to reach books on the top shelf. (JMB)

23) When subjected to a dose of insecticide, a particular breed of insect will die if the dose exceeds the tolerance of the insect. For a large population of this breed of insect, the tolerance (measured in appropriate units) is normally distributed with mean 30 and variance 9. The population is subjected to a dose of insecticide of strength 27. Assuming that the effect of each dose is independent of the previous dose, show that the probability that an insect selected at random from the survivors would die when subjected to a dose of strength 34 is approximately 0.89. Find the probability that three, or more, insects of a group of 5 insects selected from amongst the survivors would die when exposed to a dose of strength 34.
Find, for this breed of insect, the least number of units in a single dose required to ensure a 96% death rate. (U of L)

24) First class mail not delivered by the first post on the weekday following posting is considered 'late'. Assuming that 5% of all letters sent by first class mail are late, calculate the probability that, of 10 letters sent by first class post, not more than 1 will be delivered late.
State the mean and standard deviation of the probability distribution of the number of late deliveries out of 200 letters sent by first class post, and use the normal distribution to estimate the probability that not more than 5 of the letters will be delivered late. (U of L)

25) Henri de Lade regularly travels from his home in the suburbs to his office in Paris. He always tries to catch the same train, the 08.05 from his local station. He walks to the station from his home in such a way that his arrival times form a normal distribution with mean 08.00 hours and standard deviation 6 minutes.

(a) Assuming that his train always leaves on time, what is the probability that on any given day Henri misses his train?

(b) If Henri visits his office in this way 5 days each week and if his arrival times at the station each day are independent, what is the probability that he misses his train once and only once in a given week?

(c) Henri visits his office 46 weeks every year. Assuming that there are no absences during this time, what is the probability that he misses his train less than 35 times in the year? (AEB)

26) A car hire firm has 2 cars available for hire each day. Assuming that each hire is for the whole day and that the number of demands per day has a Poisson distribution with mean 2, calculate the proportion of days on which
(a) no car is hired,
(b) some demand is refused.
Show that the mean number of requests for car hire that are refused per day is equal to $4e^{-2}$.
Calculate the mean number of requests for car hire that would be refused per day if the firm had 3 cars available for hire in each day. (U of L)

27) The number of flaws in a given length of cloth may be assumed to have a Poisson distribution. The manufacturer of the cloth claims that the mean number of flaws per metre length of the cloth is equal to 0.2. To test this claim against the alternative that the mean number is greater than 0.2, it is decided to take a length of 10 metres of the cloth and to reject the claim only if this length contains 6 or more flaws.
(a) Find the probability, correct to three decimal places, that this test rule will reject the claim when, in fact, the true mean number of flaws per metre is equal to 0.2.
(b) Find the probability, correct to three decimal places, that this test rule will not reject the claim when, in fact, the true mean number of flaws per metre is equal to 0.8.
(c) Find the number which should replace the 6 in the above test rule so that the new test rule will be such that the probabilities in (a) and (b) will be approximately equal to 0.05 and 0.1, respectively. (JMB)

28) The width of a certain type of steel sheet is supposed to be not less than 19.8 m and not more than 20.3 m, and the thickness of these sheets is supposed to be not less than 0.19 cm and not more than 0.21 cm.
These sheets are produced by a machine in such a way that their widths are Normally distributed with mean 20.02 m and standard deviation 0.10 m and their thicknesses are independently Normally distributed with mean 0.200 cm and standard deviation 0.005 cm.
Find the probability that a randomly chosen sheet will
(a) meet the width specification,
(b) meet the thickness specification,
(c) fail both to meet the width specification and to meet the thickness specification,
(d) fail to meet at least one specification.
If five sheets are selected at random, what is the probability that at least four will meet the width specification?
Suppose it is required that not more than 1% of all sheets be more than 20.25 m wide. If the standard deviation of widths is kept at 0.10 m, what mean width is required? (MEI)

29) The number of incoming calls to a telephone exchange in a period of t seconds has a Poisson distribution with mean $2t$.

(a) Find, correct to three decimal places, the probability that in a period of 4 seconds the number of incoming calls will be 8 or more.

(b) Show that during a period of one second the probability of there being 3 or 4 incoming calls is *exactly* equal to the probability of there being 2 incoming calls.

(c) Find the period of time in seconds, correct to two decimal places, for which there is a probability of 0.99 that there will be at least one incoming call during that period.

(d) Let $\bar{X}$ denote the mean of the numbers of incoming calls that will occur during 40 successive periods of 5 seconds each. Use an appropriate approximation to the distribution of $\bar{X}$ to calculate, to three decimal places, the probability that the value of $\bar{X}$, to the nearest integer, will be equal to 10. (JMB)

ANSWERS

ANSWERS

Some of the answers given here are quoted in an exact form (using surds, etc.). Students who have used a calculator in their solution can check their answers by converting an exact result to decimal form.

Exercise 1a – p. 8

1) $\mathbf{r} = 3\mathbf{i} + 2\mathbf{j} + \mathbf{k} + \lambda(16\mathbf{i} + 13\mathbf{j} + 2\mathbf{k})$
2) $\mathbf{r} = 4\mathbf{i} + \mathbf{j} + 5\mathbf{k} + \lambda(3\mathbf{i} + 14\mathbf{j} + 12\mathbf{k})$
3) $4; \mathbf{r} = 3\mathbf{i} + 2\mathbf{j} - 5\mathbf{k} + \lambda(3\mathbf{i} - 2\mathbf{j} - \mathbf{k})$
4) $a = -8,\ b = 1,\ c = -2$
   or $a = 16,\ b = -7,\ c = 10$
5) a) $14\sqrt{3}$ b) $2\mathbf{i} - \mathbf{j} + 3\mathbf{k}$
   c) $\mathbf{r} = (\lambda + 2)\mathbf{i} + (\lambda - 1)\mathbf{j} + (\lambda + 3)\mathbf{k}$
6) a) No b) Yes at $\mathbf{i} - 2\mathbf{j} + \mathbf{k}$
   c) No
7) a) $(89t^2 - 190t + 150)^{1/2};\ 5\sqrt{173/89}$
   b) $(5t^2 - 18t + 18)^{1/2};\ 3\sqrt{41/5}$
8) A and C; 3 seconds later; $\mathbf{i} + 2\mathbf{j} + 3\mathbf{k}$
9) a) $\mathbf{i} - 2\mathbf{j} - 6\mathbf{k}$ b) 20 (c) $\frac{1}{2}\sqrt{205}$
11) 176, 174, 278
12) a) 76 J b) $\mathbf{i} - \mathbf{j} + 2\mathbf{k}$ c) 112 J
13) 12 J, 2 m s^{-1}

Exercise 1b – p. 14

1) Magnitude of components:
   $\frac{1}{2}\sqrt{2},\ \frac{5}{2}\sqrt{2};\ \sqrt{5},\ 2\sqrt{5};\ \frac{7}{5},\ \frac{1}{5}$
2) Magnitude of components: $\frac{10}{17}\sqrt{17}$,
   $\frac{11}{17}\sqrt{17}$
3) $2\cos t\,(3\cos 2t - \sin t)$ W; 1 J
4) a) $10\mathbf{i} - 5\sin t\,\mathbf{j} - 5\cos t\,\mathbf{k}$
   b) $20t$ W c) 160 J
5) $9t + 8t^3 + 12t^5$; 178 J;
   $\frac{3}{2}t^2\mathbf{i} - \frac{2}{3}t^3\mathbf{j} + \frac{1}{2}t^4\mathbf{k}$; 178 J
6) $36t^3 + 124t + 8$; 79 J;
   $\mathbf{v} = 4\mathbf{i} + 8\mathbf{j} + 2\mathbf{k},\ \mathbf{r} = 3\mathbf{i} + 4\mathbf{j}$, 84 J

Miscellaneous Exercise 1 – p. 15

1) $\mathbf{v} = (\sin t + t\cos t)\mathbf{i} + (\cos t - t\sin t)\mathbf{j}$;
   $\mathbf{r} = (2\cos t - t\sin t)\mathbf{i} - (2\sin t + t\cos t)\mathbf{j}$
2) $\frac{26}{17}$
3) $\mathbf{v} = \sin t\,\mathbf{i} - \cos t\,\mathbf{j} + \mathbf{k}$;
   $\mathbf{r} = (1 - \cos t)\mathbf{i} - \sin t\,\mathbf{j} + t\mathbf{k};\ m;\ m$
4) a) $8(t^2 - 2t + 2)$ b) 0
   c) $-16\sin 2t\,\mathbf{i} - 16\cos 2t\,\mathbf{j} + 8\mathbf{k}$
   d) 1
5) $20\mathbf{i} + 10\mathbf{j}$; 5; 125
6) $\frac{11}{5}, \frac{2}{5}, \frac{11}{5}, \frac{23}{5};\ \frac{22}{25}(4\mathbf{i} + 3\mathbf{j})$
7) a) 92 J b) $2\sqrt{23}$ m s^{-1}
8) $2\mathbf{i} - 5\mathbf{j} + \mathbf{k};\ F\sqrt{84}$
9) $2\mathbf{i} - \mathbf{j} + 2\mathbf{k};\ 6\mathbf{i} + 2\mathbf{j} + 4\mathbf{k};\ \sqrt{11}$,
   $\mathbf{i} + \mathbf{j} + 3\mathbf{k}$
10) 20 m s^{-1} a) 1.35 m b) 2.24 m
11) $(3 - 2T)\mathbf{i} + (4 - 5T)\mathbf{j} + (T - 2)\mathbf{k}$;
    $\sqrt{\frac{43}{15}};\ \frac{1}{15}(17\mathbf{i} - 10\mathbf{j} - 16\mathbf{k})$
12) $-6\mathbf{i} - 25\mathbf{j} - 16\mathbf{k};\ 13\mathbf{i} + 53\mathbf{j} + 34\mathbf{k}$
13) $n = 3;\ 5\mathbf{i} + \frac{4}{3}\mathbf{j} - \frac{2}{3}\mathbf{k};\ \frac{160}{3}$ J, $\frac{109}{3}$ J
14) $\mathbf{v} = 4t\mathbf{i} + (2t - 4)\mathbf{j} + 3\mathbf{k},\ \mathbf{a} = 4\mathbf{i} + 2\mathbf{j}$;
    $\frac{43}{35}\sqrt{14},\ -\frac{1}{7}\sqrt{14}$
15) $\mathbf{v} = -cn\sin nt\,\mathbf{i} + dn\cos nt\,\mathbf{j}$;
    $\mathbf{a} = -cn^2\cos nt\,\mathbf{i} - dn^2\sin nt\,\mathbf{j};\ t = \dfrac{\lambda\pi}{2n}$
16) a) $\frac{3}{2}(1 + \pi^2\cos^2\pi t + 4t^2)$
    $+ 2(\frac{1}{16}\pi^2\cos^2\{\frac{1}{4}\pi t\} + 9t^4 + 9)$
    b) $3(\pi^4\sin^2\pi t + 4)^{1/2}$
    c) $2/(5 + \pi^2)^{1/2};\ \mathbf{i} + 6\mathbf{k}$
17) $2\mathbf{i} - 3\mathbf{j} - 6\mathbf{k},\ -\mathbf{i} - 2\mathbf{j} - 2\mathbf{k}$,
    $\mathbf{i} - 5\mathbf{j} - 8\mathbf{k};\ 6\sqrt{10}$ N;
    $\mathbf{r} = (\mathbf{i} + 2\mathbf{j} + 2\mathbf{k}) + t(\mathbf{i} - 5\mathbf{j} - 8\mathbf{k})$
18) $20\mathbf{i} + 27\mathbf{j} - 34\mathbf{k};\ \sqrt{590}$ m s^{-1}; -144J
19) $6\mathbf{i} - 2\mathbf{j} + \mathbf{k}$
20) $4e^{-t}(\sin t\,\mathbf{i} - \cos t\,\mathbf{j});\ 2e^{-\pi}$J; $-2(e^{-\pi} - 1)$ J

Exercise 2a – p. 24

1) a) $x^2 + y^2 = 4$
   b) $x^2 + y^2 - 4x - 8y + 19 = 0$
   c) $x^2 - y^2 = 4$
   d) $y^2 - 8y + 4ax + 16 - 24a = 0$
2) a) $x^2 + z^2 = a^2,\ y = 0$
   b) $y^2 = z,\ x = 0$
   c) $\dfrac{x^2}{4} + \dfrac{z^2}{9} = 1,\ y = 0$
   d) $x^2 + y^2 = 1,\ z = 1$
3) a) $\mathbf{r} = \pm(\mathbf{i} + 2\mathbf{j})/\sqrt{5}$
   b) $\mathbf{r} = \frac{1}{2}(\sqrt{37} - 1)\mathbf{i} \pm \frac{1}{4}(\sqrt{37} - 1)^{1/2}\mathbf{j}$
   c) $\mathbf{r} = 3\mathbf{j} + \mathbf{k}$
4) $\mathbf{r} = (1 + a\cos\theta)\mathbf{i} + (b\sin\theta - 2)\mathbf{j}$
5) $\mathbf{r} = (1 + ap^2)\mathbf{i} + (2ap - 2)\mathbf{j}$
6) $\mathbf{r} = 3\cos\theta\mathbf{i} + 3\sin\theta\mathbf{j} + \dfrac{2\theta}{\pi}\mathbf{k}$
7) $\mathbf{r} = (3\cos\theta + 2)\mathbf{i} + \mathbf{j} + 3\sin\theta\mathbf{k}$
8) $\mathbf{r} = -\frac{21}{4}\mathbf{i} + \lambda\mathbf{j}$

Exercise 2b – p. 27

1) $\omega(\cos\theta\mathbf{i} - \sin\theta\mathbf{j} + \mathbf{k})$;
   $-\omega^2(\sin\theta\mathbf{i} + \cos\theta\mathbf{j})$
2) $-4\omega^2(\cos\theta\mathbf{i} + \sin\theta\mathbf{j})$
3) $p = \frac{1}{2}t^3$; $\mathbf{v} = 3a(2^{2/3})(p^{5/3}\mathbf{i} + p^{2/3}\mathbf{j})$,
   $\mathbf{a} = 3a(2^{1/3})(5p^{4/3}\mathbf{i} + 2p^{1/3}\mathbf{j})$
4) $\mathbf{v} = V\left\{\dfrac{1}{\sqrt{1 + 4e^{2p}}}\mathbf{i} - \dfrac{2e^p}{\sqrt{1 + 4e^{2p}}}\mathbf{j}\right\}$

Exercise 2c – p. 33

1) a) $-a\theta\omega^2,\ 2a\omega^2$
   b) $-a(1 + \theta)\omega^2,\ 2a\omega^2$
   c) $-a\omega^2(1 + 2\sin\theta),\ 2a\omega^2\cos\theta$
   d) $-2a\omega^2(\cos\theta + \sin\theta)$,
   $2a\omega^2(\cos\theta - \sin\theta)$
2) a) increases with θ
   b) increases with θ
   c) $\left(2a,\ (2n + 1)\dfrac{\pi}{2}\right)$
   d) acceleration constant.
3) $-2r\omega^2,\ -2\omega^2\sqrt{(a^2 - r^2)}$
4) $r = 2 + \cos\theta$
5) $\dfrac{a\omega\sin\theta}{(1 + \cos\theta)^2},\ \dfrac{a\omega}{1 + \cos\theta}$
6) $r = e^\theta$
7) $r = \dfrac{4}{1 - 4\sin\theta}$; -4
8) $-\dfrac{18}{r^3}$

Multiple Choice Exercise 2 – p. 34

1) c	2) b	3) d	4) c
5) a	6) a	7) b, c	8) b
9) a, b, c	10) *A*	11) *B*	12) *C*
13) *I*	14) a, b	15) *A*	16) *A*
17) c			

Miscellaneous Exercise 2 – p. 36

1) $\mathbf{r} = \left(\dfrac{t^2}{2} + vt\right)\mathbf{i} + \dfrac{1}{a}\left(t - \dfrac{1}{a}\sin at\right)\mathbf{j}$
2) $2 - \sqrt{5} < a < 2 + \sqrt{5}$
3) $-2(5\cos t\mathbf{i} + 3\sin t\mathbf{j})$, 17 units
4) $\dfrac{x^2}{9} + \dfrac{y^2}{16} = 1,\ z = 1;\ y^2 = 4x,\ z = 1;$
   $z = 1;\ \mathbf{r.k} = 1;\ \dfrac{n\pi}{\omega}$
5) $k = 2/c^2$
6) $a\mathbf{i} \pm a\sqrt{3}\mathbf{j},\ (2n + 1)\pi/2\omega$
7) $\mathbf{r} = \dfrac{8}{5 + 3\cos\theta}$; $\frac{5}{2}$; -409.6
8) $a\omega(2 + 2\sin\theta)^{1/2}$; $\frac{1}{2}\pi$
10) angular momentum per unit mass
13) $r = a(1 + 2\sin\theta)$; $(3a, \frac{1}{2}\pi)$
15) $r = ae^\theta$; $r = a(1 + \theta)$;
   $r = a(\cos\theta + \sin\theta)$
16) $2ma\omega^2\sqrt{5}$ at $\arctan(-\frac{1}{2})$ to initial line
21) $9V^2/a$ towards O
22) $-u\omega^2t,\ 2u\omega;\ 2u\omega,\ 2u\omega$

Exercise 3a – p. 47

1) $3\mathbf{k}$
2) $-5\mathbf{j}$
3) $3\mathbf{i} + 3\mathbf{j}$
4) $\mathbf{i} + 7\mathbf{j} + 4\mathbf{k}$
5) $-3\mathbf{k}$
6) $3\mathbf{i} + 3\mathbf{j} + 6\mathbf{k}$
7) $2\mathbf{i} - \mathbf{j} - \mathbf{k}$
8) $(\mathbf{a} - \mathbf{b}) \times \mathbf{F}$
9) $-2\mathbf{i} + \mathbf{j} - 3\mathbf{k},\ 2\mathbf{i} - \mathbf{j} - 4\mathbf{k},\ -7\mathbf{k}$
10) a) 5 b) $3\sqrt{3}$
11) $\mathbf{r} = \lambda\mathbf{k}$
12) $\mathbf{r} = 2\mathbf{j} - \mathbf{k} + \lambda\mathbf{i}$
13) $\mathbf{r} = -\mathbf{i} + 2\mathbf{j} - \mathbf{k} + \lambda(-\mathbf{i} + 4\mathbf{j} + 2\mathbf{k})$
14) $\mathbf{r} = -4\mathbf{j} + \lambda(\mathbf{i} + \mathbf{j})$
15) $\mathbf{r} = \mathbf{k} + \lambda(2\mathbf{i} - \mathbf{j})$
16) $\mathbf{r} = -\mathbf{j} - 2\mathbf{k} + \lambda(\mathbf{i} + \mathbf{j} - \mathbf{k})$
18) $-23\mathbf{i} - 2\mathbf{j} - 9\mathbf{k}$
19) $(\mathbf{r} - \mathbf{a}) \times \mathbf{F}$; $\pm\frac{1}{3}\sqrt{3}(\mathbf{i} + \mathbf{j} - \mathbf{k})$
20) $(-2, 2, 1)$ or $(\frac{22}{17}, 2, \frac{31}{17})$

Exercise 3b – p. 50

1) $4\mathbf{i}+\mathbf{j}-\mathbf{k};\ \sqrt{2}$
2) $3\mathbf{i}-5\mathbf{j}-8\mathbf{k};\ \sqrt{2}$
3) $-5\mathbf{i}+10\mathbf{k};\ \frac{5}{9}\sqrt{5}$
4) $-3\mathbf{i}+3\mathbf{h};\ \sqrt{6}$
5) $\sqrt{14},\ 3\mathbf{i}+2\mathbf{j}-\mathbf{k}$

Exercise 3c – p. 59

2) $-\mathbf{i}+4\mathbf{j}+2\mathbf{k}$
3) $\mathbf{r}=2\mathbf{i}-\mathbf{j}+\lambda\mathbf{i}$
4) 1
5) $\mathbf{j}+\mathbf{k},\ -4\mathbf{i}+12\mathbf{j}$
6) $7\mathbf{i}+3\mathbf{j};\ \sqrt{58}$
7) $2(p^2+q^2+r^2)^{1/2}$
8) a) a single force $4\mathbf{i}+\mathbf{j}$ along the line $\mathbf{r}=\frac{7}{2}\mathbf{j}+\lambda(4\mathbf{i}+\mathbf{j})$
   b) a couple of vector moment $\frac{1}{2}(\mathbf{a}\times\mathbf{b}+\mathbf{b}\times\mathbf{c}+\mathbf{c}\times\mathbf{a})$
   c) a single force $\mathbf{i}+3\mathbf{j}$ along the line $\mathbf{r}=-3\mathbf{j}+\lambda(\mathbf{i}+3\mathbf{j})$
   d) a single force $-\mathbf{i}+\mathbf{k}$ through 0
   e) a couple of magnitude equal to twice the area of ΔABC, in a plane perpendicular to the plane of ΔABC, together with the force $\overrightarrow{AD}$ (or equivalent using other triangles)
9) $\mathbf{F}_4=-3\mathbf{i}+4\mathbf{j}-\mathbf{k}$, $\mathbf{r}=3\mathbf{i}-\mathbf{k}+\lambda(3\mathbf{i}-4\mathbf{j}+\mathbf{k})$
10) $-\mathbf{i}+3\mathbf{j}+2\mathbf{k}$
11) $\mathbf{r}=-2\mathbf{j}+\mathbf{k}+\lambda(\mathbf{i}+\mathbf{j}-\mathbf{k})$
13) A force $\mathbf{i}+2\mathbf{j}-\mathbf{k}$ through O together with a couple $2\mathbf{i}+\mathbf{j}+4\mathbf{k}$

Multiple Choice Exercise 3 – p. 61

1) d	2) b	3) c
4) a	5) b	6) b
7) a, c	8) a, b, c	9) b, c
10) a, b	11) *E*	12) *A*
13) *A*	14) *C*	15) *D*
16) b	17) *A*	18) *I*
19) b	20) *F*	21) *T*
22) *T*	23) *F*	

Miscellaneous Exercise 3 – p. 63

2) $\frac{1}{3}\sqrt{3},\ \frac{1}{3}\sqrt{3},\ -\frac{1}{3}\sqrt{3}$ or $-\frac{1}{3}\sqrt{3},\ -\frac{1}{3}\sqrt{3},\ \frac{1}{3}\sqrt{3}$
3) $\sqrt{306}$ a) $\sqrt{35}$
   b) $\mathbf{r}=2\mathbf{i}+3\mathbf{j}-\mathbf{k}+\lambda(-5\mathbf{i}+\mathbf{j}-3\mathbf{k})$
4) $\mathbf{F}_1=-6\mathbf{i}+12\mathbf{j}+12\mathbf{k}$, $\mathbf{F}_2=6\mathbf{i}-12\mathbf{j}-4\mathbf{k}$, $\mathbf{F}_3=-2\mathbf{i}+4\mathbf{j}-4\mathbf{k};\ 28\mathbf{i}+14\mathbf{k};$ $-2;\ -2\mathbf{j}$ (units are N and N m)
5) b) $\pm\frac{1}{3}\sqrt{3},\ \mp\frac{1}{3}\sqrt{3},\ \pm\frac{1}{3}\sqrt{3}$
6) $-10\mathbf{i}-4\mathbf{j}+6\mathbf{k};\ -6\mathbf{i}-6\mathbf{j}+3\mathbf{k}$
7) a) $(\mathbf{r}-\mathbf{r}_0)\times\mathbf{F}$
   b) $\frac{1}{3}\sqrt{3},\ -\frac{1}{3}\sqrt{3},\ \frac{1}{3}\sqrt{3}$ or $-\frac{1}{3}\sqrt{3},\ \frac{1}{3}\sqrt{3},\ -\frac{1}{3}\sqrt{3}$
8) $3;\ \mathbf{r}=\mathbf{i}+3\mathbf{j}+\lambda(3\mathbf{i}+2\mathbf{j}+4\mathbf{k})$
9) $5;\ 40\sqrt{10}$
10) 6 N m; $\mathbf{r}=3\mathbf{i}-3\mathbf{j}+2\mathbf{k}+\lambda(-6\mathbf{i}-8\mathbf{k})$
11) a) $\sqrt{19}$
    b) $-10\mathbf{j}+6\mathbf{k}+\lambda(3\mathbf{i}-\mathbf{j}+3\mathbf{k})$
    c) $-24\mathbf{i}+18\mathbf{j}+30\mathbf{k};\ -3\mathbf{j}-\mathbf{k}$
12) $-\mathbf{i}+33\mathbf{j}+21\mathbf{k}$
13) $\frac{2}{5}\sqrt{6}$
14) $18\sqrt{26},\ -9\mathbf{i}+18\mathbf{j}-18\mathbf{k};$ $(4\mathbf{i}-\mathbf{j}-3\mathbf{k})/\sqrt{26}$
15) $\mathbf{i}+2\mathbf{j}+3\mathbf{k}$; a) $\sqrt{13}$
    b) $\mathbf{r}=\mathbf{i}+2\mathbf{j}+3\mathbf{k}+\lambda(3\mathbf{i}+2\mathbf{j})$
    c) $-6\mathbf{i}+9\mathbf{j}-4\mathbf{k}$ d) $-3\mathbf{k};\ 2$
16) $4\mathbf{i}+4\mathbf{j}+4\mathbf{k},\ -12\mathbf{i}-4\mathbf{j}+16\mathbf{k};\ 4\sqrt{26};$ $\mathbf{i}-3\mathbf{j}$
17) $\mathbf{R}=4\mathbf{i}-8\mathbf{j}-4\mathbf{k},\ \mathbf{G}=12\mathbf{i}+8\mathbf{j}-4\mathbf{k}$
18) Force: $3\mathbf{i}+4\mathbf{j}+4\mathbf{k}$,
    Couple: $-4\mathbf{i}+\mathbf{j}+2\mathbf{k};\ \sqrt{21}$ N m;
    $5\mathbf{i}+6\mathbf{j}+7\mathbf{k}$
19) $\mathbf{F}_4=-3\mathbf{i}+4\mathbf{j}-\mathbf{k}$,
    $\mathbf{r}=3\mathbf{i}-\mathbf{k}+\lambda(3\mathbf{i}-4\mathbf{j}+\mathbf{k});$
    $4\mathbf{i}+6\mathbf{j}+12\mathbf{k}$
20) $(3\cos t+3\sin t)\mathbf{i}+(2\sin t+\cos t)\mathbf{j};$
    $-4(\cos t+\sin t);$
    $\mathbf{r}=\lambda(3\cos t+3\sin t)\mathbf{i}+[\frac{4}{3}+\lambda(2\sin t+\cos t)]\mathbf{j}$
21) $\mathbf{F}=(2+s)\mathbf{i}+(2+t)\mathbf{j}+3\mathbf{k};$
    $\mathbf{G}=-(8+t)\mathbf{i}+(s-3)\mathbf{j}-3\mathbf{k};$
    $s=3\frac{1}{2},\ t=-2\frac{1}{2}$
22) b) (i) $\mathbf{F}_3=-\mathbf{i}-3\mathbf{k};\ \sqrt{14}$ (ii) 2
23) $18\mathbf{i}-56\mathbf{j}-20\mathbf{k};$ 160 J
24) a) 20πs
    b) 20 m; $\mathbf{v}=8\mathbf{i},\ \mathbf{a}=\frac{1}{10}(-4\mathbf{j}+\mathbf{k})$
    c) $-8(\sin\frac{1}{10}t)(\mathbf{j}+4\mathbf{k})$
    d) $y+4z=40$

Exercise 4a – p. 74

1) $2\mathbf{i}+\frac{10}{3}\mathbf{j}$
2) $\frac{1}{4}\sqrt{34}$; arctan $\frac{1}{4}$ to wall
3) $49.1°;\ \frac{2}{7}u\sqrt{7}$

4) a) $\frac{3}{4}\sqrt{13}\ \text{m s}^{-1}$ at $\arctan \frac{1}{6}\sqrt{3}$ to the wall
   b) $3\ \text{m s}^{-1}$ at 30° to the wall
   c) $\frac{3}{2}\sqrt{3}\ \text{m s}^{-1}$ along the wall
5) $\frac{1}{3}$
6) $\frac{1}{3}\sqrt{3}$; $\frac{5}{6}\sqrt{6}(\sqrt{3}+1)$
7) a) $\frac{3}{10}$ b) $\frac{3}{10}m(-5\mathbf{i}+12\mathbf{j})$
   c) $\frac{819}{200}m$

Exercise 4b – p. 79

1) $\frac{3}{8}V$
3) $\frac{1}{6}$
4) a) $-2\mathbf{i}+2\mathbf{j}; \mathbf{j}$ b) $-3\mathbf{i}+2\mathbf{j}; \mathbf{i}+\mathbf{j}$
   c) $-\mathbf{i}+2\mathbf{j}; -\mathbf{i}+\mathbf{j}$
5) $\frac{1}{2}$, $\mathbf{i}+3\mathbf{j}$
6) 1, 45° to AB
7) Along AC; $\arccos\frac{1}{8}$ to AC
8) $4\mathbf{j}$, $2\mathbf{i}+\mathbf{j}$; $5:\sqrt{13}$; 5
9) $8\mathbf{i}-\frac{11}{6}\mathbf{j}$, $2\mathbf{i}-\frac{13}{3}\mathbf{j}$

Miscellaneous Exercise 4 – p. 81

1) a) $\frac{1}{3}\sqrt{3}(1+e)u$, $\frac{1}{3}\sqrt{3}(1+e)u$, $\frac{1}{3}(2-e)u$
   b) $\frac{1}{3}\sqrt{3}m(1+e)u$, $\frac{1}{3}\sqrt{3}m(1+e)u$, $mu(1+e)$
3) $v_1=\frac{9}{5}\mathbf{n}+\frac{12}{5}\mathbf{t}$, $v_2=\frac{7}{2}\mathbf{n}+\frac{1}{5}\mathbf{t}$;
   $\frac{3}{2}\mathbf{n}+\frac{12}{5}\mathbf{t}$; $\frac{17}{10}\mathbf{n}+\frac{1}{5}\mathbf{t}$;
   $\frac{1}{50}(141\mathbf{i}-12\mathbf{j})$, $\frac{1}{50}(59\mathbf{i}+62\mathbf{j})$
5) A: $\frac{1}{2}u\{4\sin^2\alpha+(1-e)^2\cos^2\alpha\}^{1/2}$ at θ to the line of centre where $\tan\theta=(2\tan\alpha)/(1-e)$
   B: $\frac{1}{2}(1+e)u\cos\alpha$ along line of centres
6) $10\mathbf{n}+7\mathbf{t}$, $17\mathbf{n}+24\mathbf{t}$; $\frac{134}{13}\mathbf{i}+\frac{85}{13}\mathbf{j}$, $\frac{373}{13}\mathbf{i}+\frac{84}{13}\mathbf{j}$
7) $(1+e)\tan\alpha/(1-e+2\tan^2\alpha)$
8) $\frac{7}{9}$; $\frac{2}{9}u\sqrt{3}$
10) 45°
11) 4, 1
12) $\frac{4}{3}$
13) $\frac{1}{5}$, $\frac{1}{2}m$

Exercise 5a – p. 88

1) a) 1.79 s, 43.9 m b) 6.69 s, 164 m
2) $\frac{4000}{3}\{\sqrt{3}\cos\theta\sin\theta-\cos^2\theta\}$; $\frac{2000}{3}$
3) $\mathbf{r}=20t\mathbf{i}+(20t-\frac{1}{2}gt^2)\mathbf{j}$
4) a) $27\sqrt{5}$ m b) $\frac{594}{10}\sqrt{5}$ m
6) $\mathbf{v}=50\mathbf{i}+(30-10t)\mathbf{j}+50\mathbf{k}$;
   $\mathbf{r}=(5+50t)\mathbf{i}+(30t-5t^2)\mathbf{j}+(50t-3\mathbf{k})$;
   6; $\mathbf{r}.(\mathbf{i}-\mathbf{k})=8$
7) 2.29 s, 44.3 m
8) $u\sin\theta/10\cos\alpha$
10) a) 50 m b) 10 m c) 30 m

Exercise 5b – p. 97

1) $u\sqrt{21}/14$
2) a) upwards along the plane,
   b) down the plane at $\arctan\frac{37}{9}$ to the horizontal
6) a) $\dfrac{169}{2g}$ b) $\dfrac{169}{2g}$
7) $2V\sqrt{5}/5g\cos\alpha$; $\frac{1}{3}$, 1

Multiple Choice Exercise 5 – p. 99

1) d	2) c	3) d
4) d	5) b, d	6) a, b, c, d
7) a, c	8) b	9) D
10) E	11) C	12) C
13) I	14) A	15) I
16) A	17) T	18) T
19) F		

Miscellaneous Exercise 5 – p. 103

1) a) $-15\mathbf{i}+70\mathbf{k}$, $5\mathbf{i}+30\mathbf{k}$, $5\mathbf{i}+(30-10t)\mathbf{k}$
   b) $[25+(30-5t)^2]^{1/2}$; $10\mathbf{i}+95\mathbf{k}$, $25\sqrt{2}$, $3\frac{1}{12}$
3) $\arcsin\frac{1}{3}$
6) a) $\arctan 3$ b) 45°
7) $2eV\sin\theta/g\cos\alpha$; $\frac{1}{2}\cot\alpha\cot\theta-1$
8) a) 1 b) 3
9) $2\tan\beta$; $2V/g\cos\beta$
11) $(Vt\cos\alpha-\frac{1}{2}gt^2\sin\beta,\ Vt\sin\alpha-\frac{1}{2}gt^2\cos\beta)$; $2V\sin\alpha/g\cos\beta$; $V^2/g(1+\sin\beta)$; $\frac{1}{12}\pi$, $\frac{1}{4}\pi$
15) $2aV^2/2ag(V^2+2ag)^{1/2}$

Exercise 6a – p. 112

1) $m\ddot{x}+kx+R=0$
2) $m\ddot{x}+k\dot{x}+mg=0$
3) $|x|>a$, $a\ddot{x}+ka\dot{x}+gx+ga=0$; $|x|<a$, $\ddot{x}+k\dot{x}=0$
4) $M\ddot{x}=\dfrac{H}{\dot{x}}-Mk\dot{x}^2$

8) $1 - \frac{4}{9}\ln\frac{13}{4}$

9) $v = V\mathrm{e}^{-kt}$

10) $\pi/\sqrt{(n^2 - k^2)}$

11) 3

12) $\ln 2/\mathrm{s}$

13) $\ddot{x} = 5(1 - \dot{x}^2), \quad \dfrac{e^2 - 1}{e^2 + 1}$

14) $100\dfrac{\mathrm{d}v}{\mathrm{d}t} = \dfrac{2500}{v} - v, \quad 247\,\mathrm{m}$

Exercise 6b – p. 125

1) a) damped, $\frac{4}{7}\pi\sqrt{7}$ b) no
c) damped, $\pi\sqrt{21/g}$ d) simple, $\frac{2}{3}\pi$
e) no f) forced, $\frac{4}{11}\pi\sqrt{11}$

2) $\frac{3}{2}\sqrt{2}\,\mathrm{e}^{-2}\sin(2\sqrt{2})$

3) -7.16

4) $\mathrm{e}^{-2\pi/3}$

5) $\ddot{x} + \dot{x}\sqrt{\dfrac{g}{2a}} + \dfrac{2g}{a}x = 0$, yes

6) $\dfrac{3a}{4}\mathrm{e}^{-\pi/\sqrt{15}}$

7) $\frac{1}{2}u/k\,\mathrm{e}^{-\pi\sqrt{3/9}}$

Exercise 6c – p. 130

1) $(M - mt)\ddot{x} = mu - (M - mt)g$

2) $u\ln\frac{3}{2} - g;\ u\ln\frac{4}{9} + u - \frac{1}{2}g$

Miscellaneous Exercise 6 – p. 130

3) $\dfrac{g}{k^2}(1 - \ln 2)$

4) $\dfrac{\mathrm{d}v}{\mathrm{d}t} = g - kv;\ \left(\dfrac{u}{k} - \dfrac{g}{k^2}\right)(1 - \mathrm{e}^{-kT}) + \dfrac{g}{k}T$

5) $\dfrac{\mathrm{d}^2y}{\mathrm{d}t^2} = -g - k\dfrac{\mathrm{d}y}{\mathrm{d}t};\ x = \dfrac{V\cos\alpha}{k}(1 - \mathrm{e}^{-kt}),$

$y = \left(\dfrac{g}{k^2} + \dfrac{V\sin\alpha}{k}\right)(1 - \mathrm{e}^{-kt}) - \dfrac{gt}{k}$

8) $2gz = u^2\left(\dfrac{\pi}{4} - \arctan\dfrac{v^2}{u^2}\right)$

11) $y = \dfrac{g + v_0k}{ku_0}x + \dfrac{g}{k^2}\ln\left(1 - \dfrac{kx}{u_0}\right)$

12) $\dfrac{1}{k}\ln\left(g + \dfrac{ku}{g}\right);\ \dfrac{g}{k}(1 - \mathrm{e}^{-kt})$

15) $x = 4\mathrm{e}^{-3/2t}(\cos t + 3\sin t)$

16) $x = \dfrac{V}{2n}\mathrm{e}^{-nt}\sin 2nt$

18) $x = \dfrac{1}{k}V\cos\alpha(1 - \mathrm{e}^{-kt});$

$y = \dfrac{1}{k^2}(kV\sin\alpha + g)(1 - \mathrm{e}^{-kt}) - \dfrac{g}{k}t$

20) $x = -\dfrac{V\sqrt{2}}{2n}\mathrm{e}^{-2nt}\cos(2nt - \frac{1}{4}\pi) + \dfrac{V}{2n};$

$\dfrac{V}{2n}$

21) $\dfrac{V}{n}\mathrm{e}^{-nt}\cos(nt + \frac{1}{2}\pi)$

22) $x = \dfrac{\sqrt{10}}{3}\left(a - \dfrac{g}{20n^2}\right)\mathrm{e}^{-nt}\cos(3nt + \epsilon)$

$+ a + \dfrac{g}{20n^2}$ where $\tan\epsilon = -\frac{1}{3}$

24) a) $x = \frac{1}{2}a(1 + \sqrt{2})\mathrm{e}^{(1 - \sqrt{2})nt}$
$+ \frac{1}{2}a(1 - \sqrt{2})\mathrm{e}^{-(1 + \sqrt{2})nt}$
b) $x = 2a\mathrm{e}^{-nt\sqrt{3/2}}(\cos\frac{1}{2}nt + \sqrt{3}\sin\frac{1}{2}nt)$

25) $\frac{1}{8}Mu^2(1 - k)[\ln(1 - k)]^2;\ 1 - \mathrm{e}^{-2}$

26) $\mu = 43.2,\ m_1 = 1353\,\mathrm{kg},$
$a_{\max} = 79.9\,\mathrm{m\,s^{-2}},\ a_{\min} = 79.9\,\mathrm{m\,s^{-2}}$

27) $\frac{3}{2}m$

Exercise 7a – p. 141

1) $\frac{3}{2}\mathbf{i} + \frac{3}{4}\mathbf{j} + \frac{2}{3}\mathbf{k}$

2) $\frac{3}{2}\mathbf{a} + \frac{5}{6}\mathbf{b}$

3) $p = \frac{3}{2};\ q = \frac{11}{2};\ m = 2$

4) $(2, -1, -1)$

5) $\dfrac{\mathbf{a}}{2} + \dfrac{7\mathbf{c}}{10}$

6) $a = 1;\ b = -8;\ c = 1;\ (\frac{7}{3}, 1, -3)$

Exercise 7b – p. 150

1) a) $\frac{3}{5}\mathbf{i} + \mathbf{j} + \frac{7}{10}\mathbf{k}$ b) $(2\mathbf{i} + \mathbf{j} - \mathbf{k})/m$

2) a) $\frac{2}{3}\mathbf{i} + \frac{4}{3}\mathbf{j} - \frac{5}{3}\mathbf{k}$ b) $\frac{7}{3}\mathbf{i} - \frac{2}{3}\mathbf{j} + \frac{7}{3}\mathbf{k}$
c) $14\mathbf{i} - 4\mathbf{j} + 14\mathbf{k}$

3) $F/2m$ at $\arctan\frac{4}{3}$ to AB
a) increases it in ratio 5 : 3
b) none

4) $1\,\mathrm{m\,s^{-1}}$ at 60° to PQ; 10 N s; 50 J;
not at all.

5) $-\frac{1}{2}\mathbf{i} + 3\mathbf{j};\ 324\,\mathrm{J}$

6) $(\frac{8}{3}, 2);\ -7\mathbf{i} - 5\mathbf{j}$ where $\mathbf{i}$ and $\mathbf{j}$ are in the directions $\overrightarrow{\mathrm{OA}}$ and $\overrightarrow{\mathrm{OC}}$ respectively.

7) $3\mathbf{i} + 5\mathbf{j};\ \frac{8}{5}\mathbf{i} + \frac{17}{5}\mathbf{j},\ 994\frac{1}{8}\,\mathrm{J}$

Exercise 7c – p. 159

4) $\pi/3$ or 0

5) AC vertical, stable; $\widehat{ACB} = \dfrac{\pi}{3}$, unstable

8) AB horizontal; no

9) a) symmetrical, stable; at $\arccos \dfrac{a}{2b}$ to the horizontal, unstable

b) symmetrical only, unstable

10) PQ vertically downward, unstable;
PQ vertically upward, unstable;
PQ at $\dfrac{\pi}{6}$ to the horizontal (Q above P), stable

Miscellaneous Exercise 7 – p. 161

1) $\frac{5}{3}\mathbf{i} + \mathbf{j}$; $-7\mathbf{i} - 4\mathbf{j}$

2) $\dfrac{a}{8}(2\mathbf{i} + \mathbf{j})$; $3\mathbf{i} + 4\mathbf{j}$

3) $\dfrac{m_1(v_2 - v_1)}{m_1 + m_2}$

4) $\frac{11}{7}\mathbf{i} + \mathbf{j}$; $\dfrac{5\sqrt{3}}{7}$;
$[(3 + 4\sqrt{2})\mathbf{i} + (4 - 3\sqrt{2})\mathbf{j}]\,(15/\sqrt{154})$

5) a, $2a$; $\sqrt{\dfrac{ga}{3}}$, $2\sqrt{\dfrac{ga}{3}}$

6) $3(\mathbf{i} + \mathbf{j} + \mathbf{k})$; $6(\mathbf{i} + 2\mathbf{j} + 2\mathbf{k})$; 54 units

7) $128\mathbf{i} + 16\mathbf{j} - 112\mathbf{k}$; $3264m$ joules

8) a) $\frac{1}{10}(-8\mathbf{i} - 30\mathbf{j} + 39\mathbf{k})$ b) 15, 9, 25

9) $Wa[\cos\theta + \frac{1}{4}(1 + 8\sin^2\theta)^{1/2}]$ + any constant; unstable

10)

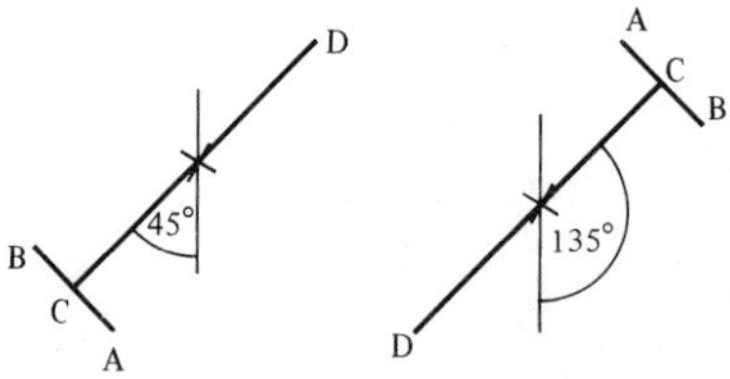

unstable stable

12) unstable

13) stable

14) stable

15) a) stable b) unstable c) unstable

16) OA vertically upward, stable;
OA vertically downward, unstable

17) stable

18) $0 < k < 2$; $k = \sqrt{2}$

19) a) $\theta = 0°$, stable; $\theta = 180°$, unstable
b) $\theta = 0°$, unstable; $\theta = 60°$, stable; $\theta = 180°$, unstable
c) $\theta = 0°$, unstable; $\theta = 180°$, stable

20) $\frac{1}{2}Wa\sqrt{2}(-1 + 6\cos\theta - 2\cos^2\theta)^{1/2} + 2Wa\sin\theta$ relative to 0

21) $\theta = 0$, unstable; $\cos\theta = \frac{3}{4}$, stable

Exercise 8a – p. 170

1) ML^2T^{-2}, MLT^{-1}, ML^2T^{-2}, T^{-1}, T^{-2}, L^2, L^3, MLT^{-2}, MLT^{-1}, ML^2T^{-3}, MLT^{-2}

2) $T = kml\omega^2$

5), 6), 7), 8), 10), 11), 12) are wrong

Exercise 9a – p. 181

1) 400 J; 200 units

2) $32\,\text{kg m}^2$

3) 2.59×10^{29} J

4) $18\,\text{kg m}^2$

5) 128π N m

6) 80π N m

7) $g/5r$

8) $\frac{1}{8}\,\text{rad s}^{-2}$; $\frac{25}{16}$ rad

9) $\dfrac{(m_1 \sim m_2)gI}{I + (m_1 + m_2)r^2}$; $\left[\dfrac{I(m_1 + m_2) + 4m_1m_2r^2}{I + (m_1 + m_2)r^2}\right]g$

10) $g/6r$

Miscellaneous Exercise 9 – p. 182

1) $\sqrt{8\pi r/g}$

2) $\dfrac{(M - m)g}{(M + m + n)a}$

3) $8\pi/\omega$; $16\pi/\omega$; 8

4) $T = \dfrac{25G}{37}$; $I = \dfrac{3G}{185\pi}$

5) $\dfrac{Wr^2(3g - 14l)}{2lg}$; $\dfrac{2Wr(6l - g)}{g}$

6) $\dfrac{2}{a}\sqrt{\left(\dfrac{gd}{3}\right)}$

7) a) $gt^2/8$ b) $2mga$

8) a) $(\omega_2 - \omega_1)I/N$
b) $(\omega_2^2 - \omega_1^2)I/2N$; $10\pi/9$; $\pi/18$; $45/8$

9) $\dfrac{g}{2}(1 - \mu\cos\alpha - \sin\alpha)$; $2Ma^2$

10) $M = 2m$; $G = \frac{1}{2}mga$; $7g/10$

12) $\frac{g}{2a}(I + 8ma^2)$
13) $4mMk^2ag/\pi n(Mk^2 + ma^2)$
14) $g/7a, 16mg/7, 18mg/7;$
$(mgaT + 7ma^2\Omega)/T$
15) $\frac{I\Omega}{2C}\ln\left(\frac{\Omega+\omega}{\Omega-\omega}\right)$
16) a) $(3G - 5L)/7a$ b) $4(G - 4L)/7I$
c) $7\Omega^2 I/8\pi(G - 2L)$

Exercise 10a – p. 195

1) a) $144a^2$ b) $20a^2$
c) $\frac{128}{5}a^2$ d) $16a^2$
2) $\frac{5}{2}Ma^2$
3) $\frac{1}{2}Ma^2$
4) $\frac{3}{10}Mr^2$
5) $\frac{53}{30}\pi a^5 m$, $\frac{153}{10}\pi a^5 m$
6) $M = 8\pi a^3 m$; $\frac{8}{3}Ma^2$
7) $\frac{7}{26}M$
8) $\frac{93}{280}Ma^2$
9) 2

Exercise 10b – p. 198

1) $\frac{50}{3}Ma^2$
2) $\frac{1}{2}Ma^2$
3) $4.08M$
4) $\frac{M}{8}\left\{\frac{12 + 8\sinh 2 + \sinh 4}{2 + \sinh 2}\right\}$

Exercise 10c – p. 203

1) $\frac{4}{3}Mb^2$
2) $\frac{1}{6}Mb^2$
3) $\frac{1}{2}Ml^2$
4) $\frac{1}{4}Ma^2$
5) $\frac{5}{6}Ma^2$
6) a) $\frac{4}{3}Ml^2$ b) $\frac{4}{3}Ml^2\sin^2\theta$
7) $\frac{1}{2}Ml^2$

Exercise 10d – p. 205

1) a) $2Ml^2$ b) $\frac{1}{3}Ml^2$
2) $\frac{2}{3}Ma^2$
3) $\frac{27}{5}M$
4) $\frac{4}{9}Ma^2$

Exercise 10e – p. 215

1) a) $\frac{4}{3}Ma^2$ b) $\frac{4}{3}M(a^2 + b^2)$
2) $\frac{3}{2}Mr^2$
3) $\frac{3}{2}Ma^2$
4) $\frac{1}{6}Ml^2$
5) $\frac{3}{2}Mr^2$
6) $7Ma^2$
7) $9Ma^2$
8) $2I$
9) $\frac{1}{6}M(3r^2 + 2h^2)$
10) a) $\frac{3}{2}Ml^2$ b) $\frac{19}{2}Ml^2$

Exercise 10f – p. 218

1) $\frac{5}{6}Ml^2$
2) $3Ma^2$
3) $\frac{Ma^2}{10}\left(\frac{15l + 8a}{3l + 2a}\right)$
4) $\frac{5}{6}Ml^2$
5) $\frac{1}{3}\left(\frac{2 + 3\pi}{2 + \pi}\right)Ma^2$
6) $\frac{214}{3}Ma^2$
7) $\frac{364}{5}Ma^2$
8) $34.4\ \text{kg m}^2$

Multiple Choice Exercise 10 – p. 219

1) c	2) d	3) a	4) c
5) e	6) a, b	7) a	8) a, c
9) c	10) b	11) *A*	12) *A*
13) *T*	14) *F*	15) *T*	

Miscellaneous Exercise 10 – p. 222

1) $5Ma^2/2$
2) $Ml^2/3$
3) $\frac{M(a^4 + b^4)}{2(a^2 + b^2)}$
5) $Ma^2/4$
6) $ml^5/5$
7) $\frac{5}{96}Mh^2$
9) $\frac{1}{3}a\sqrt{3}$; $\frac{79}{84}Ma^2$
13) $6ma^2$
15) $17Ma^2/30$
16) a) $\sqrt{(\frac{3}{10}a^2)}$ b) $\sqrt{\frac{1}{20}(3a^2 + 2h^2)}$
17) $h\sqrt{6}$, $23Ma^2/6$
18) a) $\frac{(R + r)(R^2 + r^2)}{R^2 + rR + r^2}$,
b) $[\frac{7}{8}(R^2 + r^2)]^{1/2}$

Exercise 11a – p. 230

1) $\sqrt{3g/2l}$
2) $2a$
3) $k > 3$
4) $4\sqrt{g/5r}$
5) $\left(\dfrac{12g\sqrt{2}}{5l}\right)^{1/2}$
6) $\sqrt{\dfrac{g(3\sqrt{3}-4)^{1/2}}{12a}}$
7) $\sqrt{\dfrac{(2\sqrt{3}-1)g}{6l}}$
if C is initially below AB;
$\sqrt{\dfrac{(10\sqrt{3}-13)g}{6l}}$
if C is initially above AB

Exercise 11b – p. 234

1) $4\sqrt{2g/a}$; $2mga/\pi$
2) $3\sqrt{3}\,mgr/2\pi$
3) $14mgl/3\pi$
5) $\dfrac{4\sqrt{2\pi^3}}{a}$

Exercise 11c – p. 241

1) $\sqrt{12g\sin\theta/7l}$; $6g\cos\theta/7l$
2) $4\pi\sqrt{a/3g}$
3) $4\pi\sqrt{7l/15g}$; $28l/15$
4) $2\pi\sqrt{\dfrac{3a^2+4ah+4h^2}{2g(a+2h)}}$; $h=\dfrac{a}{2}(\sqrt{2}-1)$
6) $l\sqrt{\dfrac{11}{6}}$

Exercise 11d – p. 246

1) $\frac{1}{3}mg\sin\theta$
2) $\frac{5}{2}mg\cos\theta$; $\frac{5}{2}mg$
3) $\frac{1}{4}mg\sin\theta$; $\frac{1}{2}mg(5\cos\theta-3)$;
$\frac{1}{8}mg\sqrt{487}$ at $\arctan\sqrt{3}/22$ to CA.
4) $mg\sqrt{1954}$

Multiple Choice Exercise 11 – p. 246

1) b, d	2) a, b, c	3) a, d	4) b
5) *C*	6) *B*	7) *A*	8) *C*
9) *A*	10) *A*	11) d	12) *I*
13) *F*	14) *F*	15) *T*	16) *F*
17) *T*			

Miscellaneous Exercise 11 – p. 249

1) $\sqrt{(3g/2a)}$
4) $\left(\dfrac{4mgr}{\pi n}\right)\left(\dfrac{Mk^2}{Mk^2+mr^2}\right)$
5) $\sqrt{\dfrac{2g}{3a}}$; $4mga$
6) $6g\sin\theta/7a$
7) $2<k<3+2\sqrt{2}$
8) $4mga/5\pi^2$; $\sqrt{112g/15\pi a}$
9) $\sqrt{2\pi na/g}$; $\frac{1}{4}\omega$
10) $(3\pi^3a\sqrt{2}/4g)^{1/2}$
11) $2\pi\sqrt{\dfrac{16a^2+7x^2}{(12a+7x)g}}$; $\frac{4}{7}a$
12) $2\pi\sqrt{(2a^2+3x^2)/3gx}$; $a\sqrt{\frac{2}{3}}$
15) $\frac{1}{3}mg(4-3\cos 2\theta)$, $mg\sin 2\theta$
16) $\dfrac{3g}{8a}(\sqrt{2}-1)$; $\frac{3}{8}mg$, $\frac{13}{8}mg$
17) $2(2x^2-5ax+4a^2)/(5a-4x)$
18) $\frac{1}{2}mg\sin\theta$
19) $\frac{5}{2}mg\cos\theta$; $\frac{5}{2}mg$; $\frac{1}{4}mg$; arctan 10
20) $\sqrt{g/a}$
21) $8mg$, $\frac{29}{3}mg$
22) $\cos\theta=\frac{2}{3}$; $\cos\theta=\frac{1}{3}$
23) $\frac{5}{2}mg\cos\theta$, $\frac{1}{4}mg\sin\theta$; $\frac{9}{8}mg$, $\frac{11}{8}mg$
24) a) $\dot{\theta}^2=2g(1-\cos\theta)/\lambda a$,
$\ddot{\theta}=g\sin\theta/\lambda a$
c) $\frac{5}{4}$, $\frac{1}{6}$
25) $\frac{77}{30}mg$; for $\phi>\frac{1}{2}\pi$ the sign of the moment of the weight changes but the sign of the frictional couple does not
26) $\frac{144}{17}$; $2mg$
27) $3g\cos\theta/7a$; $\frac{4}{7}mg\cos\theta$
28) a) $\frac{3}{16}mg\sqrt{2}$, $\frac{1}{16}mg(16+9\sqrt{2})$

Exercise 12a – p. 263

1) $32g/a$
2) 12.5 N m
3) a) 36 b) $12\sqrt{2/k}$
4) $\frac{4}{3}ml\,\Omega$
5) $40\sqrt{5/3}$ N s
6) a) $6m\sqrt{gl/\sqrt{3}}$
b) $6m(\sqrt{2}+1)\sqrt{gl\sqrt{3}/6}$

Exercise 12b – p. 268

1) a) $2u/3$ b) $2u/\sqrt{3}$
2) $2\sqrt{g/5a}$
3) $3J/7ma$; $J > 2m\sqrt{7ga/3}$
4) $\omega/31$
5) $4\sqrt{g/5a}$; $\dfrac{4}{9a}(u + \sqrt{5ga})$

Miscellaneous Exercise 12 – p. 269

1) $18\pi ma^2/t^2$; $18\pi ma^2/t$
2) $4ma^2\omega^2$; $8ma^2\omega$; $a\omega/g$; $a\omega^2/2g$; $16\omega/17$
3) $\frac{6}{5}Mgn$; $\frac{4}{5}M\sqrt{2gh}$
4) $2a/\sqrt{3}$; $[4ga(\sqrt{3} + 2)/\sqrt{3}]^{1/2}$
5) $v^2 > 4ag(M + 2m)(M + 3m)/3m^2$
7) $2M\sqrt{(2ga)}$; $\frac{1}{4}M\sqrt{(2ga)}$
8) $2\sqrt{3g/a}$; $\frac{4}{3}mg$, $\frac{4}{3}mg$
10) $(4g\sin\theta/3a)^{1/2}$, $2g\cos\theta/3a$; $(68g/49a)^{1/2}$
11) $\frac{20}{3}M(\frac{3}{5}ga\sqrt{2})^{1/2}$
13) $n/4a, \frac{1}{2}u$; $\frac{1}{4}mu$
15) $2m(\frac{1}{3}ga\sqrt{2})^{1/2}$
16) $2\pi\sqrt{17a/6g}$
18) $Mg\{m(1 + \mu) + I/a^2\}/\{M + m + I/a^2\}$; $V(M + m + I/a^2)$
19) $\frac{2}{3}M\sqrt{6ga}$
20) $18P/7Ma$
21) $\frac{5}{6}ml\omega$

Exercise 13a – p. 283

1) $u - d\omega\sin\omega t$, $d\omega\cos\omega t$
2) $(x - ut)^2 + y^2 = d^2$
4) $\frac{1}{2}g$; $\frac{1}{2}mg$
5) $6\sqrt{2ga}$

Exercise 13b – p. 288

1) (i) $Ma^2\omega^2$ (ii) $\dfrac{25Mv^2}{6}$
2) $\frac{20}{3}$ rad s^{-2}; $\mu \geqslant \dfrac{5}{6g}$
3) a) $\dfrac{5g\sin\alpha}{7a}$; $\mu \geqslant \frac{2}{7}\tan\alpha$
   b) $\dfrac{g\sin\alpha}{2a}$; $\mu \geqslant \frac{1}{2}\tan\alpha$
   c) $\dfrac{3g\sin\alpha}{5a}$; $\mu \geqslant \frac{2}{5}\tan\alpha$
4) a) $25g/14$ b) $\sqrt{5g/7}$
7) $\frac{14}{5}\sqrt{g/a}$; $15\sqrt{a/g}$; $\frac{7}{72}$

Exercise 13c – p. 292

3) $g\sin\theta$
4) $\dfrac{a^2\omega^2}{8\mu g}$
5) $\dfrac{8a\omega}{g}$; $\frac{22}{7}\Omega$

Exercise 13d – p. 296

1) $\dfrac{3J}{4ma}$; $\dfrac{J}{2m}$
2) $\frac{1}{6}v$
3) a) $\dfrac{V\sqrt{2}}{4l}, \dfrac{V}{2}$ b) $\dfrac{V\sqrt{2}}{8l}, \dfrac{V}{3}$
4) $\dfrac{v}{a}$, $\frac{5}{6}v$; $\frac{5}{6}\pi a$

Multiple Choice Exercise 13 – p. 301

1) a, c, d	2) d	3) a, d
4) b, d	5) *A*	6) *C*
7) *B*	8) *B*	9) *A*
10) *A*	11) b, d	12) *I*
13) *F*	14) *T*	15) *F*
16) *T*	17) *F*	

Miscellaneous Exercise 13 – p. 303

2) $(3x/g\sin\alpha)^{1/2}$
3) $\frac{2}{3}g\sin\alpha$; $\frac{1}{3}\tan\alpha$
5) $(200\,g\sin\alpha/19a)^{1/2}$
6) $\frac{1}{2}\tan\alpha$
7) $\frac{2}{21}\sqrt{3}$
8) a) $5(2a/3g\sin\alpha)^{1/2}$; b) the frictional force does no work and there is no change in PE – hence no change in KE.
9) $\frac{5}{21}\mu g$
10) 2π rad s^{-2}; $ma(g\sin\alpha - 2\pi a)/2\pi$; $\tan\alpha - 2\pi a/g\cos\alpha$
11) $a\sqrt{\frac{3}{5}}$; $\frac{3}{8}\tan\alpha$
12) a) $\frac{9}{62}g$ b) $\frac{7}{62}\sqrt{3}$
15) 0; $\frac{2}{3}g\sin\alpha$; $g\sin\alpha$
16) $\dfrac{3v}{4a}$
17) $\frac{4}{3}a$
18) $\dfrac{I\sqrt{3}}{2ml}$; $\dfrac{7I^2}{16m}$
19) $\pi a\sqrt{10}$
23) $\frac{5}{7}g\sin\alpha$; $\frac{2}{7}\tan\alpha$
24) a) $\sqrt{g/a}$ b) 0

25) vertically downward from initial position; $(\frac{3}{11}ga\sqrt{2})^{1/2}$

27) $\sqrt{\frac{24}{97}ga}$, $\frac{1}{4}\sqrt{\frac{24}{97}ga}$; a) $\frac{1}{4}a\omega\sqrt{13}$
b) $\sqrt{12g/109a}$

29) $\frac{1}{4}\sqrt{ga}$, $\frac{1}{4}\sqrt{2ga}$

31) $\frac{3}{4}mg\sin\theta\,(3\cos\theta-2)$;
$\frac{1}{4}mg(1-3\cos\theta)^2$

33) AB: $2I/3m$, BC: $I/3m$; $\frac{1}{3}I$

34) $$\int_{\alpha}^{0} 2\sqrt{\frac{l}{g}}\Big/\{\alpha^2-\theta^2+2\sqrt{1+\theta^2}-2\sqrt{1+\alpha^2}\}^{1/2}\,d\theta$$

Exercise 14a – p. 316

1) 32 million
2) $x=\frac{1}{10}(50-t)-\frac{1}{25\,000}(50-t)^3$
3) 176 s
5) $y=(e^{5/3kt}-1)/(4e^{5/3kt}+1)$
6) $x=b\Big/\left\{\dfrac{b-p}{p}e^{-bt}+1\right\}$; $x\to b$
7) 36 days, 32 days
8) $2x=N\,e^{2\alpha t}+9N\,e^{-2\alpha t}$
$4y=9N\,e^{-2\alpha t}-N\,e^{2\alpha t}$,
after $(\sqrt{13}-2)$ time units
9) $A=100\,000$, $B=4$, $\lambda=\frac{1}{2}\ln 2$
10) 63.33°; 813 s
11) $r=2r_0e^{(t-t_0)/t_0}-r_0$; $t_0(1+\ln 2)$
12) $-m\omega^2(x\mathbf{i}+y\mathbf{j})$; $m\omega^2(yz\mathbf{i}-xz\mathbf{j})$
13) $\mathbf{E}=3\mathbf{i}$, $B=\mathbf{j}$

Exercise 14b – p. 324

1) $\mathbf{r}=\{\mathbf{i}-\mathbf{j}+2t\mathbf{j}\}e^{t}$
2) $\mathbf{r}=\{\mathbf{i}+(2\mathbf{i}+\mathbf{j}+\mathbf{k})t\}e^{-t}$
3) $$\mathbf{r}=e^{-1/2t}\left\{-2\left(\cos\frac{\sqrt{3}}{2}t\right)\mathbf{i}+2\left(\sin\frac{\sqrt{3}}{2}t\right)\mathbf{j}\right\}$$
4) $\mathbf{r}=2(1-e^{-2t})\mathbf{i}$
5) $\mathbf{r}=2(\sin\sqrt{2}t)\mathbf{i}$
$+(4\cos t-4\cos\sqrt{2}t-\sin\sqrt{2}t)\mathbf{j}$
6) $\mathbf{r}=(2e^{2t}-\cos t-4\sin t)\mathbf{i}+2(\sin t)\mathbf{j}$
7) $\mathbf{r}=3\mathbf{a}\cos\lambda t+\dfrac{\mathbf{a}}{\lambda^2}\sin\lambda t-\dfrac{\mathbf{a}}{3\lambda^2}\sin 2\lambda t$
9) $\mathbf{r}=a(\mathbf{i}+\mathbf{j})\cos nt+\dfrac{b}{n}(\mathbf{i}-2\mathbf{k})\sin nt$

10) a) $\mathbf{r}=(3\mathbf{i}+\mathbf{j})e^{-2t}-(2\mathbf{i}+\mathbf{j})e^{-3t}$
b) $\mathbf{r}=\mathbf{V}te^{-nt}$
c) $\mathbf{r}=\mathbf{A}\cos\omega t+\mathbf{B}\sin\omega t-\dfrac{\mathbf{a}}{3\omega^2}\cos 2\omega t$

12) a) $\mathbf{r}=-\mathbf{i}+\mathbf{j}+e^{\theta}\mathbf{i}$
b) $\mathbf{r}=e^{\theta}(\mathbf{A}\cos 3\theta+\mathbf{B}\sin 3\theta)$; $-e^{\pi}\mathbf{j}$

13) a) $\mathbf{r}=(\mathbf{i}-\mathbf{j}+\mathbf{h})e^{4t}$
b) $\mathbf{r}=e^{-t}(\cos t-\sin t)(\mathbf{i}+\mathbf{k})$
c) $\mathbf{r}=(\sin t)\mathbf{i}+(\cos 2t)\mathbf{j}$; $y=1-2x^2$

Throughout Chapter 15, x is measured from the left-hand end of the beam.
In answers to questions where diagrams are required, the bending moment and shearing force function for each section are given.

Exercise 15a – p. 345

1) $0<x<2a$ $S=-W$; $\mathfrak{M}=Wx$
$2a<x<3a$ $S=2W$; $\mathfrak{M}=2W(3a-x)$

2) $0<x<a$ $S=-2W$;
$\mathfrak{M}=W(2x-3a)$
$a<x<2a$ $S=-W$;
$\mathfrak{M}=W(x-2a)$

3) $0<x<3a$ $S=W\left(\dfrac{x}{a}-\dfrac{1}{3}\right)$;
$\mathfrak{M}=Wx\left(\dfrac{1}{3}-\dfrac{x}{4a}\right)$
$3a<x<4a$ $S=W\left(\dfrac{x}{2a}-3\right)$;
$\mathfrak{M}=W\left(3x-8a-\dfrac{x^2}{4a}\right)$

4) $0<x<a$ $S=-W$; $\mathfrak{M}=Wx$
$a<x<3a$ $S=W\left(\dfrac{x}{a}-2\right)$;
$\mathfrak{M}=W\left(2x-\dfrac{a}{2}-\dfrac{x^2}{2a}\right)$
$3a<x<4a$ $S=W$; $\mathfrak{M}=W(4a-x)$

5) $0<x<1$ $S=-2000$ N;
$\mathfrak{M}=(2000x-2500)$ N m
$1<x<2$ $S=1000(x-)$ N;
$\mathfrak{M}=500(4x-x^2-4)$ N m

6) $0<x<a$ $S=W$; $\mathfrak{M}=-Wx$
$a<x<3a$ $S=W\left(\dfrac{x}{a}-\dfrac{5}{2}\right)$;
$\mathfrak{M}=\dfrac{W}{2}\left(5x-6a-\dfrac{x^2}{a}\right)$

7) $0 < x < 2a \quad S = \frac{W}{6}\left(\frac{3x}{a} - 13\right);$

$\mathcal{M} = Wx\left(\frac{13}{6} - \frac{x}{4a}\right)$

$2a < x < 4a \quad S = -\frac{W}{6};$

$\mathcal{M} = \frac{W}{6}(x + 18a)$

$4a < x < 6a \quad S = \frac{11W}{6};$

$\mathcal{M} = \frac{11W}{6}(6a - x)$

8) $0 < x < 0.5 \quad S = -3000$ N;
$\mathcal{M} = (3000x - 2800)$ N m
$0.5 < x < 1.0 \quad S = -1800$ N;
$\mathcal{M} = (1800x - 2200)$ N m
$1.0 < x < 1.5 \quad S = -800$ N;
$\mathcal{M} = (800x - 1200)$ N m

Miscellaneous Exercise 15 – p. 346

1) 15 N, 15 N;
$0 < x < 5 \quad S = x; \ \mathcal{M} = -\frac{1}{2}x^2$
$5 < x < 10 \quad S = -10$
$\mathcal{M} = 10x - 62.5$
S and $\mathcal{M}$ symmetrical about midpoint
$0 < x < 5 \quad S = x \ \ \mathcal{M} = -\frac{1}{2}x^2$
$5 < x < 10 \quad S = 2x - 20$
$\mathcal{M} = -x^2 + 20x - 87.5$
S and $\mathcal{M}$ symmetrical about midpoint
12.5 N m at C, D and the midpoint.

2) a) $0 < x < 4 \quad S = 500x$
$\mathcal{M} = -250x^2$
$4 < x < 8 \quad S = -1000$
$\mathcal{M} = 1000x - 8000$
$8 < x < 12 \quad S = 0$
$\mathcal{M} = 0$
b) $0 < x < 8$ as part a)
$8 < x < 12 \quad S = 250x - 3000$
$\mathcal{M} = -125x^2 + 2000x - 6000$

Greatest bending moment, 4000 N m at B in each case

3) a) $0 < x < 1 \quad S = -\frac{40}{3}; \ \mathcal{M} = \frac{40x}{3}$

$1 < x < 3 \quad S = \frac{20}{3}; \quad \mathcal{M} = 20 - \frac{20x}{3}$

$3 < x < 4 \quad S = 0; \quad \mathcal{M} = 0$
$\frac{20}{3}$ N, $\frac{40}{3}$ N; $\frac{40}{3}$ N m

b) $0 < x < 1 \quad S = 10x - \frac{80}{3};$
$\mathcal{M} = \frac{80x}{3} - 5x^2$
$1 < x < 3 \quad S = 10x - \frac{20}{3};$
$\mathcal{M} = 20 + \frac{20x}{3} - 5x^2$
$3 < x < 4 \quad S = 10x - 40;$
$\mathcal{M} = 40x - 5x^2 - 80$
$\frac{100}{3}$ N, $\frac{80}{3}$ N; $\frac{65}{3}$ N m

4) $0 < x < l \quad S = -2W; \ \mathcal{M} = 2Wx$

$l < x < 3l \quad S = \frac{4W}{3l}(x - l);$

$\mathcal{M} = 2Wl - \frac{2W}{3l}(x - l)^2$

$3l < x < 4l \quad S = \frac{4W}{3l}(x - l) - 4W;$

$\mathcal{M} = 4Wx - 10Wl - \frac{2W}{3l}(x - l)^2$

$2Wl$ at C.
$0 < x < l \quad S = 0; \ \mathcal{M} = 0$

$l < x < 3l \quad S = 2W + \frac{4W}{3l}(x - l);$

$\mathcal{M} = \frac{2W}{3l}(l - x)(2l + x)$

$3l < x < 4l \quad S = \frac{4W}{3l}(x - l) - 10W;$

$\mathcal{M} = \frac{2W}{3l}(x - 4l)(13l - x)$

$6W$

5) $0 < x < 1 \quad S = 160x - 600;$
$\mathcal{M} = 600x - 80x^2$
$1 < x < 4 \quad S = 160x - 200;$
$\mathcal{M} = 400 + 200x - 80x^2$
$4 < x < 5 \quad S = 160x - 800;$
$\mathcal{M} = 800x - 2000 - 80x^2$
525 N m; 2080 N; 2160 N m

6) $0 < x < a \quad S = \frac{Wx}{a}; \ \mathcal{M} = -\frac{Wx^2}{2a}$

$a < x < 2a \quad S = \frac{Wx}{a} - 4W;$

$\mathcal{M} = 4W(x - a) - \frac{Wx^2}{2a}$

$2a < x < 4a \quad S = \frac{Wx}{a} - 2W;$

$\mathcal{M} = 2Wx - \frac{Wx^2}{2a}$

The magnitude of the bending moment is least when zero at $x = 0, 1.17a, 4a$.
The magnitude of the bending moment is greatest when $2Wa$ at the midpoint.

7) $0 < x < 10$ $S = 10x - 8$;
$\mathfrak{M} = 8x - 5x^2$
$10 < x < 16$ $S = 10x - 200$;
$\mathfrak{M} = 200x - 1920 - 5x^2$

Greatest bending moment is 420 N m at B

8) $0 < x < 1$ $S = 500x$; $\mathfrak{M} = -250x^2$
$1 < x < 5$ $S = 500x - \frac{5800}{3}$;
$\mathfrak{M} = \frac{5800}{3}(x - 1) - 250x^2$
$5 < x < 7$ $S = 500x - \frac{3400}{3}$;
$\mathfrak{M} = 6200 + \frac{3400x}{3} - 250x^2$
$7 < x < 10$ $S = 500x - 5000$;
$\mathfrak{M} = 5000x - 250x^2 - 2500$
$\mathfrak{M} = 0$ at $x = 0, 1.18, 5.03, 10$

9) $0 < x < 1$ $S = 400x$ $\mathfrak{M} = -200x^2$
$1 < x < 5$ $S = 400(x - 3)$
$\mathfrak{M} = 1200(x - 1) - 200x^2$
$5 < x < 6$ $S = 400(x - 6)$
$\mathfrak{M} = 2400(x - 3) - 200x^2$

distant $3(\sqrt{2} - 1)$ m from each end

10) $0 \leqslant x < a$ $\mathfrak{M} = -\frac{Wx^2}{2}$
$a < x < \frac{5a}{2}$ $S = W(x - 4a)$;
$\mathfrak{M} = 4Wa(x - a) - \frac{Wx^2}{2}$
$\frac{5a}{2} < x < 4a$ $S = W(x - a)$
$4a < x \leqslant 5a$ $S = W(x - 5a)$;
$\mathfrak{M} = -\frac{W}{2}(x - 5a)^2$

11) 81 N, 63 N;
$0 < x < 3$ $S = 9x - 81$
$\mathfrak{M} = 81x - \frac{9}{2}x^2$
$3 < x < 12$ $S = 9x - 45$
$\mathfrak{M} = 108 + 45x - \frac{9}{2}x^2$
Greatest bending moment is 220.5 N m, 5 m from A
$0 < x < 3$ $S = 9x - 90$
$\mathfrak{M} = 90x - \frac{9}{2}x^2$
$3 < x < 6$ $S = 9x - 54$
$\mathfrak{M} = 108 + 54x - \frac{9}{2}x^2$
S and $\mathfrak{M}$ symmetrical about midpoint.
Greatest bending moment is 270 N m, at the midpoint

12) $0 < x < 3$ $S = 250x$; $\mathfrak{M} = -125x^2$
$3 < x < 9$ $S = 250x - 1500$;
$\mathfrak{M} = 1500(x - 3) - 125x^2$
$9 < x < 12$ $S = 250x - 3000$;
$\mathfrak{M} = 3000x - 18000 - 125x^2$

$0 < x < 3$ $S = 250x$; $\mathfrak{M} = -125x^2$
$3 < x < 8$ $S = 250(x - 8)$;
$\mathfrak{M} = -125(x^2 - 16x + 48)$
$8 < x < 9$ $S = 250(x + 4)$;
$\mathfrak{M} = -125(x^2 + 8x - 144)$
$9 < x < 12$ $S = 250(x - 12)$;
$\mathfrak{M} = -125(x^2 - 24x + 144)$
750 N; 1125 N m

13) a) $0 < x < 3$ $S = 180x$; $\mathfrak{M} = -90x^2$
$3 < x < 6$ $S = -335$;
$\mathfrak{M} = 335x - 1815$
$6 < x < 9$ $S = 65$;
$\mathfrak{M} = 585 - 65x$
b) $0 < x < 3$ $S = 180x$;
$\mathfrak{M} = -90x^2$
$3 < x < 4.5$ $S = -335$;
$\mathfrak{M} = 335x - 1815$
$4.5 < x < 7.5$ $S = \frac{400x}{3} - 935$;
$\mathfrak{M} = -\frac{200x^2}{3} + 935x - 3165$
$7.5 < x < 9$ $S = 65$;
$\mathfrak{M} = 585 - 65x$

14) bridge only. $S = 10^4(3x - 300)$
$\mathfrak{M} = 10^4\left(-\frac{3x^2}{2} + 300x\right)$
with train.
$0 < x < 50$ $S = 10^4(9x - 562.5)$
$\mathfrak{M} = \frac{10^4 x}{2}(1125 - 9x)$
$50 < x < 200$
$S = 10^4(3x - 262.5)$
$\mathfrak{M} = 10^4\left(\frac{3x^2}{2} + 262.5x + 7500\right)$
a) 43.75 m from end carrying train
b) 13:4

Exercise 16a – p. 351

1) a, c, f, -discrete
b, d, e, -continuous
2) $x = 0, 1, 2, 3$
3) $x = 0, 1, 2, \ldots, 36$
4) $x = 1, 2, 3, 4, 5, 6$

Exercise 16b – p. 353 (no answers)

Exercise 16c – p. 358

1)

x	0	1	2
$P(X = x)$	$\frac{1}{4}$	$\frac{1}{2}$	$\frac{1}{4}$

2)

x	1	2	3
$P(X=x)$	$\frac{1}{14}$	$\frac{4}{14}$	$\frac{9}{14}$

$P(X<3)=\frac{5}{14}$

3)

y	0	1	2	3	4
$P(Y=y)$	$\frac{1}{5}$	$\frac{1}{5}$	$\frac{1}{5}$	$\frac{1}{5}$	$\frac{1}{5}$

$P(Y>1)=\frac{3}{5}$

4) $\frac{1}{6}$

5) $\frac{37}{14}$

6)

x	2	3	4	5	6	7	8
$P(X=x)$	$\frac{1}{36}$	$\frac{2}{36}$	$\frac{3}{36}$	$\frac{4}{36}$	$\frac{5}{36}$	$\frac{6}{36}$	$\frac{5}{36}$

x	9	10	11	12
$P(X-x)$	$\frac{4}{36}$	$\frac{3}{36}$	$\frac{2}{36}$	$\frac{1}{36}$

$$f(x)=\begin{cases}\dfrac{x-1}{36} & x=2,3,4,5,6,7\\[2ex] \dfrac{13-x}{36} & x=8,9,10,11,12\end{cases}$$

7) $f(x)=\frac{1}{6};\ x=1,2,3,4,5,6$

8) $\frac{1}{5};\frac{4}{25};\frac{16}{125};\ f(x)=(\frac{4}{5})^{x-1}(\frac{1}{5}),$
$x=1,2,3,\ldots$

Exercise 16d – p. 361

1) a)

x	0	1	2	3	4
$P(X=x)$	$\frac{256}{625}$	$\frac{256}{625}$	$\frac{96}{625}$	$\frac{16}{625}$	$\frac{1}{625}$

b)

x	0	1	2	3	4
$P(X=x)$	$\frac{16}{81}$	$\frac{32}{81}$	$\frac{24}{81}$	$\frac{8}{81}$	$\frac{1}{81}$

c)

x	0	1	2	3	4
$P(X=x)$	$\frac{1}{16}$	$\frac{1}{4}$	$\frac{3}{8}$	$\frac{1}{4}$	$\frac{1}{16}$

d)

x	0	1	2	3	4
$P(X=x)$	$\frac{1}{625}$	$\frac{16}{625}$	$\frac{96}{625}$	$\frac{256}{625}$	$\frac{256}{625}$

e)

x	0	1	2	3
$P(X=x)$	$\frac{27}{125}$	$\frac{54}{125}$	$\frac{36}{125}$	$\frac{8}{125}$

f)

x	0	1	2
$P(X=x)$	0.078	0.259	0.346

x	3	4	5
$P(X=x)$	0.230	0.077	0.010

g)

x	0	1	2
$P(X=x)$	0.028	0.131	0.261

x	3	4	5
$P(X=x)$	0.290	0.194	0.077

x	6	7
$P(X=x)$	0.017	0.002

h)

x	0	1	2
$P(X=x)$	0.006	0.040	0.121

x	3	4	5
$P(X=x)$	0.215	0.250	0.201

x	6
$P(X=x)$	0.112

x	7	8	9
$P(X=x)$	0.042	0.011	0.002

x	10
$P(X=x)$	0.000

2) 0.010 24; 0.922 24
3) 0.136; 0.001 29 (3 s.f.)
4) (a), (c), (f)
5) 0.3125
6) 0.972 (3 s.f.)
7) 0.0625
8) 0.137 (3 s.f.)
9) 0.180 (3 s.f.)
10) 0.590; 22
11) 7

Exercise 16e – p. 364

1) a) geometric
b) rectangular
c) binomial
2) a) rectangular; $\frac{1}{13}$
b) binomial; 0.0312 (3 s.f.)
c) geometric; 0.0655 (3 s.f.)

Exercise 16f – p. 367

1) a) $\frac{19}{24}$ b) 2 c) 2.57 d) 2.5
2) $\frac{2}{3}$
3) $\frac{18}{7}$
4) 0.8

5) 12
6) 3.5
7) 1.25

Exercise 16g – p. 370

1) a) 2.9 b) 9.7 c) 24.1
2) a) 0.5 b) 0.2875 c) –0.925

Exercise 16h – p. 376

1) a) 2.7 b) 0.61
2) 3; 1
3) 2.02; 1.10 (3 s.f.)
4) 6; 1/3; 0.351 (3 s.f.)
5) 2.5; 1.25
6) $\frac{1}{2}(n+1)$; $\frac{n^2-1}{12}$
7) 5; 20
8) $1/p$; $\frac{1-p}{p^2}$
9) 2; 1.26 (3 s.f.)
11) 2; 2
12) 12.5; 143.75

Multiple Choice Exercise 16 – p. 378

1) c	2) b	3) d	4) b
5) a	6) b	7) a	
8) c (not b as $x \neq 0$)		9) c	10) C
11) A	12) B	13) A	14) E

Miscellaneous Exercise 16 – p. 379

1) 0.797 (3 s.f.)
2) 0.345 (3 s.f.)
3) a) $\frac{2}{3}$ b) $\frac{1}{14}$ c) $\frac{1}{4}$
4) a) $\frac{1}{10}$ b) $\frac{1}{25}$ c) $\frac{1}{4}$
   0.014; 0.015
5) a) 0.7; $\frac{9}{14}$
   b) (i) $\frac{2}{5}$ (ii) 0.0487 (3 s.f.)
6) a) 4.8; 0.98 (2 d.p.)
   c) 0.737 (3 s.f.)
7) a) 0.377 (3 s.f.) b) 0.889 (3 s.f.);
   1.05×10^{-4}
8) 25; 18.75; $1 - \frac{13.3^9}{4^{10}}$
9) a) 0.044 b) $\frac{5}{11}$ c) 0.121 (3 s.f.)
   d) 0.119 (3 s.f.)
10) a) $\frac{1}{10}$ b) 0.054 c) 0.056
   d) 0.098
   0.0168 (3 s.f.)
11) $\frac{1}{2}(n+1)$; 2
12) a) 100 b) 9.95 (3 s.f.)
13) 0.146 (3 s.f.); $\frac{20}{7}$; 1.56 (3 s.f.)
14) 0.0198 (3 s.f.); 0.02; 0.01996
15) 0.25; 0.5; 0.25; 1; 0.5
16) 6; $\frac{20}{3}$; $\frac{1}{3}$
17)

x	4	5	6	7	8
$P(X=x)$	$\frac{4}{36}$	$\frac{12}{36}$	$\frac{13}{36}$	$\frac{6}{36}$	$\frac{1}{36}$

$\frac{17}{3}$; 0.972 (3 s.f.)
18) $\frac{1}{4}$; 7.5; 5.625
19) a) $\frac{12}{169}$ b) $12^4/13^5$
   c) 0.48; $P(X=x) = (\frac{12}{13})^{x-1}(\frac{1}{13})$
   $x = 1, 2, \ldots; 13$
20) a) 10^{-5} b) 0.1
   0.5
21) $1-p^R$; $\left(\frac{1}{e^{\alpha}-1}\right)(e^{\alpha} - e^{p\alpha})$
22) b) θ c) $k = \frac{e^{\theta}}{e^{\theta}-1}$

$$E[Y] = \frac{\theta}{1-e^{-\theta}}$$

Exercise 17a – p. 385

1) 0.85
2) a) 0.15 b) 0.525

Exercise 17b – p. 390

1) a) $\frac{1}{2}$ b) $\frac{3}{16}$ c) 1.41 (3 s.f.)
2) a) $\frac{1}{2}$ c) $\frac{1}{2}$
3) a) $\frac{11}{16}$ b) 0.0859 (3 s.f.) c) 0
4) 2
5) $\frac{2}{3}$; 1
6) a) $\frac{e^2}{e^2-1}$ b) $\frac{e}{1+e}$
7) 4
8) $\frac{1}{2}$; $f(x) = \begin{cases} \frac{1}{2}x & 0 \leqslant x < 1 \\ \frac{1}{6}(4-x) & 1 \leqslant x < 4 \end{cases}$
9) a) $\frac{1}{2}$ b) $\frac{1}{4}$
10) 1, ln 2

Exercise 17c – p. 394

1) $f(x) = \frac{1}{6}$ $0 \leqslant x < 6$; $\frac{1}{6}$
2) $f(x) = \frac{1}{5}$ $-2 \leqslant x < 3$; $\frac{2}{5}$
3) 0.4
4) $\frac{1}{6}$
5) –4; $\frac{5}{6}$

6) Let $Y = X^2$, then $f(y) = \dfrac{1}{8y^{1/2}}$

$4 \leqslant y < 36$

7) $f(a) = \dfrac{1}{12\sqrt{(16-a)}}$ $0 \leqslant a < 12$

Exercise 17d – p. 398

1) a) 2.25; 5.4
b) $\frac{4}{3}$; 2
c) 0; $\frac{1}{3}$
2) a) $\frac{1}{36}$; 0
b) 4; $\frac{4}{5}$
c) 1; 0
3) $f(x) = 1$, $1 \leqslant x < 2$; 1.5; $\frac{7}{3}$
$f(a) = \frac{1}{2}a^{-1/2}$, $1 \leqslant a < 4$;
$\frac{7}{3}$; same

Exercise 17e – p. 401

1) a) $\frac{1}{2}$
b) $\frac{3}{4}$
2) a) $\frac{2}{3}$; $\frac{14}{9}$; 0.283 (3 s.f.)
b) e; 2; 1

Miscellaneous Exercise 17 – p. 401

1) $4^{1/3}$
2) $1 - e^{-1/3}$; $e^{-2/3}$
3) $\frac{2}{9}$; $\frac{3}{2}$
4) 2; $\frac{2}{3}$
5) 0.75; 0.3375
6) $\dfrac{2}{\pi}$; $\frac{1}{4}\pi + 1$
7) $4/(\pi - 2\ln 2)$; $(\pi - 2)/(\pi - 2\ln 2)$
9) a^{-1}; a^{-2}
10) $\frac{4}{5}$
11) a) $\frac{3}{4}$ b) $\frac{4}{3}$ c) 0.6875
d) 1.2; 0.16
12) $\frac{3}{8}$; 0.244
13) $F(x) = 4 - 2x$; $1 \leqslant x \leqslant 2$
14) a) 12 b) 0.6; 0.2
15) 0; $\frac{1}{3}$; $f(y) = \frac{1}{4}y^{-1/2}$
16) $a \geqslant 1$; $c = \dfrac{3}{2(3a-1)}$; $\frac{1}{5}$
17) a) $\frac{2}{3}$ b) $\frac{2}{3}$ c) $x/3$ d) $\frac{1}{3} + x/6$
e) 1

$$f(y) = \begin{cases} \frac{1}{3}, & 0 \leqslant y < 2 \\ \frac{1}{6}, & 2 \leqslant y < 4 \end{cases}$$

$\frac{5}{3}$

18) λ^{-1}; λ^{-2}
19) £1.15; 20p
20) a) 8×10^{-6}; 8×10^{-6}; 333, 167
c) 2 or more
d) 0.424

Answers are given to 3 s.f. unless the question requested otherwise. Note also that answers may vary slightly depending on which tables are used and on how calculators are used.

Exercise 18a – p. 413

1)	a) 0.147	b) 0.0916	c) 0.762
2)	a) 0.018	b) 0.104	
3)	a) 0.0361	b) 0.0527	c) 0.433
4)	a) 0.905	b) 0.986	c) 0.0616
5)	a) 0.990	b) 0.819	c) 0.00793
6)	a) 0.149	b) 0.271	c) 0.0404
	d) 0.0337		
7)	a) 0.0328	b) 0.0223	c) 0.272
8)	a) 0.184	b) 0.303	c) 0.354
9)	a) 0.0190	b) 0.0880	
10)	a) 0.323	b) 0.933	c) 0.003 53

Exercise 18b – p. 418

1) a) 0.002 20 b) 0.002 27
2) a) no, p too large
b) yes
c) no, n too small
d) no, n too small
3) a) 2.06×10^{-9} b) 0.005 82
4) a) 0.007 57 b) 0.125
5) a) 0.904 (binomial)
b) 0.006 74 (Poisson) or 0.006 57 (binomial)
Using binomial $n = 230$, using Poisson $n = 231$.
6) 0.271; 0.180
7) 0.000 365; 0.599
8) $\frac{1}{221}$; 0.001 22

Exercise 18c – p. 423

1)	a) 0.7257	b) 0.5199	c) 0.999 952
2)	a) 0.8531	b) 0.5871	c) 0.8078
3)	a) 0.0808	b) 0.0344	c) 0.2296
4)	a) 0.0516	b) 0.1762	c) 0.9798
5)	a) 0.0864	b) 0.2889	c) 0.11351
	d) 0.1244	e) 0.1176	f) 0.146 84
	g) 0.6294	h) 0.2358	i) 0.7032
	j) 0.890 93		
6)	a) 1	b) −0.75	c) −1.5

Exercise 18d – p. 428

1) a) 0.9772 b) 0.3085 c) 0.8413
2) a) $X < 51.048$ b) $X > 48.316$
c) $48.652 < X < 51.348$
3) a) 40.86; 22.73 b) 8.082; 3.66
c) 5.53; 2.75
4) a) 0.999 571 b) 0.9522
c) 0.6826
5) a) 0.1587 b) 0.8413 c) 0.001 35
d) 0.383
6) $X < 46.84$ g
7) 87.18 g $< X <$ 116.45 g
8) 30; 3.90 (cm)
9) 52.5 g; 5.85 g
10) 0.004 00
11) a) 0.734 b) 0.9772 c) 0.717
0.514
12) a) 0.1974 b) 0.1328 c) 0.0262

Exercise 18e – p. 434

1) 0.06
2) 0.002 36
3) 89; 58
4) a) 0.1539 b) 0.5685
5) 0.060 59
6) $\frac{1}{3}$; 0.0726 (0.0705 if use binomial)
7) 143
8) 0.1977
9) 0.51; 0.3114
10) 0.2266
11) a) 0.1097 b) 0.2696
12) 29

Miscellaneous Exercise 18 – p. 436

1) (i) 0.0228 (ii) 0.927 (iii) 0.307
a) 2.63 b) 2.34
2) a) 0.067 (2 s.f.) b) 0.42 (2 s.f.)
c) 0.081 (2 s.f.)
3) 0.91
4) a) 0.32 (2 s.f.) (Poisson)
b) 0.93 (2 s.f.)
5) 0.960
6) a) (i) 0.135 (ii) 0.947
b) (i) 0.144 (ii) 0.111 (iii) 0.0457
7) a) (i) 79 (ii) 42
b) 0.594
8) a) 10, 6, normal, 0.999
b) 2, 1.96, Poisson, 0.323
9) a) (i) 0.669 (ii) 0.841 (iii) 0.019
0.000 37
b) $\alpha^{-1}; \alpha^{-2}$
10) a) 0.303 b) 0.0902
0.282
11) a) 6 b) (i) 0.165 (ii) 0.955
12) a) (i) 0.368 (ii) 0.264
b) 21.2%; 75.5%; 3.3%
13) a) (i) 135.7 (ii) 431.5
b) (i) 0.018 (ii) 0.20 (iii) 0.000 34
14) a) 0.939 b) 0.938;
0.883
15) 42.2; 127.7
16) a) 0.0765 b) 0.2216;
0.940
17) a) 0.0892 b) 0.938
3.29×10^{-6}
18) a) (i) 0.8454 (ii) 0.2578
b) (i) 0.0164 (ii) 0.181
19) a) 0.371; 0.000 34
b) (i) 0.0030 (ii) 0.44
20) a) 0.050 b) 0.318 c) 0.423
d) $e^{-3} \sum_{r=0}^{\infty} \frac{2^r}{(r!)^2}$
21) 1/3; 20/3; 40/9; 12
22) a) 0.9599; 0.7881 b) 0.857
c) 99%
23) 0.989; 35.253
24) 0.914; 10; 9.5; 0.0721
25) a) 0.2025 b) 0.410 c) 0.0238
26) a) 0.135 b) 0.323
$9e^{-2} - 1$
27) a) 0.0166 b) 0.191 c) 5
28) a) 0.983 54 b) 0.9544
c) 7.51×10^{-4} d) 0.0613
0.997; 20.0174 m
29) a) 0.547 c) 2.30 seconds
d) 0.6826

INDEX

INDEX